Medicinal Chemistry for Practitioners

Medicinal Chemistry for Practitioners

Jie Jack Li

WILEY

This edition first published 2020

Registered Office
John Wiley & Sons, Inc., 111 River Street, Hoboken, NJ 07030, USA

Editorial Office
111 River Street, Hoboken, NJ 07030, USA

For details of our global editorial offices, customer services, and more information about Wiley products visit us at www.wiley.com.

Wiley also publishes its books in a variety of electronic formats and by print-on-demand. Some content that appears in standard print versions of this book may not be available in other formats.

Library of Congress Cataloging-in-Publication Data

Names: Li, Jie Jack, author.
Title: Medicinal chemistry for practitioners / Jie Jack Li.
Description: First edition. | Hoboken, NJ : Wiley, 2020. | Includes bibliographical references and index.
Identifiers: LCCN 2020001246 (print) | LCCN 2020001247 (ebook) | ISBN 9781119607281 (case) | ISBN 9781119607274 (adobe pdf) | ISBN 9781119607304 (epub)
Subjects: MESH: Chemistry, Pharmaceutical | Pharmaceutical Preparations–chemistry | Pharmacokinetics | Drug Delivery Systems | Drug Design
Classification: LCC RS403 (print) | LCC RS403 (ebook) | NLM QV 744 | DDC 615.1/9–dc23
LC record available at https://lccn.loc.gov/2020001246
LC ebook record available at https://lccn.loc.gov/2020001247

Cover image: [Production Editor to insert]
Cover design by [Production Editor to insert]

Set in 10/12pt TimesNewRoman by SPi Global, Chennai, India

Dedicated to Dr. Min Min Yang

Contents

Preface

This is a book that I wish I had when I became a medicinal chemist myself, fresh out of school with abundance of synthetic skills and little medicinal chemistry prowess. When I began my first medicinal chemistry job at Parke-Davis in 1997, I had to learn it "on the job".

Prof. E. J. Corey once said: "The desire to learn is the greatest gift from God". This message has resonated with me throughout my career in drug discovery and I tried to learn as much as I can. This book is the result from the "recrystallization' of all these years of learning. Here, I want to thank my mentors: Bruce Roth and Sham Nikam from Parke-Davis/Pfizer; and Nick Meanwell at BMS. I am also indebted to my fellow medicinal chemists whose papers, reviews, books, and conference presentations are cited throughout this manuscript.

I hope this book will be a good starting point for novice medicinal chemists and veteran medicinal chemists find it useful as well.

As always, I welcome your critique. You could email me your comments directly to: lijiejackli@gmail.com.

Jack Li
Nov. 1, 2019
San Mateo, California

1

Drug Targets

Following a brief discussion on selection and validation of drug targets, this chapter discusses major drug targets including enzymes, receptors, ion channels, carrier proteins, structural proteins, nucleic acids, and protein–protein interactions (PPIs).

1.1 Selection and Validation of Drug Targets

1.1.1 Factors to Consider

Choosing the right target to work on is winning half of the battle for a drug discovery project. The "right" target may be different depending upon on the size of the company, the team, the science, etc. With regard to target selection, Sir David Jack had a good approach: *The choices are made by assessing competing ideas which are invited from all the staff. The best ideas are simple, practicable with the available resources and, above all, novel enough to yield medicines that are likely to be better than probable competitors in ways that will be obvious both to doctors and their patients.*[1]

There are both business and scientific considerations for target selection. Some factors to consider are:

1. Unmet medical need;
2. Patient population;
3. Precedence, i.e., confidence in rationale (CIR);
4. Assay development;
5. Availability of animal models;
6. Biologics *versus* small molecules; and
7. Chemical matter.

1.1.1.1 Unmet Medical Need

Despite tremendous advances in drug discovery in the last decades, there are still many unmet medical needs. While the world does not need the 10th statin on the market to lower cholesterol, there are dire needs of drugs to treat cancers, viral infections, and many tropical diseases. A cure for the common flu remains elusive. As American baby boomers age, the need for geriatric medicines are on the rise. No efficacious treatment

Medicinal Chemistry for Practitioners, First Edition. Jie Jack Li.

exists for Alzheimer's disease (AD) although an astronomical amount of resources is invested in the field.

Unmet medical need is one of the key factors to consider in selecting a drug target. A case in point is when the acquired immune deficiency syndrome (AIDS) epidemic became rampant during the 1980s, no effective drugs existed and patients were dying. The urgent medical need prompted all major drug firms to join hands with the government and academia to find treatments for human immunodeficiency virus (HIV). Initial antiviral drugs discovered were older nucleotides with substantial toxicities. With the emergence of HIV protease (or peptidase) inhibitors, AIDS has become a chronic disease that can be managed rather than a death sentence.

There are times when saving lives is more important than making a profit.

1.1.1.2 Patient Population

Most blockbusters have at least one thing in common—all widely prescribed to treat a common illness such as hypertension, high cholesterol, pain, ulcer, allergies, and depression. The larger the patient population, the higher the potential for a drug to become a blockbuster drug.

Government incentives exist to discover orphan drugs to treat a condition affecting fewer than 200,000 persons in the United States.

1.1.1.3 Precedence (CIR)

Precedence of a drug target enhances its CIR. The more precedents a target has, the higher its CIR is. The gold standard in drug discovery for a drug target with regard to pharmacology in human disease is proof-of-concept (PoC).

For a particular target or a mechanism of action (MOA), it possesses CIR if it has been validated in Phase IIa clinical trials to show efficacy for this mechanism regardless of its toxicity profile. We consider such a target "known mechanism". The chances of success grow exponentially when working on a mechanism with CIR. Thanks to the lower attrition rate for targets with PoC, many competitors exist. Big pharma may throw large resources behind those projects.

The caveat is that one runs out such targets to work on fairly quickly when solely focusing on mechanisms with CIR. In addition, mechanisms with CIR cannot be sustained if everyone shies away from unprecedented mechanisms. Thankfully, most startup biotech companies focus exclusively on novel and "undruggable" targets, otherwise it would be challenging to secure funding from the venture capitals (VCs).

The drug that finishes first to gain the FDA approval becomes "first-in-class" for the MOA. Unless one of them is the "best-in-class," the others become "me-too" drugs.

1.1.1.4 Assay Development

Biochemical/enzymatic assay and cellular assay are *in vitro* assays. With flourishing biology, biochemical/enzymatic assays are relatively easy to setup. Cellular assays, granted more relevant, are more challenging. The key is to ensure the assays are reliable enough to guide the structure–activity relationship (SAR) investigations for medicinal chemists.

1.1.1.5 Availability of Animal Models

Testing a drug on animals is called *in vivo* studies. Once drugs are potent enough in both biochemical and cellular assays and possess reasonable pharmacokinetic (PK) profiles, it is then crucial to test them in animal models. The most frequently used animal models are rats, mice, and guinea pigs as they are inexpensive and easy to breed so we can have reasonably high throughput in testing our compounds. For cancers, there are many mouse species with cancer xenografts as animal models.

One of the earliest animal models was established in 1900s by Paul Ehrlich, who infected mice with *Treponema pallidum*, the bacterium that cause syphilis. Later on his associate Sahachiro Hata developed a method to infect rabbits with syphilis bacterium. These animal models greatly facilitated screening of their compounds. Without the mouse and rabbit animal models, he might not have had the capacity to experiment 605 potential drugs before his triumph with Ehrlich's 606.

Animal models can make or break a project. For instance, when Florey and Chain isolated reasonable amount of penicillin, they chose Swiss albino mice rather than guinea pigs as their animal model to test penicillin's toxicity. For unknown reasons, guinea pigs do not tolerate penicillin. If they chose guinea pigs initially to test penicillin, the emergence of this "wonder drug" could have been delayed by many years.

Furthermore, rodents are very different from *homo sapiens*. Therefore, it is always prudent to test investigational drugs on higher species such as rabbits, dogs, and monkeys before testing on humans.

When working on MMP-13 specific inhibitors as potential treatment for osteoarthritis (OA), our team at Parke–Davis used Sprague–Dawley rats and New Zealand white male rabbits as animal models. The artificial arthritis was created by either surgically (manually damaging some cartilage between the animal's joints) or treating the joints with sodium iodoacetate to damage the joints so that they mimic arthritis.[2]

Last century, zoologists have created many invaluable animal models to emulate human diseases. Unfortunately, not all diseases have suitable animal models. For example, it is challenging to make animals "depressed." To create an animal model for depression, scientists fed reserpine to animals so they became sedated, thus *partially mimicking* some symptoms of depression. Animal models for other psychiatric illnesses are even more challenging, so is the animal model for stroke.

One popular genus of monkey used in drug discovery is the cynomolgus monkey. A known animal model for hair growth is stumped-tailed monkey. Amusingly, the only viable animal model for studying leprosy, decades ago, was the nine-banded armadillo. Thankfully, they were plentiful then. For testing a topical drug's skin penetration, adorable mini-pigs are used as the animal model—you have an animal model and a pet at the same time.

There have been times that the drug's target is not known, and the only means of gauging the SAR is using animal models. Anticoagulant clopidogrel (Plavix) and cholesterol absorption inhibitor ezetimibe (Zetia) were both discovered using animal models because their MOAs were unknown at the time of their discovery.

1.1.1.6 Biologics versus Small Molecules

In today's business environment, one has to wrestle with the question of small molecule versus biologic approaches once a target is chosen. Oral small molecule drugs penetrate cell membrane and modulate the targets inside the cell. Biologics can only interact with cell-surface or secreted proteins and are never given orally. If given the choice of convenience, oral small molecule drugs are certainly easier to swallow, so to speak.

Another disparity between small molecule drugs and biologics is Chemistry, Manufacturing, and Control (CMC). When a patent expires for a small molecule drug, making a generic version is relatively straightforward. Everyone can make generic drugs with relative ease; but making biosimilars is not a walk in the park even though patents for some biologics have expired. The CMC for biologics are so challenging that even the innovators occasionally have difficulty making each batch consistent for their own original biologics. This also explains why biosimilars do not significantly lower the prices of biologics, whereas generic drugs decimate the price for small molecule brand name drugs.

The debate of small molecule drugs versus biologics is not new. Ehrlich's magic bullet theory represented the side for championing small molecules. On the opposite side, Emil von Behring was a staunch supporter of serum therapy. Despite their initial preference of small molecules versus serum therapy, both won the Nobel Prize: von Behring in 1901 and Ehrlich in 1908. Now, more than a century has gone by, the great debate is still raging on. Certainly, last century was the century of small molecules, garnering greater than 90% of the blockbuster drugs, but their golden age draws to a close today. However, blockbuster drugs for biologics are on the rise. Because of difficulties in producing monoclonal antibodies (many biologics are monoclonal antibodies), herculean efforts are undertaken by big Pharma to develop biosimilars. Subsequently, many biosimilars are not available even long after the original patent for the brand name biologic has expired.

1.1.1.7 Chemical Matter

Unless one decides to pursue biologics exclusively, chemical matter always matters if one chooses to pursue small molecule drugs to treat the diseases.

It is always beneficial to have some chemical matter as precedents. Regardless from either competitors or academia, known chemical matter is invaluable as a tool compound to evaluate the fidelity of the biochemical assays and to help target validation.

Any given drug target will fall into one of the four quadrants as depicted in Figure 1.1. Quadrant 1 hosts projects with known mechanism and known chemotype. These projects have the highest possibility of success. If one chooses to be truly innovative and solely work on frontier targets, the projects fall into quadrant 4 with both unknown mechanism and unknown chemotype. While there are no solid statistics, the success rate for the project to reach the market is low.

With a known mechanism and unknown chemotype, quadrant 2 has the second best chance of success because medicinal chemists are now savvy enough to create our own chemical matter. For quadrant 3 with known chemotype and unknown mechanism, it is still challenging although its chance of succeeding is exponentially higher than that of quadrant 4.

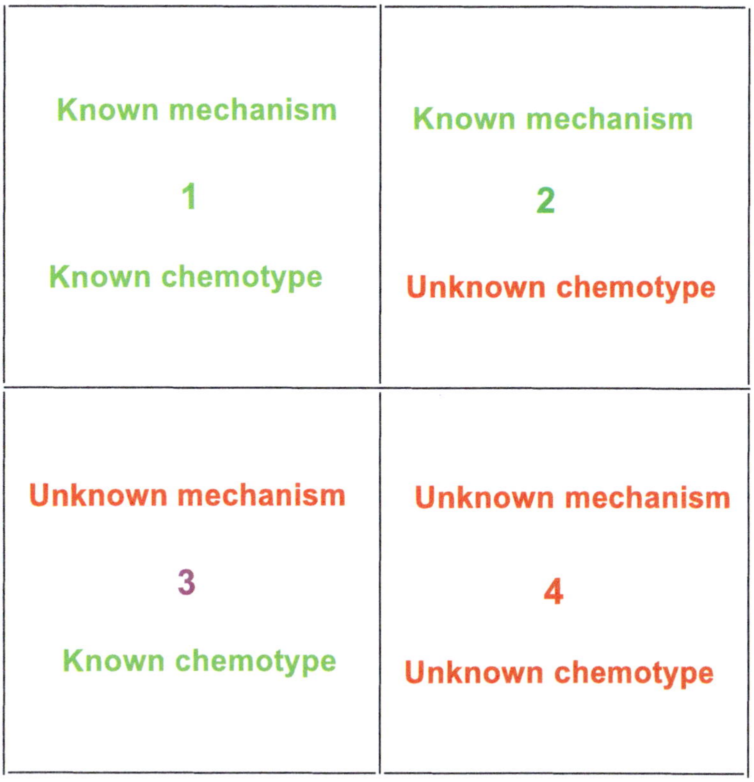

Figure 1.1 The drug target quadrants.

Patient population, CIR, and unmet medical needs are big pictures one has to ponder. When everything is about equal, the tie-breaker would be logistics in executing the project. If a mechanism already has existing chemical matter in the literature, chances of success are better than another mechanism that has no chemical matter at all. In addition, if an assay has been published for one mechanism, the project could get jump-started more rapidly in comparison to another one without known assays. Furthermore, animal models are critical to the success of a project. It is mandatory to have an animal model for any given project.

1.1.2. Target Selection

Target selection in drug discovery is the decision to focus on finding an agent with a particular biological action that is anticipated to have therapeutic utility. According to a 2006 article, there are at least 324 drug targets.[3]

Target selection is influenced by a balance of complex scientific, medical, and strategic considerations. In selecting a drug target, the reliability of its genesis decreases as the following:

- Pharmacology in human disease. Targets with PoC have the highest chance of success;
- Pharmacology in animal model;
- Animal model availability;
- Cell model;
- Literature precedent;
- Anecdotal findings; and
- Hunch.

There are several major classes of drug targets.[4]

The largest biochemical class as drug targets are enzymes—nearly half of oral drugs on the market work on enzymes. Enzymes are body's biological catalysts that accelerate biological reactions. The most fruitful enzymes for drug discovery today are kinases, a class of enzymes that catalyze the transfer of a phosphate group from adenosine triphosphate (ATP) to a specific molecule, such as tyrosine, threonine, and serine. More than thirty kinase inhibitors have been approved by the FDA thus far to treat cancers and other diseases. Phosphatases, a class of enzymes that catalyze the transfer of a phosphate group from a specific molecule, such as tyrosine, threonine, and serine, are extremely hard to inhibit with small molecules. But significant progress has been made on these "undruggable" targets. More on this later.

The second largest class of drug targets are receptors on the surface of the cell membranes. One-third of marketed drugs modulate receptors. One subtype of receptors, G-protein-coupled receptors (GPCRs) have been especially successful for drug discovery. Antihistamines such as fexofenadine (Allegra), loratadine (Claritin), desloratadine (Clarinex), and cetirizine (Zyrtec) treat allergy through the mechanism of blocking histamine H_1 receptors. Cimetidine (Tagamet) and ranitidine (Zantac) are selective histamine H_2 receptor inhibitors (β-blockers).

The third class of drug targets (7% drugs on the market) are ion channels. They include sodium, potassium, calcium ion channels, etc. Nifedipine (Adalat) and amlodipine (Norvasc) are both calcium channel blockers (CCBs). Puffer fish is extremely poisonous because its toxin tetrodotoxin is a sodium channel blocker. Gabapentin (Neurontin) and pregabalin (Lyrica) are binders of the $\alpha_2\delta_1$ unit of the calcium channel, affecting Ca^{++} currents.

The fourth and fifth classes of drug targets are transporters (4%) and nuclear hormone receptors (4%), respectively. The remaining drug targets are relatively trivial with market shares less than 3%.

1.1.3. Target Validation

Once a target is identified, target validation is the next essential step. Drug target validation is to ascertain that the drugs are hitting the intended targets rather than off-targets. As Manning wisely pointed out, drug discovery is such a long, arduous, and expensive enterprise, it makes sense to focus your resources on validated targets rather than poorly understood ones.[4]

As mentioned before, a target having a drug with proven efficacy in Phase IIa gains PoC and CIR. Indeed, *clinical validation* is a solid vote of confidence on the validity the targets′ function and pathway.

Innovative drug targets, by definition, often do not have clinical validations. Preclinical validation is the second best thing for a target. This includes

- Genetic evidence, human genetic evidence is superior than that from other species;
- Association with disease state, again, human genetic evidence is superior than that from other species;
- Biochemical activation of function induces signs of disease, in the same vein, human genetic evidence provides higher confidence than that from other species;
- Function modulation in preclinical models improves disease signs; and
- Reason to expect mechanism-based adverse effects.

Target validation should determine the function of the target as well as the disease pathway it regulates. It is also helpful to gauge the importance of the disease pathway and projected therapeutic index (TI) if a drug interacts with the target. Ideally, target validation is optimally carried out in humans. In reality, this is rarely done for obvious reasons. Rather, target validation is divided into *in vitro* tests in test tubes and *in vivo* tests in animals.

Target validation with an *in vitro* test examines the biochemical functions of the target when it is bound to a ligand. Some of the cutting-edge gene manipulation techniques are used such as gene knockouts (KOs), antisense technology, and RNA interference (RNAi). To compensate some drawbacks associated with genetic techniques, proteomics, the study and manipulation of the protein make-up of a cell, has become a preferred approach in target validation.

Since these *in vitro* experiments are done using enzymes, cells, or tissues in test tubes, the outcome cannot be closely translated to human. A negative result is *surely* a bad sign; but a positive outcome is not necessarily always a sure bet. It merely means that one small hurdle has been overcome, and many more are ahead. In contrast, a positive result from *in vivo* target validation with animal model is a great leap forward for the project.

In the absence of a valid animal model, KOs, in which genes are deleted or disrupted to halt their expression, have been useful in predicting drug actions. Many important biologic targets have been validated using the gene knocked out method.

Due to the fact that both *in vitro* and *in vivo* target validations are expensive, many *in silico* techniques have been developed for target validation. Several companies have been founded to specialize in creating software to model drug–receptor interactions.

Whereas animal models are indispensable to drug discovery, humans are different from animals. An animal model is just a model. How the drug behaves in humans would have to be ultimately tested in humans.

According to some,[5] target validation likely consists of the following six steps:
1. Discovering the biomolecule of interest;
2. Evaluating its potential as a target;
3. Designing a bioassay to measure biological activity;
4. Constructing a high-throughput screen (HTS);
5. Performing screening to find hits; and
6. Evaluating hits.

1.2 Enzymes

Like catalysts in chemical reactions, enzymes are catalysts in the body. They are agents that accelerate chemical reactions in a biological system without being consumed themselves. There are six major classes of enzymes.

1. Oxidoreductases: Oxidations and reductions;
2. Transferase: Group transfer reactions;
3. Hydrolase (Protease): Hydrolysis reaction;
4. Lyase: Addition or removal of groups to form double bonds;
5. Isomerase: Isomerization and intramolecular group transfer; and
6. Ligase: Joining two substrates at the expense of ATP.

How does an enzyme work?

More than a hundred years ago, Emil Fischer put forward the "lock-and-key" hypothesis. In essence, enzyme is like a lock and substrate is like a key. Once a matching pair binds to each other tightly, bonds on the substrates are weakened. Subsequently, the reaction takes place and the substrate is converted to the product. The enzyme, unchanged, is again available for the next reaction. As our knowledge accumulates, the simplistic "lock-and-key" model is not adequate to explain many results observed in real life. In 1960, Daniel Koshland at UC Berkeley proposed the "induced fit" theory.[6] Briefly, when the active site on the enzyme makes contact with the proper substrate, the enzyme molds itself to the shape of the molecule.

Today, enzymes are the most consequential class of drug targets. Overall, more than 52% of recent small molecule clinical drug candidates (DCs) come from enzymes. Conspicuously, 30% of all DCs come from kinases, which are discussed more extensively later in this chapter, and 23% of all DCs come from other enzymes. In contrast, 17% of all DCs come from GPCR's and 9% of all DCs come from epigenetics.[7]

1.2.1 Competitive Inhibitors

Competitive inhibitors act at the substrate-binding site and therefore compete with the substrate. They are also known as orthostatic inhibitors, and many of them are *reversible inhibitors.* In this section, we discuss six classes of competitive inhibitors including

1. ACE inhibitors;
2. Kinase inhibitors;
3. HCV NS3/4A serine protease inhibitors;

4. DPP-4 inhibitors;
5. Proteasome inhibitors; and
7. Novel α-cyano-acrylamide reversible inhibitors.

1.2.1.1. ACE Inhibitors

As shown in Figure 1.2, renin transforms angiotensinogen (a peptide with 450 amino acids) to angiotensin I (a peptide with 10 amino acids). Subsequently, angiotensin-converting enzyme (ACE) converts its substrate angiotensin I to the more active angiotensin II, a peptide with eight amino acid, the product. ACE is a chloride-dependent zinc metallopeptidase that is present in both membrane-bound and soluble forms. ACE is a hydrolase (protease), and its action is to speed up hydrolysis reaction of the Phe–His peptide bond. ACE contains two homologous active sites in which a zinc ion catalyzes the cleavage of C-terminal dipeptides (e.g., His–Leu) from various proteins, including angiotensin I and bradykinin. Cleavage of angiotensin I by ACE generates the octapeptide angiotensin II. Angiotensin II activates membrane-bound angiotensin receptors, the end result of which is vasoconstriction and sodium retention leading to a net increase in blood pressure. Since vasodilation and salt excretion are among the many processes mediated by bradykinin, the destruction of circulating bradykinin by ACE is also believed to contribute to hypertension.[8]

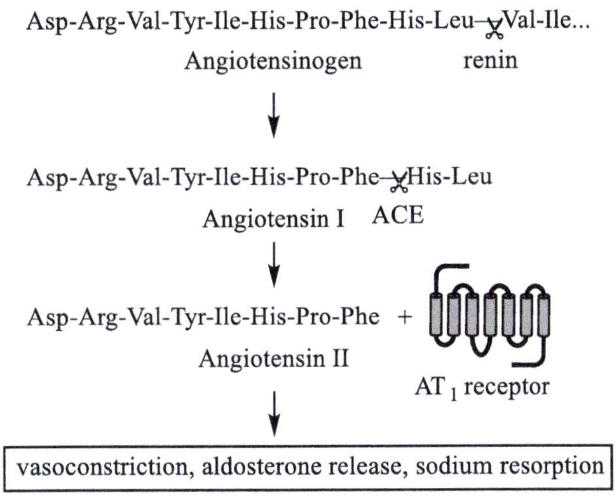

Figure 1.2 The renin-angiotensin system (RAS).

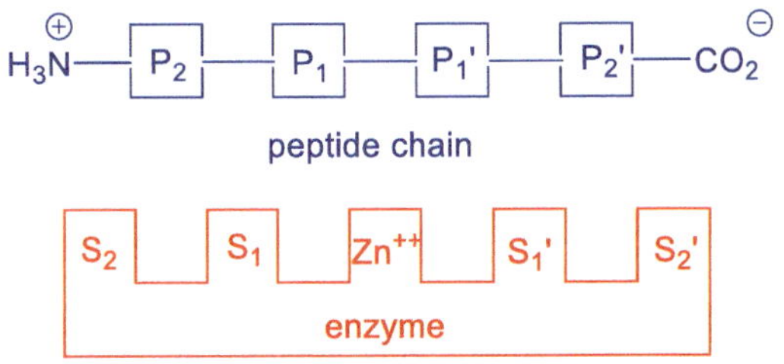

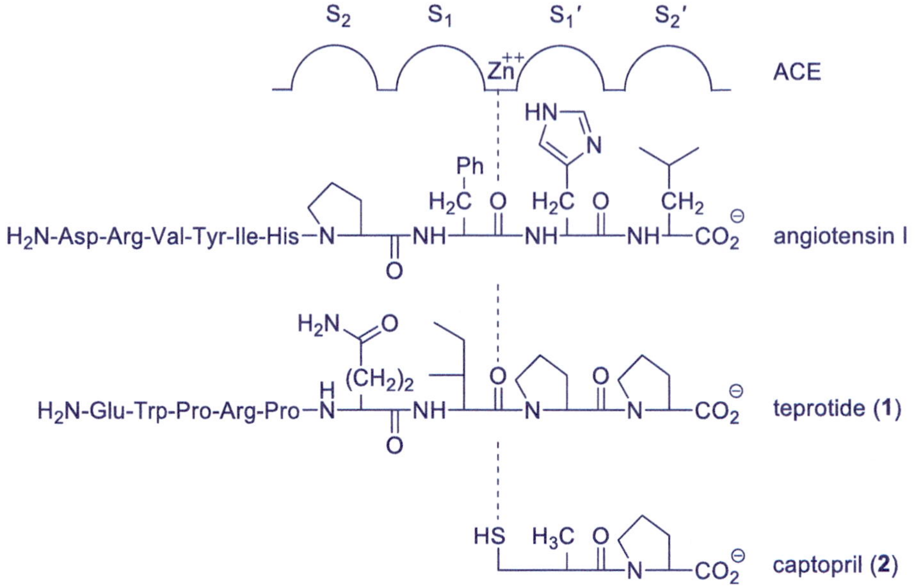

Figure 1.3 Definition of binding sites for zinc-containing proteases and proposed binding of angiotensin I and early ACE inhibitors.

As shown in Figure 1.3, biochemist David Cushman and organic chemist Miguel A. Ondetti at Squibb isolated a nonapeptide, teprotide (**1**) from a poisonous venom extract of the Brazilian pit viper *Bothrops jararaca*. Using teprotide (**1**) as a starting point, they curtailed the molecule and replaced its carboxylate group with a thiol (–SH) and achieved a significant increase in potency in ACE inhibition. The resulting drug became the first oral ACE inhibitor, captopril (Capoten, **2**). Captopril (**2**) mimics the functions of angiotensin I and its thiol group chelates with the Zn^{++} cation at the active site on ACE. Since its binding to ACE competes with the substrate angiotensin I, captopril (**2**) is a *bona fide* competitive inhibitor. It is also a reversible inhibitor (Figure 1.4).

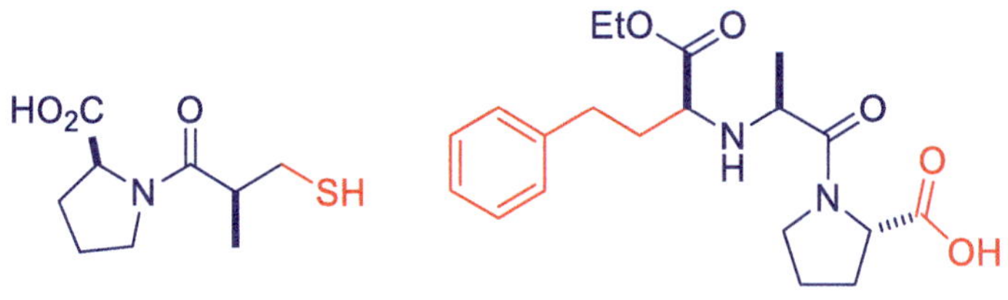

captopril (Capoten, **2**)
ACE inhibitor

enalapril (Vasotec, **3**)
ACE inhibitor

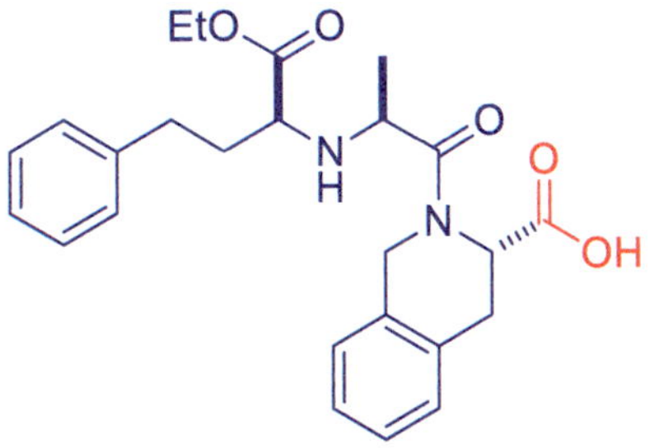

quinapril (Accupril, **4**)
ACE inhibitor

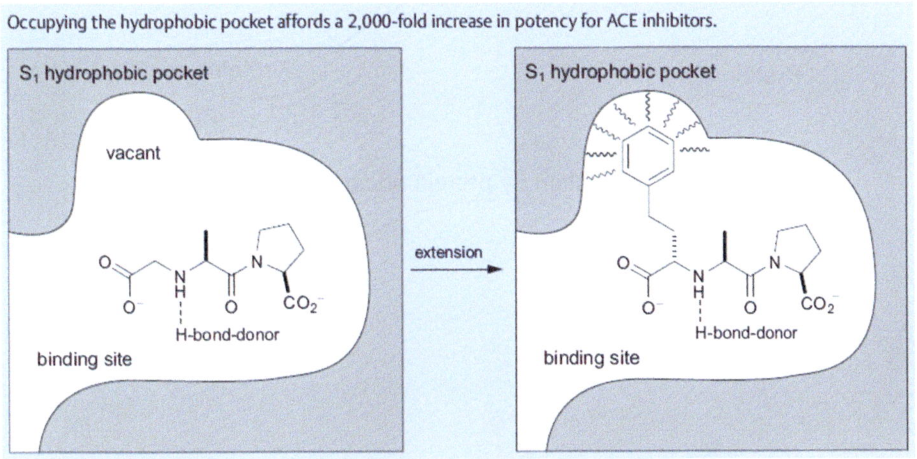

Figure 1.4 S_1 hydrophobic binding pocket inside ACE.

Captopril (**2**) suffers from a trio of side effects associated with the thiol group, including bone marrow growth suppression due to a decrease in circulating white blood cells, skin rash, and loss of taste. To improve on captopril (**2**), Merck scientists led by Arthur A. Patchett replaced the thiol group with carboxylate to avoid thiol's liabilities. Their greatest discovery was addition of the phenylethyl group, which took advantage of a previously unexplored hydrophobic pocket at the binding site but not utilized by captopril (**2**), which resulted in a 2,000-fold increase of potency![9] They arrived at enalaprilat, which suffered poor oral bioavailability. They simply converted the acid into its corresponding ethyl ester, creating enalapril (Vasotec, **3**), a prodrug of enalaprilat, with excellent oral bioavailability. Another popular ACE inhibitor is quinapril hydrochloride (Accupril, **4**), discovered by Parke–Davis. The tetrahydroisoquinoline to replace proline not only provided novel IP, but also boosted its bioavailability. Both enalapril (**3**) and quinapril (**4**) are competitive inhibitors. Their carboxylate groups are the chelating group that bind reversibly to the zinc cation at the active site.

1.2.1.2. Kinase Inhibitors

The field of kinase inhibitors is a *tour de force* of medicinal chemistry. Tremendous successes have been achieved during the last two decades as evidenced by more than 30 kinase inhibitors on the market. The majority of kinase reversible inhibitors are competitive inhibitors. What do they compete with? Adenosine triphosphate (ATP)!

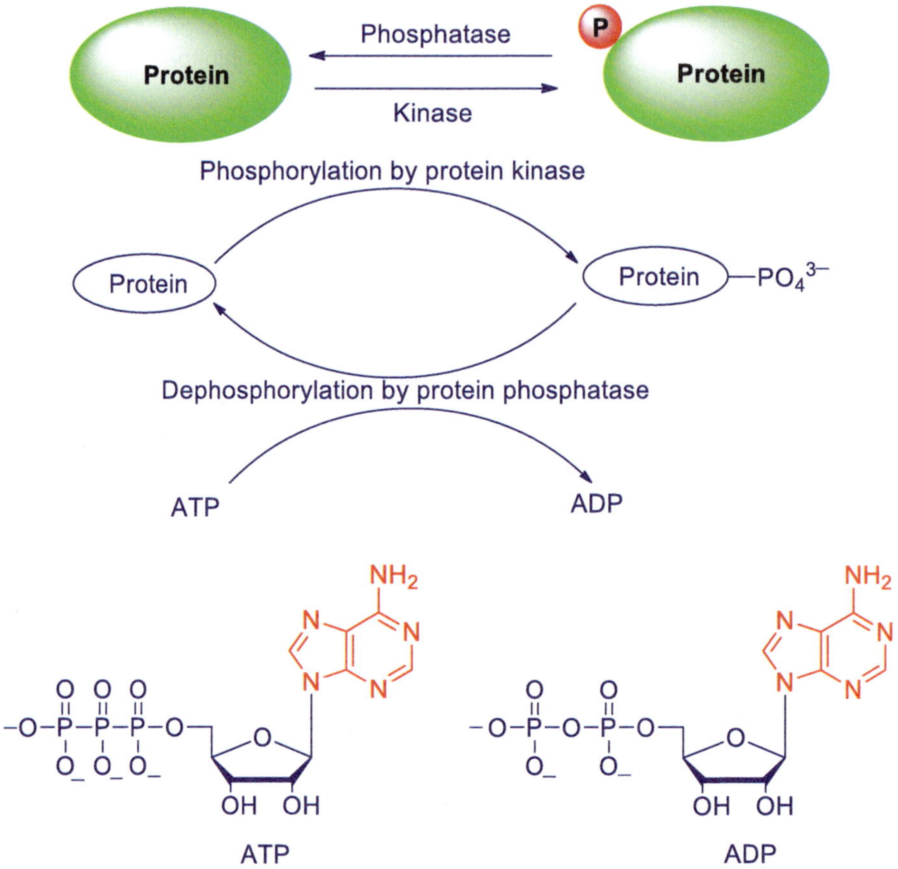

Figure 1.5 The function of protein kinases and phosphatases.

As far as an enzyme goes, a kinase's task is very straightforward. As shown in Figure 1.5, protein kinases modulate intracellular signal transduction by catalyzing the phosphorylation of specific proteins. In contrast, protein phosphatases function through the dephosphorylation of proteins to modulate biological activities. Many cellular processes result from the interplay between phosphorylation by protein kinases and dephosphorylation by protein phosphatases. Therefore, protein kinase inhibition is an area for therapeutic intervention against a variety of diseases such as cancer, inflammatory disorders, and diabetes. At some point 10 years ago, more than 25% research projects in the drug industry was focused on kinases. It is worth noting that although there are over 5,000 protein kinases in human, there are only a few phosphatases. Most kinase inhibitors are flat aromatic compounds that mimic the adenine portion of ATP. Since most kinase inhibitors bind reversibly the ATP binding pocket on the kinase and block ATP binding (thereby inhibiting kinase activities), they are ATP-competitive inhibitors.

Indeed, most protein kinase inhibitors are competitive inhibitors—they occupy the enzymes' ATP pocket (loop) and prevent phosphorylation of key amino acids on kinases. Those amino acids include serine (Ser, S), threonine (Thr, T), and tyrosine (Tyr, Y). The fact that most kinase competitive inhibitors are flat is a reflection of their mimicry of the adenosine moiety on ATP. The two MEK inhibitors on the market, in contrast, are noncompetitive, reversible (allosteric) inhibitors.

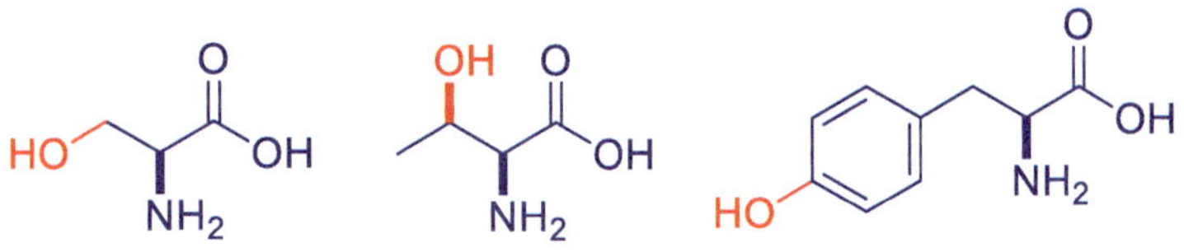

serine (Ser, S) threonine (Thr, T) tyrosine (Tyr, Y)

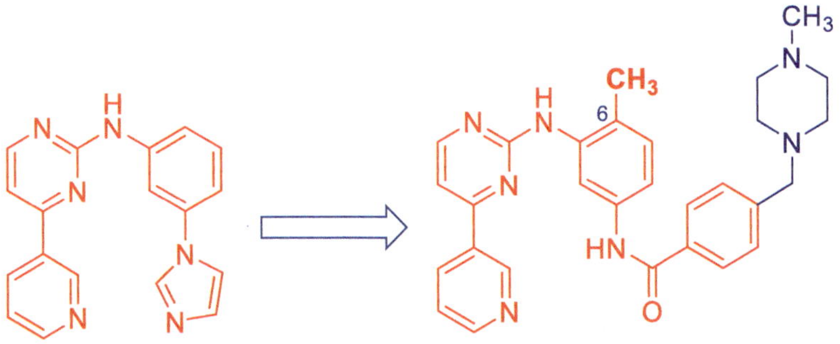

phenylaminopyrimidine **5**

imatinib (Gleevec, **6**)
Novartis, 2001
Bcr-Abl kinase inhibitor

The first kinase inhibitor on the market was Novartis' protein-tyrosine kinase inhibitor imatinib (Gleevec, **6**) for the treatment of chronic myeloid leukemia (CML) and gastrointestinal stromal tumors (GISTs). From an HTS campaign, Novartis' chemists identified phenylaminopyrimidine **5** as a PKC-α inhibitor with an IC_{50} of approximately 1 μM. It did not inhibit either an abnormal protein tyrosine kinase Abl or platelet-derived growth factor receptor (PDGFR), another kinase. Using phenylaminopyrimidine **5** as a starting point, intensive SAR studies involving more than 300 analogs eventually led to imatinib (**6**). Introduction of the "flag-methyl group" at the 6-position of the phenyl ring (highlighted in bold) abolished the activity against PKC-α. Attachment of the piperazine was intended to boost its aqueous solubility. The fact that it also binds to the kinase's active site was an added bonus. Imatinib (**6**) was obtained as a "selective" Bcr-Abl-tyrosine kinase inhibitor.[10] Bcr-Abl is an abnormal protein tyrosine kinase produced by the specific chromosomal abnormality called the Philadelphia chromosome that is a marker CML. The drug also inhibits another tyrosine kinase receptor, the c-kit receptor that is associated with GIST. Later studies revealed that imatinib (**6**) actually blocks a panel of at least eight protein kinases, including aforementioned Bcr-Abl, PDGFR, and c-kit.

The success of imatinib (**6**) was momentous. It revolutionized cancer therapy. Since kinase inhibitors only target cancer cells and leave normal cells alone, they are known as targeted cancer therapies with substantially fewer side effects. Before kinase inhibitors, toxicities were always closely associated with chemotherapies.

Again, the majority of kinase inhibitors are reversible, ATP-competitive inhibitors, especially the early ones. A representative of ATP-competitive inhibitors is sunitinib maleate (Sutent, **7**), an inhibitor of vascular endothelial growth factor receptors (VEGFR1 and VEGFR2) and platelet-derived growth factor receptors (PDGFR-α and PDGFR-β); fetal liver tyrosine kinase receptor 3 (Flt3) and stem-cell factor receptor (c-KIT). Another example is palbociclib (Ibrance, **8**), a selective cyclin-dependent kinase (CDK)-4/6 inhibitor.

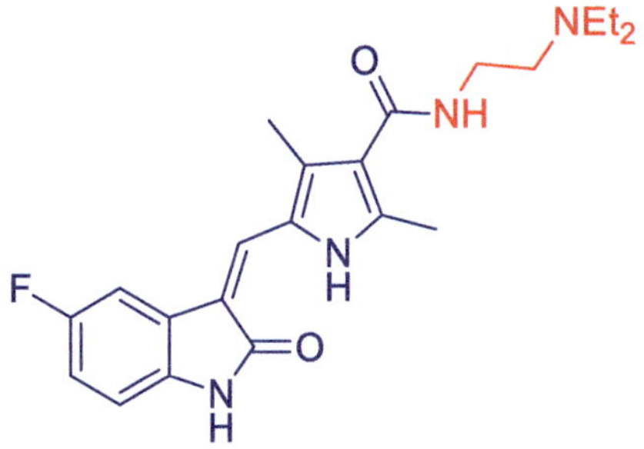

sunitinib (Sutent, **7**)
Sugen, 2007
VEGFR and PDGFR inhibitor

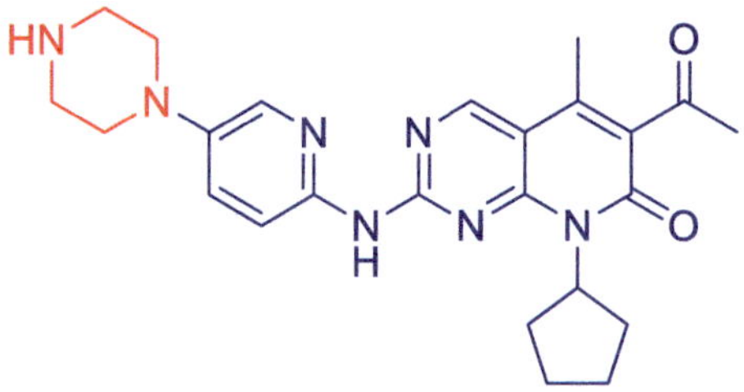

palbociclib (Ibrance, **8**)
Pfizer, 2015
CDK4/6 inhibitor

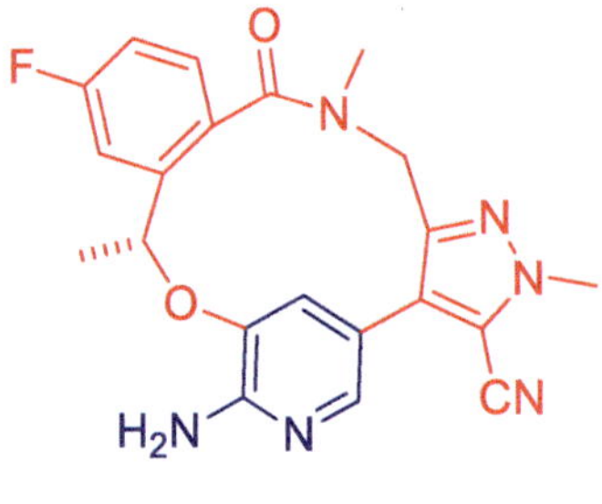

lorlatinib (Lorbrena, **9**)
Pfizer, 2018
ALK/ROS1 Inhibitor

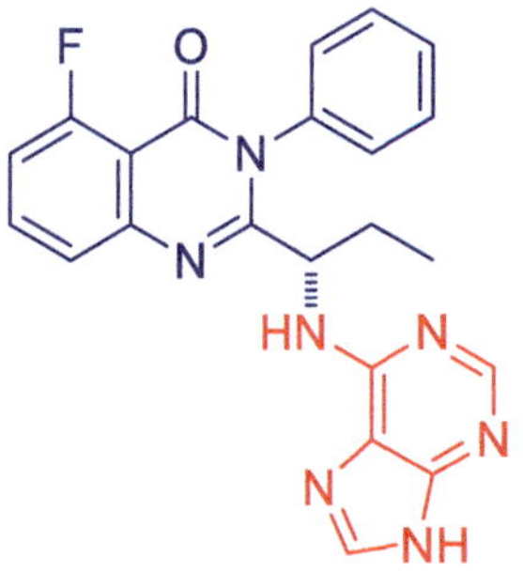

idelalisib (Zydelig, **10**)
Gilead, 2014
PI3Kδ inhibitor

Additional ATP-competitive inhibitors include lorlatinib (Lorbrena, **9**), an inhibitor of anaplastic lymphoma kinase (ALK) and c-Ros oncogene 1 (ROS1) fusion kinase and idelalisib (Zydelig, **10**), an inhibitor of phospoinositide-3-kinase-δ (PI3Kδ) kinase.

1.2.1.3. HCV NS3/4A Serine Protease Inhibitors

Both telaprevir (Incivek, **11**) and boceprevir (Victrelis, **12**), hepatitis C virus (HCV) NS3/4A serine protease inhibitors, are competitive inhibitors. They are also covalent reversible inhibitors.

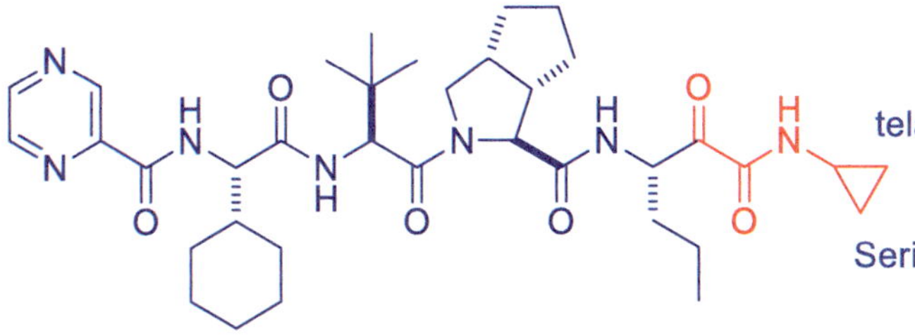

telaprevir (Incivek, **11**)
Vertex, 2011
HCV NS3/4A
Serine Protease Inhibitor

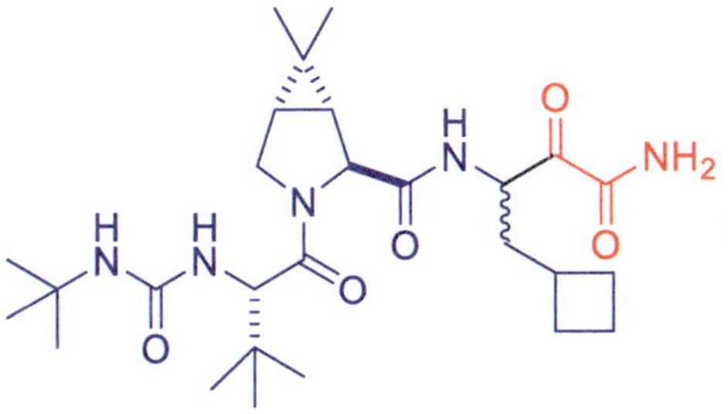

boceprevir (Victrelis, **12**)
Schering–Plough/Merck, 2011
HCV NS3/4A
Serine Protease Inhibitor

The HCV genome encodes a polyprotein of structural and nonstructural (NS) proteins. HCV NS3/4A serine protease is a noncovalent heterodimer of the N-terminal ~180 residue portion of the 631-residue NS3 protein with the NS4A co-factor. Similar to ACE, HCV NS3/4A serine protease also has a zinc cation. But unlike the catalytic Zn^{++} on ACE that is key to its activity, the Zn^{++} cation at the C-terminal domain distal from the active site on HCV NS3/4A serine protease may merely play a structural rather than catalytic role. Its *bona fide* catalytic site is composed of a catalytic triad of Ser_{139}, His_{57}, and Asp_{81}. Unfortunately, the active site is flat and open (solvent exposed) without any deep binding pockets. That was why no viable hits emerged from Vertex's HTS campaign, although they were the first to solve the first crystal structure of the active form of the NS3/4A protease in 1996.[11]

James W. Black famously said, "The most fruitful basis for the discovery of a new drug is to start from an old drug." Vertex ended up choosing the minimal length of the enzyme's natural substrate, a decapeptide with 10 amino acid residues (EDVVCC-SMSY), as their starting point of their chemical matter. During their SAR investigations, Vertex chemists realized that the covalent and reversible inhibitors were 10–1,000 times more potent than inhibitors that relied solely on noncovalent interactions. Among the *electrophilic warheads* including aldehydes, carboxylic acids, trifluoromethyl ketones, chloromethyl ketones, and α-keto-amides as serine protease inhibitors, α-keto-amide warhead resulted in the highest potency. Eventually, telaprevir (Incivek, **11**) emerged and was approved by the FDA in 2011.[11] Similarly, Schering–Plough/Merck's HCV NS3/4A serine protease inhibitor boceprevir (Victrelis, **12**) was also approved by the FDA in 2011.[12]

As far as the MOA is concerned, working in concert with His_{57} and Asp_{81}, the Ser_{139} on HCV NS3/4A serine protease adds to the ketone warheads on inhibitor **11**/**12** to form a covalent tetrahedral hemiacetal intermediate **13**, which closely mimics the transition state of the hydrolysis processes by the serine protease.

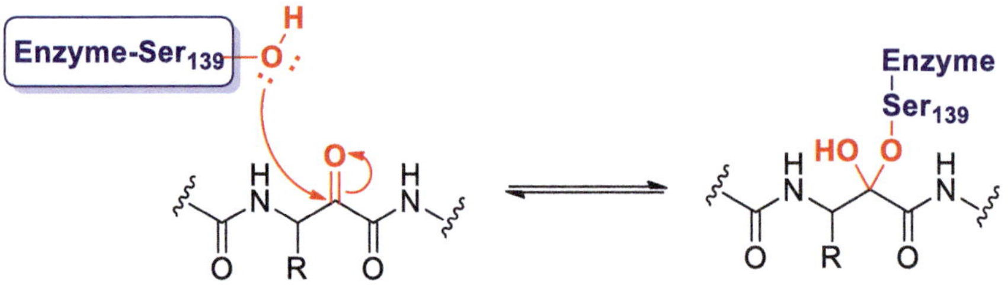

1.2.1.4. DPP-4 Inhibitors

Dipeptidyl peptidase IV (DPP-4) inhibitors emerged a decade ago as an important new therapy for the treatment of type 2 diabetes. As shown in Figure 6, DPP-4 is a non-classical serine protease. Its enzymatic function is to cleave incretin hormone (a 30 amino acid peptide) glucagon-like peptide 1 (GLP-1). Therefore, DPP-4 derived its name from the fact that the enzyme cleaves off two amino acids: dipeptide histamine–alanine (His–Ala).

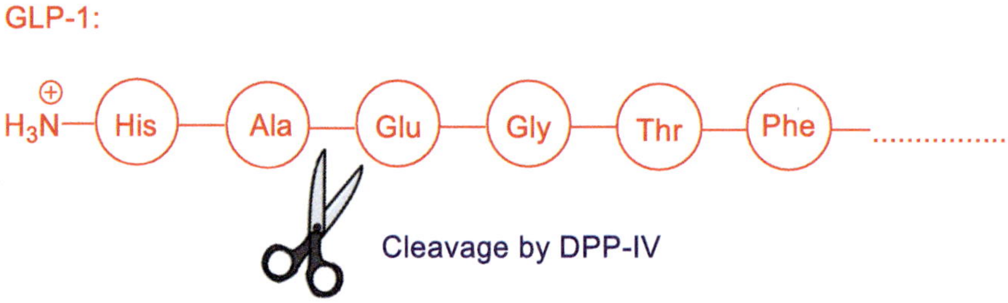

Figure 1.6 DPP-4 cleaves GLP-1 at the penultimate position from the N-terminus.

Meanwhile, as shown in Figure 1.7, DPP-4 inhibitors stimulate insulin secretion indirectly by enhancing the action of GLP-1 and glucose-dependent insulinotropic polypeptide (GIP). GLP-1 and GIP stimulate insulin secretion in a glucose-dependent manner, thus posing little or no risk for hypoglycemia. In addition, GLP-1 stimulates insulin biosynthesis, inhibits glucagon secretion, slows gastric emptying, reduces appetite, and stimulates the regeneration and differentiation of islet β-cells. DPP-4 inhibitors increase circulating GLP-1 and GIP levels in humans, which leads to decreased blood glucose levels, hemoglobin A_{1c} (H_bA_{1c}) levels, and glucagon levels. DPP-4 inhibitors possess advantages over alternative diabetes therapies, including a lowered risk of hypoglycemia, a potential for weight loss, and the potential for the regeneration and differentiation of pancreatic β-cells.[13] The first two "gliptin" DPP-4 inhibitors on the market were cyanopyrrolidines, vildagliptin (Galvus, **14**) and saxagliptin (Onglyza, **15**).[14] Both of them are reversible covalent inhibitors and the nitrile group serves as the electrophilic warhead. The P_1 cyanopyrrolidines occupy the S_1 sub-pocket of the peptidase.

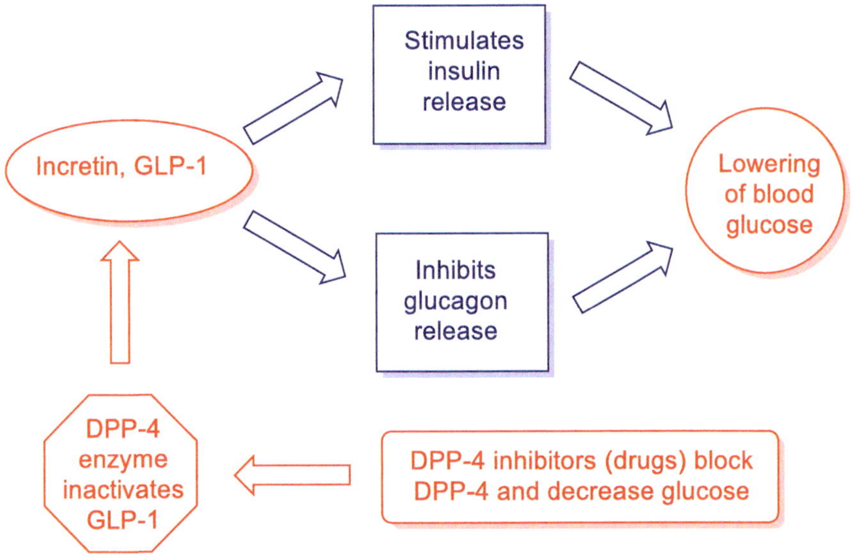

Figure 1.7 Functions of DPP-4 inhibitors.

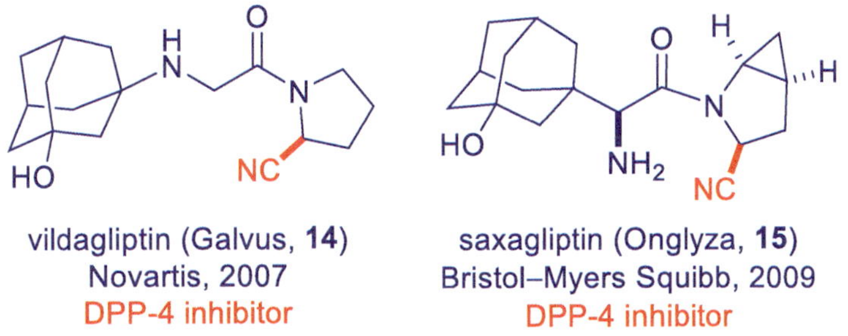

How does the reversible covalent bond form between the DPP-4 enzyme and its inhibitor vildagliptin (**14**) and saxagliptin (**15**)?

The Pinner reaction!

In organic chemistry textbooks, the Pinner reaction is the partial solvolysis of a nitrile to yield an iminoether (imidate, imidoester). Coincidently, this is exactly what happens between DPP-4 and its reversible covalent inhibitors. Thus, the hydroxyl group from the active site serine$_{630}$ of DPP-4 reacts with the nitrile group on the inhibitors and reversibly forms the corresponding imidate product.

The reversible covalent MOA of nitrile-containing gliptins may be depicted as shown below.[14] In essence, vildagliptin (**14**) binds to the DPP-4 enzyme and forms a noncovalent complex inside the S$_1$ pocket. Under the synergistic functions of serine$_{630}$ and tyrosine$_{547}$ at the active site, covalent bonding of the catalytic serine$_{630}$ hydroxyl with the pendant nitrile group forms covalent complex via a Pinner-like reaction. Collapse of the covalent complex then gave rise to imidate **16**, which is not stable and is reversibly converted to vildagliptin (**14**) in due course.

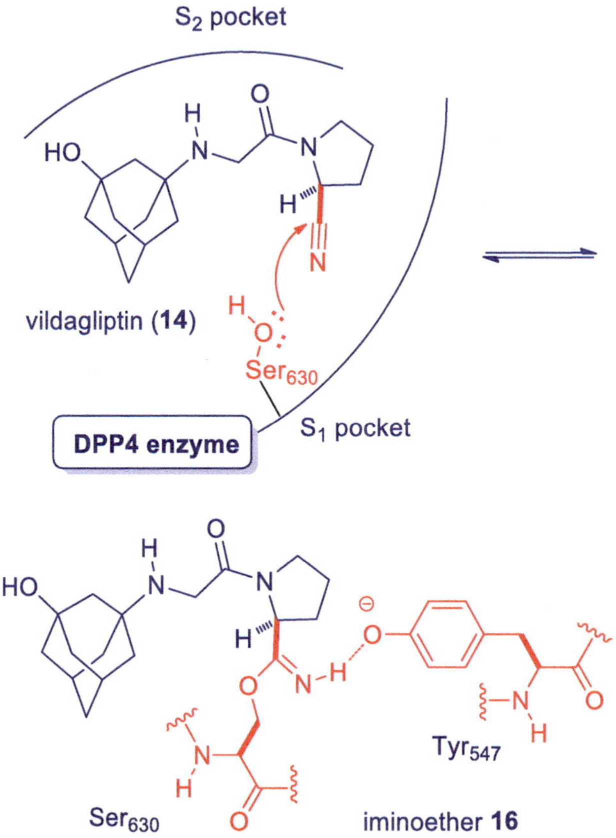

Saxagliptin (**15**) is a slow tight-binding DPP-4 inhibitor. Its MOA is similar to that of vildagliptin (**14**), namely, a reversible covalent inhibitor. Interestingly, one additional experiment often ruins a beautiful story. In an effort to look for backups to saxagliptin (**15**), BMS arrived at compound BMS-538305 (**17**) as the result of "wholesale removal" of its nitrile warhead. BMS-538305 (**17**) turned out to be very potent (Ki = 10 nM).[15] Since it cannot be a covalent inhibitor, it is most likely a "garden variety" reversible competitive inhibitor.

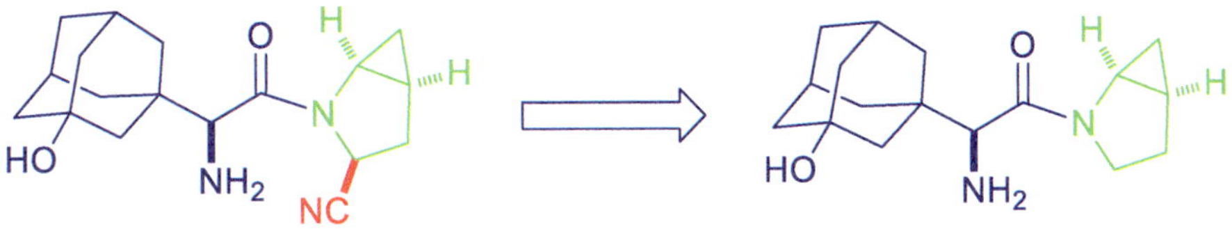

saxagliptin (**15**)
DPP-4 inhibitor

BMS-538305 (**17**)
DPP-4 inhibitor

1.2.1.5. Proteasome Inhibitors

The proteasome is a large protein complex that has a central role in the regulation of cellular function by catalyzing the ATP-dependent degradation of cellular proteins. In the normal healthy cell, the majority of intracellular proteins are labeled through poly-ubiquitination, targeting them for proteolysis within the multicatalytic 26S proteasome. The 26S proteasome, in turn, is a cylindrical structure comprised of a 20S catalytic core, with caspase-like, trypsin-like, and chymotrypsin-like (CT-L) activities. The catalytic core is capped by two 19S regulatory subunits that are involved in directing the entry of poly-ubiquitin tagged proteins into the enzyme complex.[16]

Proteasome inhibitors bortezomib (Velcade, **18**) and ixazomib (Ninlaro, **19**) preferentially inhibits the CT-L activity of the proteasome, resulting in accumulation of pro-apoptotic proteins in the cell, ultimately leading to apoptosis and cell death. They exert their proteasome inhibition through the boronic acid, which forms covalent bonds with an active-site threonine (Thr, T). Through proteasomal inhibition, bortezomib (**18**) and ixazomib (**19**) stabilizes IκB, activates c-Jun-terminal kinase, and stabilizes the CDK inhibitors p21 and p27, the tumor suppressor p53 and pro-apoptotic proteins.[16]

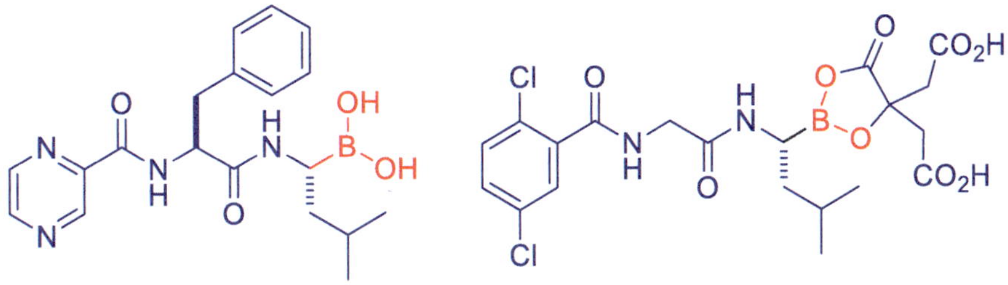

bortezomib (Velcade, **18**)
Millennium, 2003
proteasome inhibitor

ixazomib (Ninlaro, **19**)
Millennium, 2015
proteasome inhibitor

On the molecular level, as shown below, crystal structure of the boronic acid-based proteasome inhibitor bortezomib (**18**) in complex with the yeast 20S proteasome sheds light to illuminate the covalent nature of the inhibitor.[17] The β5 (chymotrypsin-like) sites are the most important sites in protein breakdown. Its N-terminal threonine (Thr 1) adds to bortezomib (**18**) and forms tetrahedral adduct **20**, as detected by the X-ray structure of the cocrystal.

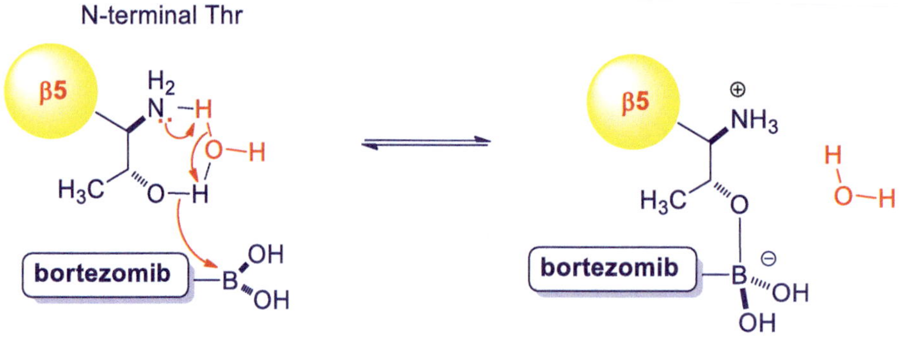

bortezomib (**18**) tetrahedral adduct **20**

Many proteasome inhibitors work via forming covalent bonds with proteasome as their MOA. The warheads include aldehydes, epoxyketones (*vide infra*, see Section 1.2.5), α-ketoaldehydes, vinylsulfones, and boronates. Among these warheads, boronic acid moiety is special because it ensures increased specificity for the proteasome, as opposed to earlier generations of synthetic inhibitors. For instance, peptide aldehydes showed cross-reactivity toward cysteine proteases and low metabolic stability. Furthermore, the boronic acid core ensures high affinity for hard oxygen nucleophiles in contrast to soft cysteine nucleophiles, according to the Lewis′ hard–soft acid–base (HSAB) principle. As expected, the boron atom covalently interacts with the nucleophilic oxygen lone pair of Thr1O$^{\gamma}$. Tetrahedral adduct **20** is further stabilized by a hydrogen bond between the N-terminal amino group of threonine and one of the hydroxyl groups of the boronic acid.[17] This hydrogen bond explains why boronates are more potent inhibitors of proteasomes than of serine proteases, a group of enzymes that they were originally developed to inhibit.

1.2.1.6. Novel α-Cyano-acrylamide Reversible Inhibitors

Irreversible covalent inhibitors are historically avoided because of potential adverse effects. Although great strides have been made on the front of irreversible covalent kinase inhibitors to address mutation issues, concerns about off-target modifications motivate the development of reversible cysteine-targeting kinase inhibitors.

In 2012, Taunton cleverly invented the concept of reversible conjugate addition to kinase drug discovery.[18] Ordinary acrylamides form covalent bonds with cysteine. But adding an additional electron-withdrawing group such as a nitrile group at the α-position makes the acrylamides reversible. As shown below, hetero-Michael addition of cysteine to α-cyano-acrylamide gives rise to the β-thioether. Since the α-nitrile adds much elevated resonance stabilization of the corresponding anion, the resultant adduct favors the reverse reaction. These covalent reversible inhibitors remain covalently linked to their target protein as long as they are stabilized by additional noncovalent interactions with the binding pocket, which drastically increases target residence times. Upon target degradation or after nonspecific cysteine labelling, however, the free and unmodified inhibitor is released. The utility of this concept was demonstrated by designing Bruton′s tyrosine kinase (BTK) reversible covalent inhibitors. The cysteine-reactive cyanoacrylamide electrophile offered those BTK inhibitors with elongated biochemical residence time spanning from minutes to 7 days.[19]

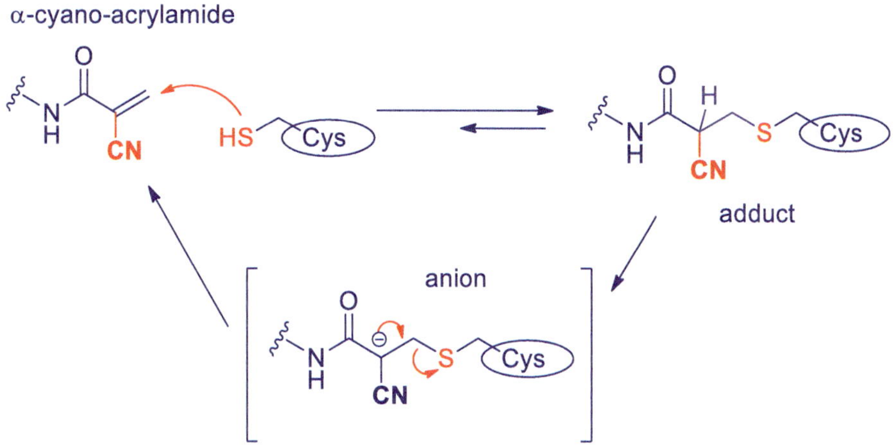

Many additional classes of enzymes have competitive inhibitors as drugs on the market. They include

1. HIV protease, integrase, and reverse transcriptase inhibitors to treat AIDS;
2. Histone deacetylase (HDAC) inhibitors to treat cancer;
3. 3-Hydroxy-3-methylglutaryl coenzyme A (HMG-CoA) reductase inhibitors (statins) to lower cholesterol;
4. Poly adenosine diphosphate (ADP) ribose polymerase (PARP) inhibitors for treating cancer;
5. Phosphodiesterase 5 (PPD-5) inhibitors to treat erectile dysfunction (ED); and
6. Dipeptidyl peptidase IV (DPP-4) inhibitors for the treatment of diabetes.

1.2.2 Allosteric Inhibitors

Many *allosteric inhibitors* are non-competitive, reversible inhibitors. In this section, we will discuss allosteric kinase inhibitors, allosteric phosphatase inhibitors, and allosteric nonnucleoside reverse transcriptase inhibitors (NNRTIs). In this section, three classes of allosteric inhibitors are exemplified:

1. Allosteric Kinase Inhibitors;
2. Allosteric Phosphatase Inhibitors; and
3. Allosteric NNRTIs.

1.2.2.1. Allosteric Kinase Inhibitors

In the 1990s, my former colleagues at Parke–Davis fortuitously discovered from HTS hits, several mitogen-activated protein kinase/extracellular signal-regulated kinase (MEK) inhibitors that were not ATP or extracellular signal-regulated kinase (ERK) competitive. In another word, they inhibited protein kinase by blocking phosphorylation without directly targeting ATP binding. The serendipitous discovery led to the discovery of PD0184352 (CI-1040, **21**) and PD0325901 (**22**). It was learned several years later that the original MEK screening protein was an intrinsically active truncation mutant rather than a partially phosphorylated form, which, unlike activated phosphoMEK, was susceptible to the original HTS hits, **21** and **22**.[20]

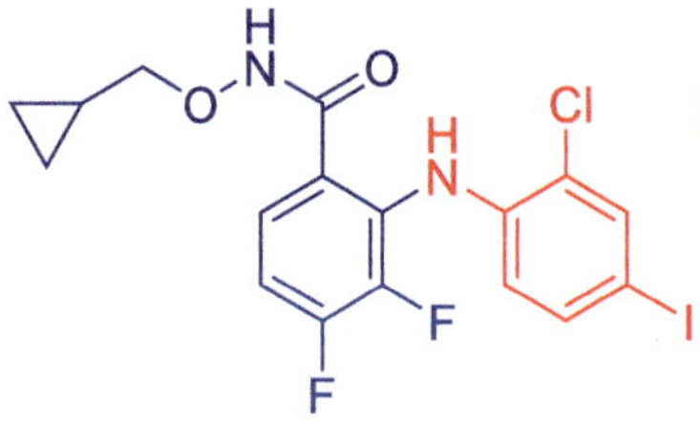

PD-0184352 (CI-1040, **21**)
MEK1/2 inhibitor

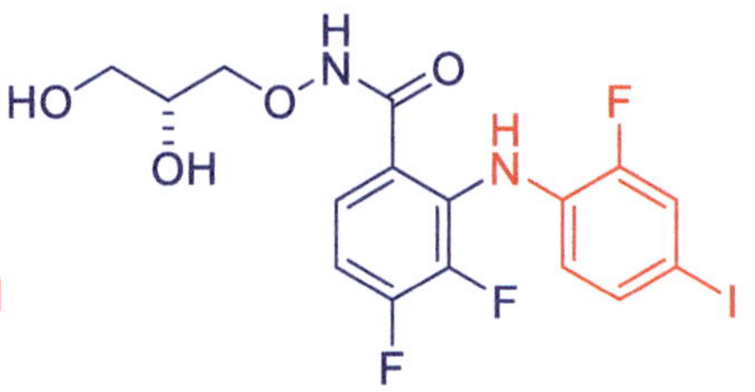

PD-0325901 (**22**)
MEK1/2 inhibitor

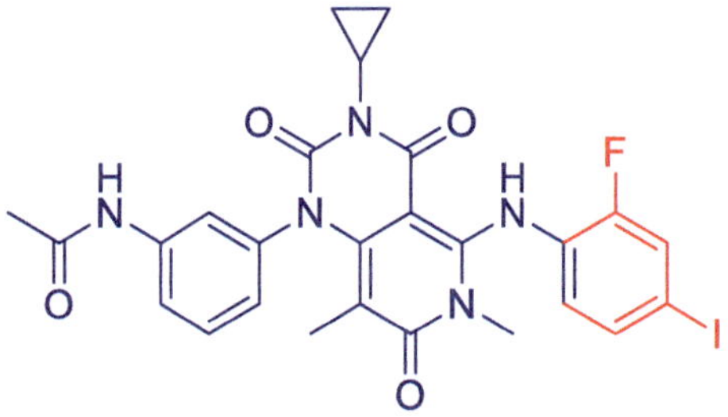

trametinib (Mekinist, **23**)
GSK, 2013
MEK1/2 inhibitor

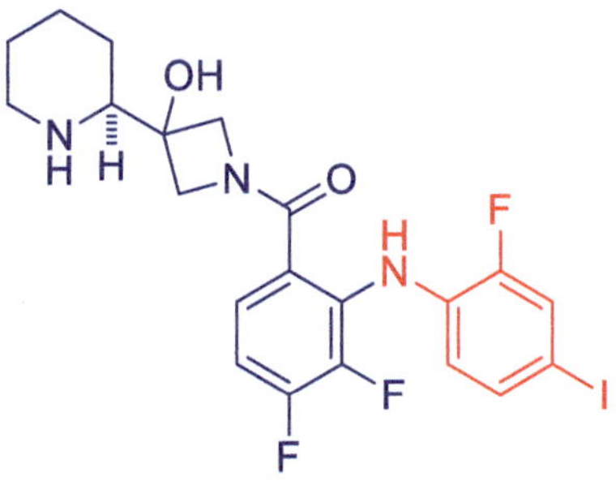

cobimetinib (Cotellic, **24**)
Exelixis/Genentech, 2015
MEK1/2 inhibitor

Although **21** and **22** never reached the market due to business, political, and other reasons (for instance, the hydroxamate functionality is a structural alert), they paved the road for the success of two other selective MEK1/2 allosteric inhibitors. One of them is GSK's trametinib (Mekinist, **23**)[21] and the other is Exelixis' cobimetinib (Cotellic, **24**, see Figure 1.8 for its allosteric binding site at the MEK1 enzyme).[22] Both of them are MEK1/2 inhibitors approved by the FDA to treat metastatic melanoma carrying the BRAF V600E mutation. In 2018, the third inhibitor, Array Biopharma's binimetinib (Mektovi) was approved by the FDA, in combination with encorafenib (Braftovi, a

BRAF inhibitor), for the treatment of patients with un-resectable or metastatic melanoma with a BRAF V600E or V600K mutation.

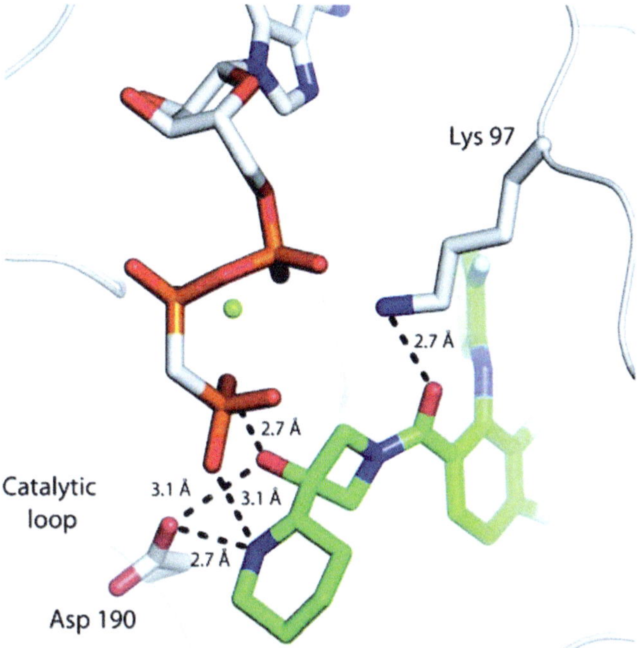

Figure 1.8 MEK1:AMP–PCP ternary complex co-crystal structure for **24**. Dash lines indicate key contacts for the carboxamide and aminoethanol fragments. The allosteric MEK inhibitor **24** does not occupy the ATP catalytic loop. Source: Sebolt–Leopold and Bridges 2009.[20] Reprinted with permission.

Apart from MEK1/2 inhibitors **23** and **24**, it has been long believed that allosteric inhibitors are elusive for other kinases. Indeed, thus far, **23** and **24** are only two allosteric kinase inhibitors on the market approved by regulatory agencies. However, our knowledge gained over the last two decades has aided us to make headway in allosteric inhibitors for kinases. Recently, Novartis reported discovery of asciminib (ABL001, **28**), an allosteric tyrosine kinase inhibitor of BCR-ABL1.[23]

Asciminib (**28**) was discovered employing the fragment-based drug discovery (FBDD) strategy.[23] Since only allosteric inhibitors at the myristate pocket of ABL1 kinase was of interest, the ATP site was blocked by imatinib (**6**). The ABL1–imatinib complex was screened by nuclear magnetic resonance (NMR) T1ρ and waterLOGY ligand experiment. Primary aniline **25** and secondary aniline **26** were identified as initial fragment hits. Neither of the two initial fragment hits was active in a biochemical ABL1 kinase assay, as expected. Unexpectedly, although fragment hits are generally not expected in cellular assay, follow-up compounds such as **27** exhibited low micro-molar K_d values and had a GI_{50} value of 8 μM. Further structure-based optimization for potency, physicochemical, pharmacokinetic, and drug-like properties, culminated in asciminib (**28**), undergoing Phase III clinical studies in CML patients since 2017.

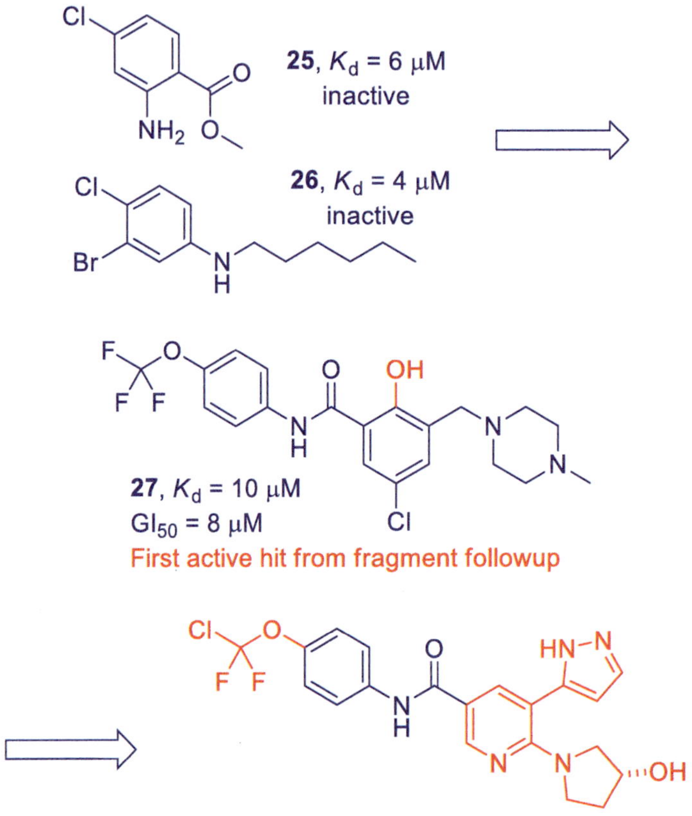

asciminib (ABL001, **28**)
allosteric Bcr-Abl1 inhibitor

1.2.2.2. Allosteric Phosphatase Inhibitors

Nowadays, allosteric inhibition is increasingly applied toward difficult drug targets. Recent achievement in the field of phosphatase inhibitors is a good example. Unlike kinases, phosphatases do not have a defined substrate like ATP to guide drug design. Src homology region 2-containing protein tyrosine phosphatase (SHP2) is a nonreceptor protein tyrosine phosphatase and scaffold protein. It is comprised of three domains: N-SH2, C-SH2, and PTP, where the active site resides (PTP is short for PTPase, i.e., *p*rotein *t*yrosine *p*hosphat*ase*, see Figure 1.9). Since the phosphate group binding site is highly positively charged and often does not have a distinctive small molecule pocket, competitive SHP2 inhibitors mimicking phosphate are very challenging. The initial competitive SHP2 inhibitors discovered during the last two decades invariably possessed ionizable functional groups, and thus had difficulty crossing cell membranes or enter bloodstream.[24,25]

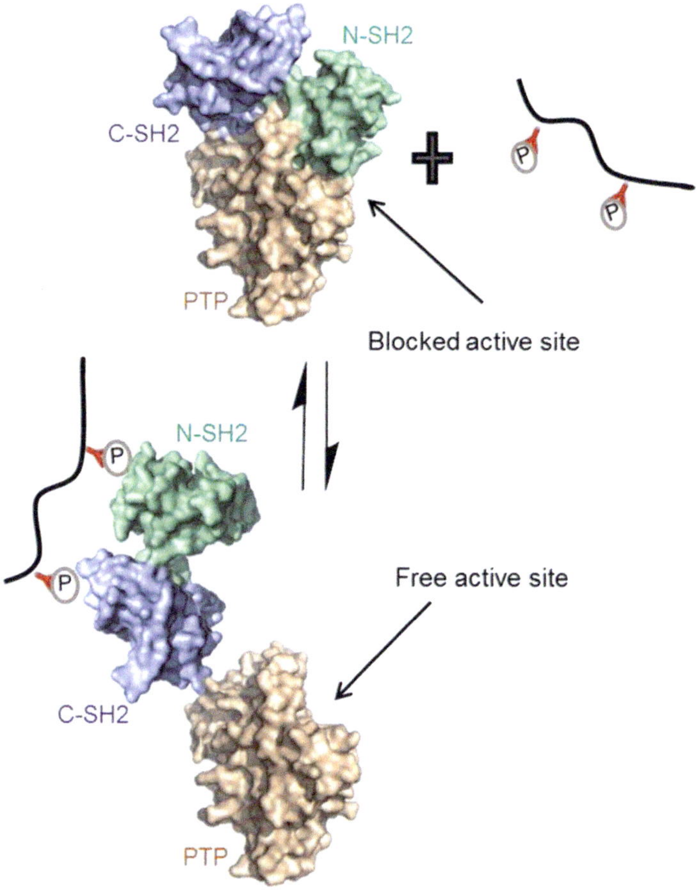

Figure 1.9 Equilibrium of SHP2 in closed (blockade active site) and open (free active site). Source: Garcia Fortanet et al. 2016.[25] Reprinted with permission of American Chemical Society.

In 2016, Novartis reported an allosteric SHP2 inhibitor SHP099 (**29**), which occupies a tunnel-like binding site (a pocket formed by the confluence of the three domains) in SHP2′s closed conformation. Because SHP2 is only active when adopting the open conformation, SHP099 (**29**) behaves like a molecule glue that prevent the opening of SHP2 (see Figure 1.9). As an allosteric inhibitor, SHP099 (**29**) does not need to be phosphate-like. It has appropriate affinity, cell permeability, and other properties that enable oral administration.[25] One of its analogs, TNO155, is now in clinical trials. If successful, a surge of drug discovery on phosphatases will ensue, similar to the way that the approval of imatinib (**6**) influenced the development of drugs on protein kinases.[26]

Expanding their initial success, Novartis described identification of a second, distinctive, previously unexplored binding site. Allosteric site-2 is a cleft formed at the surface of the N-terminal SH2 and PTP domains. They also identified SHP244 (**30**) as a weak SHP2 inhibitor with modest thermal stabilization of the enzyme (see Figure 1.10).[27]

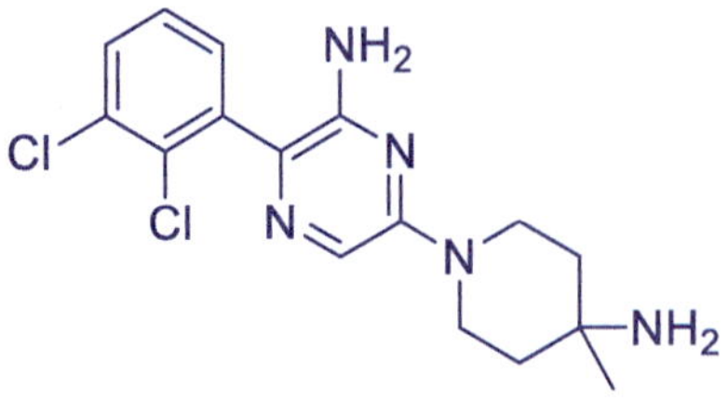

SHP099 (**29**)
SHP2 inhibitor

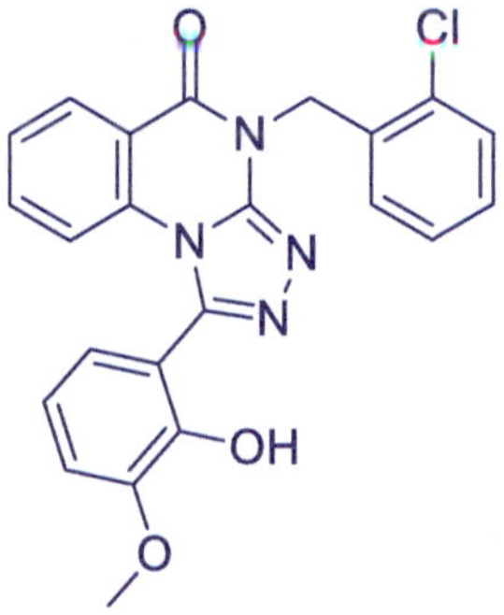

SHP244 (**30**)
site-2 SHP2 inhibitor

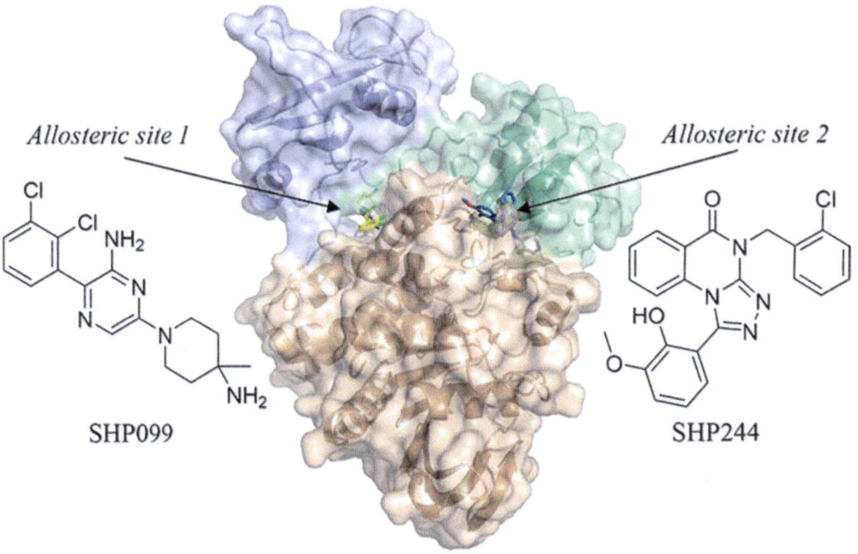

Figure 1.10 Dual allosteric inhibition of SHP2 phosphatase. Source: Fodor et al. 2018.[27] Reprinted with permission of American Chemical Society.

1.2.2.3. Allosteric NNRTIs

NNRTIs stands for nonnucleoside reverse transcriptase inhibitors.

HIV reverse transcriptase (RT) transforms the single-strained ribonucleic acid (RNA) viral genome to double-strained deoxyribonucleic acid (DNA) before entering the cell nucleus. Initial nucleoside reverse transcriptase inhibitors (NRTIs) such as azidothymine (AZT, Retrovir), lamivudine (3TC, Epivir), and abacavir (Ziagen) are considered prodrugs because they become active only after phosphorylation by kinases. They bind directly to the RT enzyme and are plagued by toxicities.

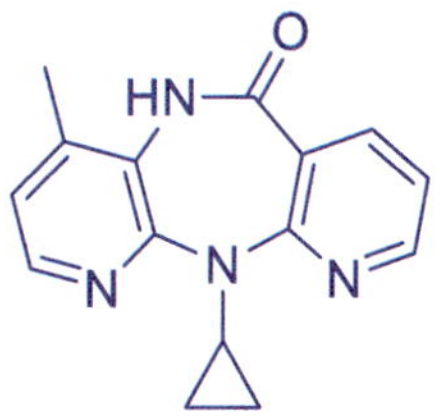

nevirapine (Viramune, **31**)
Boehringer Ingelheim, 1996
NNRTI

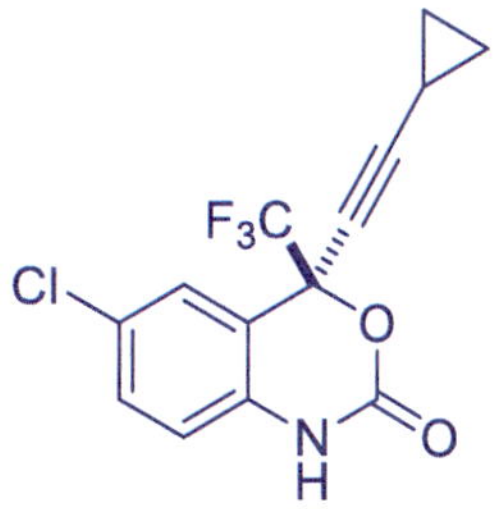

efavirenz (Sustiva, **32**)
Bristol-Myers Squibb/Merck, 1998
NNRTI

On the other hand, NNRTIs allosterically bind to reverse transcriptase to inhibit transcription. Unlike NRTIs, they do not require phosphorylation via intracellular metabolism to be active. The first-generation NNRTIs are nevirapine (Viramune, **31**), efavirenz (Sustiva, **32**), and delavirdine (Rescriptor, **33**). They adopt a "butterfly-like" conformation when bound to the allosteric site of the RT enzyme. Regrettably, the rapid emergence of drug resistance (at least nine amino acid mutations) dramatically reduced their potency, thus compromising the patient's clinical compliance.[28]

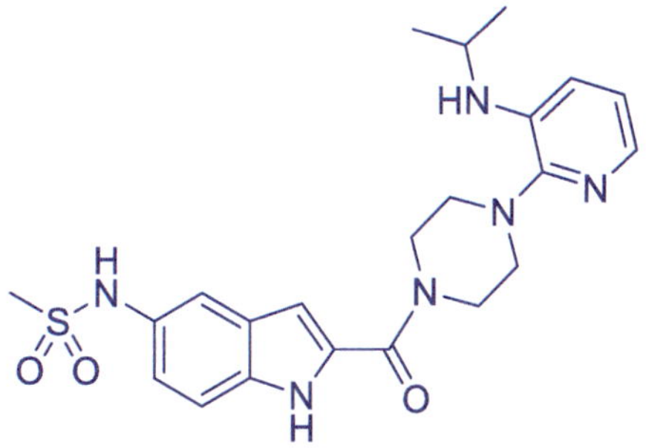

delavirdine (Rescriptor, **33**)
Upjohn/Pfizer, 1997
NNRTI

The second-generation NNRTIs etravirine (Intelence, **34**) and rilpivirine (Edurant, **35**) have a high genetic barrier to resist various clinically relevant mutations. This is because of their intrinsic structural flexibility. Although structurally very different, both the first-generation NNRTIs **31**–**33** and the second-generation NNRTIs etravirine (**34**) and rilpivirine (**35**) bind to the same hydrophobic binding site, namely the NNRTI binding pocket (NNIBP) that is located a short distance of 10 Å from the catalytic site.[29] Therefore, NNRTIs **31**–**35** are all allosteric inhibitors. It is worth mentioning that the NNIBP has high flexibility as it does not exist until binding with NNRTI, the formation of which is related to torsional rotations of the flexible side chains of some relevant amino acids.

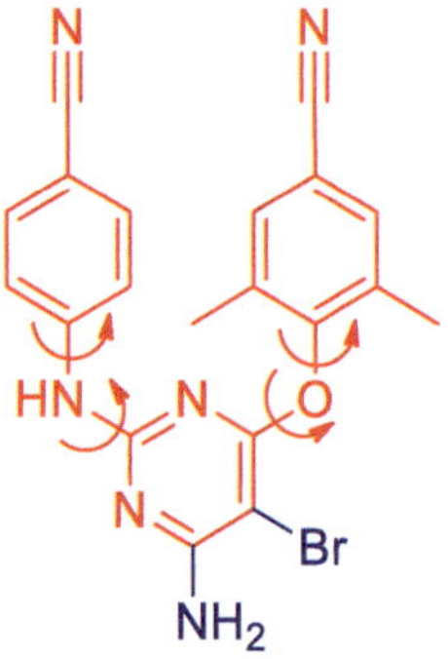

etravirine (Intelence, **34**)
Tibotec/J&J, 2008
NNRTI

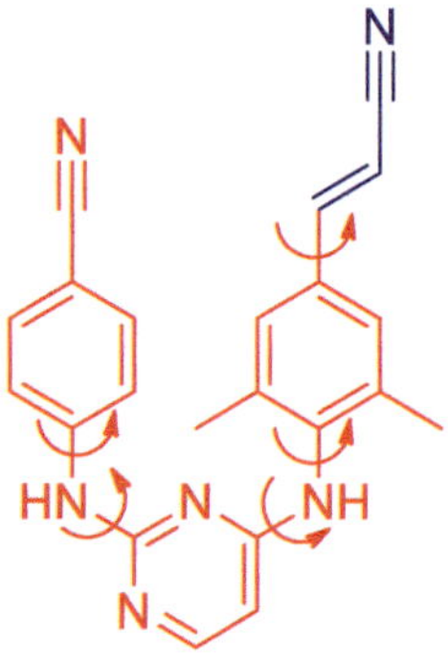

rilpivirine (Edurant, **35**)
Tibotec/J&J, 2011
NNRTI

Unlike the first-generation NNRTIs **31**–**33**, which adopt a "butterfly-like" conformation, the second-generation NNRTIs etravirine (**34**) and rilpivirine (**35**) tend to bind to HIV-1 RT in a "U"-shaped conformation (also named "horseshoe mode") in order to adapt to the changeable hydrophobic pocket. The conformational adjustments ("wiggling") and rotational and translational shifts ("jiggling") of the inhibitors within the binding pocket help them to retain potency against mutant HIV-1 viruses.[30] Interestingly, rilpivirine (**35**)'s behavior at low pH and its intrinsic flexibility may facilitate drug aggregation to spherical nanoparticles (100–200 nm in diameter at low pH) and favor the oral bioavailability.

1.2.3 Covalent Irreversible Inhibitors

Several advantages exist for covalent inhibitors such as increased biochemical efficacy, longer duration of action on the target, and lower efficacious doses. Historically, the quintessential classic drugs such as aspirin and penicillin G are both covalent inhibitors. Aspirin is an irreversible inhibitor of cyclooxygenase (COX) and operates via acylation of an active site serine (Cys, C) residue. Penicillin covalently binds to the active site serine of DD-transpeptidase. In addition, clopidogrel (Plavix) and omeprazole (Prilosec) are both prodrugs of covalent inhibitors after bio-activations.

In recent years, there has been a renewed interest in drug development of covalent inhibitors. Several of these inhibitors have been approved by the FDA and EMA and more covalent inhibitors are in early and late stage development. Covalent inhibitors can have superior affinity toward target proteins compared to their non-covalent counterparts, potentially resulting in improved therapeutic benefit. As of 2019, there are over 40 FDA-approved drugs with a proven covalent MOA.

Many amino acids could react with electrophilic warheads to form covalent bonds with covalent inhibitors. Cysteine (Cys, C) is the most frequently used nucleophile residue on the target proteins. This is not surprising because the thiol group is the strongest nucleophile present on all amino acids. Cysteine is a unique amino acid that possesses an aliphatic thiol group. It is also the least common amino acids in proteins, but most of the 518 human kinases have an accessible noncatalytic cysteine within the reach of the active site. In protein kinases, cysteines are not involved in catalysis but some regulatory functions such as sulfenylation or oxidation. Targeting cysteine located in close proximity to the kinase ATP binding sites has been the most fruitful strategy for covalent kinase inhibitors.

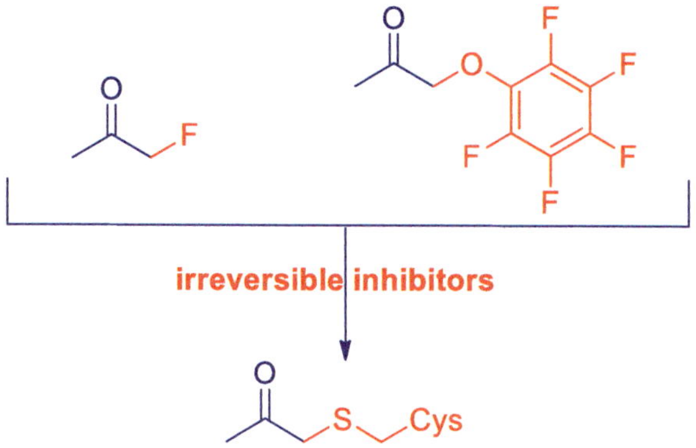

Since cysteine is not always available at the protein binding site, targeting lysine (Lys, K) would be the second best choice.

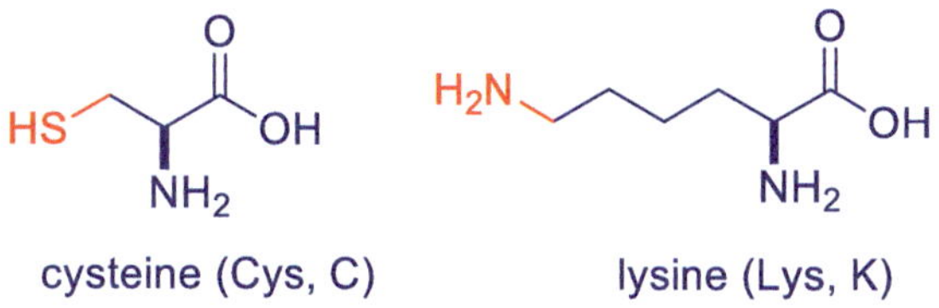

Nowadays, we are more open-minded toward noncatalytic residues including serine (Ser, S), threonine (Thr, T), tyrosine (Tyr, Y), methionine (Met, M), glutamate (Glu, E), and aspartate (Asp, D).

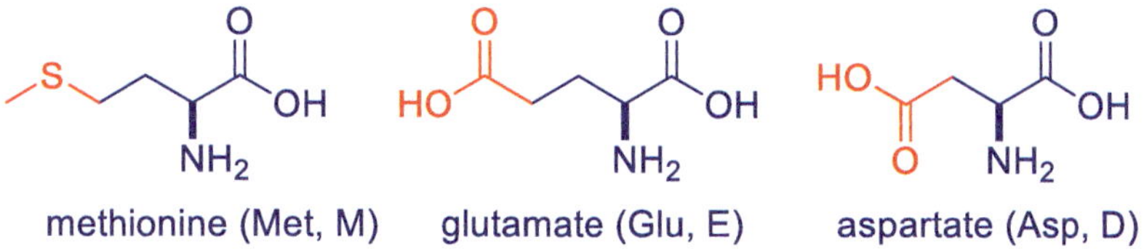

With regard to the "warheads," they have to be sufficiently stable against ubiquitous nucleophiles such as glutathione to reach their targets in cellular or *in vivo* settings. Thus far, acrylamides have been the most fruitful electrophiles for targeting cysteine in kinases due to their relative low intrinsic activity.

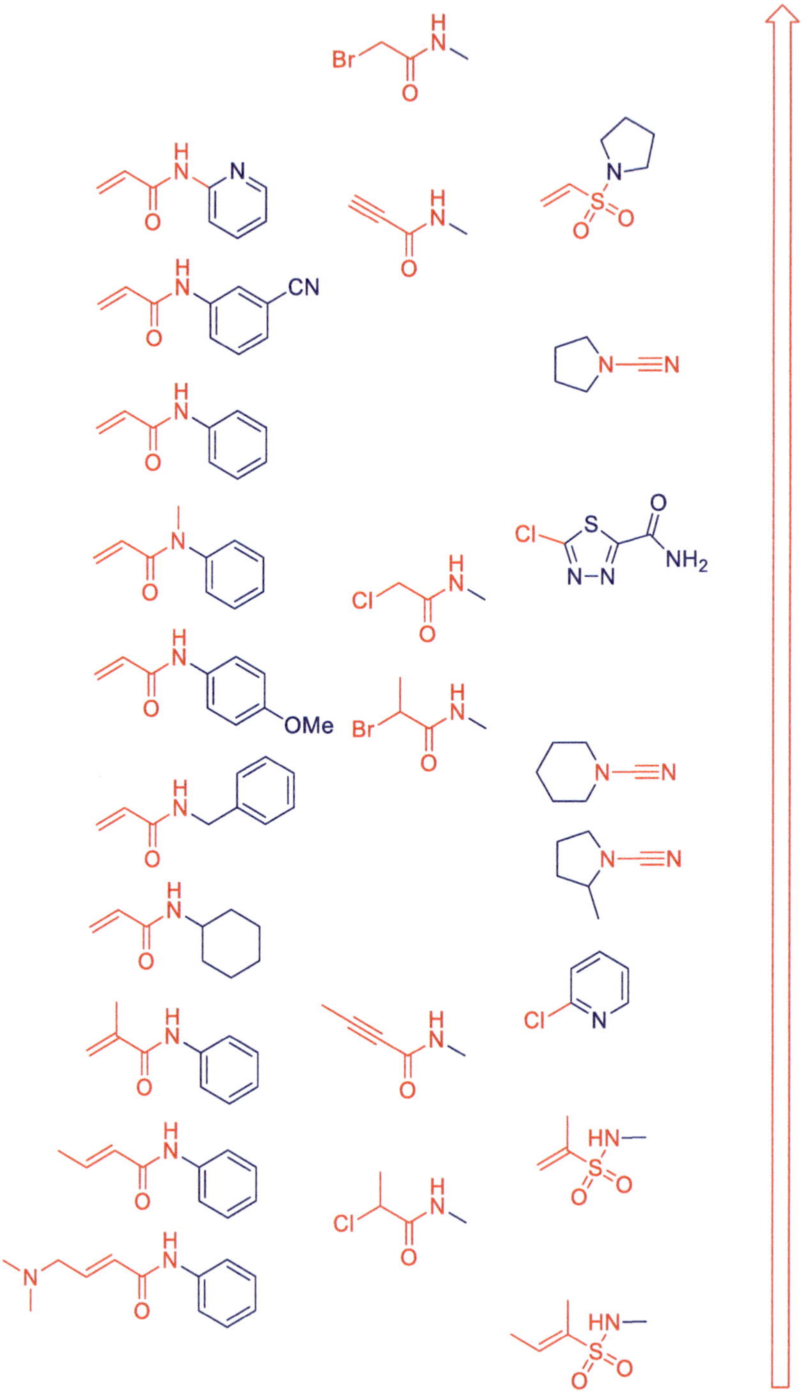

Figure 1.11 Warhead reactivity.

Warheads′ reactivity may be fine-tuned by varying electronic effects, strength of leaving group, and steric hindrance. As shown in Figure 1.11, as much as 500-fold change of reactivity may be achieved for modulation of the warhead′s reactivity.

As we will see in Chapter 4, an unselective Michael acceptor on a drug is a structural alert—it is possible to be a safety liability since it may form covalent bonds with nucleophiles in the physiological environment if the drug is not selective against the intended target. However, off-target effects associated with covalent inhibitors are typically due to covalent binding to other proteins, resulting in cell damage or immunological response.

Nowadays, covalent inhibitors are undergoing a reconnaissance. Unlike the garden variety Michael acceptor-containing compounds, these covalent inhibitors use functional groups with relatively low reactivity, combined with a highly selective reversible binding motif. Such inhibitors are commonly referred to as targeted covalent inhibitors (TCIs). Covalent inhibition is attractive in that it offers the potential for extended duration of pharmacodynamic modulation relative to pharmacokinetic profile of the inhibitor.

Five targets with covalent irreversible inhibitors will be highlighted in this section including:

1. Bruton′s tyrosine kinase (BTK) inhibitors;
2. Covalent EGFR inhibitors;
3. KRAS inhibitors;
4. Fatty acid amide hydrolase (FAAH) inhibitors; and
5. Monoacylglycerol lipase (MAGL) inhibitors.

1.2.3.1 Bruton′s Tyrosine Kinase (BTK) Inhibitors

Like most covalent kinase inhibitors to follow, BTK inhibitors, bind to the ATP binding pocket in protein kinases. The covalent bond is formed between the reactive functional group, often referred to as the "warhead," and a cysteine residue located in the binding pocket.

BTK inhibitor ibrutinib (Imbruvica, **36**) was the first covalent inhibitor for kinases on the market. Bruton′s tyrosine kinase (BTK), also known as agammaglobulinemia tyrosine kinase (ATK) or B-cell progenitor kinase (BPK), is a non-receptor tyrosine kinase. It has a pleckstrin-homology (PH) domain, SH3 and SH2 (Src homology) domains, and a kinase domain. The BTK polypeptide has 659 amino acid residues with a molecular weight of 76 kDa. Inhibition of BTK activity prevents downstream activation of the BCR pathway and subsequently blocks cell growth, proliferation, and survival of malignant B cells (Figure 1.12).[31]

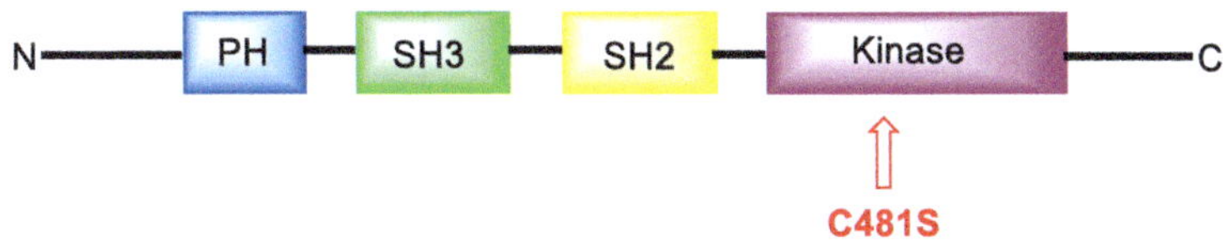

Figure 1.12 The structure of BTK. The C481S mutation in the kinase domain mediates resistance to ibrutinib (**36**).[32]

Ibrutinib (Imbruvica, **36**) is the first-in-class BTK inhibitor for mantle cell lymphoma, chronic lymphocytic leukemia (CLL), and Waldenstrom's macroglobulinemia. It is a selective small molecule inhibitor with subnanomolar activity against BTK. It reacts with Cys-481 at the rim of the ATP binding site of BTK, leading to inhibition of BTK enzymatic activity.[32] As a covalent irreversible inhibitor, ibrutinib (**36**) displays a decoupling between its pharmacokinetic and pharmacological properties. It is quickly cleared but still able to sustain inhibition of BTK activity because that once irreversibly bound with ibrutinib (**36**), BTK loses catalytic activity until regeneration by protein synthesis. In 2017, the second BTK inhibitor, acalabrutinib (Calquence, **37**), gained approval from regulatory agencies.[33] While it works through the same MOA as progenitor ibrutinib (**36**), acalabrutinib (**37**)'s warhead is a methylpropargyl amide. They are known as the first-generation BTK inhibitors.

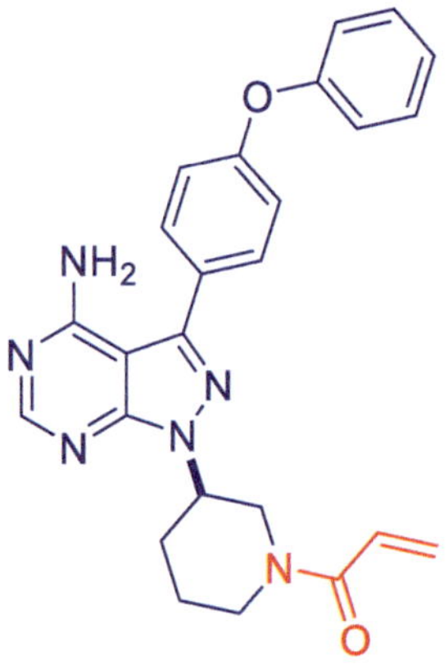

ibrutinib (Imbruvica, **36**)
Pharmacyclics/Jansen, 2013
BTK Inhibitor

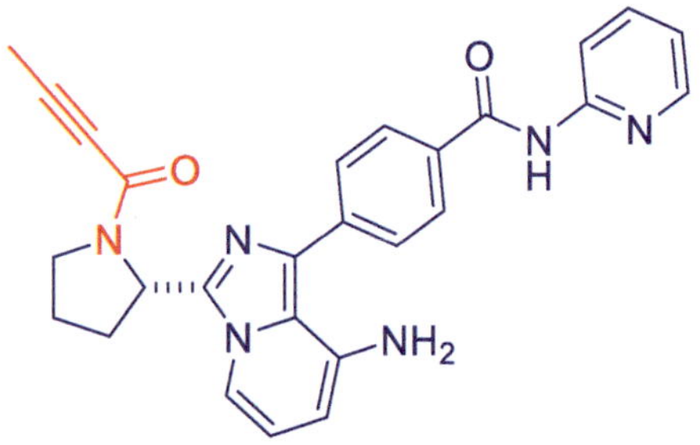

acalabrutinib (Calquence, **37**)
AZ/Acerta, 2017
BTK inhibitor

Cancer is indeed a formidable and cunning adversary. Cancer cells would do whatever it takes to survive and often it is exhausting to outsmart the enemy. Although ibrutinib (**36**) and acalabrutinib (**37**) can do wonders to cancer patients afflicted with B-cell malignancies, 75% of them develop resistance to them within 2 years. Close scrutiny revealed that a substitution of serine for cysteine at residue 481 (C481S, see Figure 1.11) took place. Such a mutation led to a less nucleophilic serine so that the first-generation BTK inhibitors are no longer effective. Efforts are underway to discover the second-generation BTK inhibitors.[34]

1.2.3.2 Covalent EGFR Inhibitors

Epidermal growth factor receptor (EGFR) inhibitors are one of the earliest kinase inhibitors on the market. Gefitinib (Iressa, **38**), erlotinib (Tarceva), and vandetanib (Caprelsa) are considered the first-generation EGFR inhibitors. They are all ATP-competitive inhibitors. After gefitinib (**38**) received the FDA approval in 2003, patients on EGFRs for over approximately one year acquired resistance through specific mutation within the binding pocket. Approximately 60% of patients treated with gefitinib (**38**) and erlotinib developed a T790M (a gatekeeper residue) single-point mutation responsible for the resistance to therapy. To combat the T790M mutation, gefitinib (**38**) was modified to afford covalent inhibitor afatinib (Gilotrif, **39**). It was designed to undergo a hetero-Michael addition with Cys_{797} at the active site of EGFR as shown in Figure 1.13.[35]

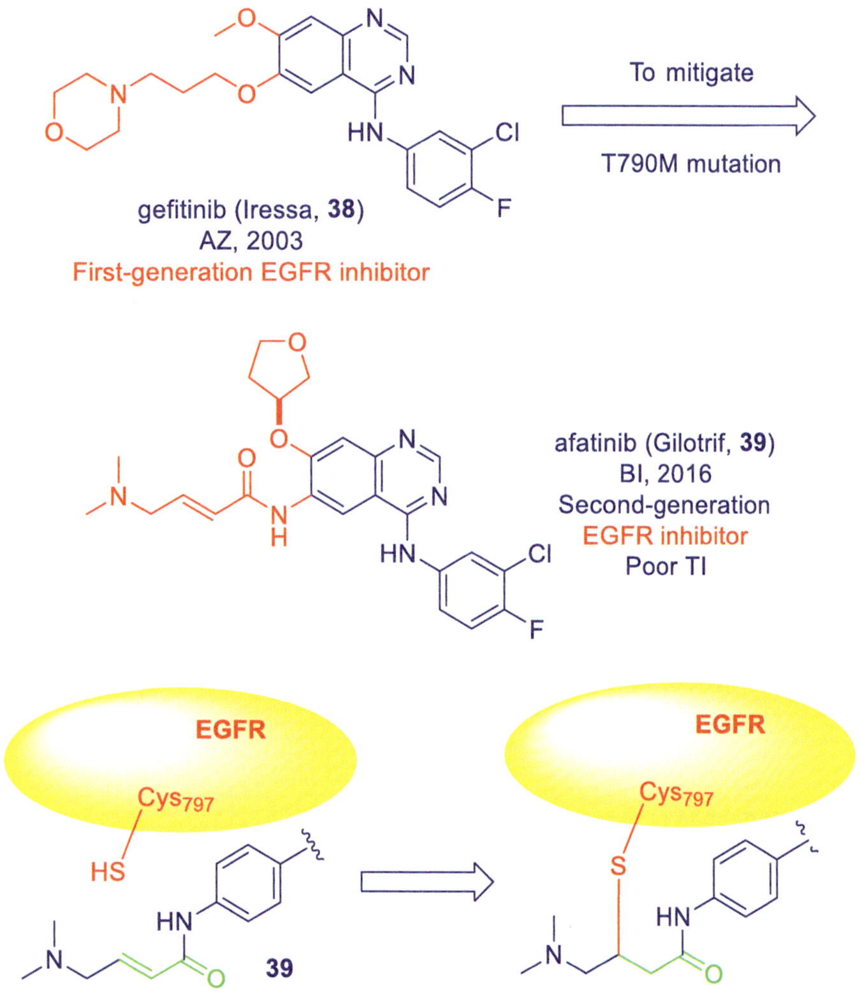

Figure 1.13 Afatinib (**39**), a second-generation EGFR inhibitor, was designed to undergo a hetero-Michael addition with Cys_{797} in the active site of EGFR.

The T790M mutation of afatinib (**39**), a second-generation EGFR inhibitor, lowers the affinity of the initial binding event of afatinib (**39**) before covalent linkage to Cys_{797} based on X-ray cocrystal analysis. This may have led to toxicity and lack of efficacy in clinics. Again, to outsmart cancer, the third-generation of covalent EGFR inhibitors have been invented and two of them have been approved by the FDA. They are neratinib (Nerlynx, **40**) and osimertinib (Tagrisso, **41**), respectively.[36]

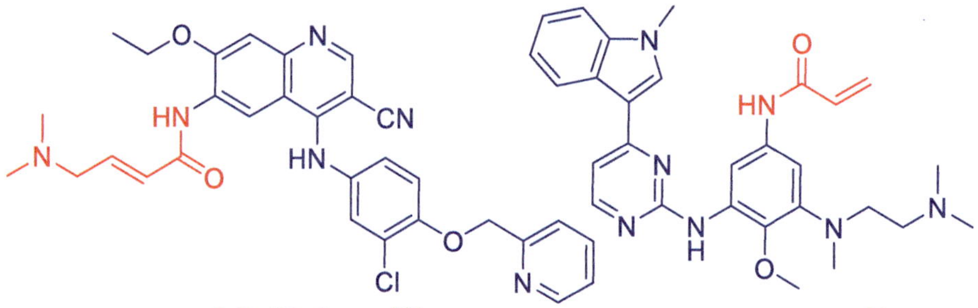

neratinib (Nerlynx, **40**)
Puma, 2017
EGFR inhibitor (third-generation)

osimertinib (Tagrisso, **41**)
AstraZeneca, 2017
EGFR inhibitor (third-generation)

As a testimony of how wily are cancer cells, the first evidence of osimertinib (**41**) resistance mediated by the EGFR mutation C797S was reported in 2015. As shown beneath, since Cys_{797} has been mutated to serine (Ser, S), its hydroxyl tail is not nearly as nucleophilic as the thiol group on cysteine.[37] The consequence of Cys_{797} mutation is high toxicity and decreased binding affinity to mutant kinase, possibly because a more hydrophilic character of Ser_{797} could lead to repulsion with the acrylamide moiety. Since then, a tremendous amount of effort has been devoted to discover the fourth-generation EGFR inhibitors to overcome the EGFR C797S mutation.[38] This time around, it seems to be even more challenging, and there is a long road ahead of us.

osimertinib (**41**)

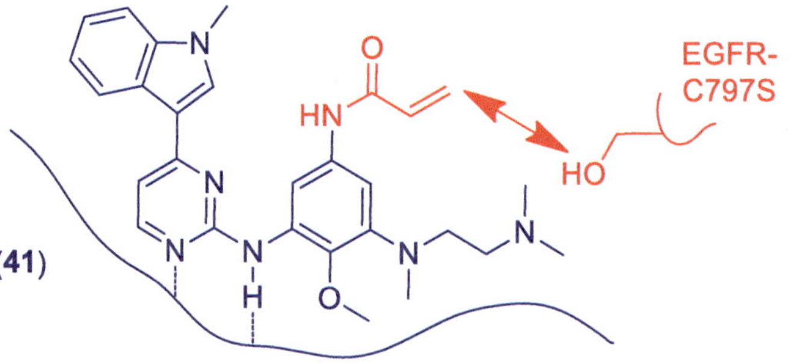

1.2.3.3 KRAS Inhibitors

The RAS proteins are members of a large superfamily of low molecular-weight guanosine triphosphate (GTP)-binding proteins. Approximately 30% of all human cancers contain activating Ras mutations. RAS gene family are the most frequently mutated oncogenes in human cancers. Mutant RAS appears in 90% of pancreatic, 45% of colon, and 35% of lung cancers. Among the three Ras genes, K-Ras is the most frequently mutated isoform (86%), followed by N-Ras (11%) and H-Ras (3%). With regard to KRAS proteins, there are several mutations. $KRAS^{G12C}$ mutations, in particular, predominate in nonsmall cell lung cancer (NSCLC, 45–50% of mutant KRAS^{G12C}). KRAS^{G12D} is important in pancreatic cancer (61%), colon cancer (42%), and NSCLC (22%).[39]

RAS has long been viewed as undruggable due to its lack of deep pockets for binding of small molecule inhibitors. Recent successes with RAS direct inhibitors, KRAS^{G12C} in particular, are a testimony of scientists' ingenuity and perseverance. Covalent guanosine mimetic inhibitors have been discovered. Some difluoromethylene bisphosphonate analogs are devoid of liability of the hydrolytic instability of the diphosphate moiety present in SML-8-71-1 and provide the foundation for development of prodrugs.[40]

Wellspring has brought a covalent KRAS^{G12C}-specific inhibitor ARS-1620 (**42**) to clinical trials in 2018.[41a] On its heel, Mirati and Array Biopharma also announced their covalent KRAS^{G12C}-specific inhibitor MRTX1257 (**43**) is in Phase I clinical trials in 2018. Both ARS-1620 (**42**) and MRTX1257 (**43**) are being tested to treat NSCLC.[41b] More excitingly, in 2019,[42] Amgen reported their Phase I clinical trial outcome of their first-in-class covalent KRAS^{G12C}-specific inhibitor **AMG 510**: 54% of 13 evaluable NSCLC patients experienced a partial response (PR) at the target dose of 960 mg in the ongoing Phase 1 study. Meanwhile, 46% of patients had stable disease for a disease control rate of 100% at the target dose. The FDA granted **AMG 510** fast track designation for previously treated metastatic NSCLC with KRAS^{G12C} mutation. The KRAS^{G12C}, the previously undruggable target has now become one of the most popular targets.

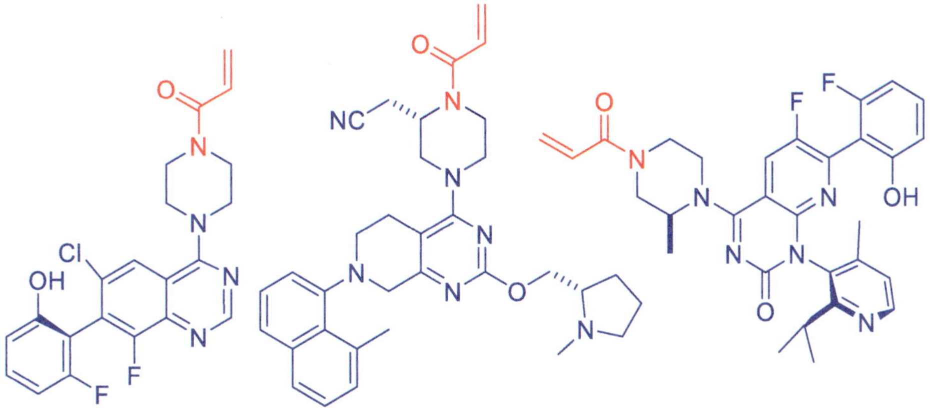

ARS-1620 (**42**) KRAS^{G12C} covalent inhibitor

MRTX1257 (**43**) KRAS^{G12C} inhibitor

AMG 510 KRAS^{G12C} inhibitor

1.2.3.4 Fatty Acid Amide Hydrolase (FAAH) Inhibitors

FAAH is an integral membrane serine hydrolase responsible for the degradation of fatty acid amide signaling molecules such as endocannabinoid anandamide (AEA), which has been shown to possess cannabinoid-like analgesic properties. FAAH belongs to the amidase class of enzyme, a subclass of serine hydrolases that has an unusual Ser–Ser–Lys catalytic triad. The Ser–His–Asp catalytic triad is more common among hydrolyses.

Pfizer discovered PF-04457845 (**44**), a highly potent and selective FAAH inhibitor that reduces inflammatory and noninflammatory pain.[43] Mechanistic and pharmacological characterization of PF-04457845 (**44**) revealed that it is a covalent irreversible inhibitor involving carbamylation of FAAH's catalytic Ser_{241} nucleophile. The mechanism of FAAH inhibition of **44** involving the Ser_{241}–Ser_{217}–Lys_{142} catalytic triad is shown below: nucleophilic attack of the carbamate on **44** by the hydroxyl nucleophile on Ser_{241} of the Ser_{241}–Ser_{217}–Lys_{142} catalytic triad gives rise to tetrahedral intermediate **45**. Collapse of intermediate **45** then results in four products including inactive covalently modified FAAH (**46**), a serine, a lysine, and by-product pyridazin-3-amine.

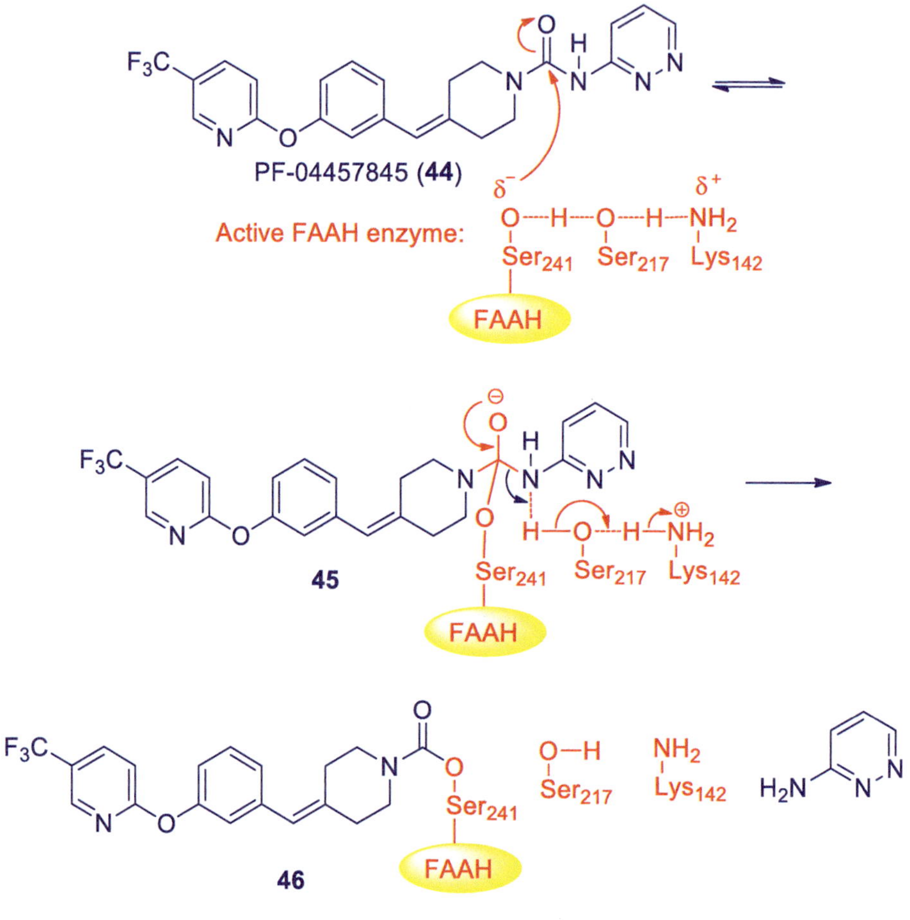

1.2.3.5 Monoacylglycerol Lipase (MAGL) Inhibitors

Like FAAH, MAGL is also a serine hydrolase with an active site containing a classical Ser–His–Asp triad. It is the main enzyme responsible for degradation of the endocannabinoid 2-arachidonoylglycerol (2-AG) in the central nervous system (CNS). MAGL catalyzes the conversion of 2-AG to arachidonic acid (AA), a precursor to the proinflammatory eicosannoids such as prostaglandins. The characteristic reactivity of the active site enables the opportunity for covalent modification of the key serine residue (Ser_{122}).

Azetidine carbamate **47** is an irreversible covalent MAGL inhibitor discovered by Pfizer.[44] The hexafluoroisopropanol (HFIP) group here serves as the leaving group when attacked by the key serine residue (Ser_{122}) at the enzyme's active site. The products are covalently modified MAGL (**48**) and HFIP.

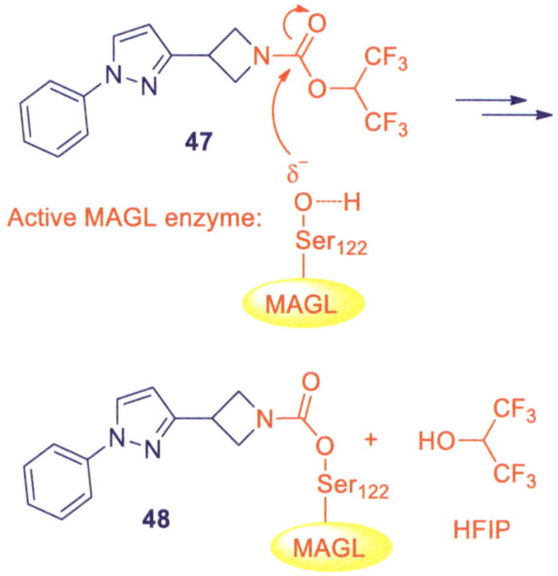

1.2.4 Transition-state Mimetics

A transition-state mimetic is an inhibitor that mimics the transition-state structure of the substrate of an enzyme, which, by definition, has the highest energy. They have the least stable conformation. Transition-state mimetics also take advantage of the better binding according to Koshland's "induced fit" theory.[45]

Transition state theory teaches us that chemically stable mimetics of enzymatic transition states will bind tightly to their associated enzymes. Inhibitors that resemble the transition state of substrates can provide confirmation for the proposed mechanism of an enzyme. If such an inhibitor binds tightly to the enzyme, the enzyme is stabilizing a substrate in a similar conformation during catalysis.

Drugs as transition-state mimetics (mimics, or analogs) are often competitive inhibitors because they almost invariably bind to the active sites. Many examples of transition-state mimetics exist in the aspartic protease field where inhibitors were designed to mimic the proposed transition-state structure during cleavage of the substrate. Human aspartic proteases include renin, pepsin, gastricsin, cathepsins D and E, and HIV-1 protease.

Three classes of drugs are discussed in this section:

1. HIV-1 protease inhibitors;
2. Influenza virus neuraminidase inhibitors; and
3. DD-transpeptidase inhibitors.

1.2.4.1 HIV-1 Protease Inhibitors

The best-known examples of transition-state mimetics are probably HIV-1 protease inhibitors such as ritonavir (Norvir, **49**) and darunavir (Prezista, **50**). In fact, ritonavir (Norvir, **49**) was designed while cognizant of the structure of the proposed transition-state hydrolysis intermediate of the viral polypeptide substrate by HIV protease and the symmetry of the protease.

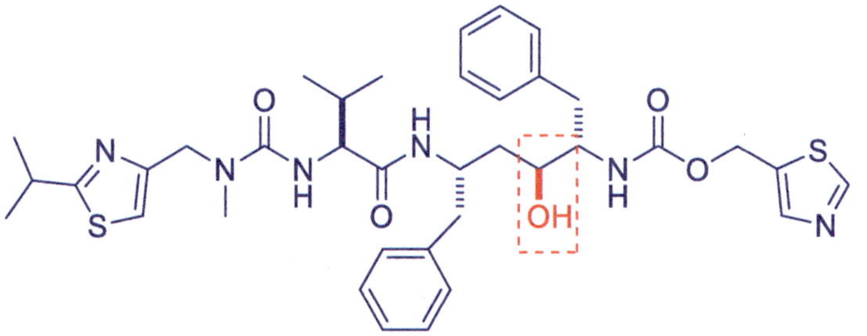

ritonavir (Norvir, **49**)
Abbott, 1996
HIV protease inhibitor

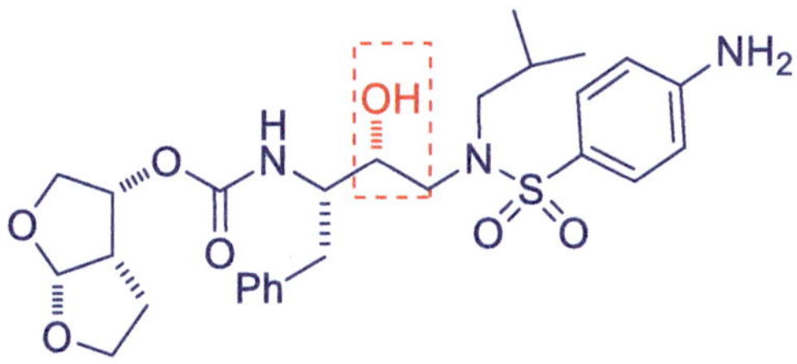

darunavir (Prezista, **50**)
Tibotec, 2006
HIV protease inhibitor

HIV protease is an aspartic protease. It is so named because its substrate's scissile peptide bond is hydrolyzed by two aspartic acids on the protease. Fully functional HIV-1 protease is a homodimer creating a single catalytic site with two essential Asp residues, one from each subunit, whose carboxylate groups are involved in catalysis. The HIV protease dimer consists of two identical, noncovalent subunit of 99 amino acid residues associated in a two-fold (C-2) symmetric fashion. This explains why some HIV protease inhibitors are somewhat symmetrical. The active site of HIV protease is actually formed at the dimer interface and contains two conserved catalytic aspartic acid residues, one from each monomer. The substrate binding cleft is composed of equivalent residue from each subunit and is bound on one side by the active side aspartic acids, Asp25 and Asp125.

As shown later, the MOA for HIV protease inhibitors is through mimicry of the transition-state hydrolysis intermediate. The amino acids Asp25 and Asp25′ in the cleavage site are located between S1 and S1′. Under the synergistic action of Asp25 and Asp25′, water adds to substrate **51** to deliver the tetrahedral transition state hydrolysis intermediate **52**. Since the energy level for transition state is the highest, it readily collapses to deliver the hydrolyzed products **53** as an acid and an amine.[46,47]

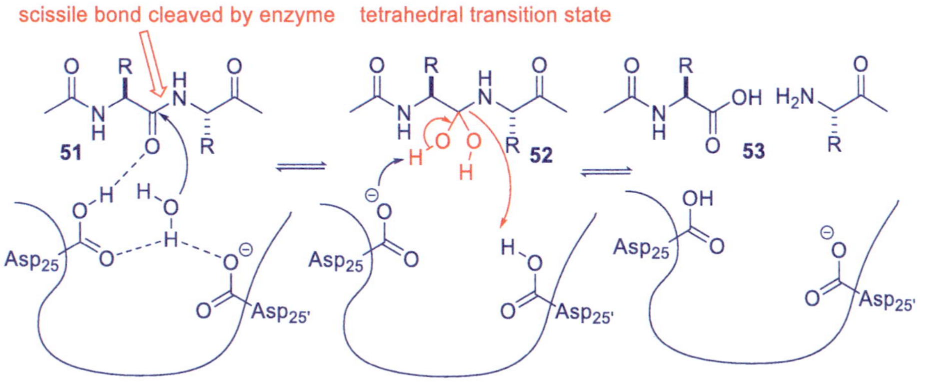

All marketed HIV protease inhibitors, including ritonavir (**49**) and darunavir (**50**), work as transition state mimetics. To further improve the bioavailability of these HIV protease inhibitors, tertiary alcohol transition-state mimetics have been prepared as aspartic protease inhibitors.[48] Furthermore, silanediol **54** was prepared as a serine protease inhibitor. It inhibits the serine protease chymotrypsin with a K_i of 107 nM. Inhibition of the enzyme may involve exchange of a silane hydroxyl with the active site serine nucleophile, contrasting to previous silanediol protease inhibitors.[49]

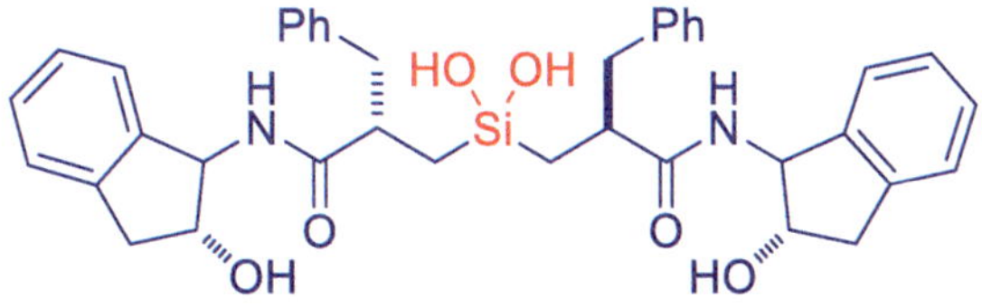

serine protease inhibitor **54**
chymotrypsin K_i = 107 nM

1.2.4.2 Influenza Virus Neuraminidase Inhibitors

Flu drugs zanamivir (Relenza, **59**),[50] oseltamivir (Tamiflu, **61**),[51] and peramivir (Rapivab, **62**)[52] as neuraminidase inhibitors are quintessential transition state mimetics.

Influenza virus neuraminidase (sialidase) is attached to the viral surface by a single hydrophobic sequence of 29 amino acids. Its active site is located in a deep pocket and the 18 amino acids making up the active site itself are constant. The active site of the enzyme will bind and stabilize the transition-state **56**[53] more effectively than it will stabilize the substrate (**55**) itself, thus resulting in an overall decrease in activation energy for the chemical transformation. Neuraminidase functions to remove terminal sialic acid (**57**) residues and promote release of virus particles from the cells.

As shown later, under the influence of neuraminidase, cell surface sialic acid (**55**) is cleaved to give rise to transition state **56** as an oxonium cation, in addition to glycoprotein. Nucleophilic addition of water to transition state **56** produces sialic acid (**57**), which itself is not a very potent neuraminidase type A inhibitor. In the early 1990s, the high-resolution crystal structure of neuraminidase/sialic acid (**57**) complex became available. It was a great boon to aid structure-based drug design (SBDD) that was instrumental to the success of discovering neuraminidase inhibitors.

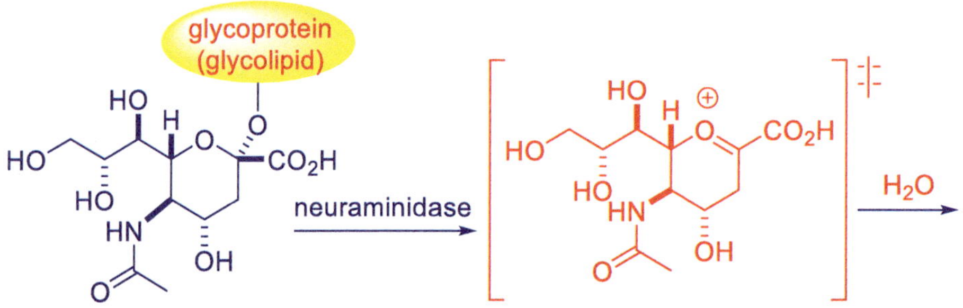

cell surface sialoside (**55**) transition state **56**

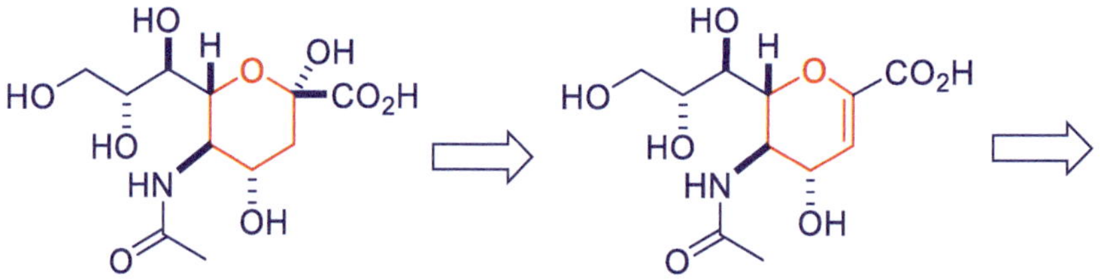

sialic acid (**57**)
K_i = 1 x 10^{-3} M (A/N2)

DANA (**58**)
K_i = 4 x 10^{-6} M (A/N2, IC_{50} = 7 μM)

Indeed, 2-deoxy-2,3-didehydro-*N*-acetyl-neuraminic acid (DANA, **58**), a mimetic of transition state **56**, is close to 4,000-fold more potent than sialic acid (**57**). The crystal structures of neuraminidase/DANA (**58**) complex revealed their interactions are characterized by both strong charge–charge interactions and limited hydrophobic contacts. Even though replacing the bottom hydroxyl group with a primary amine already offered an approximately 100-fold boost of potency, the guanidine isostere zanamivir (Relenza, **59**) proved to be the most potent analog. Due to its highly polar nature, zanamivir (**59**) can only be administered topically via inhalation.[50]

To discover an orally active neuraminidase inhibitors, Gilead chose to replace the pyranose (dihydropyran) core structure with a chemically and enzymatically more stable cyclohexene scaffold. Using cyclohexene as a suitable bioisostere was a wise choice because it mimics the proposed flat oxonium cation in the transition state of sialic acid cleavage. Paradoxically, Gilead discovered that wholesale replacement of the left-hand glycerol side-chain with hydrophobic substituents to increase the molecule's lipophilicity is favored. The X-ray crystal structure revealed that the lipophilic side chains bound to the hydrophobic pocket consisted of Glu_{276}, Ala_{246}, and Ile_{222} of the enzyme active site. The increase in potency due to interaction with pockets 1 and 2 was sufficient so that it was not necessary to incorporate the guanidinium group that was critical to the potency of zanamivir (**59**). GS-4071 (**60**) emerged as one of the most potent influenza neuraminidase inhibitors against both influenza A and B strains.[51] Its ethyl ester prodrug oseltamivir (Tamiflu, **61**) proved to be safe and efficacious for the oral treatment and prophylaxis of human influenza infection and was approved by the FDA in 1999.

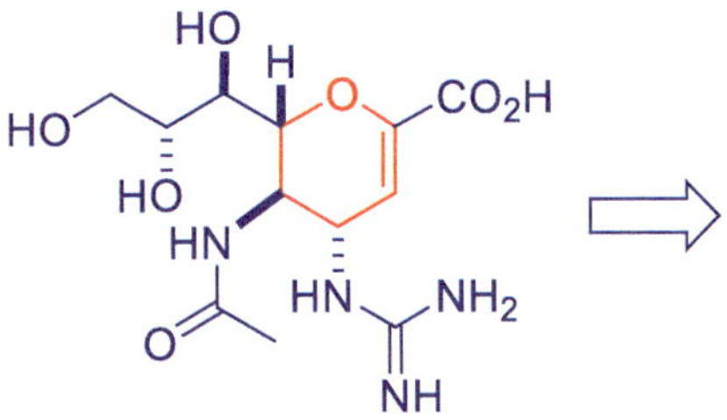

zanamivir (Relenza, **59**)
$K_i = 2 \times 10^{-10}$ M (A/N2)

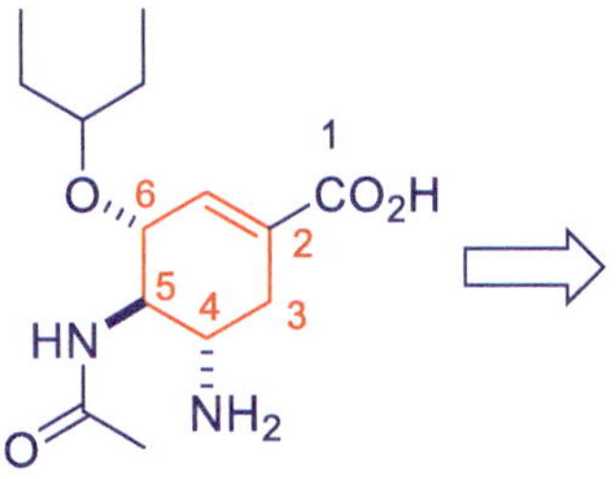

GS-4071 (**60**)
IC_{50} = 1.3 nM (A/N2)

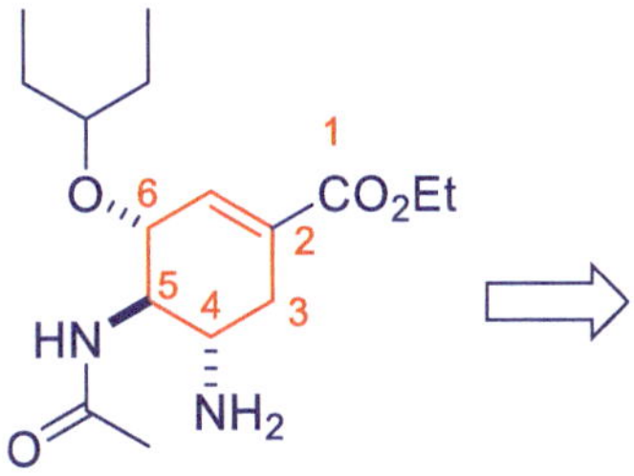

oseltamivir (Tamiflu, **61**)
Gilead, 1999

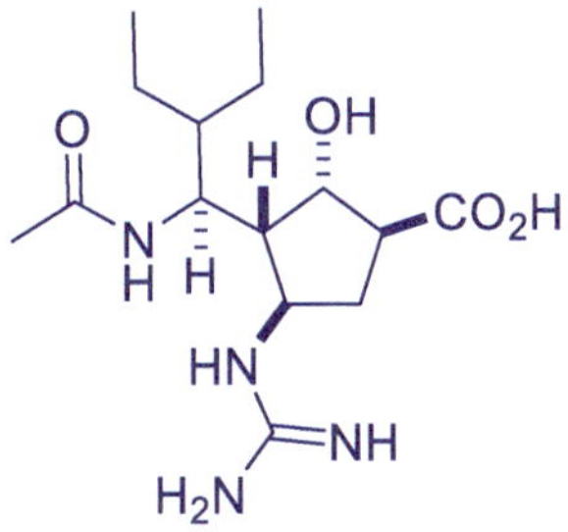

peramivir (Rapivab, **62**)
IC_{50} = 1.1 nM (A/N2)

Extensively guided by crystal structures, BioCryst discovered that a cyclopentane ring was a suitable scaffold for a novel class of neuraminidase inhibitors. They eventually arrived at peramivir (Rapivab, **62**),[52] which takes advantage of the two previous successful drugs. On the one hand, peramivir (**62**) retains the guanidine motif for potency since it occupies the fourth binding pocket replacing the existing water and involves in charge-based interactions with residues Asp_{151}, Glu_{119}, and Glu_{227}. On the other hand, mimicking oseltamivir (**61**), its aliphatic 3-pentyl group binds to the hydrophobic pocket of neuraminodase. The interactions between the 3-pentyl group and the hydrophobic pocket were previously not observed in the crystal structure of neuraminidase/sialic acid (**57**) complex.

1.2.4.3 DD-Transpeptidase Inhibitors

Penicillin has saved numerous lives, and we have learned much about how it works. We know that it is a covalent irreversible inhibitor because its β-lactam ring is opened by the serine alcohol of β-lactamase. Here, penicillin G (**64**) may be considered as a transition-state mimic of tripeptide Gly–D-Ala–D-Ala (**63**) at the glycoprotein DD-transpeptidase. The tripeptide Gly–D-Ala–D-Ala (**63**) is involved in the cross-linking of glycol-peptides constituting the cell walls of bacteria.[54]

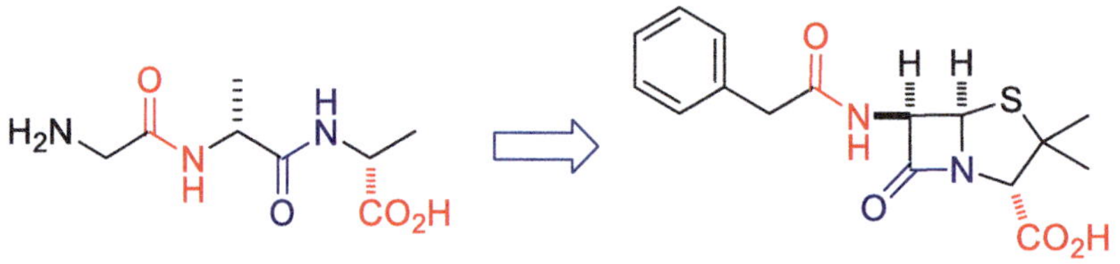

Gly–D-Ala–D-Ala (**63**) penicillin G (**64**)

1.2.5 Suicide Substrates

Suicide substrates are also known as suicide inhibitors, or mechanism-based inactivators (MBIs). They are modified substrates that are partially processed by the normal catalytic mechanism of an enzyme. The enzyme itself participates in the irreversible modification, causing it to lose efficacy. This enzyme participation results in even greater specificity by these inhibitors at the active site.

We noted that proteasome inhibitors bortezomib (Velcade, **18**) and ixazomib (Ninlaro, **19**) function by forming reversible covalent bonds between their boronic acids and the N-terminal nucleophile (Ntn) threonine (Thr 1) at the active site. Another proteasome inhibitor carfilzomib (Kyprolis, **67**) forms irreversible covalent bond with the N-terminal threonine (Thr 1) at the active site of the target protein.[55] This MOA makes carfilzomib (**67**) a suicide substrate. The outcome is similar to those of its brethren **18** or **19**, i.e., carfilzomib (**67**) binds and inhibits the CT-L activity of the proteasome of 20S protease. Inhibition of proteasome-mediated proteolysis provides poly-ubiquitinated proteins, which ultimately causes cell cycle arrest and apoptosis of cancer cells.

Historically, the genesis of carfilzomib (Kyprolis, **67**) traces back to the discovery of an α′,β′-epoxyketone tetrapeptide natural product epoxomicin (**65**). It potently and irreversibly inhibits the catalytic activity of 20S proteasome. It is specific for the proteasome and does not inhibit other proteases such as calpain, papain, cathepsin B, chymotrypsin, and trypsin.

Crews at Yale elucidated its MOA as a protease inhibitor[56] and improved its potency with YU-101 (**66**). Taking over the baton, Proteolix at South San Francisco added a morpholine motif at the left-hand side of the molecule and obtained carfilzomib (**67**) as a more water-soluble analog.[57] After its FDA approval in 2006, it is given to patients intravenously (iv) biweekly or more frequently dosing schedules. Further optimizations resulted in an orally bioavailable protease inhibitor tripeptide PR-047 (ONX-0912, oprozomib, **68**), which has an absolute bioavailability up to 39% in rodents and dogs.[58] The key modifications consist of removal of the left-hand amino acid and replacement of the morpholine terminus with an aromatic 2-methylthiazole.

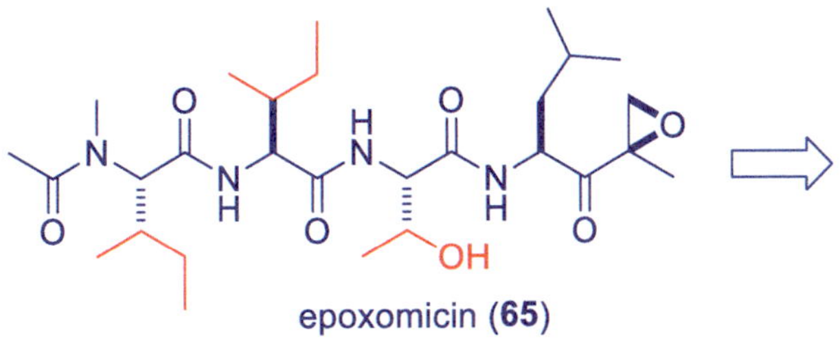

epoxomicin (**65**)

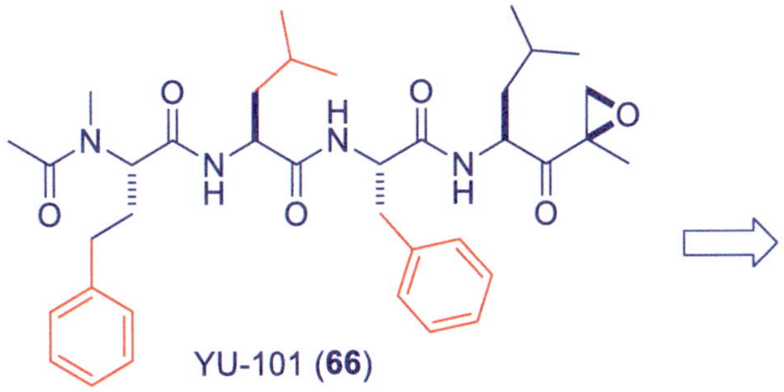

YU-101 (**66**)

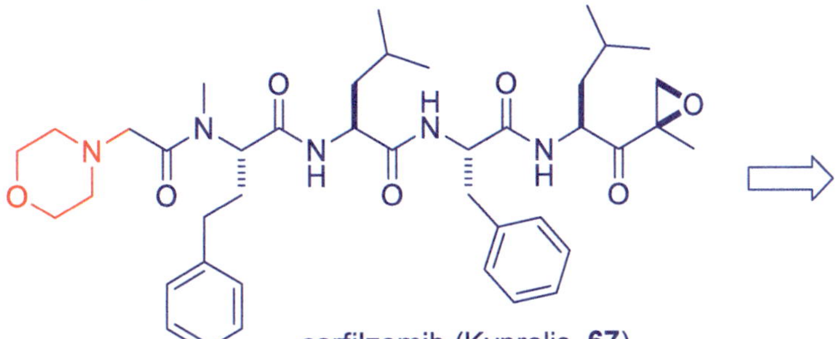

carfilzomib (Kyprolis, **67**)
Onyx, 2012
proteasome inhibitor

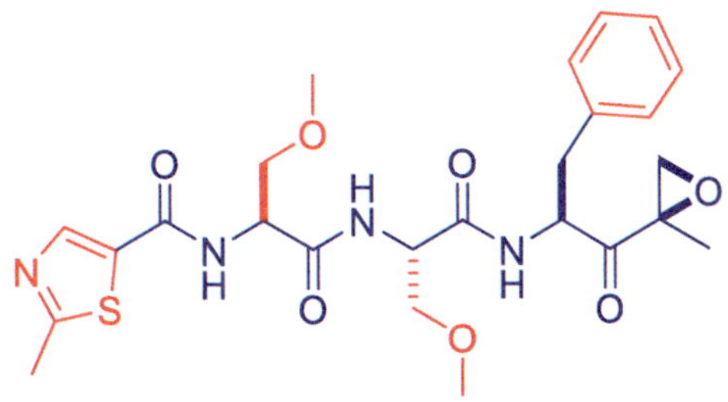

PR-047 (ONX-0912, oprozomib, **68**), *F*% = 39%

Back to the MOA of inhibitors **65–68** on a molecular level, it is exemplified here employing the reaction between epoxomicin (**65**) and the N-terminal nucleophile (Ntn). With the aid of a water molecule, the hydroxyl group on threonine 1 (Thr 1, **70**) adds to the carbonyl at the C-terminus of epoxomicin (**65**) to give a tetrahedral hemiacetal intermediate **69**. Subsequently, the amine group on Thr 1 attacks the epoxide moiety on epoxomicin (**65**) to afford the intramolecular cyclization product as morpholine **71**.[56] Activation of the epoxide may be facilitated by an intramolecular hydrogen bond formation between the N5 hydrogen and the oxygen of the epoxide. The β-methyl substituent on the epoxide moiety is key to attenuate the reactivity of the keto-epoxide "warhead." Without the methyl, the keto-epoxide "warhead" is too reactive.

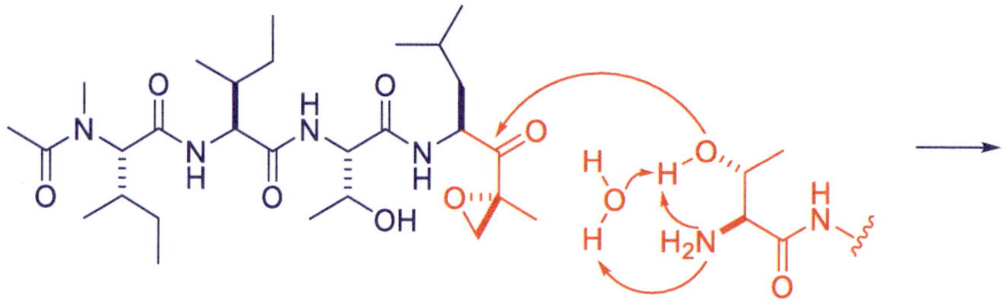

epoxomicin (**65**) Thr 1 (**70**)

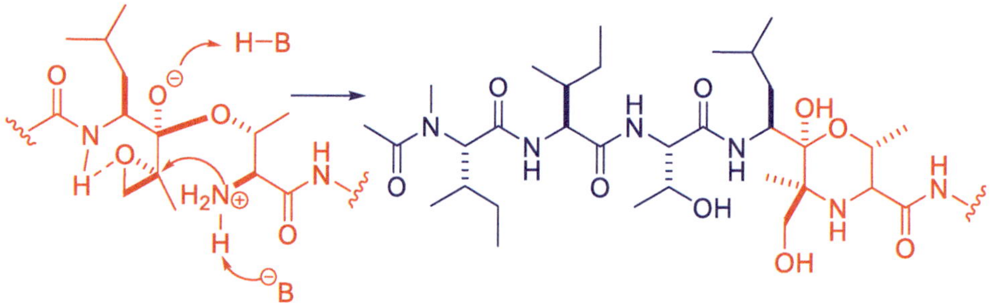

intermediate **69** morpholine **71**

Glycopeptide transpeptidase catalyzes the formation of cross-links between D-amino acids [tripeptide Gly–D-Ala–D-Ala (**63**)] in the cell walls of bacteria. This enzyme also catalyzes the reverse reaction, the hydrolysis of peptide bonds. During the course of hydrolyzing the strained lactam bond in penicillin G (**64**), the enzyme activates the inhibitor (penicillin), which then covalently modifies an active site serine in the enzyme to afford covalent adduct **72**. In effect, the enzyme loses its functions by cleaving the strained peptide bond in penicillin.

MBIs have important ramifications on cytochrome 450 enzymes.[59] There are some common features with suicide substrates. We will look at MBIs in depth in Chapter 4 on Structural Alerts.

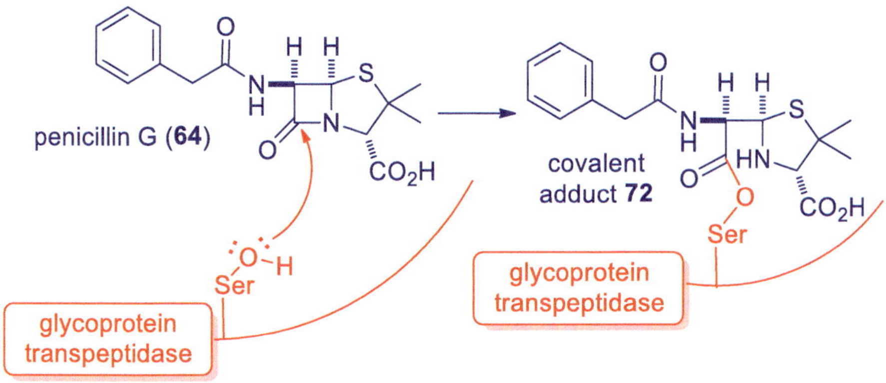

1.2.6 Isozyme Selectivity of Inhibitors

Human body is a complicated machinery. Many enzymes have isoforms numbering from two to hundreds. Whereas cyclooxygenase has COX-1 and COX-2 isoforms, there are over 500 isoforms in the kinase family. To achieve specific or selective inhibitors of single protein isoform targets within the complex, multi-component biological systems requires highly sophisticated drug design strategies.[60]

We have come a long way in appreciating and targeting isoforms. Here, we discuss selective modulators for isoforms of four classes of enzymes:

1. COX-2 selective inhibitors;
2. HDAC selective inhibitors;
3. PDE selective inhibitors; and
4. IDH selective inhibitors.

1.2.6.1 COX-2 Selective Inhibitors

Nonsteroid anti-inflammatory drugs (NSAIDS) such as Aspirin (**73**), ibuprofen (Advil, **74**), and naproxen (Aleve, **75**) are cyclooxygenase inhibitors. In the mid-1980s, evidence began to emerge that there are two subtypes of cyclooxygenases. Between the two isoforms of cyclooxygenases, one is inducible and the other is constitutive. The inducible enzyme associated with the inflammatory process was named cyclooxygenase II (COX-2), whereas the constitutive enzyme was called COX-1. COX-2 is localized mainly in inflammatory cells and tissues and becomes up-regulated during the acute inflammatory response; COX-1 is mainly responsible for normal physiological processes such as protecting the gastric mucosa and maintaining dilation of blood vessels. In general terms, COX-1 is the good enzyme and COX-2 is the bad one; hence, it was envisioned that a COX-2 selective inhibitor would be beneficial in both inhibiting prostaglandin production and reducing adverse gastrointestinal and hematologic side effects.

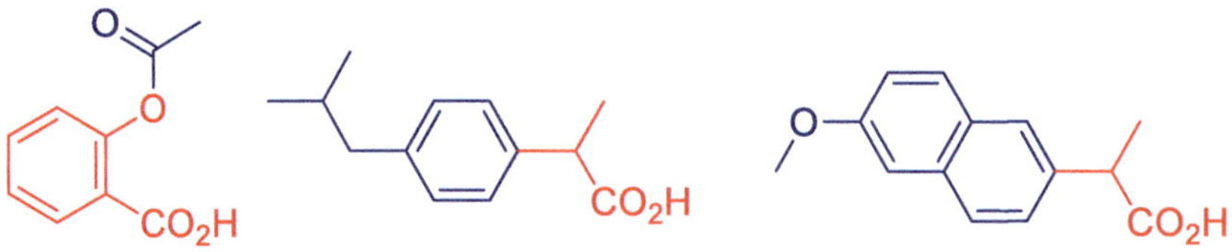

Aspirin (**73**) ibuprofen (Advil, **74**) naproxen (Aleve, **75**)

The first COX-2 selective inhibitor on the market, celecoxib (Celebrex, **76**) was approved in 1999 and the second, rofecoxib (Vioxx, **77**), was approved in 1998. Both celecoxib (**76**) and rofecoxib (**77**) quickly became blockbuster drugs for treating osteoarthritis (OA) and rheumatoid arthritis (RA). The third and the most selective (see Table 1.1)[61] COX-2 selective inhibitor, valdecoxib (Bextra, **78**), emerged in 2001. Regrettably, both rofecoxib (**77**) and valdecoxib (**78**) had to be withdrawn from the market place when post-market clinical trials for rofecoxib (**77**) showed that there was a five-fold increase in myocardial infarction (heart attack) among patients treated with rofecoxib (**77**) over those treated with naproxen (**75**).

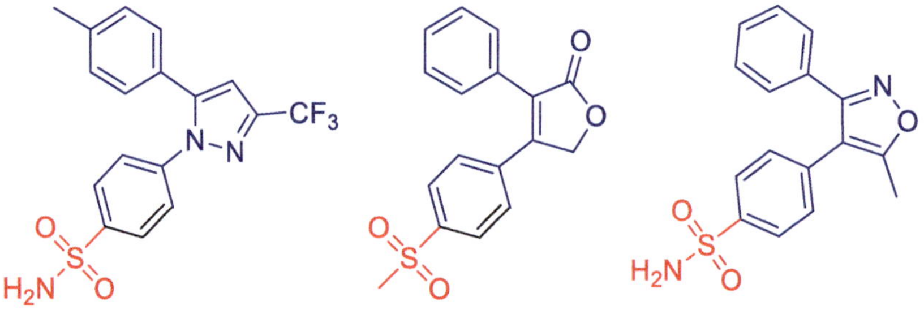

celecoxib (Celebrex, **76**) Pfizer, 1998 COX-2 selective inhibitor

rofecoxib, (Vioxx, **77**) Merck, 1999 Withdrawn, 2004

valdecoxib, (Bextra, **78**) Pfizer, 2001 Withdrawn, 2005

Table 1.1. Selectivity of NSAIDS.[61]

Drug	IC_{50} (μM), COX-1	IC_{50} (μM), COX-2	Ratio
Aspirin (**73**)	1.2	15.8	13.1
Ibuprofen (**74**)	3.3	37	11.4
Naproxen (**75**)	1.1	36	32.7
Celecoxib (**76**)	1.2	0.83	0.7
Rofecoxib (**77**)	15	0.018	0.0012
Valdecoxib (**78**)	150	0.005	0.00003

Source: https://jasbsci.biomedcentral.com/articles/10.1186/s40104-018-0259-8#rightslink.

1.2.6.2 HDAC Selective Inhibitors

The 18 isoforms of histone deacetylases (HDACs) are divided into four classes according to their sequence homology to their yeast analogs. They are viable epigenetic drug targets. HDAC inhibitors, vorinostat (SAHA, Zolinza, **79**), romidepsin (Istodax, **80**), belinostat (Beleodaq, **81**), and panobinostat (Farydak, **82**), are all pan-HDAC inhibitors without significant selectivity among four classes of 18 isoforms. With the exception of romidepsin (**80**), all HDAC inhibitors possess hydroxamate as their chelating group to bind to the catalytic zinc cation. The hydroxamate's strong chelating power may be the root of their lack of selectivity. Today, significant resources have been deployed to discover selective inhibitors.[62]

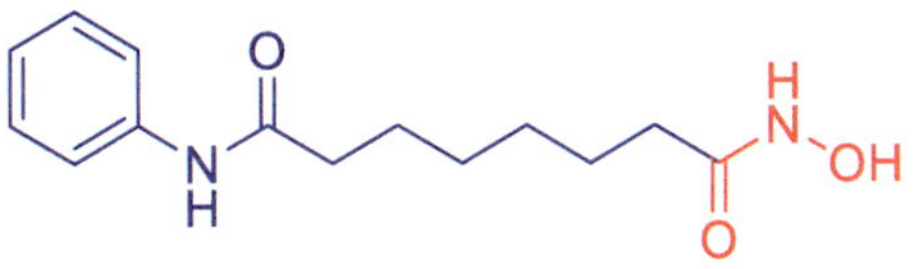

vorinostat (SAHA, Zolinza, **79**)
Merck, 2006
HDAC inhibitor

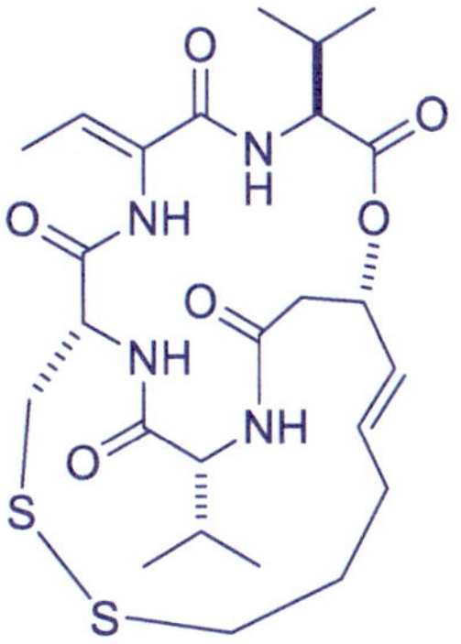

romidepsin (Istodax, **80**)
Gloucester, 2009
HDAC inhibitor

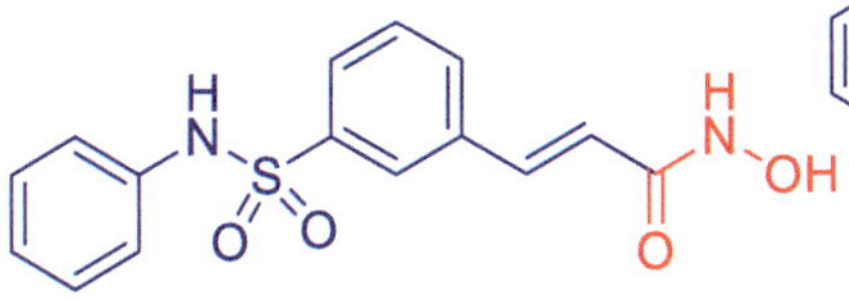

belinostat (Beleodaq, **81**)
TopoTarget, 2014
HDAC inhibitor

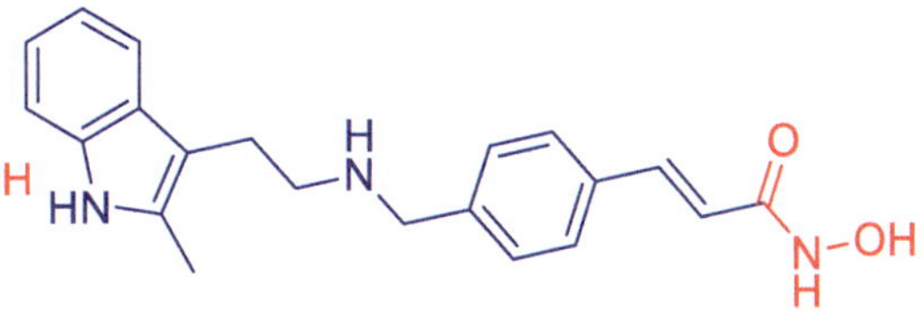

panobinostat (Farydak, **82**)
Novartis, 2015
HDAC inhibitor

A somewhat selective HDAC inhibitor chidamide (Epidaza, **83**) was approved by Chinese regulatory agencies for marketing in 2017. It functions as a potent HDAC1,2,3, and 10 subtype selective inhibitor. Its *o*-aminoaniline moiety serves as the zinc binding group (ZBG). Addition of a 4-fluoro substituent on the ZBG led to a better pharmacokinetic profile for the drug.[63]

HDAC1,2,3, and 8 belong to Class I HDACs. Employing *o*-aminoaniline moiety as the ZBG, RGFP966 (**84**) was shown to a potent selective HDAC3 inhibitor.[64]

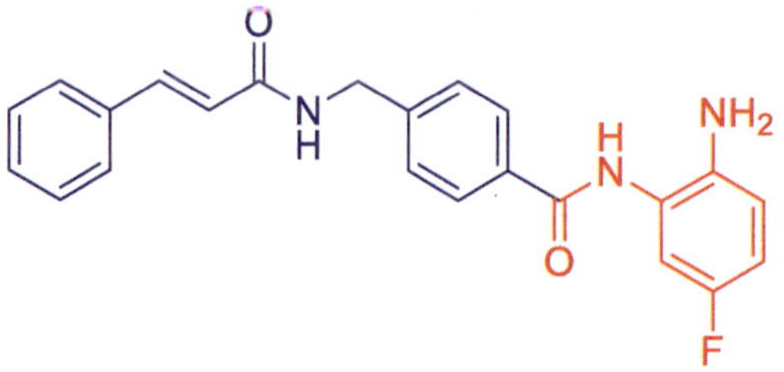

chidamide (Epidaza, **83**)
HDAC1, IC_{50} = 70 nM
HDAC2, IC_{50} = 110 nM
HDAC3, IC_{50} = 50 nM
HDAC10, IC_{50} = 60 nM

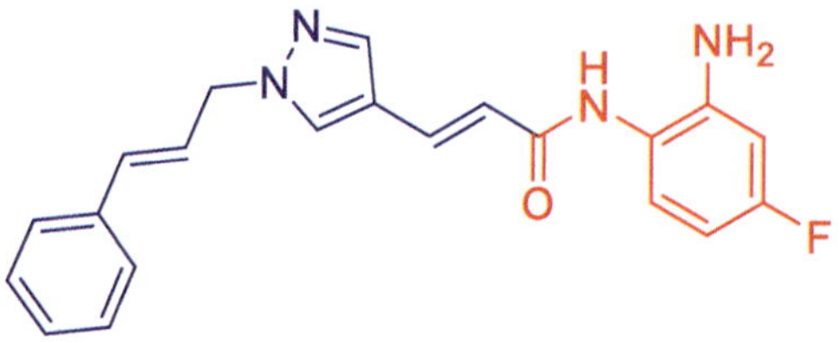

RGFP966 (**84**)
HDAC1, IC_{50} = 2,800 nM
HDAC2, IC_{50} = 1,700 nM
HDAC3, IC_{50} = 80 nM
HDAC8, IC_{50} >100,000 nM

It seems prudent to stay away from the hydroxamate as the ZBG in order to achieve selectivity against HDACs. In an effort to find an HDAC6 selective inhibitor, a thiol was chosen as the ZBG. The resulting compound **85** showed selective inhibitory activity for HDAC6 over HDAC1 and HDAC4.[65]

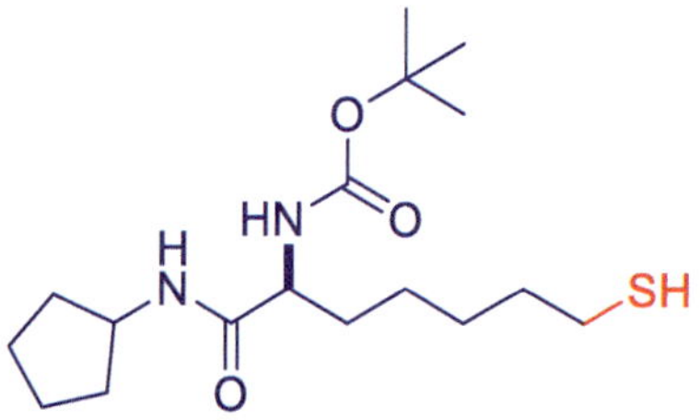

thiol **85**
HDAC1, IC_{50} = 1,210 nM
HDAC4, IC_{50} = 1,030 nM
HDAC6, IC_{50} = 29 nM

1.2.6.3 PDE Selective Inhibitors

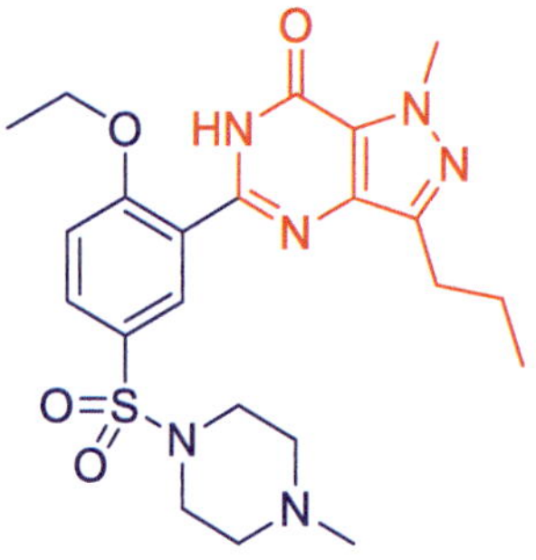

sildenafil citrate (Viagra, **86**)
PDE5 inhibitor

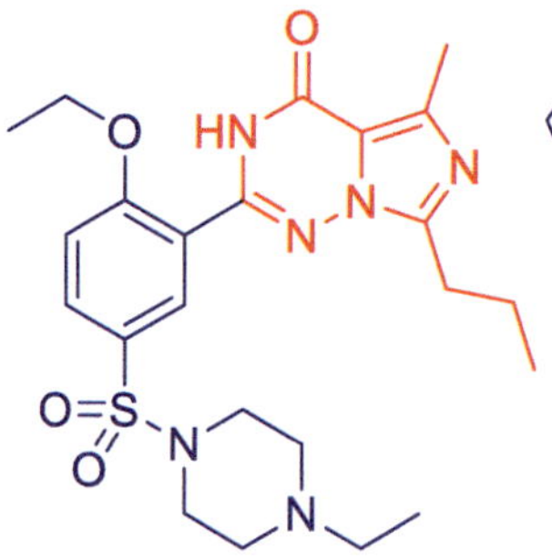

vardenafil (Levitra, **87**)
PDE5 inhibitor

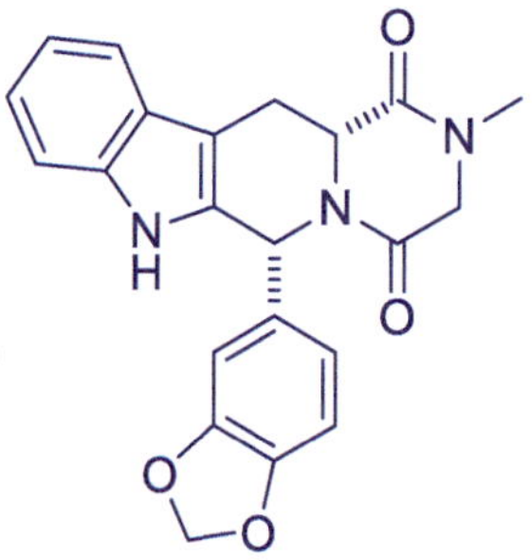

tadalafil (Cialis, **88**)
PDE5 inhibitor

We now have better appreciation of the functions of phosphodiesterases after the success of selective inhibitors of cyclic guanosine monophosphate (cGMP)-specific type 5 (PDE5). The emergence of sildenafil (Viagra, **86**), vardenafil (Levitra, **87**), and tadalafil

(Cialis, **88**) for the treatment of erectile dysfunction in the market place provided more incentive in the field.

As shown in Figure 1.14, levels of cGMP are controlled by phosphodiesterases (PDEs), which convert cGMP into GMP. A PDE inhibitor would prevent the breakdown of cGMP and the subsequent increase in cGMP concentration would allow smooth muscle cells in the kidney and blood vessels to relax thus lowering blood pressure. Nitric oxide production may be impaired in patients suffering from erectile dysfunction leading to low levels of cGMP, which can be quickly degraded by PDE5. Inhibition of PDE5 by sildenafil (**86**) slows the breakdown of cGMP allowing for higher concentrations to build up in the corpus cavernosum leading to an erection.[66]

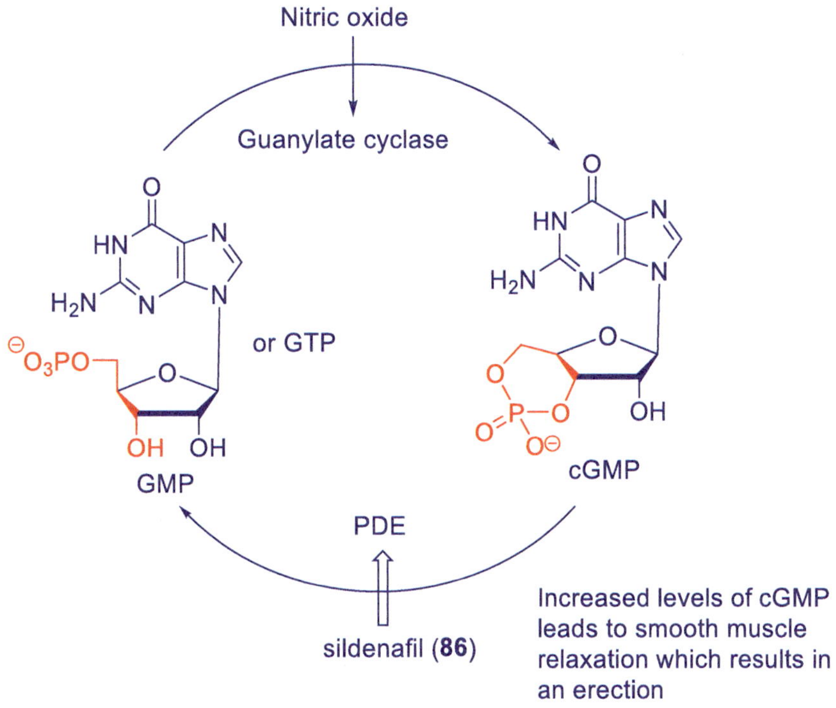

Figure 1.14 Roles of nitric oxide, cGMP, phosphodiesterase, and sildenafil (**86**) in erectile function

Among the cyclic nucleotide phosphodiesterases, only PDE4, PDE7, and PDE8 are selective for cAMP hydrolysis. Three selective PDE4 inhibitors have been approved for the treatment of immune-mediated and inflammatory diseases. Their beneficial effects in reducing antigen-inducing bronchoconstriction as well as airway smooth muscle construction have been well documented. AstraZeneca's phthalimide-derivative roflumilast (Daliresp, **89**) and its major metabolite, roflumilast *N*-oxide (IC_{50} = 2.0 nM), are selective PDE4 inhibitors prescribed to treat chronic obstructive pulmonary disease (COPD). Celgene's apremilast (Otezla, **90**) was approved by the FDA to treat psoriasis and psoriatic arthritis. Another selective PDE4 inhibitor ibudilast (Ketas, **91**) has been used only in Japan since the 1980s for the treatment of bronchial asthma, post stroke dizziness, and conjunctivitis. It is later found to be a selective PDE4 inhibitor.[67] In 2017, Anacor/Pfizer's crisaborole (Eucrisa, **92**) was approved for the treatment of atopic dermatitis. Meanwhile, many additional selective PDE4 inhibitors are in clinical trials for non-pulmonary indications.

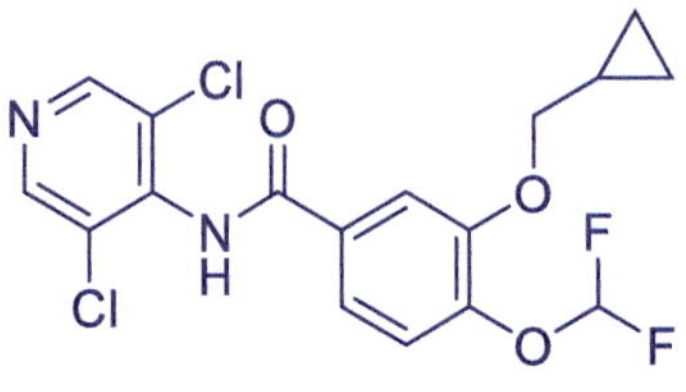

roflumilast (Daliresp, **89**)
Nycomed/AZ, 2011
selective PDE_4 inhibitor
IC_{50} = 0.8 nM

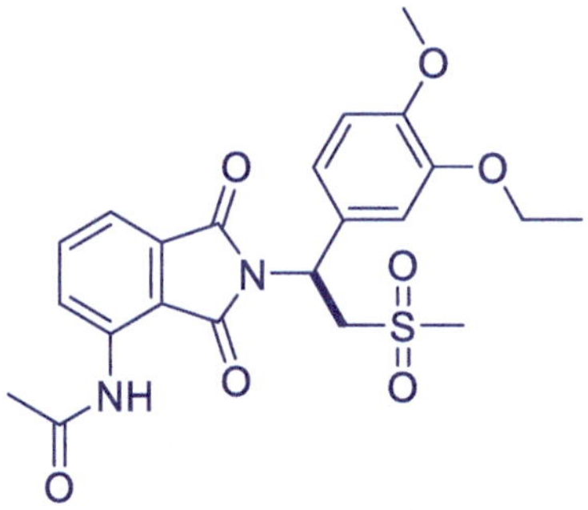

apremilast (Otezla, **90**)
Celgene, 2014
selective PDE_4 inhibitor
IC_{50} = 80 nM

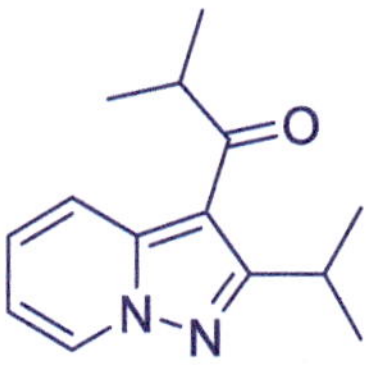

ibudilast (Ketas, **91**)
Japan, 1980s
selective PDE_4 inhibitor
IC_{50} = 1–10 μM

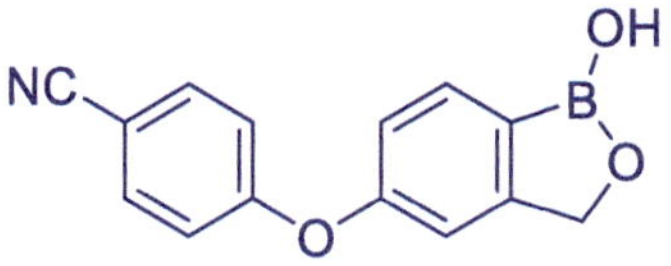

crisaborole (Eucrisa, **92**)
Anacor/Pfizer, 2017
selective PDE_4 inhibitor

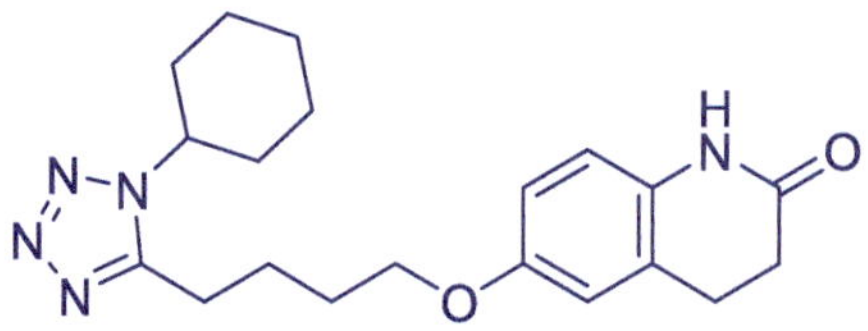

cilostazol (Pletal, **93**)
Otsuka, 1999
selective PDE_3 inhibitor

Selective PDE3 inhibitor cilostazol (Pletal, **93**) was approved in Japan in 1999 as an antiplatelet agent. It has been mostly evaluated in the Asian population and its use has been limited by tolerability due to headache, diarrhea, dizziness, or increased heart rate.[68] Both milrinone (Primacor, **94**) and enoximone (Perfan, **95**) are also found to be selective PDE3 inhibitors.

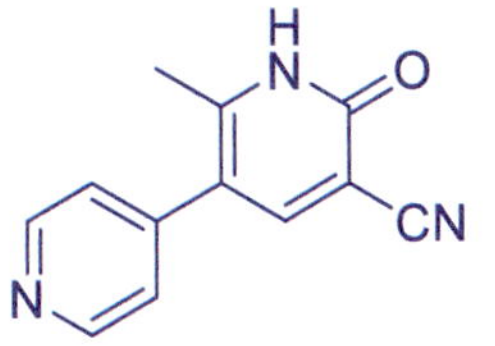

milrinone (Primacor, **94**)
Bedford, 2002
selective PDE_3 inhibitor

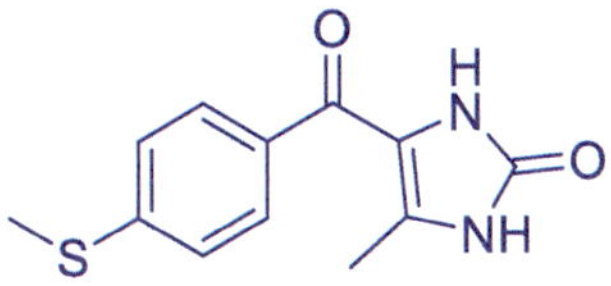

enoximone (Perfan, **95**)
Richardson–Merrell, 1980
selective PDE_3 inhibitor

1.2.6.4 IDH Selective Inhibitors

One of the latest achievements in treating leukemia is Agios′ success in both IDH2 and IDH1 inhibitors. IDH stands for isocitrate dehydrogenase. In the tricarboxylic acid cycle, wild-type IDH1/2 catalyze the conversion of isocitrate (**96**) to the corresponding ketone α-ketoglutarate (α-KG, **97**). However, mutant IDH1/2 reduce α-KG (**97**) to the oncometabolite 2-(*R*)-hydroxyglutarate (D-2-HG, **98**). Since D-2-HG (**98**) induces a block in cell differentiation and causes cells to lose the ability to advance from an immature to a fully differentiated state, somatic mutation m-IDH1/2 are found in ~20% of adults with acute myeloid leukemia (AML). Inhibitors of m-IDH1/2 provide viable therapeutic effect to treat IDH1/2 mutated tumors.[69]

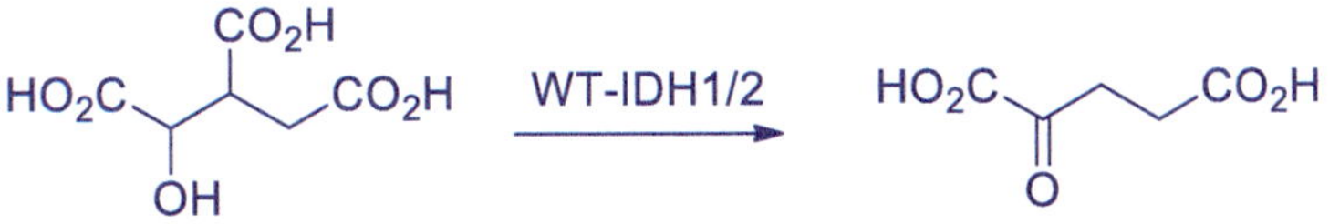

isocitrate (**96**) α-KG (**97**)

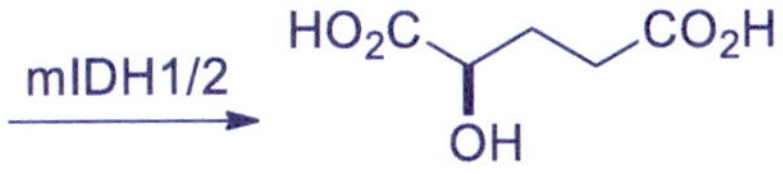

2-(*R*)-HG (**98**)

mIDH inhibitors represent a novel class of targeted cancer metabolism therapy that induces differentiation of proliferating cancer cells. Agios′ first-in-class selective IDH2 inhibitor enasidenib (Idhifa, **99**) was approved in 2017 for the treatment of adults with relapsed or refractory AML and an IDH2 mutation as detected by an FDA-approved test. Interestingly, the Abbott RealTime™ IDH2 companion diagnostic test (CDx), which received FDA approval concurrently with enasidenib, is indicated for the identification of AML patients with IDH2 mutations who may be candidates for treatment with enasidenib (**99**).[70]

In 2018, Agios′ encore success came in the form of its selective IDH1 inhibitor ivosidenib (Tibsovo, **100**). Somatic point mutations at a key arginine residue (R132) within the active site of the metabolic enzyme isocitrate dehydrogenase 1 (IDH1) confer a novel gain of function in cancer cells, resulting in the production of D-2-HG (**98**). Elevated 2-HG levels are implicated in epigenetic alterations and impaired cellular differentiation. IDH1 mutations have been described in an array of hematologic malignancies and solid tumors.[71] Ivosidenib (**100**) has been approved for the treatment of adult patients with relapsed or refractory AML with a susceptible IDH1 mutation.

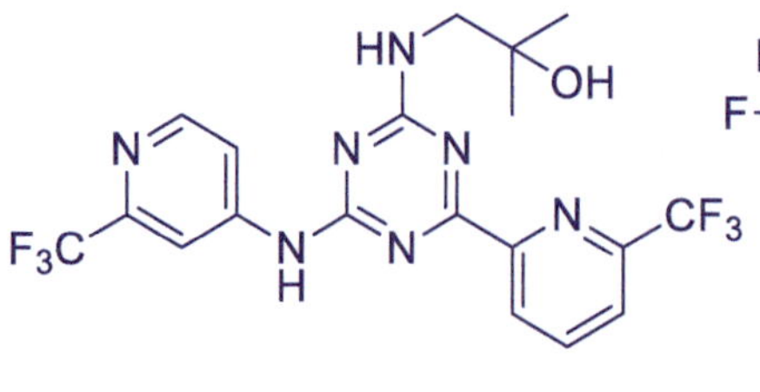

enasidenib (Idhifa, **99**)
Agios/Celgene, 2017
IDH2 inhibitor

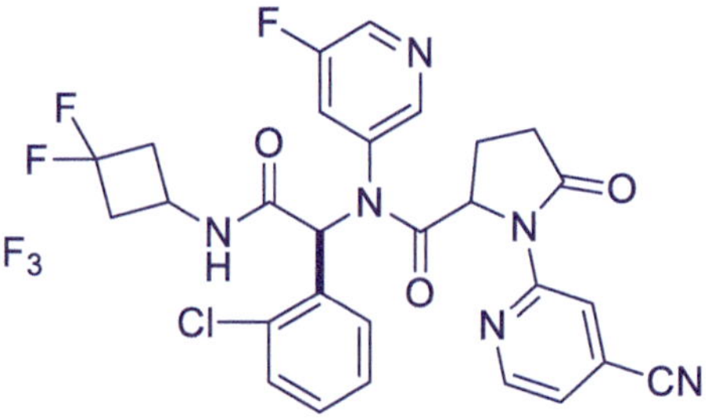

ivosidenib (Tibsovo, **100**)
Agios, 2018
IDH1 inhibitor

1.3 Receptors

At the end of nineteenth century, the concept of the drug receptor theory began to germinate largely thanks to Paul Ehrlich's side chain theory and John Newport Langley's concept of receptive substances.[72,73] Today, the cornucopia of drug receptors include no fewer than 130 GPCRs, along with many nuclear hormone receptors, growth factor receptors, and ionotropic receptors. Most receptors are membrane-bound. The only type of intracellular receptors are nuclear hormone receptors.

While an enzyme catalyzes a transformation of a substrate to a product, a receptor only serves as a receptacle, allowing the ligand to bind but without changing it chemically.

Molecules that bind to receptors are known as ligands. Receptor *modulator* is an even more sweeping concept, covering both ligands and molecules that modulate downstream effects. For a given receptor, its endogenous (natural) ligand binds to the receptor and produces an effect. For drugs that bind to receptors, they may be characterized as

1. Antagonists;
2. Agonists;
3. Partial agonists, etc.

1.3.1 Antagonists

Receptors' biological functions begin to manifest after ligand binding. If a compound that binds to a receptor and then blocks its biological functions, the ligand is called an antagonist. An antagonist is sometimes called a blocker. Many drugs on the market are receptor antagonists. There are more receptor antagonist drugs than agonist drugs.

β-blockers are good examples of receptor antagonists.

After adrenaline's discovery in the early 1900s, it was believed that there was one adrenaline receptor. But in the 1940s, evidence began to surface suggesting that there might be two different isoforms of adrenaline receptors. In 1948, Ahlquist proposed that there are two types of receiving mechanisms, or sites in the cardiovascular system: α, prevailing in the heart, β, in the blood vessels. Because they are receptors for adrenaline and adrenaline-like substances, they are known as "adrenergic" receptors. The two types of adrenergic receptors (adrenoceptors in short) are termed α-adrenoceptor and β-adrenoceptor. Black helped ICI to discover the first β-blocker (antagonist) propranolol (Inderal, **101**) in 1964, employing the SBDD approach for the first time.

Since then, dozens of β-blockers entered the market place. They include oxprenolol (Trasacor, **102**), pindolol (Visken, **103**), and betaxolol (Kerlone, **104**). Scrutiny of the structures revealed that they all have exactly the same pharmacophore (in red), which is likely to provide key interactions with the β-adrenergic receptor. In 2014, Boehringer–Ingelheim's olodaterol (Striverdi, **105**), a selective β_2-adrenergic receptor *agonist* (activator), was approved to treat COPD.

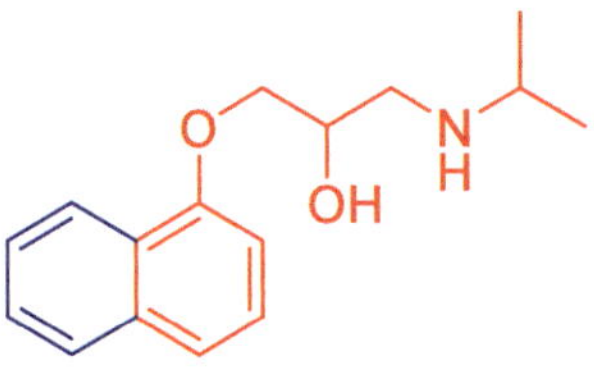

propranolol (Inderal, **101**)
ICI
β-adrenoceptor antagonist

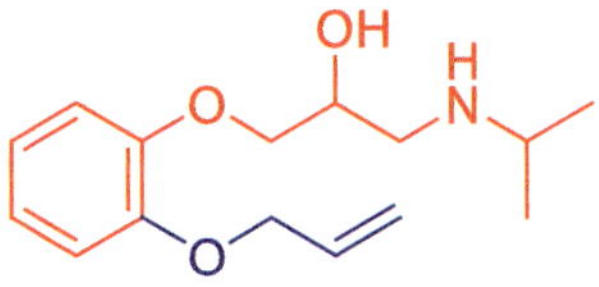

oxprenolol (Trasacor, **102**)
Ciba
β-blocker

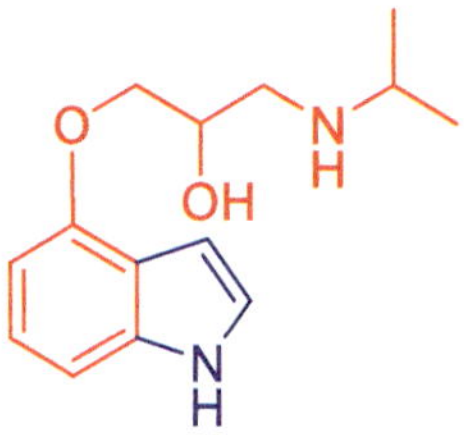

pindolol (Visken, **103**)
Sandoz
β-blocker

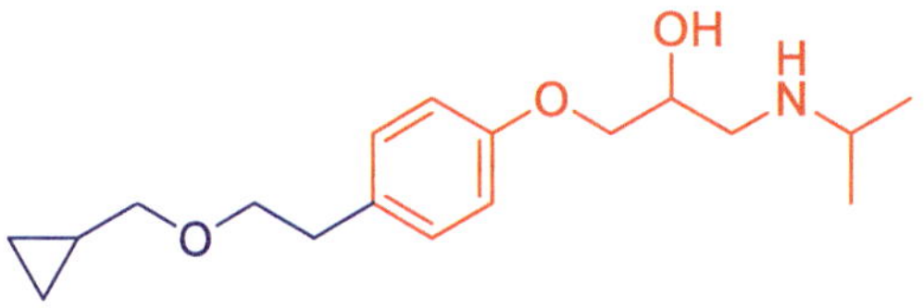

betaxolol (Kerlone, **104**)
Synthelabo-Searle, 1985
β-adrenoceptor **antagonist**

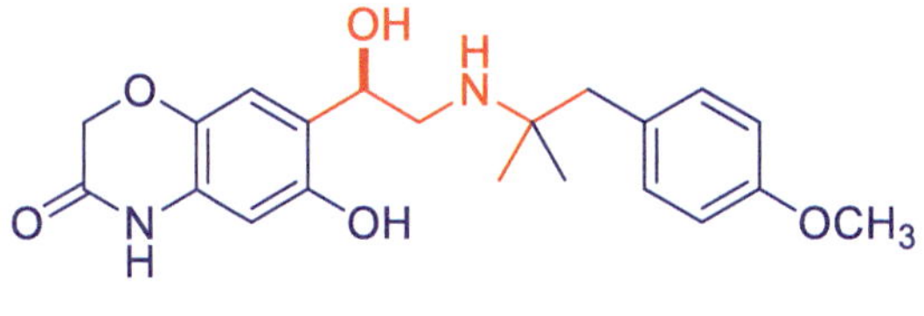

olodaterol (Striverdi, **105**)
Boehringer-Ingelheim, 2014
β-adrenoceptor **agonist**

Ironically, the crystal structure of human β_2-andrenergic receptor was not solved until 2007.[74] Solving crystal structures of GPCRs for hormones and neurotransmitters are very challenging because of their low natural abundance, inherent structural flexibility, and instability in detergent solutions. So it is not surprising that several Nobel Prizes were bestowed to researches associated with solving GPCR crystal structures such as Raymond Stevens and Brian Kobilka in 2012 for solving the crystal structure of human β_2-andrenergic receptor. Here is a list of GPCRs and chronology of when their structures were solved: bovine rhodopsin, 2000; human A_{2A} adenosine receptor, 2008; human β_1-andrenergic receptor, 2008; human CXCR4 receptor, 2010; and human dopamine D_3 receptor, 2010. Today, there are over 120 GPCR crystal structures deposited in the Protein Data Bank (PDB) of 32 different receptor from families scattered across the phylogenetic tree, including class A, B, C, and frizzled GPCRs.[75]

1.3.2 Full and Partial Agonists

An agonist binds to a receptor and produces a pharmacological effect. The majority of agonists may be divided as either full agonists or partial agonists. Full agonists have 100% of the effect as the endogenous ligand, whereas partial agonists have less than 100% of the effect as the endogenous ligand. Partial agonists often act as both antagonists and agonists.

We can also look at partial agonists from biological assay's perspective. In a given assay, partial agonists are drugs that bind to and activate a given receptor, but have only partial efficacy at the receptor relative to a full agonist. For instance, a partial agonist in assay A can be a full agonist in assay B, but an antagonist in assay C.

Besides steroids, which are nuclear hormone receptor agonists, the most common class of drugs that function as receptor agonists are probably opioids. Morphine (**106**)'s pharmacological effects have been known for millennia. Even heroin (3,6-di-acetyl-morphine, **107**) has been in use (misuse might be a better description) since 1897. The proposal and models of opioid receptor were forwarded in the 1960s. In 1975, the endogenous ligands for opioid receptors, also known as endophines (for *endo*genous mor*phine*), were isolated as pentapeptide methionine-enkephalin (**108**, Tyr-Gly-Gly-Phe-Met) and leucine-enkephalin (H-Tyr-Gly-Gly-Phe-Leu).

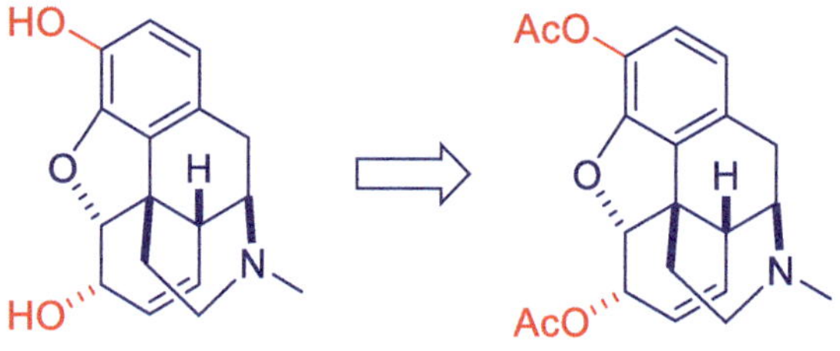

morphine (**106**) heroin (**107**)
μ receptor agonist μ receptor agonist

It soon became evident that multiple opioid receptors exist. Today, four types of opioid receptors have been identified: mu (μ, activated by morphine), delta (δ, for deferens because it was first discovered in mouse), kappa [κ, activated by ketocyclazocine (**109**)], and opioid-receptor like-1 (ORL-1). These four opioid receptors, all GPCRs, have now been cloned and characterized. They are known to form both homomeric and heteromeric receptor complexes. In fact, opioid receptors can form heteromeric receptor complexes with nonopioid receptors as well, for instance, μ and α_{2a} adrenoceptors.[76]

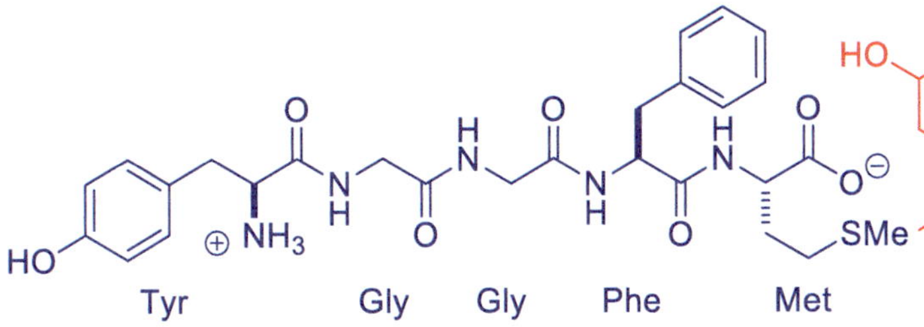

H-Tyr-Gly-Gly-Phe-Met-OH, or YGGFM
endogenous opioid: met-enkephalin (**108**)

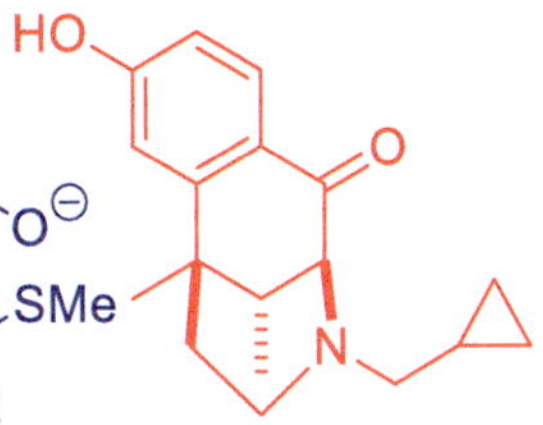

ketocyclazocine (**109**)
κ receptor agonist

Subtle variations of a molecular structure could readily convert an opioid receptor agonist to an antagonist. Morphine (**106**) is an agonist for the μ-opioid receptor, with good selectivity for the μ-receptor over related κ- and δ-opioid receptors. Simply switching morphine (**106**)'s *N*-methyl group to an *N*-allyl substituent, the resulting nalorphine (**110**) is a μ-opioid antagonist. Similarly, naloxone (**111**) is both a μ-opioid antagonist and κ opioid antagonist. As a nonselective opioid antagonist, naloxone (**111**) is less effective in blocking encephalin-induced inhibition of the nerve-evoked contraction of the mouse vas deferens.[77]

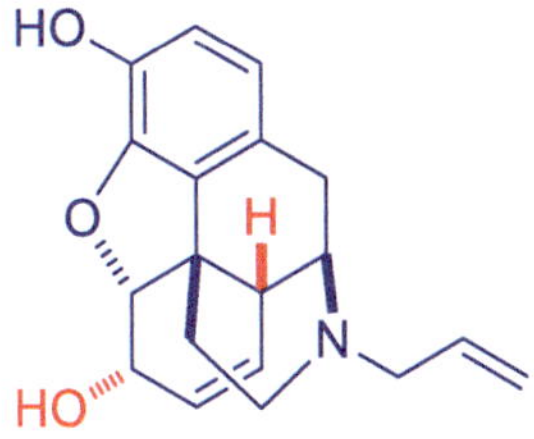

nalorphine (**110**)
μ opioid antagonist
κ opioid partial agonist

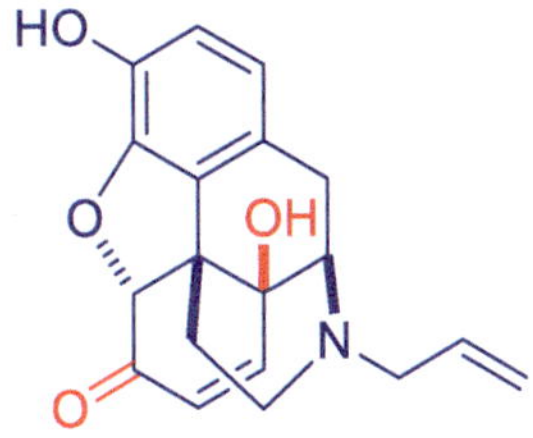

naloxone (**111**)
μ opioid antagonist
κ opioid antagonist

Many drugs were discovered without the benefit of understanding the drug targets. In 1960, Janssen synthesized fentanyl (Duragesic, **112**), which is 80–100-fold more potent than morphine (**106**) as an analgesic. Later on, it was found that fentanyl (**112**) is a μ opioid agonist. In fact, fentanyl (**112**) and sufentanil (Sufenta, **113**) are among the most potent μ agonists known and generally have been used as adjuncts for surgical anesthesia. Remifentanil (Ultiva, **114**), also a μ agonist, is rapidly metabolized by plasma and tissue esterases, has a terminal elimination half-life of less than 10 min and does not accumulate in tissues. Therefore, it is ideal for use by continuous infusion in perioperative setting.[78]

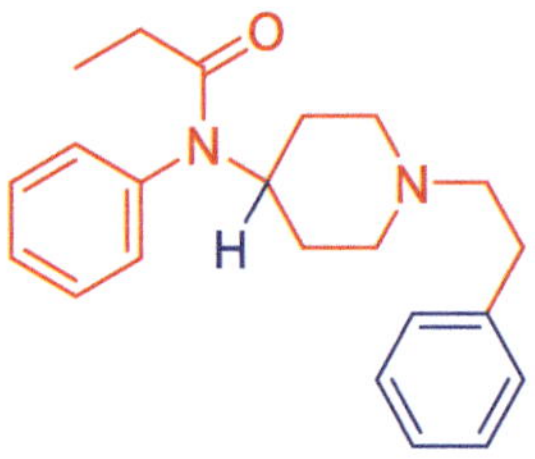

fentanyl (Duragesic, **112**)
μ opioid agonist
$t_{1/2}$ = ~3 h

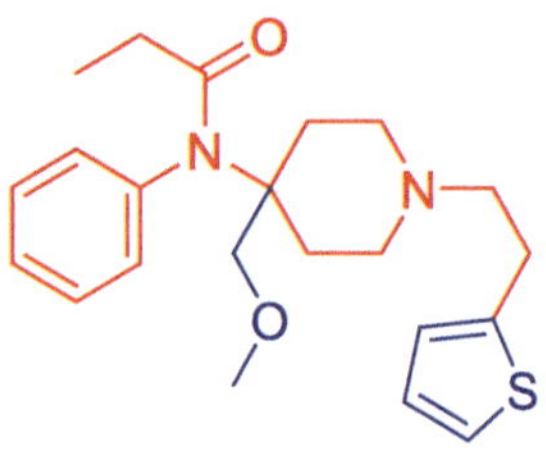

sufentanil (Sufenta, **113**)
μ opioid agonist
$t_{1/2}$ = 97 min

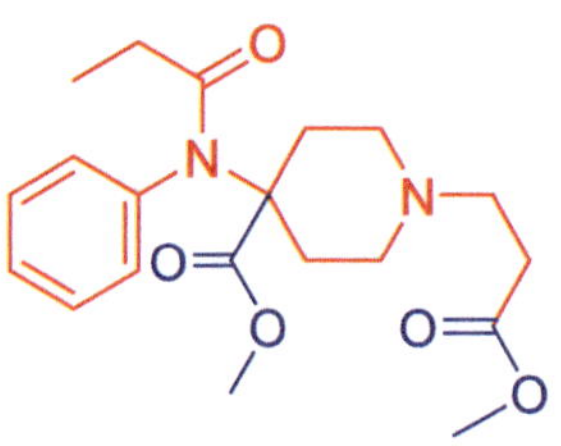

remifentanil (Ultiva, **114**)
μ opioid agonist
$t_{1/2}$ < 10 min

1.3.3 Antagonist and Agonist Interconversion

We just learned that a small structural change, replacing the *N*-methyl group on potent μ-opioid agonist morphine (**106**) with an *N*-allyl, resulted in the potent μ-opioid antagonist nalorphine (**110**).

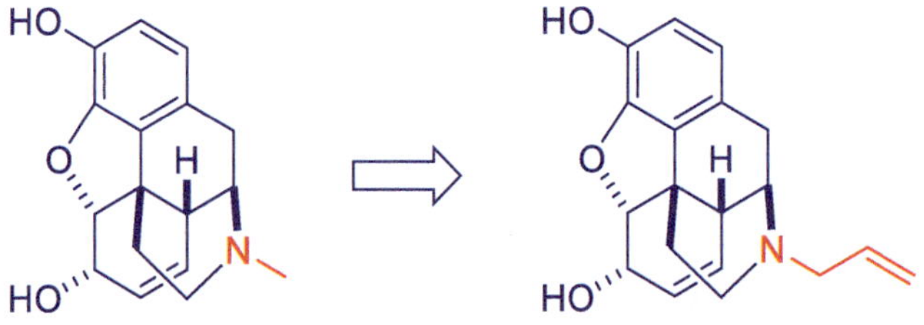

morphine (**106**)
μ receptor **agonist**
K_i = 0.53 nM

nalorphine (**110**)
μ receptor **antagonist**
K_i = 0.36 nM

In the same vein, small modifications to the structure of GPCR ligands can lead to major changes in functional activities, switching agonists to antagonists or vice versa.[79] For instance, histamine-2 (H_2) receptor antagonists may suppress gastric acid, whereas H_2 receptor agonists would stimulate gastric acid secretion. Histamine (**115**), as the endogenous ligand, itself an H_2 receptor agonist, causes stimulation of gastric acid secretion. During the process of discovery of the first blockbuster drug cimetidine (Tagamet, **118**) as an H_2 receptor antagonist to treat peptic ulcer, James Black at SmithKline and French, along with his chemistry colleagues, prepared 2-methylhistamine (**116**). Instead of suppressing gastric acid secretion, 2-methylhistamine (**116**) stimulated acid secretion, a manifestation as an H_2 receptor agonist. The next key compound guanylhistamine (**117**) acted as both an antagonist and an agonist, also known as a partial agonist. Many reiterations and extensive SAR investigations eventually produced cimetidine (**118**) as a full antagonist.[80]

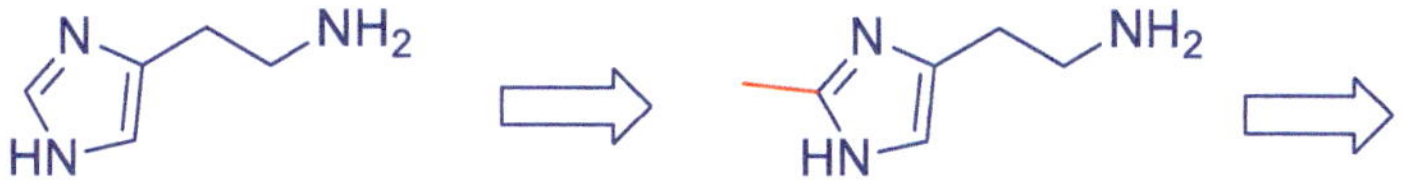

histamine (**115**) — H_2 receptor agonist

2-methylhistamine (**116**) — H_2 receptor agonist

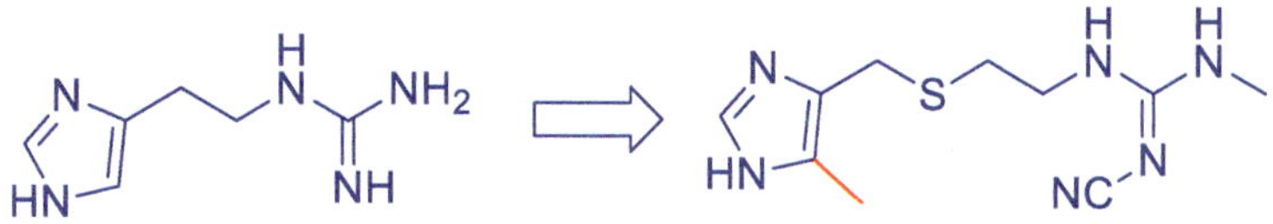

guanylhistamine (**117**) — H_2 receptor agonist/antagonist

cimetidine (Tagamet, **118**) — H_2 receptor antagonist

Activation of the GPCR 5-HT_{2B} receptor has dire consequences. Among the 14 serotonin receptor subtypes, 5-HT_{2B} receptor is notorious, known as the "death receptor". A 5-HT_{2B} receptor agonist drug is responsible for the cardiovascular side effect of Wyeth's obesity drug Fen–Phen: *fen*fluramine (**119**)/*phen*termine (**120**).[81] More precisely, the major metabolite of fenfluramine (**119**), norfenfluramine, is a potent 5-HT_{2B} agonist as tested in Gq calcium release assays, whereas fenfluramine (**119**) itself exhibits little activation. It has been shown that activation of 5-HT_{2B} receptors is necessary to produce valvular heart disease and that serotonergic medications that do not activate 5-HT_{2B} receptors are unlikely to produce valvular heart disease. Therefore, all clinically available medications with serotonergic activity and their active metabolites should be screened for agonist activity at 5-HT_{2B} receptors and that clinicians should consider suspending their use of medications with significant activity at 5-HT_{2B} receptors.

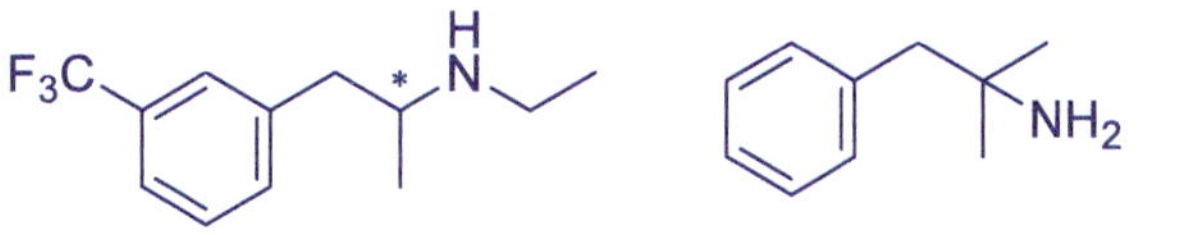

fenfluramine (**119**) — $5HT_{2B}$ receptor agonist

phentermine (**120**) — $5HT_{2B}$ receptor agonist

Ergot alkaloid pergolide (Permax, **121**) is a dopamine receptor (D_2) agonist prescribed for treating Parkinson's disease before it was withdrawn from the market due to toxicities involving heart valve damage (valvulopathy). This potentially deadly side effect has been observed with many, but not all, ergot alkaloids (e.g., ergotamine) used in the clinic and has been linked to activation of the 5-HT_{2B} receptor. Pergolide (**121**) is tested as a 5-HT_{2B} agonist. Replacing the *N*-propyl fragment with an *N*-methyl group gave rise to 6-methyl-pergolide (**122**), which is now a 5-HT_{2B} receptor antagonist while maintaining to be a D_2 agonist.[82]

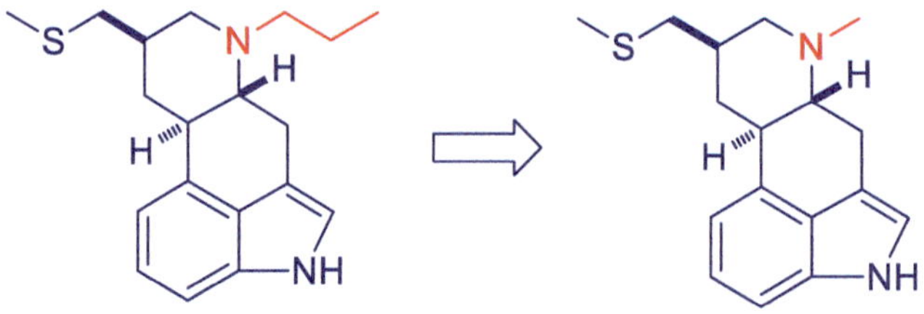

pergolide (Permax, **121**)
5-HT_{2B} receptor agonist
D_2 agonist

6-methyl-pergolide (**122**)
5-HT_{2B} receptor antagonist
D_2 agonist

Numerous additional examples exist where small structural alterations resulted in interconverting GPCR agonists and antagonists.[79]

In addition to receptor antagonists, agonists, and partial agonists, there are also receptor *inverse agonists*, *biased agonists*, etc. An inverse agonist binds to the same receptor as an agonist but induces a pharmacological response opposite to that agonist. On the other hand, biased agonists belong to the category of biased ligands. They are ligands showing distinct potency and/or efficacy toward different signaling pathways engaged by a given receptor. This concept, also known as "functional selectivity," presents an evident advantage for the development of therapeutics with better safety profiles. Indeed, targeting only the pathway(s) involved in disease modification without affecting other functions of the receptor enables the development of drug candidates with reduced risk of adverse effects.

All of these jargons are indeed very confusing to us medicinal chemists (to some biologists as well, I secretly suspect). When encountering these terms, it might be a good time to talk to your biology/pharmacology colleagues. For rocket science, it is probably wise to converse with rocket scientists rather than attempting launching a rocket on your own.

1.3.4 G-Protein-Coupled Receptors

GPCRs are the most fruitful targets for drug discovery—134 GPCRs are targets for drugs approved in the United States or European Union.[83] The family of GPCRs have been classified according to their pharmacological properties into four main sub-families (Figure 1.15):

1. Classes A, rhodopsin-like receptors;
2. Class B, secretin-like receptors;
3. Class C, metabotropic glutamate/pheromone receptors; and
4. Class F, frizzled receptors.

As mentioned before, GPCR agonists mimic the endogenous ligand (e.g., dopamine (DA) in Parkinson's disease), resulting in synthetic activation of the GPCR signaling pathway. In contrast, GPCR antagonists block the signaling of a receptor. β-blockers **101–105** for the treatment of cardiac disorders such as hypertension are GPCR antagonists.

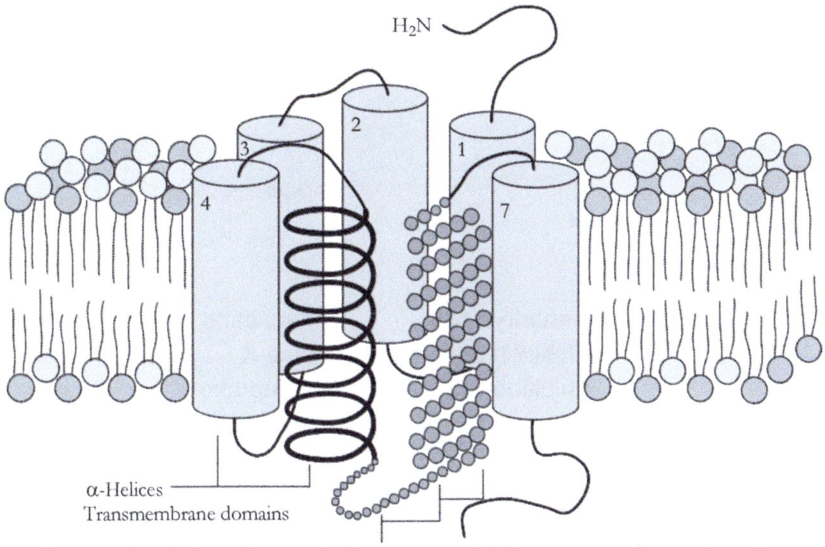

Figure 1.15 G-Protein-coupled receptor with 7 trans-membrane domains.
Source: Vivien H. Li.

1.3.4.1 Serotonin Receptors

In the human body, serotonin (5-hydroxytryptamine, 5-HT, **124**) is generated from natural amino acid tryptophan (**123**) via a two-step process involving the tryptophan hydroxylase and the decarboxylase. As a neurotransmitter, serotonin has a profound impact on our physiology, especially on our CNS, namely, mood. Hence, serotonin is also known as a happy molecule. Lower than normal levels of serotonin are often associated with depression. As a side, it is interesting to know that fruits like banana, tomatoes, nuts, plums, venoms of bees, and wasps also contain serotonin.

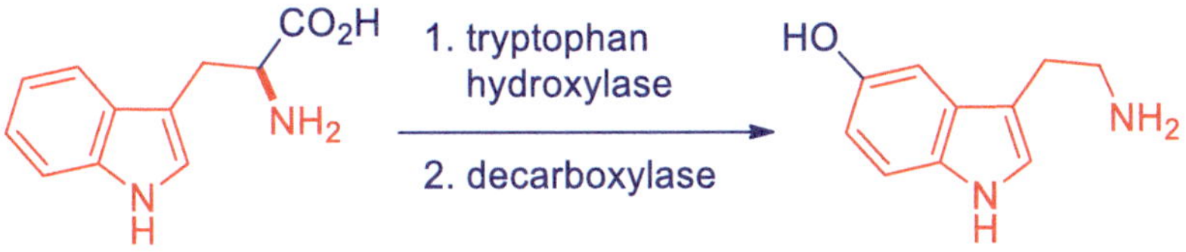

tryptophan (**123**) serotonin (5-hydroxytryptamine, 5-HT, **124**)

As the endogenous ligand of serotonin receptors, serotonin (**124**) travels from presynaptic neurons to postsynaptic neurons, where serotonin (5-HT) receptors reside. The process would be a happy ending except that presynaptic neurons are also able to "reuptake" serotonin molecules and lower the levels of serotonin. The old monoamine oxidase inhibitors (MAOIs) could block the oxidation of serotonin to biologically inactive 5-hydroxyindole acetic acid, but they were not selective, causing severe side effects. Using an old anti-histamine, i.e., a histamine receptor-1 (H_1) receptor antagonist), diphenhydramine (Benadryl, **125**) as a starting point, Lilly discovered the selective serotonin reuptake inhibitor (SSRI), fluoxetine (Prozac, **126**).[84]

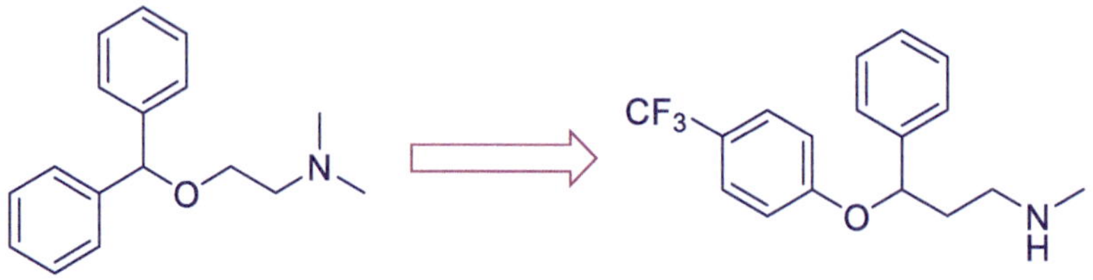

diphenhydramine (Benadryl, **125**) fluoxetine (Prozac, **126**)

inhibition of serotonin uptake K_i 17 nM

inhibition of norepinephrine uptake K_i 2,703 nM

Today, many SSRIs are generic drugs now including paroxetine (Paxil, **127**) and sertraline (Zoloft, **128**). The success of fluoxetine (**126**) opened a floodgate for efficacious and safe antidepressants. For example, trazodone (Desyrel, **129**), a phenyl-piperazine compound, is a combined 5-HT_{2A} receptor antagonist and a weak serotonin reuptake inhibitor (SRI). Merck KGa's vilazodone (Viibryd, **130**), another phenyl-piperazine compound, is a combined SRI and a 5-HT_{1A} receptor partial agonist. Lundbeck's vortioxetine (Brintellix, **131**) is the latest entry of antidepressant. It has a complicated poly-pharmacology. But it is an inhibitor of 5-HT_{1A} and 5-HT_{3A} receptors and serotonin transporter (SERT), among several other receptors.

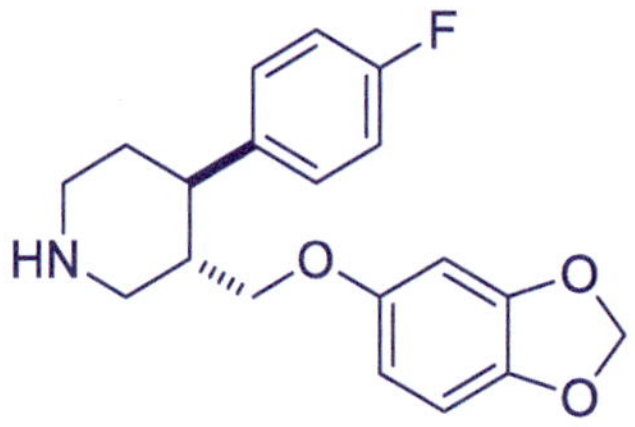

paroxetine (Paxil, **127**)
GSK, 1992
SSRI

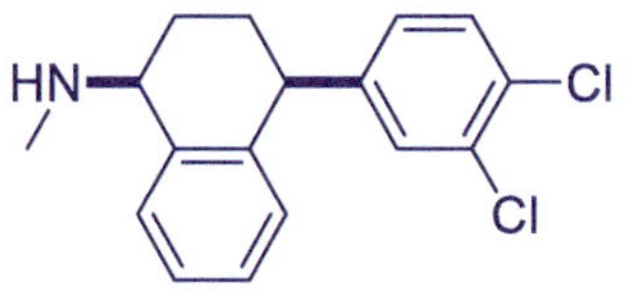

sertraline (Zoloft, **128**)
Pfizer, 1997
SSRI

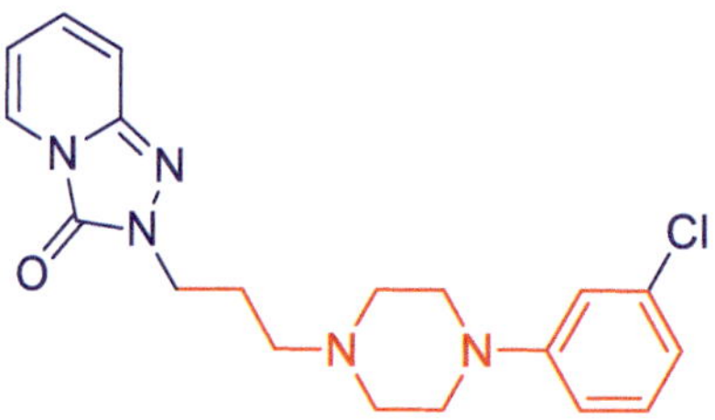

trazodone (Desyrel, **129**)
Angelini, 1981
5-HT_{2A} antagonist /SRI

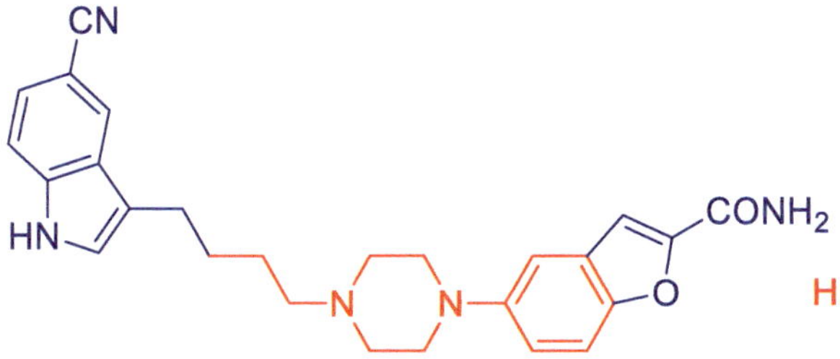

vilazodone (Viibryd, **130**)
Merck KGa, 2011
5-HT_{1A}/5-HT_{3A} antagonist

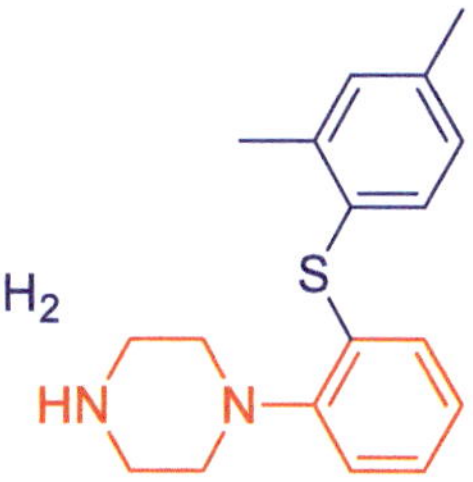

vortioxetine (Brintellix, **131**)
Lundebck/Takeda, 2013
Polypharmacology

We now have better understanding of the many subtypes of the serotonin receptors. As shown in Figure 1.16,[85] serotonin receptors may be divided to seven major subtypes. Six of them are all GPCRs except 5-HT_3 is a black sheep in the family, being a cation-selective ligand-gated ion channel. Subtype 5-HT_1 is also further divided to three mini-subtypes: 5-HT_{1A}, 5-HT_{1B}, and 5-HT_{1D}. Subtype 5-HT_2 is further divided to three mini-subtypes as well: 5-HT_{2A}, 5-HT_{2B}, and 5-HT_{2C}., among which the "death receptor" 5-HT_{2B} was mentioned before. Furthermore, Subtype 5-HT_5 is further divided to two mini-subtypes: 5-HT_{5A} and 5-HT_{5B}.

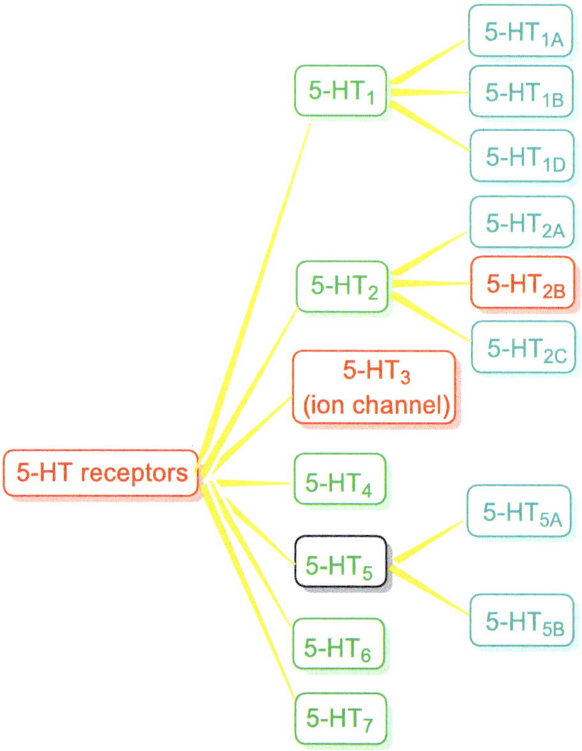

Figure 1.16 Subtypes of serotonin receptors.

Among these 14 5-HT receptor subtypes, 5-HT_{1A} receptors are the most studied and best characterized. Many drugs are found to be its ligands, modulating the 5-HT_{1A} receptors as either agonists or antagonists, etc.[86]

Many triptans for the treatment of migraines are 5-HT_{1D} agonists. Ergot alkaloids, such as ergotamine (a 5-HT_{2A} antagonist), were used to treat migraine for over a century. It was recognized that their beneficial effects resulted from activation of 5-HT_1-like receptors, specifically 5-$HT_{1B/1D}$ receptors. This led to the development of sumatriptan succinate (Imitrex, **132**), a selective 5-$HT_{1B/1D}$ agonist, as the first specific antimigraine medication. It is believed that 5-$HT_{1B/1D}$ agonists elicit their antimigraine action by selective vasoconstriction of excessively dilated intracranial, extracerebral arteries and/or inhibiting the release of inflammatory neuropeptides from perivascular trigeminal sensory neurons.[87]

Once the pharmacophore was identified, many "me-too" triptans emerged to offer relives to the migraine patients. Representative triptans include zolmitriptan (Zomig, **133**), eletriptan hydrobromide (Relpax, **134**), and frovatriptan succinate (Frova, **135**), among others.

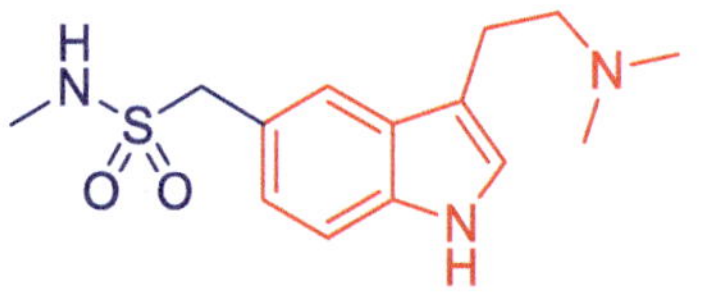

sumatriptan succinate (Imitrex, **132**)
GSK, 1995, 5-$HT_{1B/1D}$ agonist

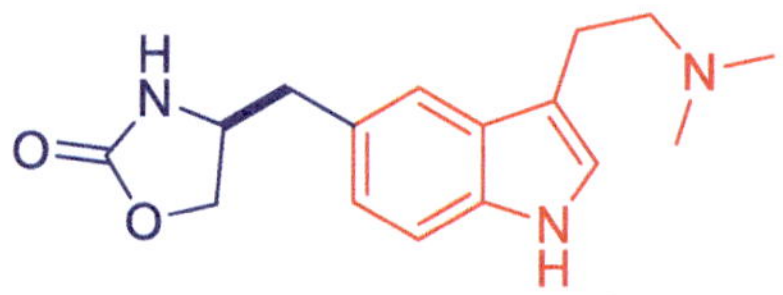

zolmitriptan (Zomig, **133**)
Wellcome, 1997, 5-$HT_{1B/1D}$ agonist

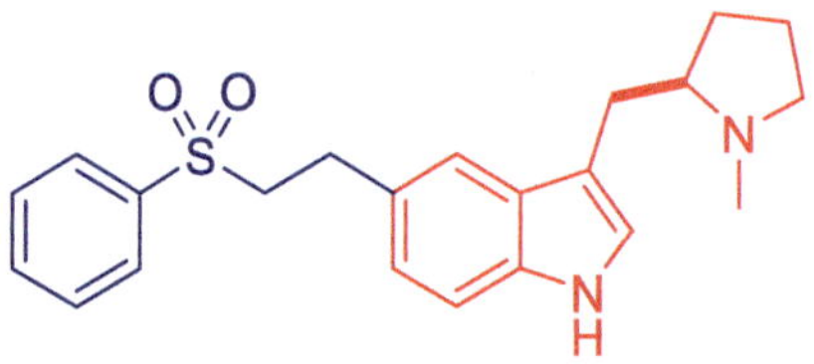

eletriptan hydrobromide (Relpax, **134**)
Pfizer, 2002, 5-$HT_{1B/1D}$ agonist

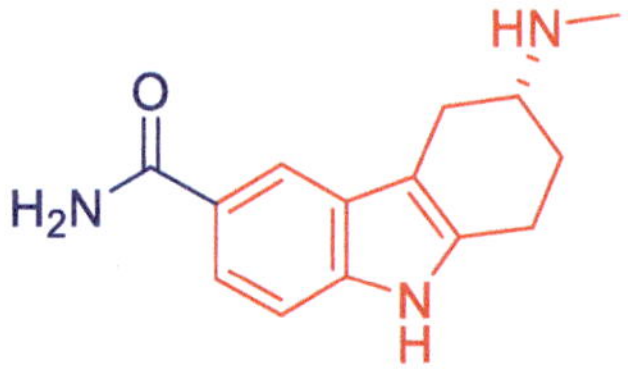

frovatriptan succinate (Frova, **135**)
SmithKline/Elan, 2001
5-$HT_{1B/1D}$ agonist

It has been suggested that 5-HT_{1B} receptor activation results in vasoconstriction of intracranial vessels, while inhibition of neuropeptide release is mediated via the 5-HT_{1D} receptor. Selective 5-HT_{1D} agonists have recently been identified and are being studied to determine the relative importance of these receptor mediated events on the antimigraine activity.[88]

In addition to cimetidine (Tagamet, **118**), histamine H_2 receptor antagonists have been fruitful as anti-ulcer drugs. Ranitidine (Zantac, **136**) and famotidine (Pepcid, **137**) are just two additional examples.

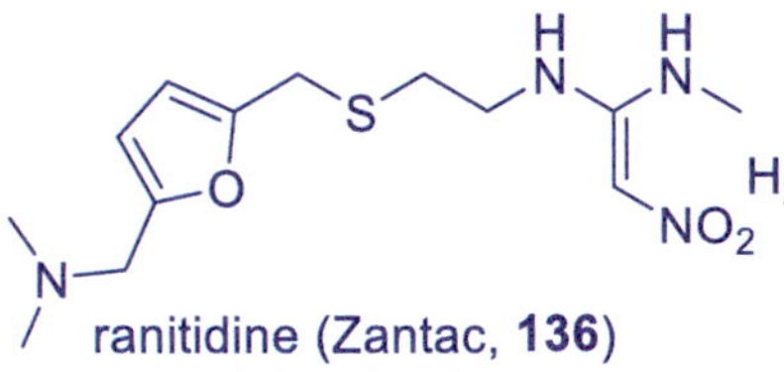

ranitidine (Zantac, **136**)
GSK, 1983
H_2 antagonist

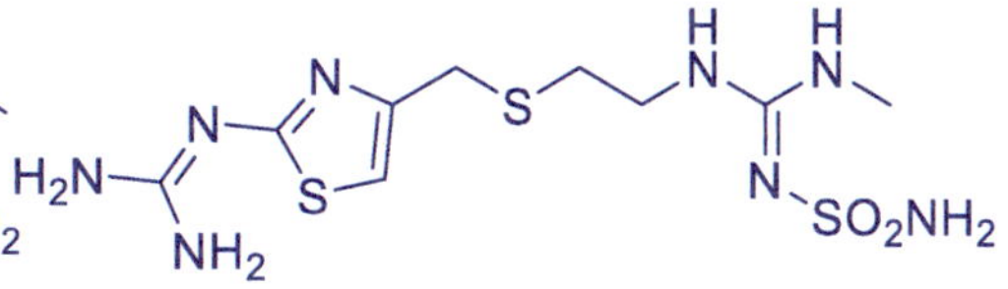

famotidine (Pepcid, **137**)
Yamanouchi/Merck, 1986
H_2 antagonist

Arena Pharmaceuticals' lorcaserin (Belviq, **138**) is a 5-HT_{2C} receptor agonist for the treatment of obesity.[89] Meanwhile 5-HT_{2A} receptor is of paramount importance as antipsychotic drugs in view of the "D_2/5-HT_{2A}" theory. All atypical antipsychotics are potent antagonists of serotonin 5-HT_{2A} and dopamine D_2 receptors including Lilly's olanzapine (Zyprexa, **139**) as one of the more popular atypical antipsychotics for treating schizophrenia. More on this in the section on dopamine receptors.[90]

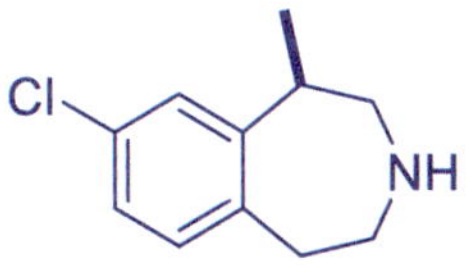

lorcaserin (Belviq, **138**)
Arena, 2012
5-HT_{2C} agonist

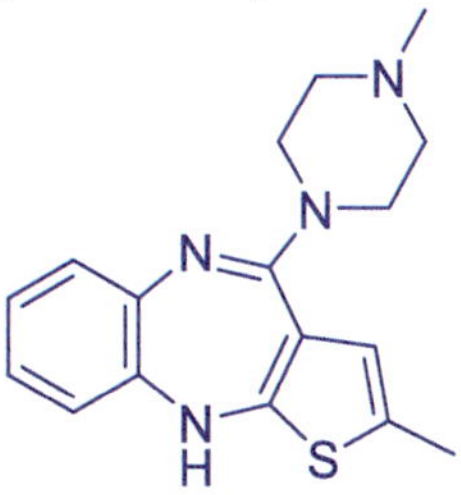

olanzapine (Zyprexa, **139**)
Eli Lilly, 1996
5-HT_{2A} antagonist
D_2 antagonist

The latest entry of selective 5-HT_2 antagonist is Sprout Pharmaceuticals' flibanserin (Addyi, **140**) for treating female sexual dysfunction. Its major pharmacology is both a 5-HT_{2A} receptor antagonist and a 5-HT_{1A} receptor agonist.[91] A 5-HT_{2A} inverse agonist, Acadia's pimavanserin (Nuplazid, **141**) is approved for the treatment of Parkinson's disease.[92]

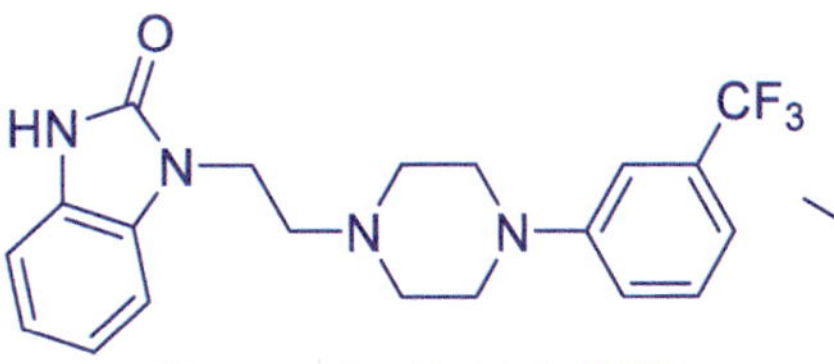

flibanserin (Addyi, **140**)
Sprout/BI, 2016
5-HT_{2A} antagonist
5-HT_{1A} agonist

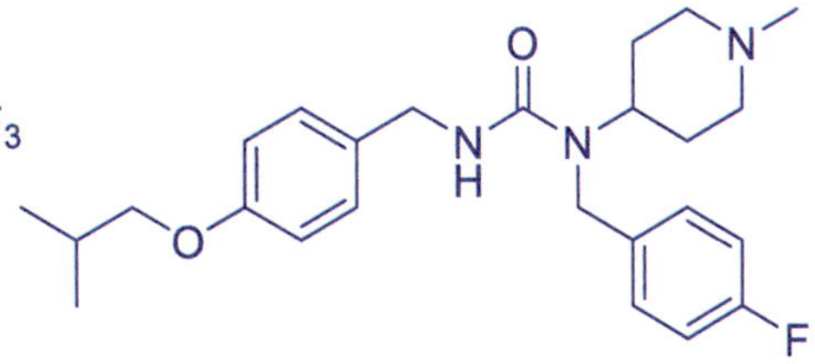

pimavanserin (Nuplazid, **141**)
Acadia, 2017
5-HT_{2A} inverse agonist

5-HT_3 receptor antagonists ondansetron (Zofran, **142**), granisetron (Kytril, **143**), tropisetron (Navoban, **144**) are used as anti-emetics for nausea and vomiting linked to chemotherapies in cancer treatments.[93]

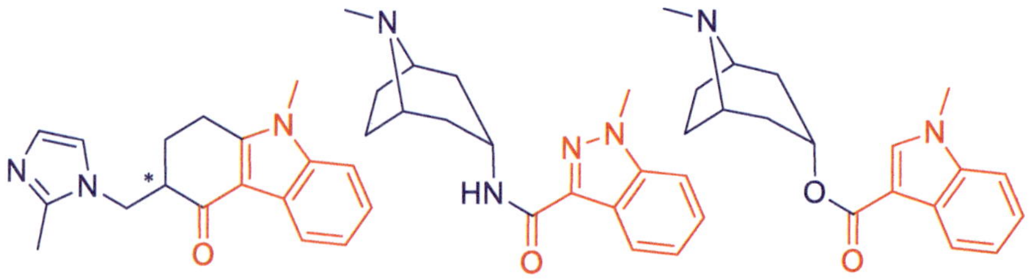

ondansetron (Zofran, **142**) granisetron (Kytril, **143**) tropisetron (Navoban, **144**)
GSK, 1990 Roche, 1994 Novartis, 1992
5-HT_3 antagonist 5-HT_3 antagonist 5-HT_3 antagonist

Knowledge gained from **142–144** and similar 5-HT_3 receptor antagonists turned out to be ligands for both 5-HT_3 and 5-HT_4 receptors. Novartis′ tegaserod (Zelnorm, **145**), a partial agonist of 5-HT_4 receptor, is prescribed for irritable bowel syndrome (IBS) and constipation.[94] The presence of a guanidinium cation in carbazimidamide **145** at physiological pH limited crossing the blood–brain barrier (BBB), thereby avoiding CNS side effects of the drug. At the end of 2018, the FDA approved Shire′s prucalopride (Motegrity, **146**) for the treatment of chronic idiopathic constipation (CIC) in adults. It functions as 5-HT_4 receptor agonist.[95]

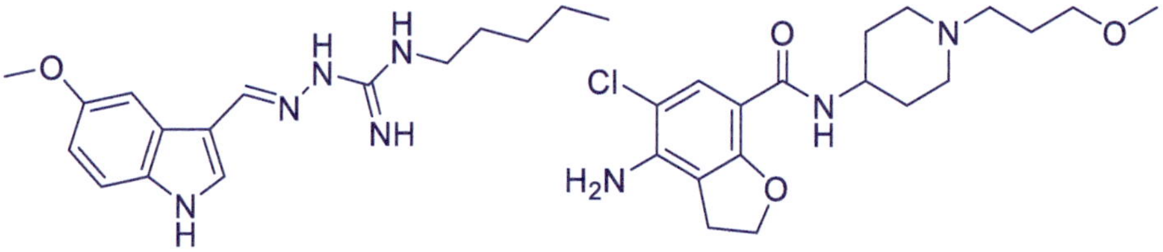

tegaserod (Zelnorm, **145**)
Novartis, 2002
partial agonist of 5-HT_4 receptor

prucalopride (Motegrity, **146**)
Shire, 2018
5-HT_4 agonist

The 5-HT_6 receptor is predominantly expressed in the CNS. Therefore, its agonists and antagonists have been investigated for a variety of CNS diseases. Axovant′s SB-742457 (intepirdine, **147**), a selective 5-HT_6 receptor antagonist,[94(b)] failed Phase III clinical trials for Alzheimer′s disease (AD) in 2017. Previously, Lundbeck′s selective 5-HT_6 receptor antagonist Lu AE58054 (idalopirdine, **148**)[97] also failed to meet its primary endpoint for AD in 2013.

SB-742457 (intepirdine, **147**)
5-HT_6 receptor antagonist

Lu AE58054 (idalopirdine, **148**)
5-HT_6 receptor antagonist

Regrettably, it seems that no 5-HT_6 receptor modulator has gained regulatory approval thus far despite tremendous amount of resources invested. Literature on other more obscure 5-HT receptor subtypes is even sparser.

1.3.4.2 Dopamine Receptors

Dopamine (**151**), a catecholamine neurotransmitter, is the endogenous ligand for dopamine receptors. It is biosynthesized in dopaminergic neuron terminals from the essential amino acid L-tyrosine (**149**). With the aid of cytosolic enzyme tyrosine hydroxylase (TH), L-tyrosine (**149**) is oxidized to L-dihydroxyphenylalanine (L-dopa, **150**). Aromatic L-amino acid decarboxylase (AADC) decarboxylates L-dopa (**150**) to dopamine (**151**).

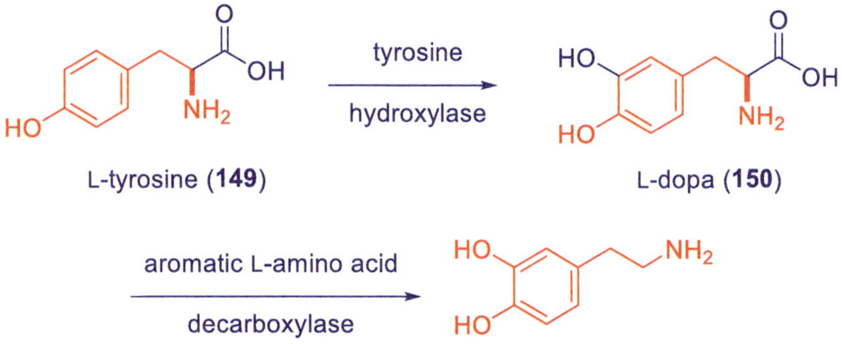

L-tyrosine (**149**)

L-dopa (**150**)

dopamine (**151**)

Dopamine is intimately associated to many CNS conditions. L-Dopa is prominently featured in the movie *Awakening*. It temporarily restored Parkinson's disease patients′ "normalcy" before they relapsed.

The family of dopamine (DA) receptors may be divided to two subgroups: D_1-like receptors (D_1 and D_5 subtypes) and D_2-like receptors (D_2, D_3, and D_4 subtypes, Figure 1.17).[97]

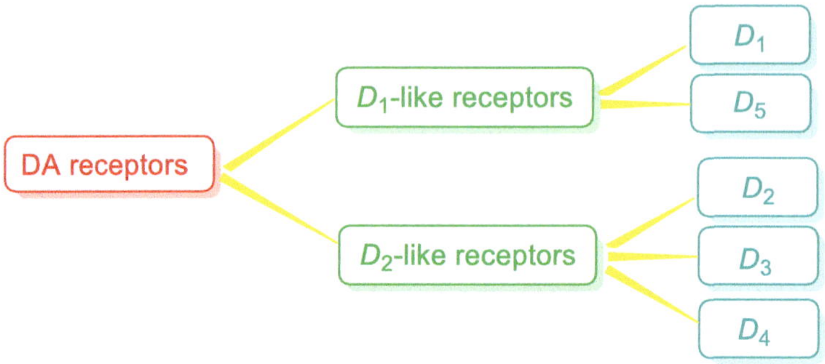

Figure 1.17 Subtypes of dopamine receptors.

Just like serotonin receptors, many neurotransmitters and CNS drugs bind to more than one types of receptors (e.g., both 5HT and DA receptors) and several subtypes of each receptor. Much research has been devoted to teasing out the intricacy of the role of each subtype of particular receptor. In addition to 5HT and DA receptors, adrenoceptors and *N*-methyl-D-aspartic acid (NMDA) receptors are frequently responsible for CNS functions.

In drug discovery, a selective compound is often preferred because of lower off-target toxicity. Selective dopamine receptor drugs are considered breakthroughs since they are difficult to achieve. SKF 39393 (**152**) is a partial agonist of the D_1 receptor. Schering–Plough's benzazepines SCH 23390 (**153**) and its conformationally more restraint ethylene-bridged analog SCH 39166 (ecopipam, **154**) are D_1/D_5 selective antagonists. Ecopipam, (**154**) advanced to Phase III clinical trials for Tourette syndrome after failing in trials for schizophrenia, addiction, obesity, and other D_1-dependent neurological disorders. All three drugs **152**–**154** possess one or two phenol functionality, which is susceptible to rapid first-pass metabolism and results in low plasma levels and poor bioavailability. Efforts have been taken to replace the culprit phenol group with more lipophilic isosteres such as indole, indazole, benzotriazole, benzimidazlolones, and benzothiazolones.[98]

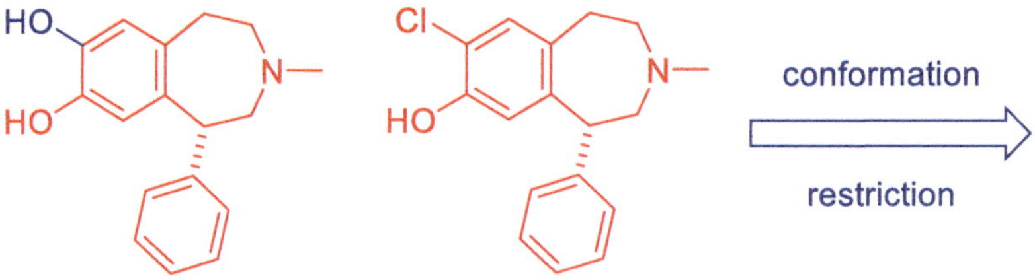

SKF 39393 (**152**)
D_1 partial agonist

SCH 23390 (**153**)
selective D_1/D_5 antagonist

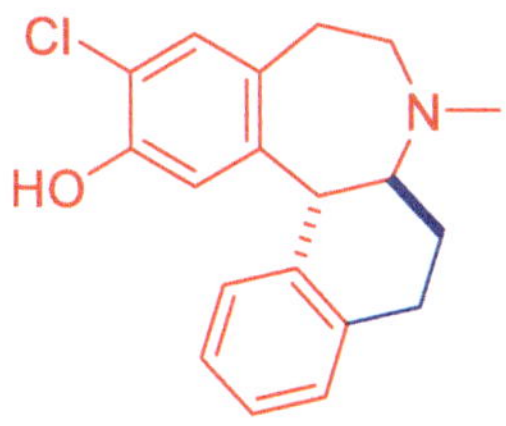

SCH 39166 (ecopipam, **154**)
selective D_1/D_5 antagonist

The D_2 dopamine receptor is one of the most validated drug targets for neuropsychiatric and endocrine disorders. But selective D_2 receptor ligands are hard to come by. L741,626 (**155**) was a selective D_2 receptor antagonist versus the D_3 receptor with a D_2/D_3 ratio of 6.5. Simple replacement of the chlorine atom with an iodine atom gave rise to its analog **156** with exponentially improved selectivity with a D_2/D_3 ratio of 49. Another selective D_2 antagonist JNJ-37822681 (**157**) had a D_2/D_3 ratio of 7.3 whereas the latest selective D_2 antagonist **158** had a D_2/D_3 ratio of 41.[99]

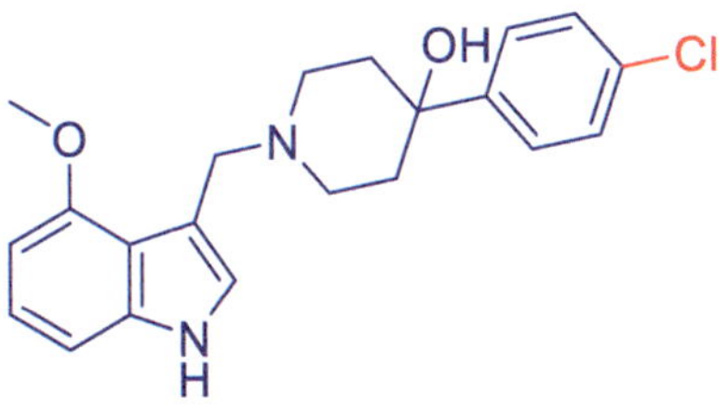

L741,626 (**155**)
K_i: D_2 = 9.5 nM, D_3 = 100 nM
D_2/D_3 = 6.5

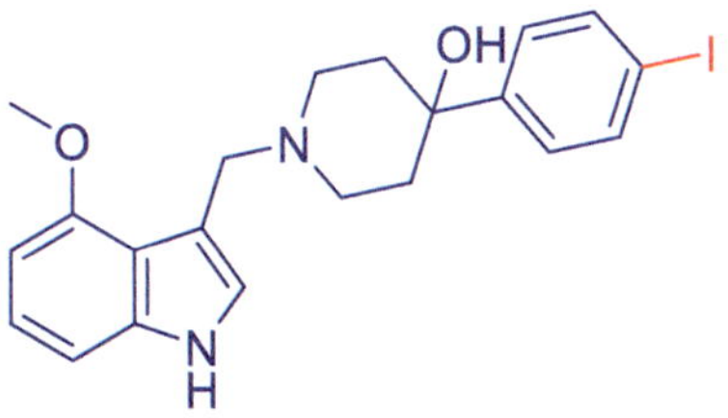

iodo-analog of L741,626 (**156**)
D_2 = 1.3 nM, D_3 = 100 nM
D_2/D_3 = 49

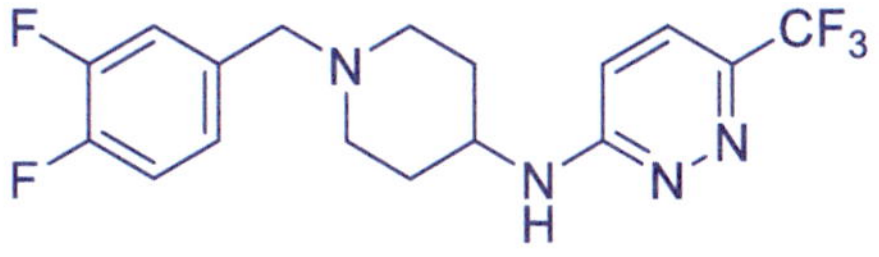

JNJ-37822681 (**157**)
K_i: D_2 = 158 nM, D_3 = 1,159 nM
D_2/D_3 = 7.3

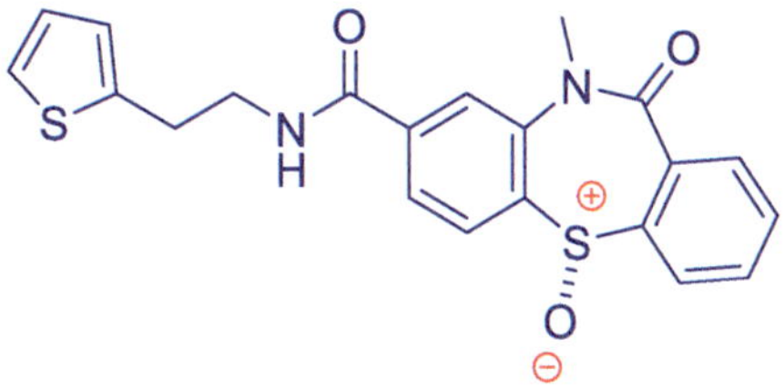

158
D_2 = 100 nM, D_3 = 4,100 nM
D_2/D_3 = 41.0

Atypical antipsychotics, also known as serotonin–dopamine antagonists, effectively reduce extrapyramidal symptoms (EPS) that plagued older typical antipsychotics. As mentioned before, atypical antipsychotics such as Pfizer's ziprasidone (Geodon, **159**) and Otsuka's aripiprazole (Abilify, **160**) are potent antagonists of dopamine D_2 and serotonin 5-HT_{2A} receptors. Because aripiprazole (**160**) was so successful, in 2015, Alkermes gained approval to sell its prodrug aripiprazole lauroxil (Aristada, **161**). Meanwhile, Otsuka garnered FDA's nod to market its "me-too" drug to Abilify: brexpiprazole (Rexulti, **162**).

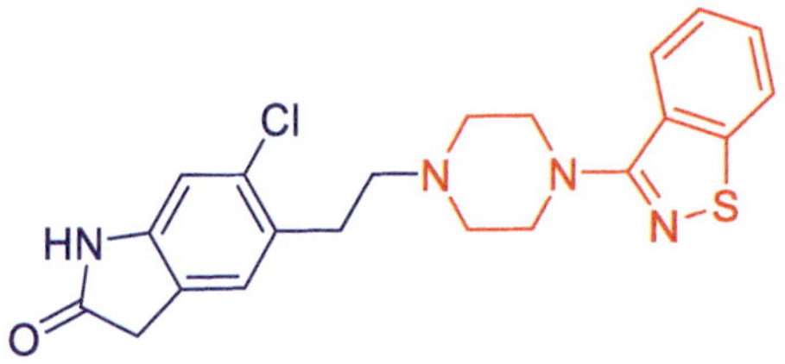

ziprasidone (Geodon, **159**)
Pfizer, 2001
full D_2 antagonist
5-HT_{2A} antagonist

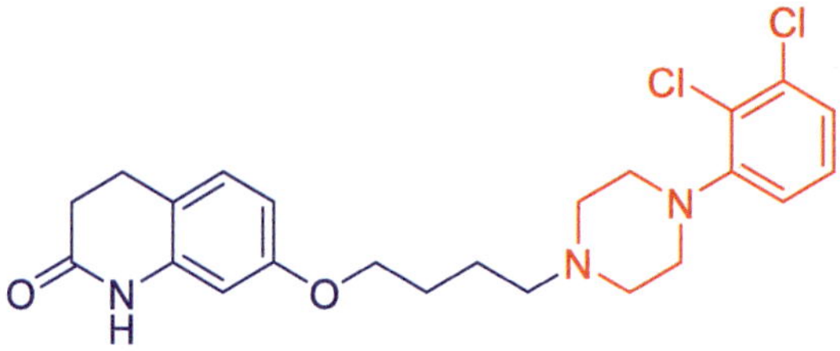

aripiprazole (Abilify, **160**)
Otsuka/BMS, 2002
D_2 antagonist
5-HT_{2A} antagonist

In addition to binding to D_2 and 5-HT_{2A} receptors, those atypical antipsychotics also act on many other receptors including multiple serotonin receptors (5-HT_{1A}, 5-$HT_{1B/1D}$, 5-HT_{2C}, 5-HT_3, 5-HT_6, and 5-HT_7), the noradrenergic system (α_1 and α_2), the muscarinic acetylcholinergic system (M_1), and the histamine receptors (H_1). It has been postulated that the additional 5-HT_{1A} agonist activity shown by several atypical antipsychotic agents could reduce EPS and alleviate the anxiety that often precipitates psychotic episodes in schizophrenia patients.

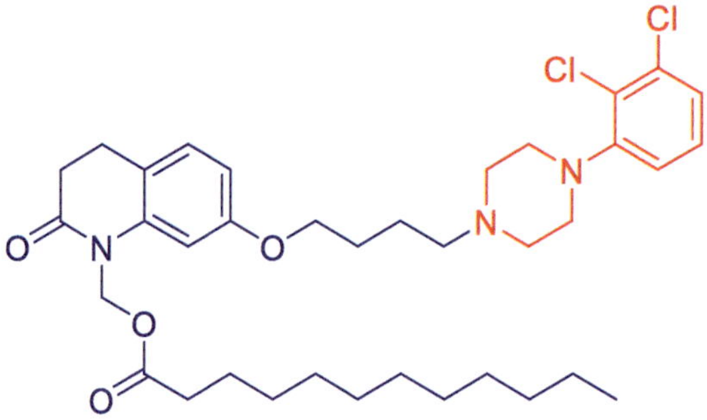

aripiprazole lauroxil
(Aristada, **161**)
Alkermes, 2016
partial agonist of
D_2, 5-HT_{1A} and
5-HT_{2A} receptors

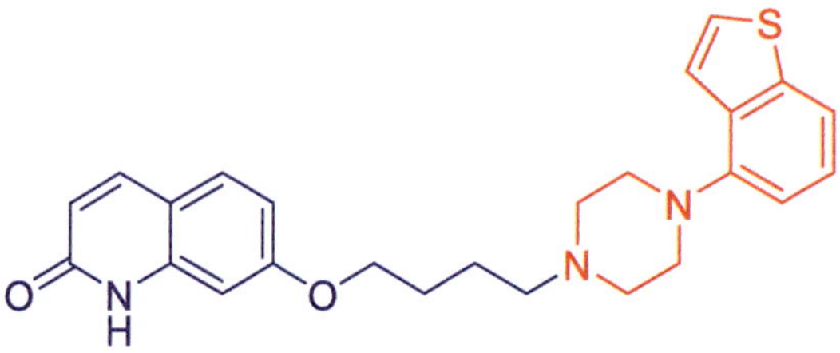

brexpiprazole
(Rexulti, **162**)
Otsuka/Lundbeck, 2016
partial agonist of
D_2, 5-HT_{1A} and
5-HT_{2A} receptors

Introduction of buspirone (Buspar, **163**) by Bristol–Myers Squibb in 1988 was considered revolutionary for the treatment of general anxiety disorder (GAD). Buspirone (**163**), the first member of the azapirone family of drugs, is an antagonist of D_2, D_3, and D_4 receptors. It is also a partial serotonin agonist at the 5-hydroytryptomine (5-HT_{1A}) receptor in the brain, effectively treating anxiety without concomitant sedative, muscle relaxant, or anticonvulsant activities.[100] Unlike benzodiazepines, it does not significantly interact with γ-aminobutyric acid type A ($GABA_A$) receptors.

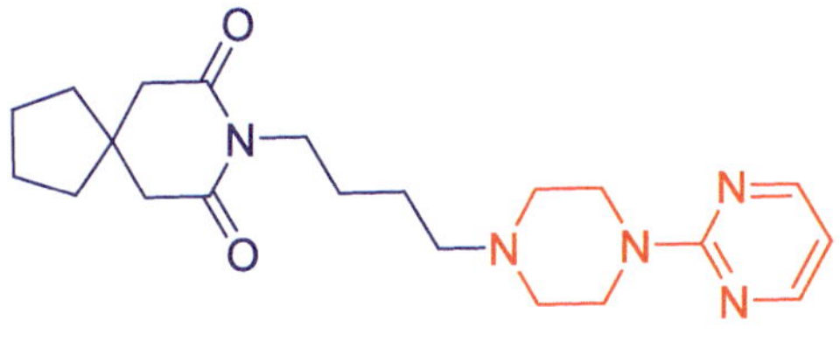

buspirone (Buspar, **163**)
BMS, 1988
$D_{2,3,4}$ antagonist
5-HT_{1A} partial agonist

Selective D_3 receptor antagonists and/or partial agonists have shown efficacy in animal models for drug abuse and other CNS disorders.[101] Regrettably, GSK598,809 (**164**), a selective D_3 antagonist caused significant hypertension in dogs in the presence of cocaine, may preclude further development of these agents toward cocaine addiction.[102] Another selective D_3 antagonist, (*R*)-PG648 (**165**), is 200-fold more selective against the D_2 receptor. An atypical antipsychotic cariprazine (Vraylar, **166**) is a partial agonist of D_3 and D_2 receptors with high selectivity for the D_3 receptor. The recent entry of a D_3 selective antagonist **167** is highly selective with a 1,700-fold selective over the D_2 receptor. It inhibited oxycodone-induced hyperlocomotion in mice and reduced oxycodone-induced locomotor sensitization. The evidence support the D_3 receptor as a target for opioid dependence treatment.[103]

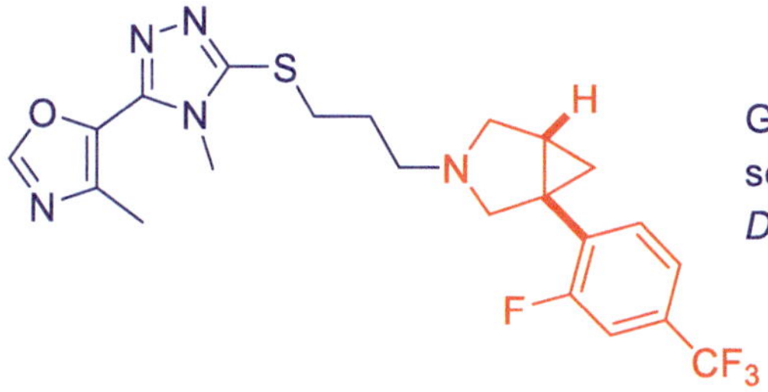

GSK598,809 (**164**)
selective D_3 antagonist
D_3 K_i = 6.2 nM, D_2/D_3 = 119

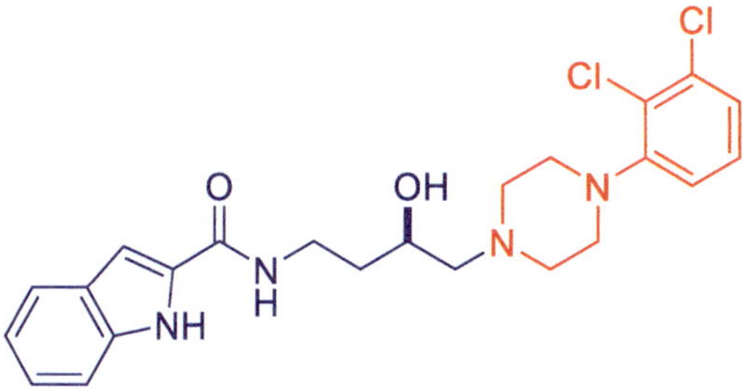

(*R*)-PG648 (**165**)
selective D_3 antagonist
D_3 K_i = 2.9 nM, D_2/D_3 = 200

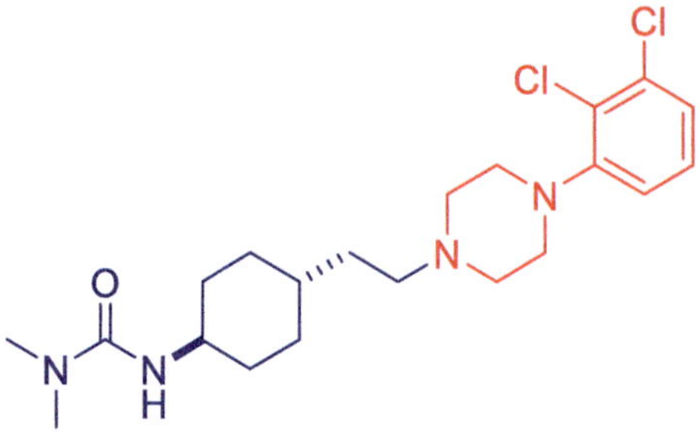

cariprazine (Vraylar, **166**)
Gedeon Richter/Actavis, 2015
partial agonist of
D_2, 5-HT_{1A} and 5-HT_{2A} receptors

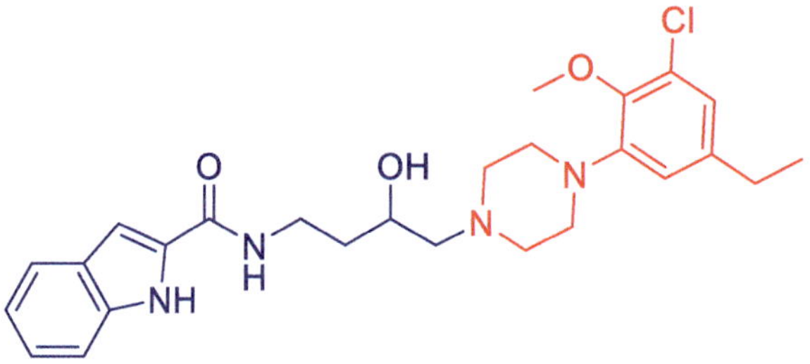

167
selective D_3 antagonist
D_3 K_i = 6.84 nM
D_2/D_3 = 1,700

It is hard to believe that the human D_4 receptor was not cloned and identified until 1991. Since the atypical antipsychotic clozapine (Clozaril, **168**) was found to possess higher affinity for D_4 relative to other dopamine receptor subtypes D_1, D_2, D_3, and D_5, great enthusiasm was generated to find unique clinical efficacy of selective D_4 antagonists. Unfortunately, in 1997, Merck's selective D_4 antagonist L-745,870 (**169**) failed to show sufficient therapeutic response for acutely psychotic inpatients with schizophrenia. Thus, D_4 receptor as a drug target fell out of favor for decades until recently the target is experiencing a reconnaissance. Many selective D_4 receptor ligands have been discovered and we will observe how they fare in clinics, especially for treating the Parkinson's disease.[104]

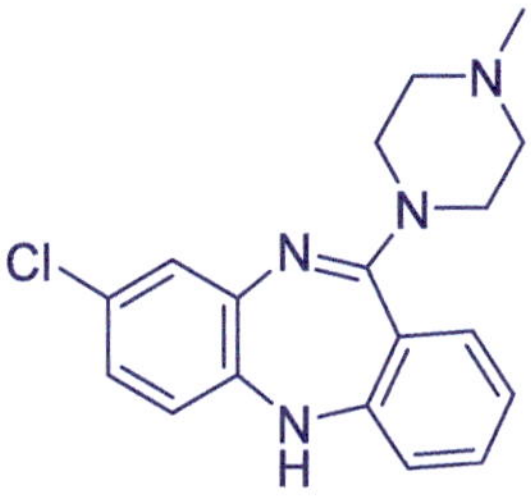

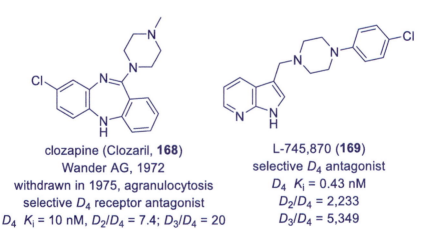

clozapine (Clozaril, **168**)
Wander AG, 1972
withdrawn in 1975, agranulocytosis
selective D_4 receptor antagonist
D_4 K_i = 10 nM, D_2/D_4 = 7.4; D_3/D_4 = 20

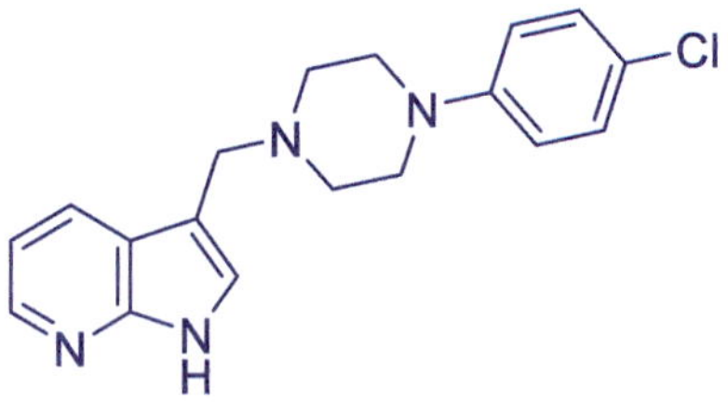

L-745,870 (**169**)
selective D_4 antagonist
D_4 K_i = 0.43 nM
D_2/D_4 = 2,233
D_3/D_4 = 5,349

An excellent and comprehensive review on selective dopamine receptor agents was published in 2013.[97a]

1.3.4.3 The Latest GPCR Drugs

Histamine H_1 and H_2 receptors have been fruitful targets, yielding antihistamines to treat allergy and ulcer, respectively. The latest entry of H_3 receptor drug is Bioproject's pitolisant (Wakix, **170**) for treating narcolepsy. It is an H_3 receptor inverse agonist.[105]

Merck's suvorexant (Belsomra, **171**) is the first-in-class orexin receptor antagonist for treating insomnia.[106] Orexin receptors (OX_1R and OX_1R) are orphan GPCRs with a pair of peptides, orexins, as their endogenous ligands. The drug suvorexant (**171**) offers an alternative to positive allosteric modulators (PAMs) of the GABA receptor as treatment of insomnia.

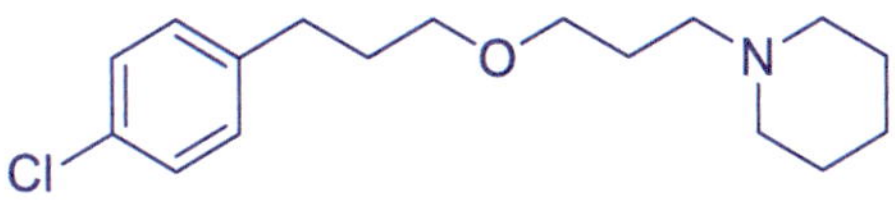

pitolisant (Wakix, **170**)
Bioproject, 2017
H_3 receptor inverse agonist

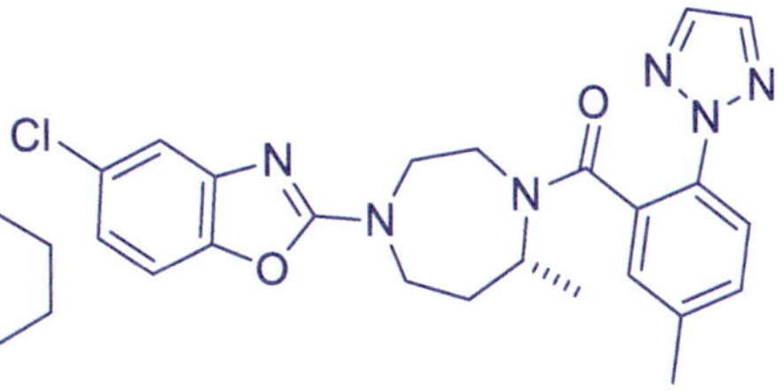

suvorexant (Belsomra, **171**)
Merck, 2014
orexin receptor antagonist

Recently, GPCR antagonists provided novel therapeutics for viral infections and cancers.

Pfizer's maraviroc (Selzentry, **172**) is the first-in-class CCR5 receptor antagonist approved for the treatment of HIV.[107] The chemokine receptor CCR5 has been demonstrated to be the major co-receptor for the fusion and entry of macrophage tropic (R5-tropic) HIV-1 into cells. Approximately 50% of individuals are infected with strains that maintain their requirement for CCR5. Moreover, CCR5-deficient individuals are apparently fully immunocompetent, indicating that absence of CCR5 function may not be detrimental and that a CCR5 antagonist should be well tolerated.

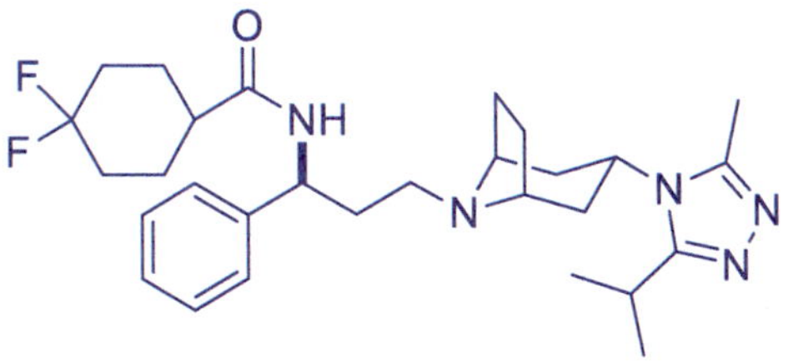

maraviroc (Selzentry, **172**)
Pfizer, 2011
CCR5 receptor antagonist

The FDA's approvals of Genentech's vismodegib (Erivedge, **173**) in 2012 and Novartis' sonidegib (Odomzo, **174**) in 2015 heralded a new era of basal cell carcinoma

(BCC) treatments with smoothened (SMO) inhibitors by interrupting the hedgehog (Hh) signaling pathway.[108] As shown in Figure 1.18, patched (PTCH1) is a 12-transmembrane domain protein located on the surface of the responding cell. It suppresses the activity of SMO, a seven-transmembrane GPCR-like receptor. Since there is no endogenous ligands for the SMO receptor, it is considered to be an orphan GPCR. Activated SMO would initiate a downstream signaling cascade leading to activation of transcription factors for the Glioma-associated oncogenes (Gli) via suppressor of the fused homolog (Sufu).

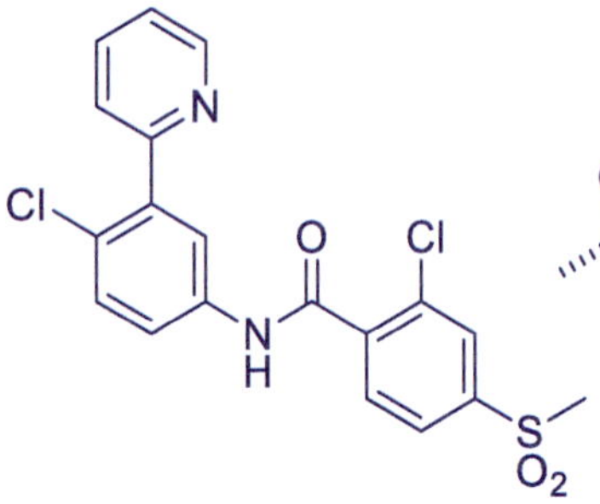

vismodegib (Erivedge, **173**)
Genentech, 2012
hedgehog pathway inhibitor

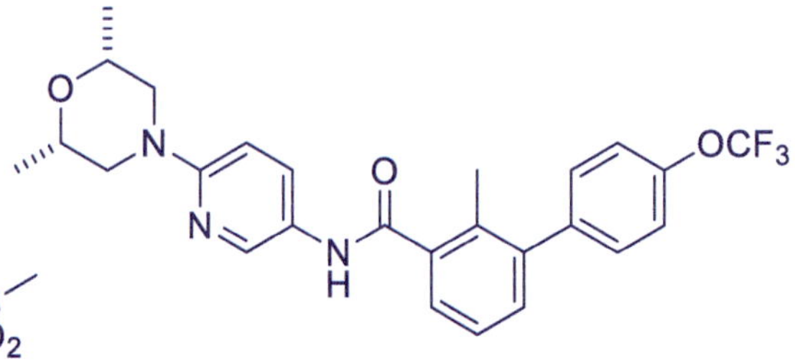

sonidegib (Odomzo, **174**)
Novartis, 2015
hedgehog pathway inhibitor

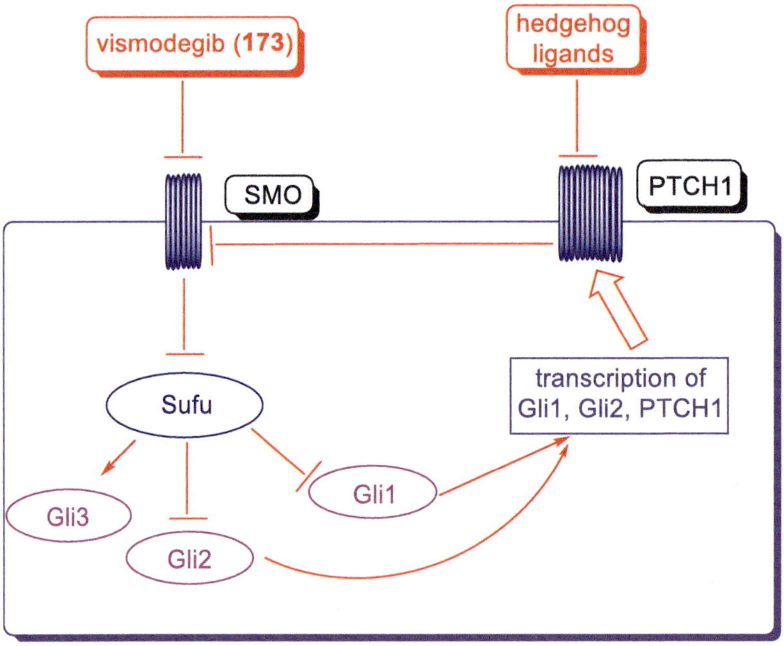

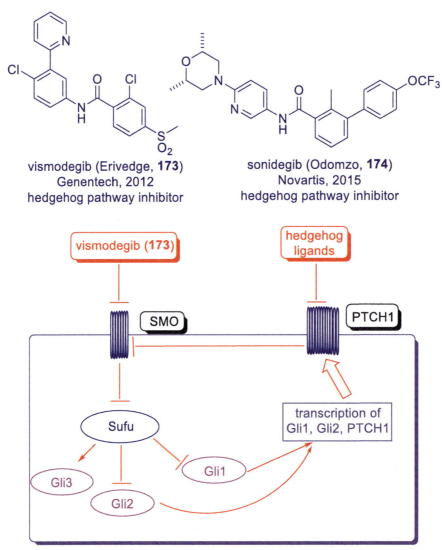

Figure 1.18 The Hedgehog pathway.

Hedgehog pathway inhibitors vismodegib (**173**) and sonidegib (**174**) may be considered as GPCR antagonists as well. Their approval provided clear PoC of the validity of using small molecule inhibitors to treat BCC. Unlike the conventional chemotherapies, there cancer-target therapies achieved great selectivity between cancer cells and normal cells thus exhibiting improved therapeutic efficacy and decreased toxicological events.

1.3.5 Nuclear Receptors

Nuclear receptors (or nuclear hormone receptors) are the only type of intracellular receptors, while most other receptors are membrane-bound. Like GPCRs and protein kinases, nuclear receptors are a rich source of pharmaceutical targets. Over 80 nuclear receptor-targeting drugs have been approved for 18 nuclear receptors.

1.3.5.1 Androgen Receptor

Androgen (**175**), the male hormone, is also known as testosterone. In contrast, estrogen (**176**) is the female hormone. Conversion of androgen (**175**) to estrogen (**176**) is facilitated by the CYP450 aromatase.

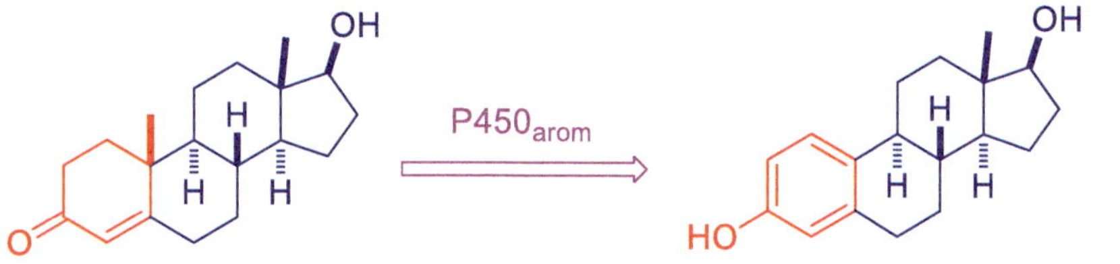

androgen (**175**), C_{19} steroid estrogen (**176**), C_{18} steroid

Charles B. Huggins won the 1966 Nobel Prize for delineating the impacts of hormones on prostate and breast cancers. Whereas elevated levels of androgen (**175**) correlates to more incidents of prostate cancers in man, estrogen (**176**) may serve as fuel for breast cancer (BRCA). Since then on, great achievements have been made in employing anti-androgens to treat prostate cancer and estrogen receptor (ER) modulators to treat breast cancer.

Androgen receptor (AR) is a ligand-activated nuclear hormone receptor and its endogenous ligand androgen stimulates the growth of prostate cancer. Therefore, AR antagonists (anti-androgens) compete with endogenous ligands androgens for the androgen receptor. When an antagonist binds to AR, it induces a conformational change of AR that impedes transcription of key androgen-regulated genes and therefore inhibits the biological effects of androgens, such as testosterone and dihydrotestosterone. AR antagonists, also referred to as anti-androgen agents, can be categorized as steroidal or nonsteroidal. Steroidal anti-androgens were first developed in the late 1960s and are distinguished from the nonsteroidal agents by their physiological progestational effects. Examples of steroidal anti-androgens are megestrol acetate and medroxyprogesterone.

Nonsteroidal AR antagonists have been the subject of extensive investigation during the past three decades due to the fact that they are generally better tolerated by patients. As a result, there are several marketed drugs that are nonsteroidal AR antagonists. In 1983, flutamide (Eulexin, **177**) was launched by Schering–Plough. It is actually a prodrug and the active metabolite is hydroxyflutamide where the isopropyl group is oxidized *in vivo*). Sanofi–Aventis launched nilutamide (Nilandron, **178**) in 1987 and AstraZeneca gained approval of bicalutamide (Casodex, **179**) in 1993. The addition of bicalutamide (**179**) to standard of care, either as mono-therapy or as adjuvant treatment, improved progression-free survival in men with locally advanced prostate cancer, which has spread to the area just outside the prostate. Regrettably, after a period of two to four years, the cancer becomes resistant to such treatments using non-steroidal AR antagonists **177**–**179**. In the castration-resistant (formerly called hormone refractory or androgen-independent) stage, former AR antagonists **177**–**179** become partial agonists and their use in cancer treatment must be discontinued.[109]

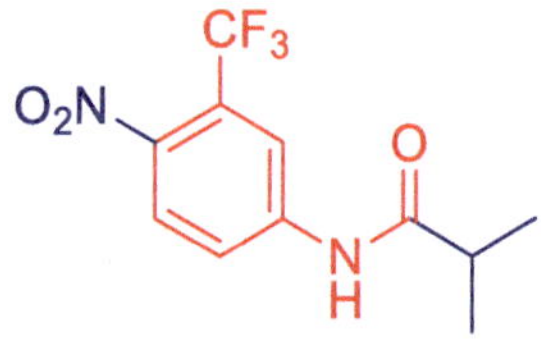

flutamide (Eulexin, **177**)
Schering-Plough, 1983
androgen receptor antagonist

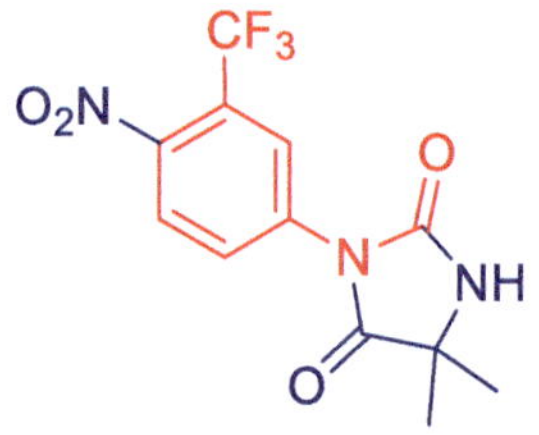

nilutamide (Nilandron, **178**)
Sanofi, 1987
AR antagonist

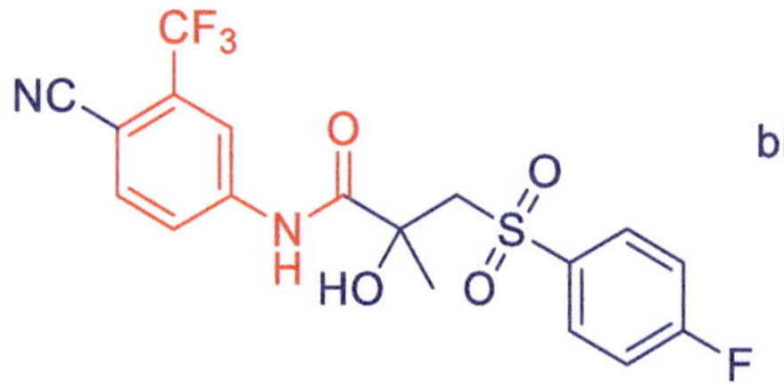

bicalutamide (Casodex, **179**)
AZ, 1993
AR antagonist

New AR antagonists enzalutamide (Xtandi, **180**) and apalutamide (Erleada, **181**) are the answer to this need.[110] In murine xenograft models of metastasized-castration-resistant prostate cancer (mCRPC), apalutamide (**181**) demonstrated greater antitumor activity than enzalutamide (**180**). Furthermore, apalutamide (**181**) penetrates less effectively the BBB (blood-brain barrier) than enzalutamide (**180**), suggesting that the chance of developing seizures may be less than with enzalutamide (**180**).

Here it might be an opportune time to comment on structural alerts. Both flutamide (**177**) and nilutamide (**178**) have the nitro group as a structural alert. On the other hand, bicalutamide (**179**) is devoid of the nitro group yet its anilide is no longer in a cyclic ring. While enzalutamide (**180**) and apalutamide (**181**) possess neither a nitro nor a linear anilide. However, both of them have the thiohydantoin moiety as a structural alert. At the end of the day, it comes down to the therapeutic widow. For the treatment of potentially fatal prostate cancer, a drug with potential of hepatotoxicity is still a viable drug to save lives.

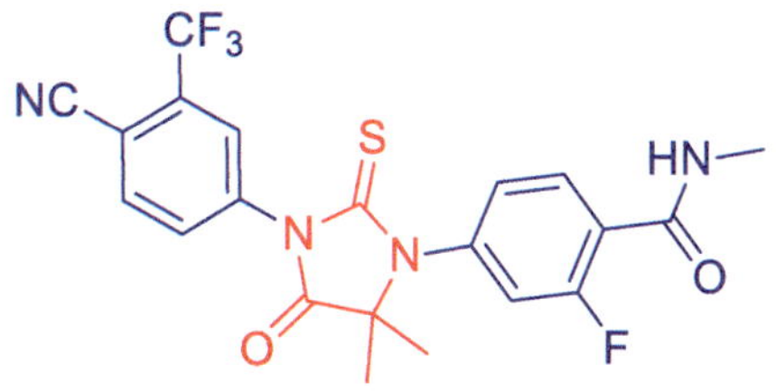

enzalutamide (Xtandi, **180**)
Medivation/Astellas, 2012
androgen receptor antagonist

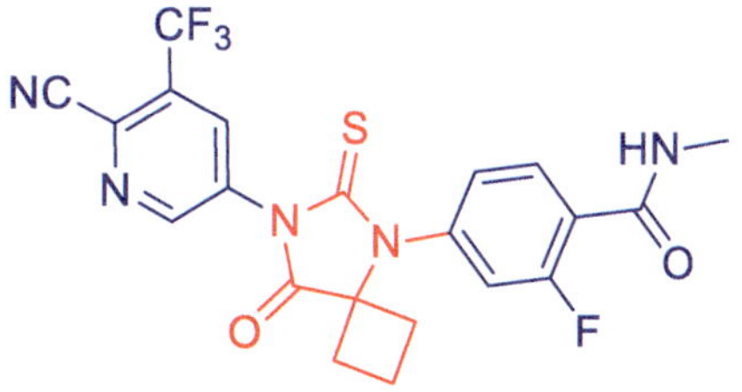

apalutamide (Erleada, **181**)
Janssen, 2018
androgen receptor antagonist

1.3.5.2 Estrogen Receptor

Endogenous estrogens such as 17β-estradiol (**182**) and estrone are the main hormones involved in the development and maintenance of the female sex organs and mammary glands. They play a pivotal role in the growth and function of a number of other tissues, both in males and females, such as the skeleton, cardiovascular system, and central nervous system. Natural estrogens such as 17β-estradiol (**182**) and estrone function by first binding to intracellular estrogen receptors (ERs), of which two types have been described (ERα and ERβ). Subsequently, the ERs modulate transcription of target genes on different tissues, resulting in the overall physiological effects.

The linkage between estrogen hormone and invasive breast cancer has long been established. On a molecular level, the rationale for estrogen hormones' causing breast cancer may be attributed to its metabolism. The phenol functional group on estradiol (**182**) may be oxidized by CYP450 to the corresponding catechol 4-hydroxyestradiol (**183**), which is readily further oxidized to the *ortho*-quinone, estradiol-3,4-quinone (**184**). An excellent Michael acceptor, the mutagenic species *ortho*-quinone **184** can trap the guanine fragment on DNA, resulting in depurinized DNA in addition to the purine-estratriol adduct **186** via the intermediacy of *o*-hydroquinone **185**.[111]

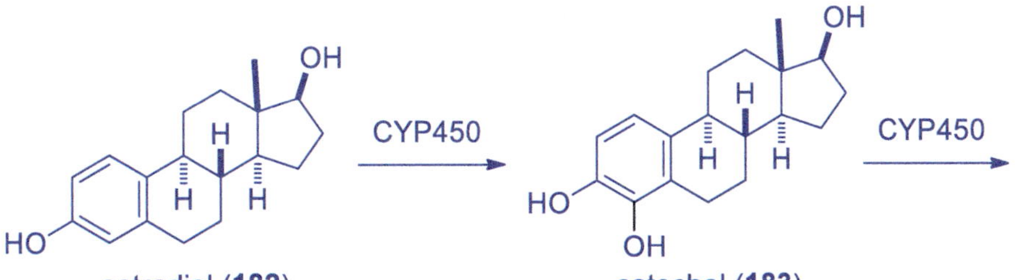

estradiol (**182**) catechol (**183**)

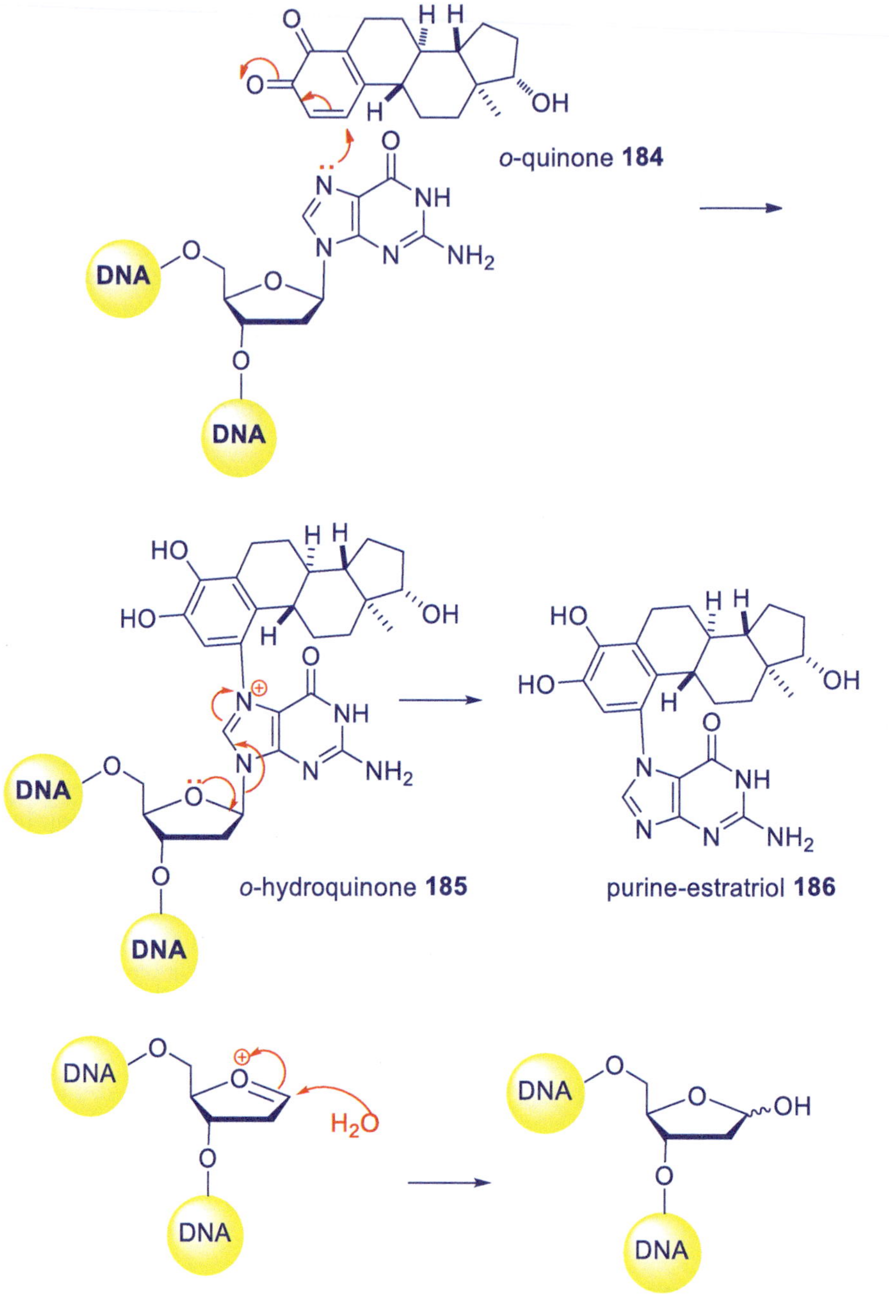

o-quinone **184**

o-hydroquinone **185**

purine-estratriol **186**

depurinized DNA

ICI serendipitously discovered chlorotrianisene (**187**), the prototype of the triphenylethylene (TPE) structural class, as an ER modulator. To boost its solubility, a dimethylaminoalkyl group was attached to produce clomiphene (Clomid, **188**), the first ER modulator studied clinically. It has been used to treat infertility in women who do not ovulate since the mid-1960s.

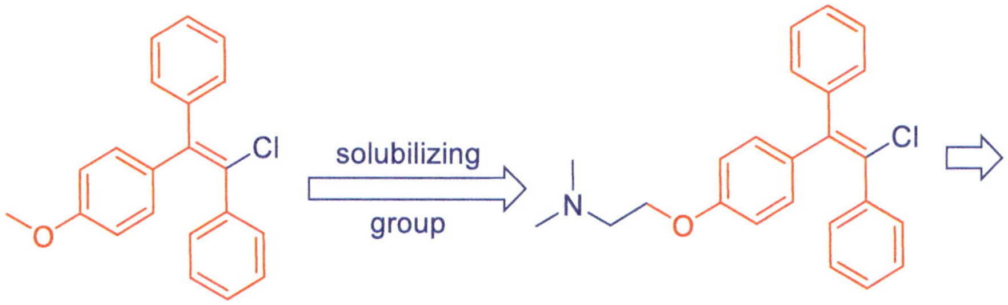

chlorotrianisene (**187**) clomiphene (Clomid, **188**)

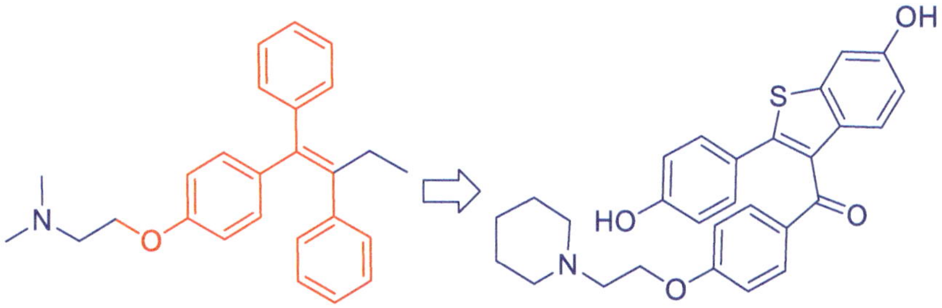

tamoxifen (Nolvadex, **189**)
ICI, 1977, SERM

raloxifene (Evista, **190**)
Lilly, 1997, SERM

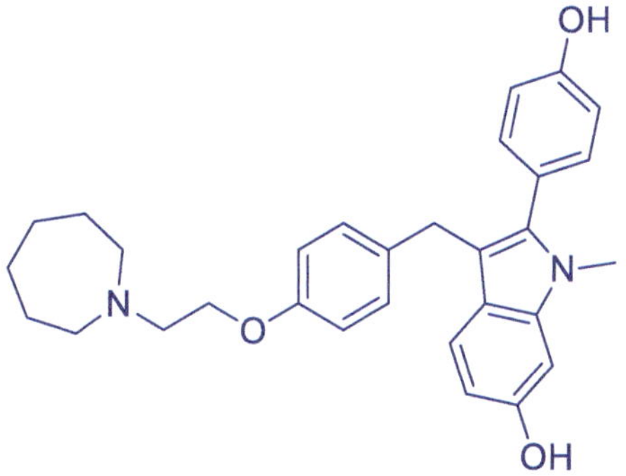

bazedoxifene (**191**)
Pfizer, 2013
SERM

Tamoxifen (Nolvadex, **189**), since its emergence in 1977, had been the gold standard for the treatment of breast cancer for decades. It displayed strong antagonism of estrogen in mammary tissue. In terms of its MOA, in addition to being a selective estrogen receptor modulator (SERM), tamoxifen (**189**) binds high affinity with several other targets such as microsomal antiestrogen binding site (AEBS), protein kinase C, calmodulin-dependent enzymes, and acyl coenzyme A:cholesterol acyl transferase (ACAT).[112]

The discovery of compounds being able to mimic the effects of estrogen in skeletal and cardiovascular systems while producing virtually complete antagonism in breast and uterine tissues led to the coining of the term selective estrogen receptor modulator (SERM), of which raloxifene (Evista, **190**) is a representative and exemplifies the benzothiophene structural class. The discovery of raloxifene (**190**) spurred further investigation around the benzothiophene core, but it also encouraged development of related scaffolds. Thus indole-based SERM bazedoxifene (**191**, Duavee in combination with Premarin) obtained the FDA approval in 2013 for the prevention and treatment of postmenopausal osteoporosis.[113]

Another structural class of SERMs ormeloxifene (**192**) and lasofoxifene (Fablyn, **193**) can be considered as conformational variants of the TPE scaffold. Ormeloxifene (**192**) was launched in India as a contraceptive in the early 1990s,[114] and lasofoxifene (**193**) was approved by the EMA as an osteoporosis therapy and a treatment of vaginal atrophy.[115]

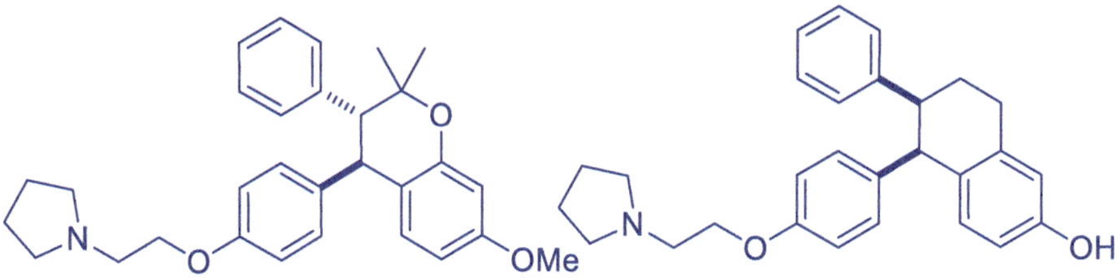

ormeloxifene (**192**)

lasofoxifene (Fablyn, **193**)
Ligand/Pfizer, 2009, SERM

A steroid-like scaffold was successful as a drug with fulvestrant (Faslodex, **194**), which is a full ER antagonist that displays no agonistic effects. It works by down-regulating and degrading the ER,[116] and is the first-in-class selective estrogen receptor degrader (SERD) approved by the FDA in 2002 used in breast cancer therapy.

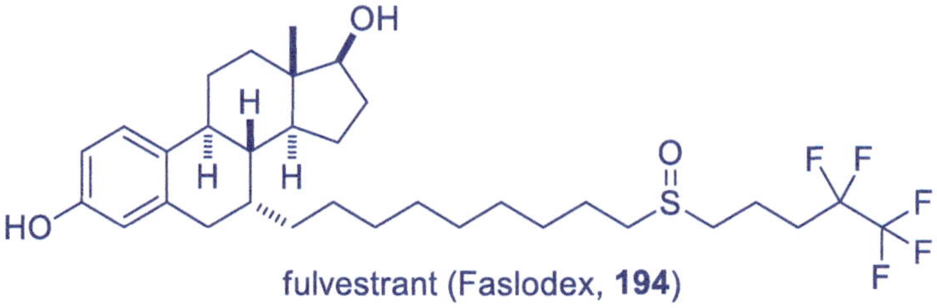

fulvestrant (Faslodex, **194**)

1.3.5.3 PPAR Receptor

Peroxisome proliferators-activated receptor-gamma (PPAR-γ) is a nuclear receptor. The thiazolidinediones (TZDs) or "glitazones" are PPAR-γ agonists that improve metabolic control in patients with type 2 diabetes through the improvement of insulin sensitivity.[117] Sankyo/Parke–Davis' troglitazone (Rezulin, **195**) was marketed in 1997 but was withdrawn in 2000 due to liver toxicities. GSK's rosiglitazone (Avandia, **196**) was approved in 1998 and was withdrawn in 2010 due to cardiovascular toxicities. The only PPAR-γ agonist on the market in the United States is Takeda/Lilly's pioglitazone (Actos, **197**).[118]

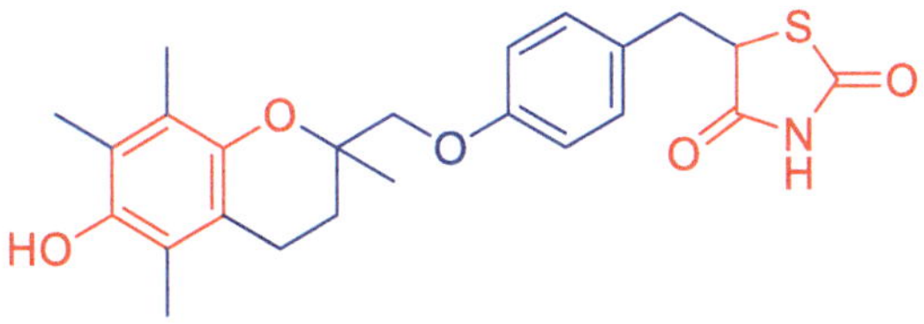

troglitazone (Rezulin, **195**)
Sankyo/Parke–Davis
1997–2000
liver toxicities

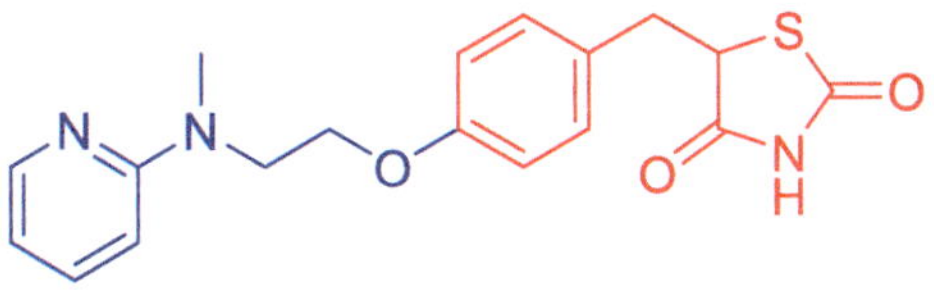

rosiglitazone (Avandia, **196**)
GSK, 1998–2010
cardiovascular toxicities

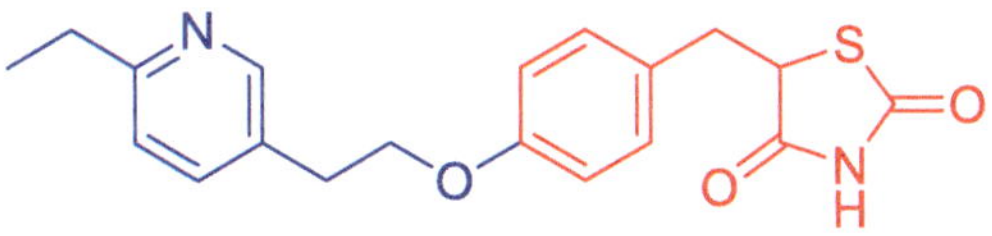

pioglitazone (Actos, **197**)
Takeda/Lilly, 1999

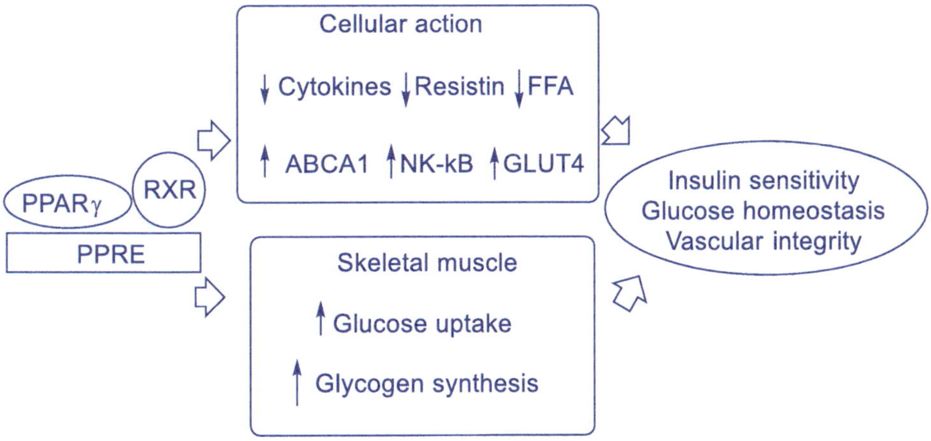

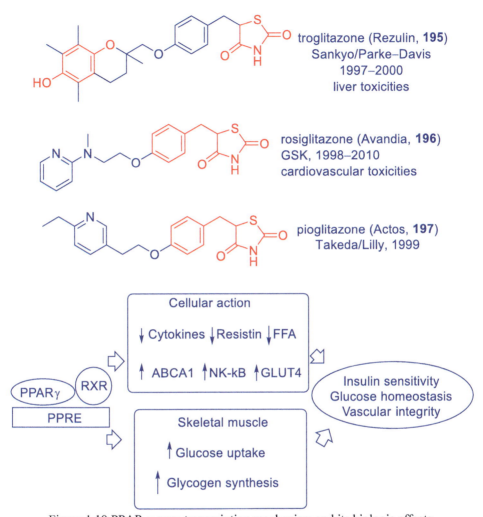

Figure 1.19 PPAR-γ gene transcription mechanism and its biologic effects.

TZDs are PPAR-γ agonists. They exert their anti-diabetic effects through a mechanism that involves activation of the γ-isoform of the TZD-induced activation of PPAR-γ. The process alters the transcription of several genes involved in glucose and lipid metabolism and energy balance. They include those that code for lipoprotein lipase, fatty acid transporter protein, adipocyte fatty acid binding protein, fatty acyl-CoA synthase, malic enzyme, glucokinase, and the GLUT4 glucose transporter. TZDs reduce insulin resistance in adipose tissue, muscle and the liver. However, PPAR-γ is predominantly expressed in adipose tissue. It is possible that the effect of TZDs on insulin resistance in muscle, and liver is promoted via endocrine signaling from adipocytes. Potential signaling factors, include free fatty acids (FFAs, well-known mediators of insulin resistance linked to obesity) or adipocyte-derived tumor necrosis factor-α (TNF-α), which is over expressed in obesity and insulin resistance (Figure 1.19).[119]

1.3.6 Growth Factor Receptors

Growth factors bind to growth factor receptors locate outside the cell membrane and trigger a cascade of biological responses. Well-known growth factor receptors include epidermal growth factor receptor (EGFR), fibroblast growth factor receptor (FGFR), vascular endothelial growth factor receptor (VEGFR), and platelet-derived growth factor receptor (PDGFR), and insulin-like growth factor receptor (IGFR). These tyrosine kinase-linked receptors are important to cell growth and cell division. They have dual action in that they act both as a receptor and as an enzyme (tyrosine kinase). More than nine tyrosine kinase inhibitors of VEGFR are on the market. Interestingly, VEGFR antagonists are investigated anti-angiogenic agents or as a treatment of psoriasis.[120]

Sunitinib maleate (Sutent, **7**) is an inhibitor of VEGFR-1 and VEGFR-2 and PDGFR-α and PDGFR-β. Gefitinib (Iressa, **38**) is the first-generation EGFR inhibitors and afatinib (Gilotrif, **39**), a second-generation EGFR inhibitor. To combat EGFR mutations such as T790M and C797S, the third-generation EGFR inhibitors neratinib (Nerlynx, **40**), and osimertinib (Tagrisso, **41**) are designed as covalent inhibitors.[121] Boehringer–Ingelheim's nintedanib (Ofev, **198**) is a FGFR/Flt3 inhibitor.[122] Although many kinase inhibitors inhibit FGFR, selective FGFR inhibitors are far and in between. Finally, IGFR inhibitors have shown promise as a treatment of diabetes.

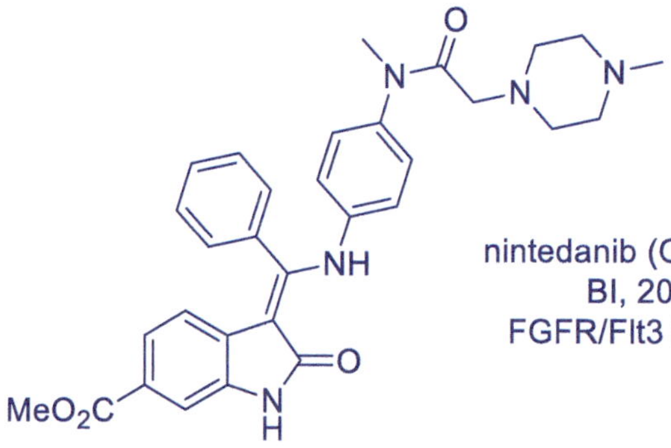

nintedanib (Ofev, **198**)
BI, 2014
FGFR/Flt3 inhibitor

1.3.7 Ionotropic and Metabotropic Receptors

Neurotransmitters such as glutamate and GABA bind to receptors first before exerting their biological effects. Glutamate is an excitatory neurotransmitter that acts upon the glutamatergic receptors, comprised of two classes: metabotropic (mGluRs) and ionotropic (iGluRs).[123] Ionotropic receptors encompass receptors and ion channels will be the focus of next section (Section 1.3.8). Metabolic receptors, on the other hand, do not have their binding sites associated with channel. Whereas ionotropic receptors are ligand-gated ion channels, metabolic receptors are GPCRs.

The iGluRs are classified further by their selective agonists: kainate, *N*-methyl-D-aspartate (NMDA), and α-amino-3-(5-methyl-3-oxo-1,2-oxazol-4-yl)propionic acid (AMPA). The discovery and characterization of mGluR and iGluR have led to rigorous efforts to identify novel drugs that selectively inhibit these receptors. Regrettably, after decades of intense research, no drug targeting an mGluR has yet received marketing approval. Recent work has focused on AMPA receptors as they have been shown to play a key role in the generation and spread of epileptic seizures, and AMPA receptor antagonists have a better safety profile in addition to a broader spectrum of activity than NMDA receptor antagonists. Eisai's perampanel (Fycompa, **199**) is an AMPA receptor antagonist for the treatment of seizure.[124]

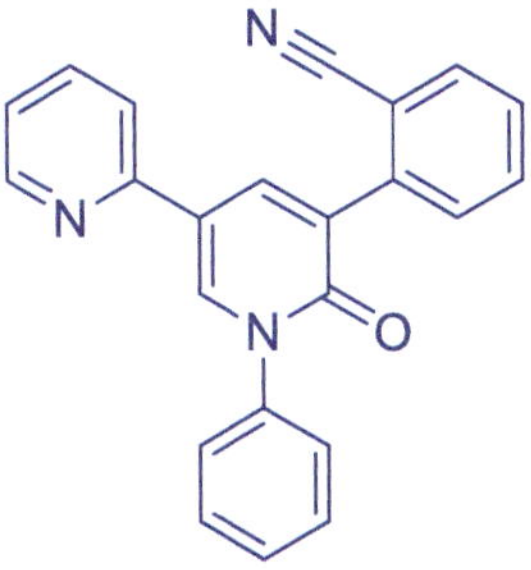

perampanel (Fycompa, **199**)
Eisai, 2012
AMPA receptor antagonist

Ionotropic receptors include nicotinic acetylcholine receptors and GABA receptors, etc. Another important specific of ionotropic glutamate receptors are NMDA receptors. Drugs like esketamine (Spravato, **200**) for treatment-resistant depression (TRD) and memantine (**201**) for treating Alzheimer's disease bind to NMDA receptors in addition to several other receptors.[125]

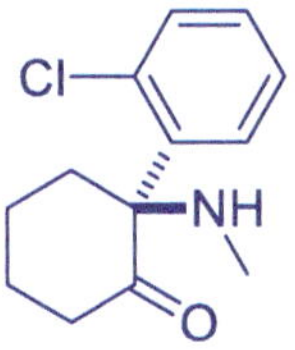

esketamine (Spravato, **200**)
nasal spray CIII, Janssen, 2019

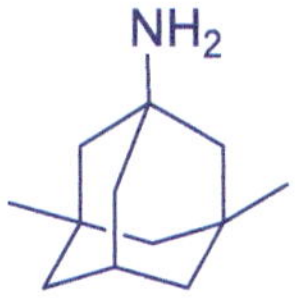

memantine (Namenda, **201**)

Well-known drugs to treat insomnia such as zolpidem (Ambien, **202**) and eszopiclone (Lunesta, **203**) are $GABA_A$ receptor agonists. $GABA_A$ receptors are sometimes referred to as benzodiazepine receptors, are the most abundant of the inhibitory neurotransmitter receptors. They are pentameric membrane proteins which are ligand gated chloride ion channels that can be modulated by multiple binding sites. Currently seven $GABA_A$ subunits with multiple isoforms and at least eight $GABA_A$ receptor subtypes comprised of various subunits are known.

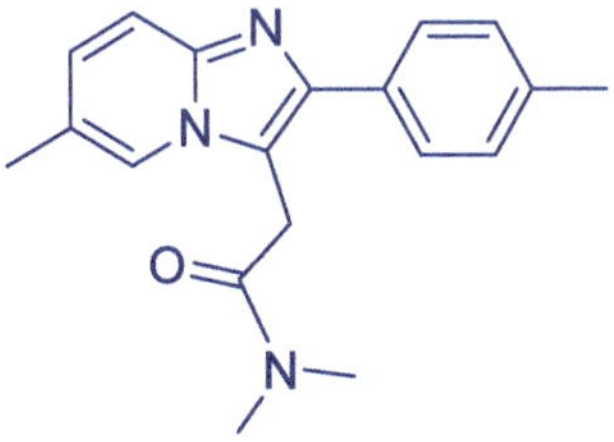

zolpidem (Ambien, **202**)
Sanofi-Aventis, 1999
$GABA_A$ receptor agonist

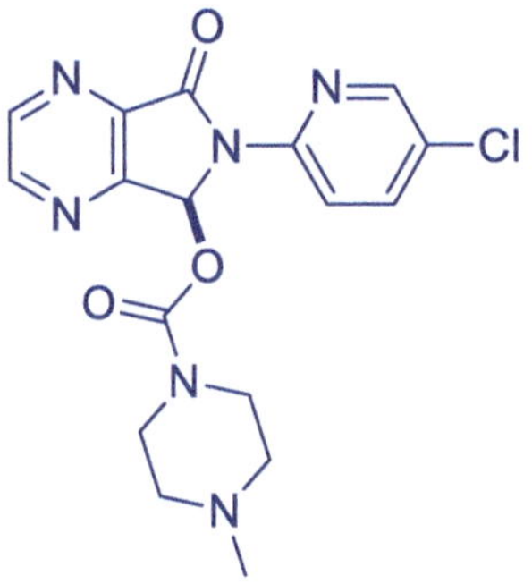

eszopiclone (Lunesta, **203**)
Sepracor, 2005
$GABA_A$ receptor agonist

1.4 Ion Channels

1.4.1 Calcium Channel Blockers

In the 1970s, several molecules including verapamil, perhexiline, diltiazem, and prenylamine all showed promise in animal models to lower blood pressure. It was Albrecht Fleckenstein who found out the common ground for the four drugs: they are all calcium channel blockers (CCBs).

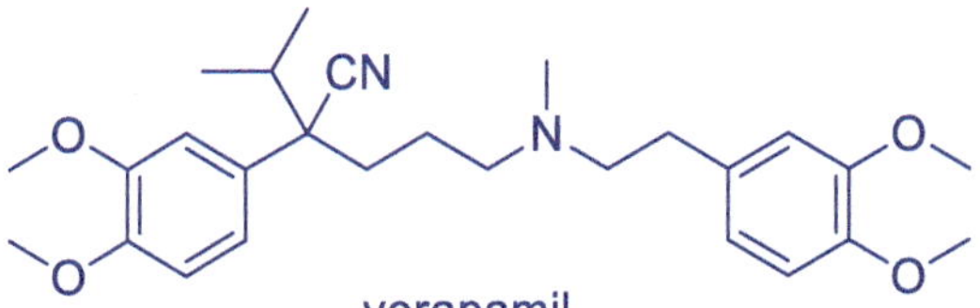

verapamil

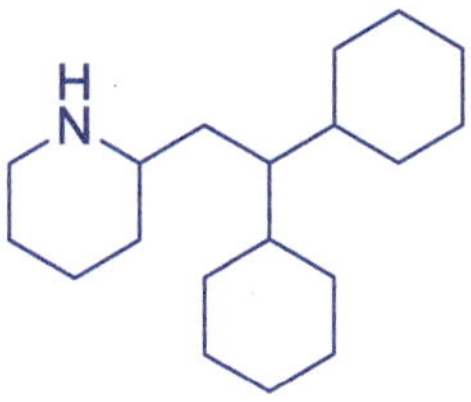

perhexiline

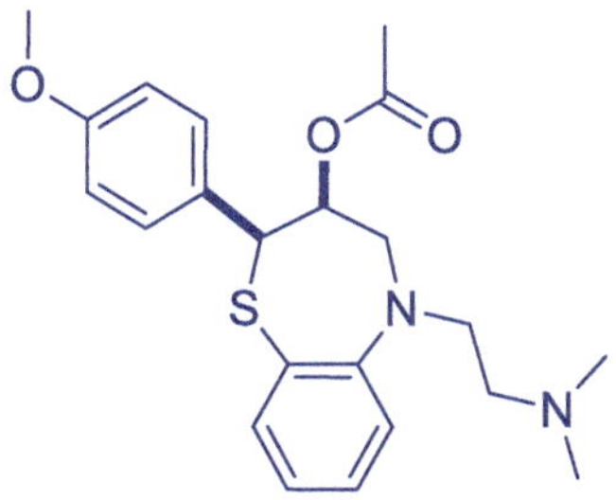

diltiazem

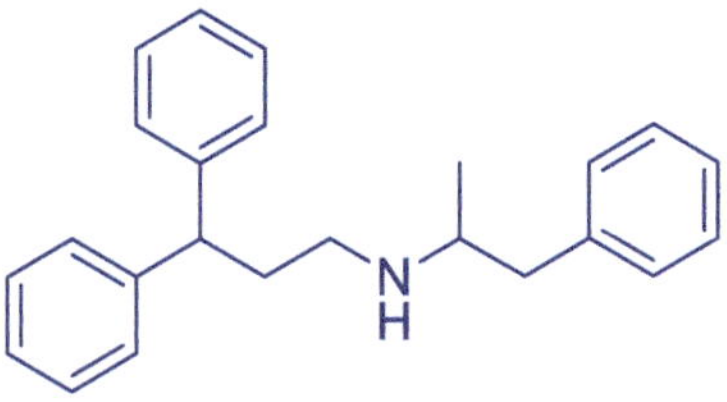

prenylamine

Although none of the four drugs made it to the market, Bayer succeeded with nifedipine (Adalat, **204**), the first-generation CCB. Nifedipine (**204**) suffers from a short half-life and the nitrophenyl group is a structural alert for toxicity. Pfizer's amlodipine besylate (Norvasc, **205**) belongs to the third generation CCB. An intramolecular hydrogen bond helps stabilizing the molecule. The primary amine-ether promotes a long half-life and enables a qd regimen. Finally, the chlorophenyl group is not as toxic as the nitrophenyl group.

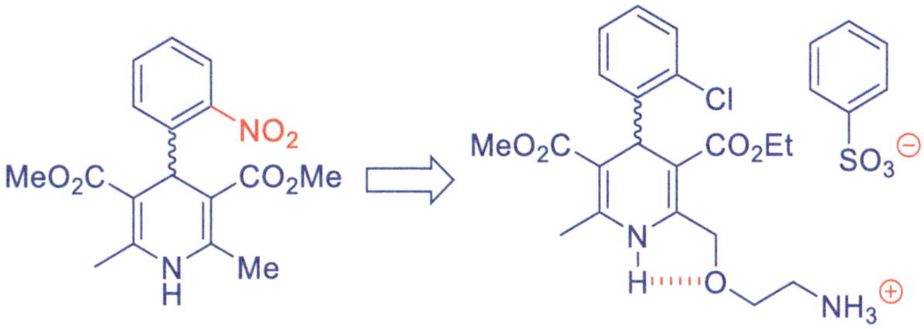

nifedipine (Adalat, **204**)
$t_{1/2}$ ~ 1 h

amlodipine besylate (Norvasc, **205**)
$t_{1/2}$ ~ 34 h, *F* 74%, V_d 16 L/kg

Curiously, although Pfizer's both gabapentin (Neurontin, **206**) and pregabalin (Lyrica, **207**) owe their genesis to GABA and GABA pharmacophore is imbedded in their structures, they do not directly interact with GABA receptors. Rather they are ligands of the $\alpha_2\delta$ subunit of voltage gated calcium channels.[126]

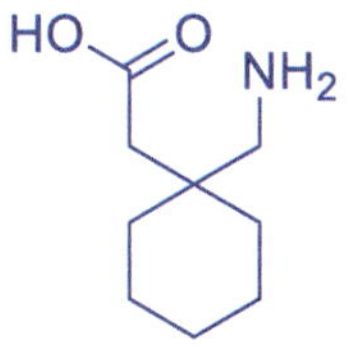

gabapentin (Neurontin, **206**)
Pfizer, 1993
α2δLigands

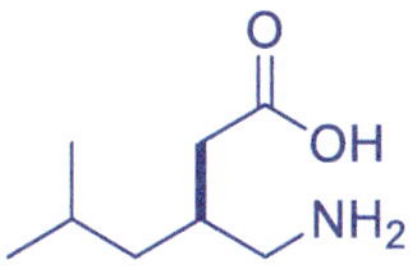

pregabalin (Lyrica, **207**)
Pfizer, 2004
α2δLigands

1.4.2 Sodium Channel Blockers

The famous Japanese Fugu fish's neurotoxin tetrodotoxin is a sodium channel blocker. Benzocaine, ketamine, and phenytoin all block sodium channel among their polypharmacology. Flecainide (Tambocor, **208**) and propafenone (Rythmol, **209**), also sodium channel blockers, have been prescribed to treat atrial fibrillation (AF, abnormal heart rhythm) since the early 1980s. Notice that the structure of propafenone (**209**) bears uncanny resemblance to β-blockers.

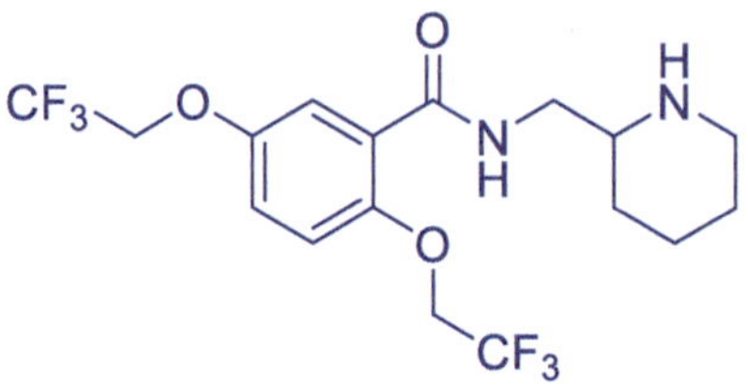

flecainide (Tambocor, **208**)
3M, 1982
Class Ic antiarrythmic

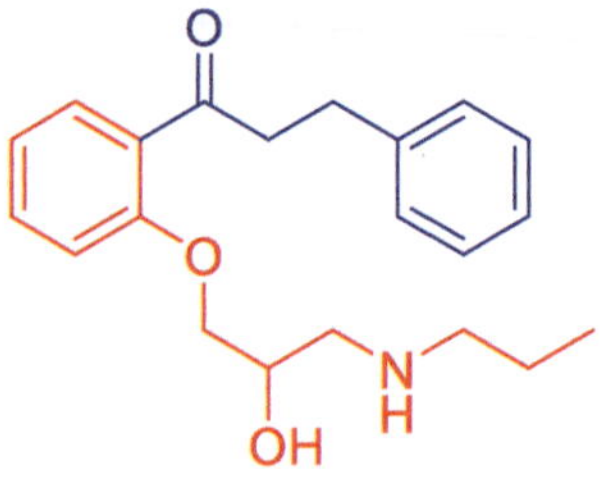

propafenone (Rythmol, **209**)
Abbott, 1982
Class Ic antiarrythmic

1.4.3 Potassium Channel Blockers

Bee's venom apamin, a cyclic peptide, is a potassium channel blocker. An old heart drug amiodarone (Cordarone, **210**) is a potassium channel blocker that has been prescribed to treat arrhythmia. It is not a clean (selective) drug, also inhibiting sodium and calcium channels. A potassium ion (K^+) channel known as human ether-a-go-go (hERG, Kv11.1), plays a central role in cardiac repolarization. Drugs that are hERG substrates tend to have QTc prolongation and consequently cardiotoxicities. Pfizer's dofetilide (Tikosyn, **211**), a class III antiarrhythmic agent was withdrawn due to hERG channel inhibition and implication torsades de pointes (TdP), a rare but serious condition manifested as QT prolongation on ECG. More discussion on this topic in Chapter 3.

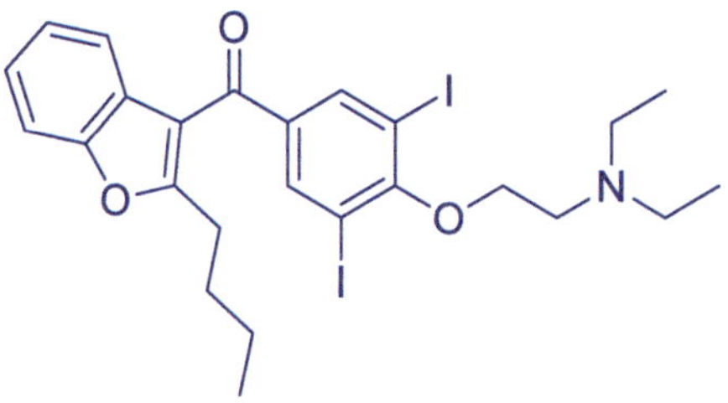

amiodarone (Cordarone, **210**)
K channel blocker

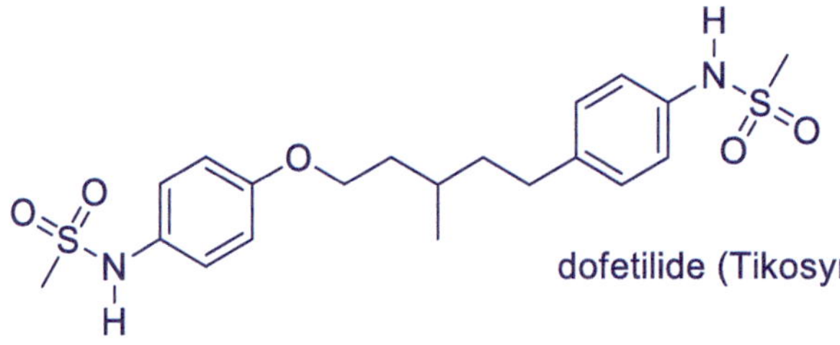

dofetilide (Tikosyn, **211**)

1.5 Carrier Proteins

Carrier proteins, also known as carrier transporters or transporter proteins, actively transport molecules across membranes. They have been targets of several classes of drugs. SSRIs, such as fluoxetine (Prozac, **126**), paroxetine (Paxil, **127**), and sertraline

(Zoloft, **128**), bind to the transporter of serotonin, preventing its uptake into the cell. Other active transporters responsible for reuptake of neurotransmitters are dopamine, glycine, and GABA transporters.

Sodium-glucose cotransporter-2 (SGLT2) is a fruitful drug target. Several SGLT2 inhibitors are on the market for treating diabetes including canagliflozin (Invokana, **212**), dapagliflozin (Farxiga, **213**), and empagliflozin (Jardiance, **214**).[127]

canagliflozin (Invokana, **212**)
Mitsubishi Tanabe/
Janssen, 2013
SGLT2 inhibitor

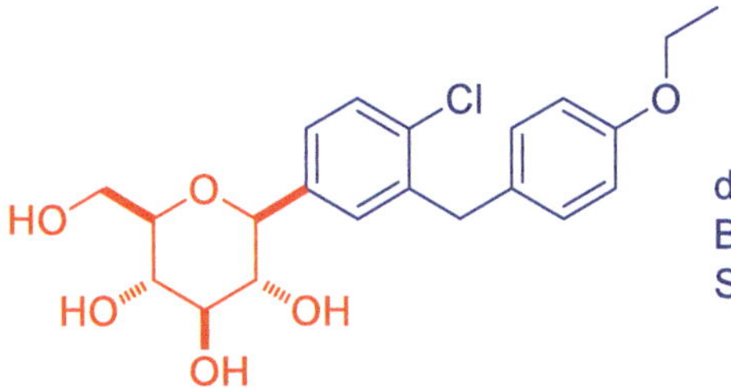

dapagliflozin (Farxiga, **213**)
BMS/AstraZeneca, 2014
SGLT2 inhibitor

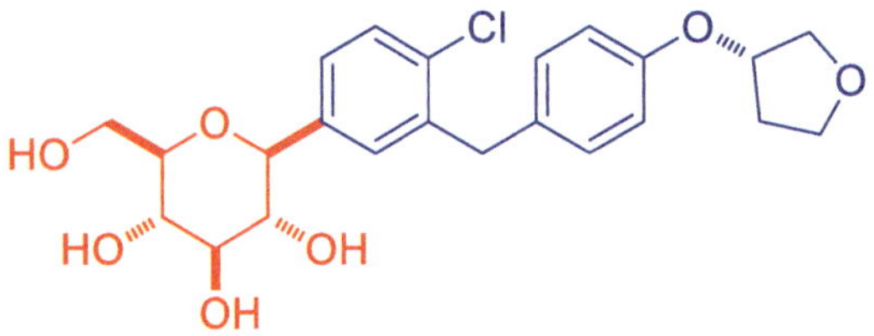

empagliflozin (Jardiance, **214**)
Boehringer Ingelheim, 2014
SGLT2 inhibitor

Efflux transporters include permeability glycoprotein (Pgp), organic anion transporter (OAT) family; multidrug resistance-associated protein (MRP); and breast cancer resistance protein (BCRP). We will discuss Pgp more extensively in Section 3.3.3.

1.6 Structural Proteins

Tublin is probably the most fruitful drug targets among structural proteins. A protein, tublin polymerizes itself into long filaments that form microtubules, which promote formation of the mitotic spindle responsible for separation of chromosomes and change formations to regulate intracellular transport.

Colchicine was probably the first known ligand to bind tublin and inhibit microtubule polymerization. Vinca alkaloids vinorelbine (Navelbine) and vinblastine (Velban) promote depolymerization of tubulins thus disrupt cancer cell growth. In contrast, paclitaxel (Taxol, **215**) promotes stabilization of microtubules, thus preventing cancer cell growth. In essence, paclitaxel (**215**) interferes with the protein microtubules in the cell, which pulls apart the chromosomes before cell division (mitosis). In the presence of paclitaxel (**215**), cells can no longer divide two daughter cells, and the tumor gradually dies.[128] Similarly, ixabepilone (Ixempra, **216**), a lactam analog of lactone epothilone B, also works as a microtubule stabilizer.[129]

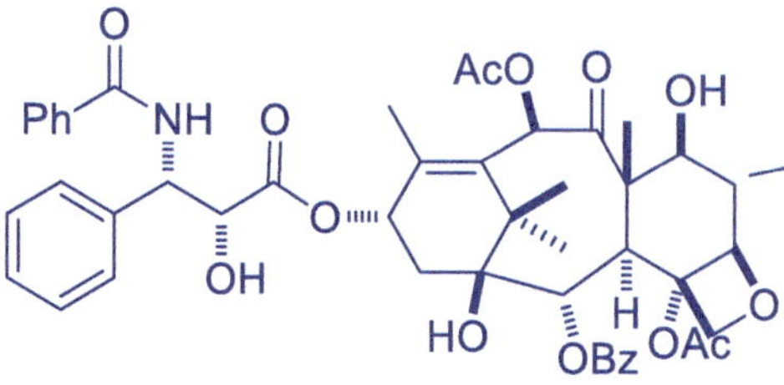

paclitaxel (Taxol, **215**)
BMS, 1993
microtubule stabilizer

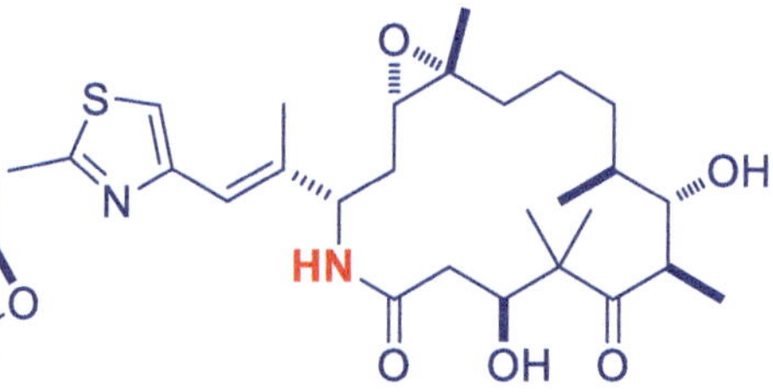

ixabepilone (Ixempra, **216**)
BMS, 2007
microtubule stabilizer

1.7 Nucleic Acids

Nucleic acids may be divided to DNAs and RNAs. DNA alkylating agents and intercalators have been employed as cancer treatments, but they are generally associated with serious toxicities.

Sulfa drugs are the earliest antimetabolites which interfere with normal cellular function, particularly the synthesis of DNA that is required for replication. As shown beneath, sulfonamide "disguises" itself as a building block for folic acid (**217**) synthesis: *para*-amino-benzoic acid (PABA). After incorporating into the DNA structures, the fake PABA disrupts DNA synthesis.

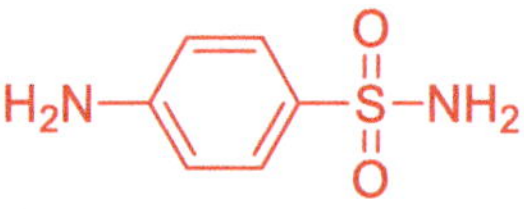

sulfonamide

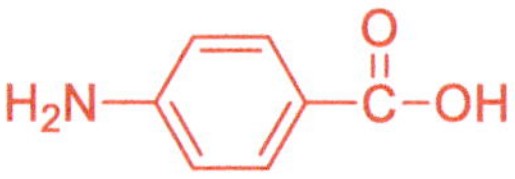

para-amino-benzoic acid (PABA)

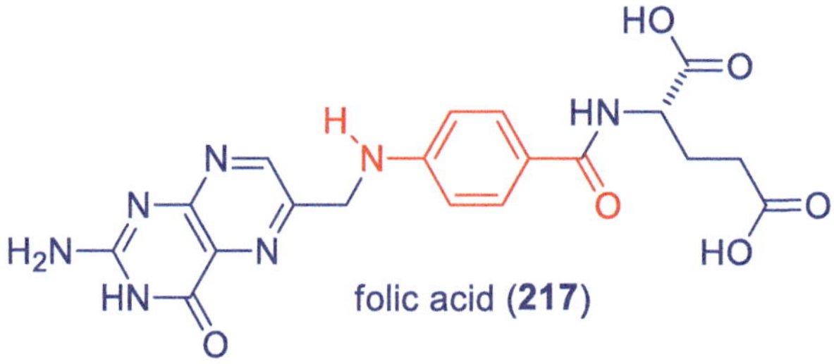

folic acid (**217**)

Another anti-metabolite drug methotrexate (**218**) was the first effective drug to treat childhood leukemia. Although its direct target is dihydrofolate reductase, the end-result is interrupting DNA synthesis.

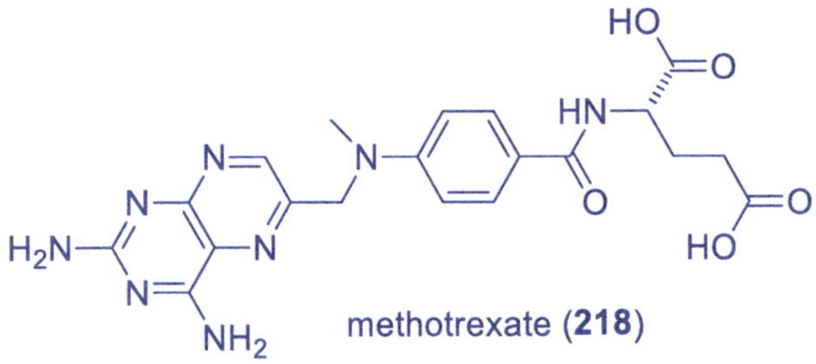

methotrexate (**218**)

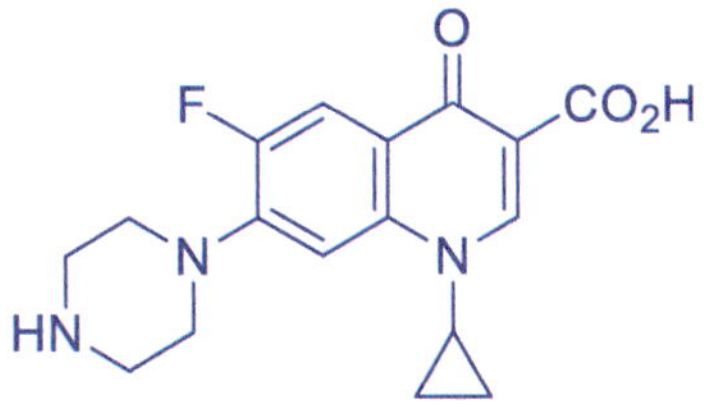

ciprofloxacin (Cipro, **219**)
Bayer, 1987
topoisomerase II inhibitor

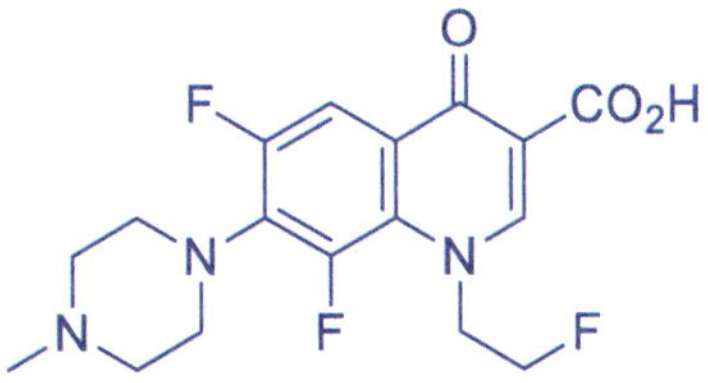

fleroxacin (Quinodis, **220**)
topoisomerase IV inhibitor

Fluoroquinolone antibacterials are DNA topoisomerase inhibitors. Ciprofloxacin (Cipro, **219**) is an inhibitor of DNA topoisomerase II and DNA gyrase, and fleroxacin (Quinodis, **220**) is a DNA topoisomerase IV and gyrase inhibitor.

A natural product, bleomycin used to treat skin cancer, is a DNA chain cutter.

Antisense drugs are oligonucleotides that contain the sequence of bases complementary to those found in a short section of the target nucleic acids. The drug and the nucleic acid bind to each other by extensive hydrogen bonding network.[130] Mipomersen (Kynamro), approved in 2013 for lowering low-density lipoprotein (LDL), is an orally bioavailable, second-generation antisense oligonucleotide.

RNA interference (RNAi) is an endogenous mechanism for controlling gene expression. It results in the cleavage of target messenger RNA (mRNA) by small interfering RNAs bound to the RNA-induced silencing complex. Recently, Alnylam's RNAi therapeutic patisiran [Onpattro, a large molecule with a molecular weight (MW) of 13,424] was approved to treat a rare genetic disorder called hereditary transthyretin-mediated amyloidosis. In 2018, Ionis/Akcea receive the FDA approval for their RNAi therapy inotersen (Tegsedi), also for the treatment of hereditary transthyretin amyloidosis (hATTR).

1.8 Protein–Protein Interactions

We have been successful targeting DNA, RNA, and proteins, but less so for PPIs. The reality is that many biological processes are mediated by proteins that act in a cooperative manner. Therefore, drugs targets grow exponentially if modulating PPIs is successful. Unfortunately, the contact surface is large even for the simplest binary PPIs. The dogma was that small molecules cannot offer enough affinity to become efficacious drugs.

Historically, a couple of drugs targeting integrin PPIs crossed the finish-line and gained the FDA approval. One is Medicure's antiplatelet tirofiban (Aggrastat, **221**), which works by inhibiting the PPI between fibrinogen and the platelet integrin receptor GP IIb/IIIa. Not bioavailable orally, it is given via IV. The other is Shire's eye drop lifitegrast (Xiidra, **222**) for treating dry eye by inhibiting an integrin, lymphocyte function-associated antigen-1 (LFA-1), from binding to intracellular adhesion molecule 1 (ICAM-1). Both drugs are not orally bioavailable and their impact is not significant.

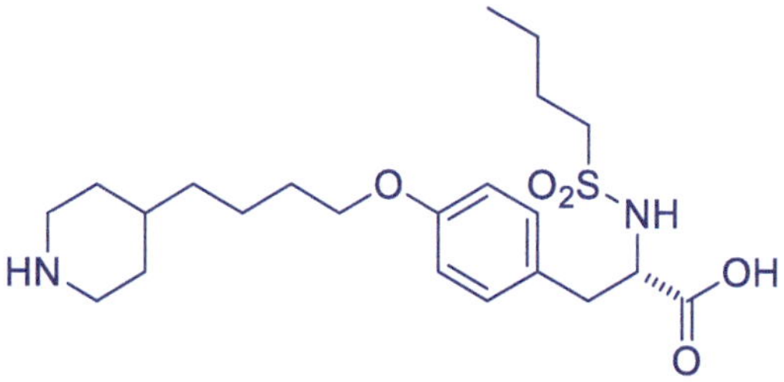

tirofiban (Aggrastat, **221**)
Medicure, 2000
PPI inhibitor

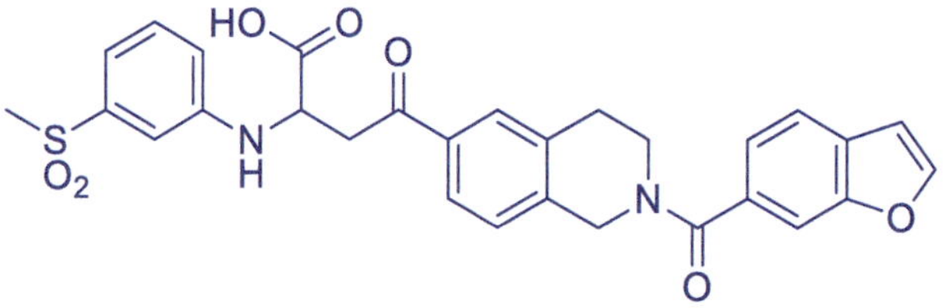

lifitegrast (Xiidra, **222**)
Shire, 2016
PPI inhibitor

A recent spectacular success in Abbvie's venetoclax (Venclexta, **226**) ignited great interests in PPIs as new drug targets. It is a B-cell lymphoma 2 (Bcl-2) inhibitor discovered by employing the FBDD strategy. Instead of the co-crystallography tactic, "SAR by NMR" method was key to generate their fragment hits. From initial screening a 10,000 compound library with MW <215 at 1 mM concentration, *p*-fluorophenyl-benzoic acid (**223**) emerged as one of the first-site (P1) ligands. Later on, screening a 3,500 compound library with MW ~ 150 at 5 mM concentration identified the second site (P2) ligand 5,6,7,8-tetrahydro-naphthalen-1-ol (**224**).[131] A protracted and winding road consisting of identifying the third binding site (P3), designing away from serum deactivation from domain II of human serum albumin (HSA-III) binding, boosting oral bioavailability, and removing the nitrophenyl structural alert cumulated to the discovery of navitoclax (**225**) as a potent and orally bioavailable Bcl-2 inhibitor (not selective against Bcl-x_L).[132] Eventually, the fourth binding site (P4) was replaced with 7-azaindole ether and its N atom captured an additional hydrogen bond with Arg104 on the target, giving rise to venetoclax (Venclexta, **226**) as a potent, selective (against Bcl-x_L, Bcl-w, and Bcl-1), and orally bioavailable Bcl-2 inhibitor. In 2016, it was approved by the FDA for treating CLL with the 17p deletion.[133]

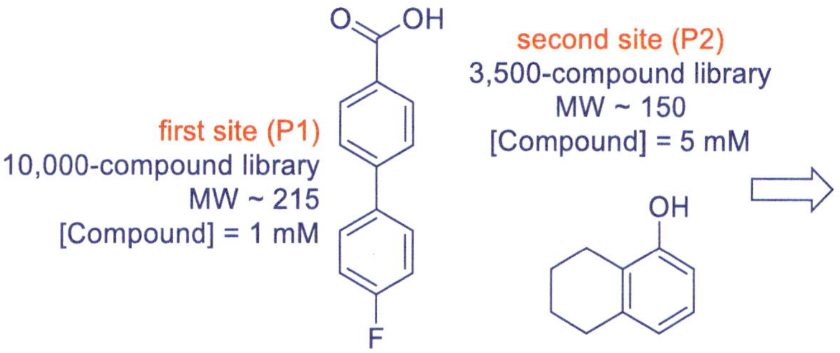

223, K_d = 300 μM **224**, K_d = 2,000 μM

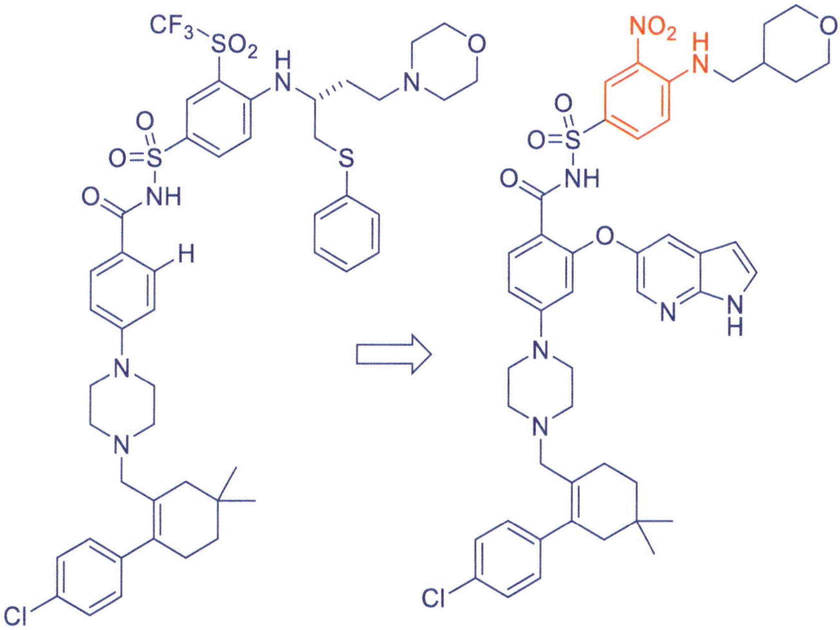

navitoclax (**225**)
K_i = 0.04 nM, F ~ 30%, LE = 0.2

venetoclax (Venclexta, **226**)
K_i = 0.01 nM, F ~ 29%, LE = 0.2

1.9 Further Reading

Botana, L. M.; Loza, M. Eds. Therapeutic Targets: Modulation, Inhibition, and Activation. Wiley: Hoboken, NJ, **2012**.

Copeland, R. A. Evaluation of Enzyme Inhibitors in Drug Discovery: A Guide for Medicinal Chemists and Pharmacologists. *2nd Ed.* Wiley: Hoboken, NJ, **2013**.

Dosa, P. I.; Amin, E. A. *J. Med. Chem.* **2016,** *59,* 810–840.

Lu, C.; Li, A. P., Eds. Enzyme Inhibition in Drug Discovery and Development: The Good and the Bad. Wiley: Hoboken, NJ, **2010**.

Ye, N.; Zhang, A.; Neumeyer, J. L.; Baldessarini, R. J.; Zhen, X.; Zhang, A. *Chem. Rev.* **2013,** *113*, PR123–PR178.

1.10 References

1. Jack, D. In Walker, S. R., Ed. Creating the Right Environment for Drug Discovery. Quay Publishing: Lancaster, UK, **1991**, pp. 67–71.
2. Li, J. J.; Yue, W.-S.; Ortwine, D. F.; Johnson, A. R.; Man, C.-F.; Baragi, V.; Kilgore, K.; Dyer, R. D.; Han, H.-K. *J. Med. Chem.* **2008**, *51*, 835–841.
3. Overington, J. P.; Al-Lazikani, B.; Hopkins, A. L. *Nat. Rev. Drug Discov.* **2006**, *5*, 993–996.
4. Manning, A. M. Target Identification and Validation. In Li, J. J.; Corey, E. J., Eds. Drug Discovery, Practices, Processes, and Perspectives. Wiley: Hoboken, NJ; **2013**, pp. 43–66.
5. Chen, X.-P.; Du, G.-H. *Drug Discov. Ther.* **2007**, *1*, 23–29.
6. Koshland, D. E., Jr. *Angew. Chem. Int. Ed. Engl.* **1994**, *33*, 2475–2478.
7. Brown, D. G.; Boström, J. *J. Med. Chem.* **2018**, *61*, 9442–9468.
8. Cee, V. J.; Olhava, E. J. Leading ACE Inhibitors for Hypertension. In Johnson, D. S.; Li, J. J., Eds. The Art of Drug Synthesis. Wiley: Hoboken, NJ, **2007**, pp. 143–158.
9. Patchett, A. A.; Harris, E.; Tristram, E. W.; Wyvratt, M. J.; Wu, M. T.; Taub, D.; Greenlee, N. S.; Hirschmann, R.; et al. *Nature* **1980**, *288*, 280–283.
10. Zimmermann, J.; Buchdunger, E.; Mett, H.; Meyer, T.; Lydon, N. B. *Bioorg. Med. Chem. Lett.* **1997**, *7*, 187–192.
11. Rao, B. G.; Murcko, M.; Tebbe, M. J.; Kwong, A. D. Discovery and Development of Telaprevir (Incivek™)—A Protease Inhibitor to Treat Hepatitis C Infection. In Fischer, J.; Rotella, D. P., Eds. Successful Drug Discovery. Wiley-VCH: Weinheim, **2015**, pp. 195–212.
12. Howe, A. Y.; Venkatraman, S. *J. Clin. Transl. Heptol.* **2013**, *1,* 22–32.
13. Zettl, H.; Schubert-Zsilavecz, M.; Steinhilber, D. *ChemMedChem* **2010**, *5,* 179–185.
14. Kim, Y. B.; Kopcho, L. M.; Kirby, M. S.; Hamann, L. G.; Weigelt, C. A.; Metzler, W. J.; Marcinkeviciene, J. *Arch. Biochem. Biophys.* **2006**, *445,* 9–18.
15. Simpkins, L. M.; Bolton, S.; Pi, Z.; Sutton, J. C.; Kwon, C.; Zhao, G.; Magnin, D. R.; Augeri, D. J.; Gungor, T.; Rotella, D. P.; et al. *Bioorg. Med. Chem. Lett.* **2007**, *17*, 6476–6480.
16. Kisselev, A. F.; van der Linden, W. A.; Overleeft, H. S. *Chem. Biol.* **2012**, *19*, 99–115.

17. Groll, M.; Berkers, C. R.; Ploegh, H. L.; Ovaa, H. *Structure* **2006**, *14*, 451–456.
18. Serafimova, I. M.; Pufall, M. A.; Krishnan, S.; Duda, K.; Cohen, M. S.; Maglathlin, R. L.; McFarland, J. M.; Miller, R. M.; Froedin, M.; Taunton, J. *Nat. Chem. Biol.* **2012**, *8*, 471–476.
19. Bradshaw, J. M.; McFarland, J. M.; Paavilainen, V. O.; Bisconte, A.; Tam, D.; Phan, V. T.; Romanov, S.; Finkle, D.; Shu, J.; Patel, V; et al. *Nat. Chem. Biol.* **2015**, *11*, 525–531.
20. Sebolt-Leopold, J. S.; Bridges, A. J. Road To PD0325901 and Beyond: The MEK Inhibitor Quest. In Li, R.; Stafford, J. A., Eds. Kinase Inhibitor Drugs. Wiley: Hoboken, NJ; **2009**, pp. 205–227.
21. Abe, H.; Kikuchi, S.; Hayakawa, K.; Iida, T.; Nagahashi, N.; Maeda, K.; Sakamoto, J.; Matsumoto, N.; Miura, T.; Matsumura, K.; et al. *ACS Med. Chem. Lett.* **2011**, *2*, 320–324.
22. Rice, K. D.; Aay, N.; Anand, N. K.; Blazey, C. M.; Bowles, O. J.; Bussenius, J.; Costanzo, S.; Curtis, J. K.; Defina, S. C.; Dubenko, L.; Koltun, E. S.; et al. *ACS Med. Chem. Lett.* **2012**, *3*, 416–421.
23. Schoepfer, J.; Jahnke, W.; Berellini, G.; Buonamici, S.; Cotesta, S.; Cowan-Jacob, S. W.; Dodd, S.; Drueckes, P.; Fabbro, D.; Gabriel, T.; et al. *J. Med. Chem.* **2018**, *61*, 8120–8135.
24. Chen, Y.-N. P.; LaMarche, M. J.; Chan, H. M.; Fekkes, P.; Garcia-Fortanet, J.; Acker, M. G.; Antonakos, B.; Chen, C. H.-T.; Chen, Z.; Cooke, V. G.; et al. *Nature* **2016**, *535*, 148–152.
25. Garcia Fortanet, J.; Chen, C. H.-T.; Chen, Y.-N. P.; Chen, Z.; Deng, Z.; Firestone, B.; Fekkes, P.; Fodor, M.; Fortin, P. D.; Fridrich, C.; et al. *J. Med. Chem.* **2016**, *59*, 7773–7782.
26. Salamoun, J. M.; Wipf, P. *J. Med. Chem.* **2016**, *59*, 7771–7772.
27. Fodor, M.; Price, E.; Wang, P.; Lu, H.; Argintaru, A.; Chen, Z.; Glick, M.; Hao, H.-X.; Kato, M.; Koenig, R.; et al. *ACS Chem. Biol.* **2018**, *13*, 647–656.
28. (a) Namasivayam, V.; Vanangamudi, M.; Kramer, V. G.; Kurup, S.; Zhan, P.; Liu, X.; Kongsted, J.; Byrareddy, S. N. *J. Med. Chem.* **2019**, *62*, 4851–4883. (b) Li, D.; De Clercq, E.; Liu, X. *J. Med. Chem.* **2012,** *55*, 3595–3613.
29. Kohlstaedt, L. A.; Wang, J.; Friedman, J. M.; Rice, P. A.; Steitz, T. A. *Science* **1992**, *256*, 1783–1790.
30. Gu, S.; Lu, H.-H.; Liu, G.-Y.; Ju, X.-L.; Zhu, Y.-Y. *Eur. J. Med. Chem.* **2018**, *158*, 371–393.
31. Liang, C.; Tian, D.; Ren, X.; Ding, S.; Jia, M.; Xin, M.; Thareja, S. *Eur. J. Med. Chem.* **2018**, *158*, 315–326.
32. Pan, Z.; Scheerens, H.; Li, S.; Schultz, B. E.; Sprengeler, P. A.; Burrill, L. C.; Mendonca, R. V.; Sweeney, M. D.; Scott, K. C.; Grothaus, P. G.; et al. *ChemMedChem* **2007**, *2*, 58–61.
33. Akinleye, A.; Chen, Y.; Mukhi, N.; Song, Y.; Liu, D. *J. Hematol. Oncol.* **2013**, *6*, 59(1–9).
34. Wu, J.; Liu, C.; Tsui, S. T.; Liu, D. *J. Hematol. Oncol.* **2016**, *9*, 80(1–7).
35. Cheng, H.; Nair, S. K.; Murray, B. W. *Bioorg. Med. Chem. Lett.* **2016**, *26,* 1861–1868.
36. Cheng, H.; Planken, S. *ACS Med. Chem. Lett.* **2018**, *9*, 861–863.

37. Grabe, T.; Lategahn, J.; Rauh, D. *ACS Med. Chem. Lett.* **2018**, *9*, 779–782.
38. Chen, L.; Fu, W.; Zheng, L.; Liu, Z.; Liang, G. *J. Med. Chem.* **2018**, *61*, 4290–4300.
39. Stephen, A. G.; Esposito, D.; Bagni, R. K.; McCormick, F. *Cancer Cell* **2014**, *25*, 272–281.
40. Xiong, Y.; Lu, J.; Hunter, J.; Li, L.; Scott, D.; Choi, H. G.; Lim, S. M.; Manandhar, A.; Gondi, S.; Sim, T.; et al. *ACS Med. Chem. Lett.* **2017**, *8*, 61–66.
41. (a) Janes, M. R.; Zhang, J.; Li, L.-S.; Hansen, R.; Peters, U.; Guo, X.; Chen, Y.; Babbar, A.; Firdaus, S. J.; Darjania, L.; et al. *Cell* **2018,** *172*, 578–589. (b) Fell, J. B.; Fischer, J. P.; Baer, B. R.; Ballard, J.; Blake, J. F.; Bouhana, K.; Brandhuber, B. J.; Briere, D. M.; Burgess, L. E.; Burkard, M. R.; et al. *ACS Med. Chem. Lett.* **2018**, *9*, 12301234.
42. Anon, *Cancer Discov.* **2019**, *9*, 988–989.
43. Ahn, K.; Smith, S. E.; Liimatta, M. B.; Beidler, D.; Sadagopan, N.; Dudley, D. T.; Young, T.; Wren, P.; Zhang, Y.; Swaney, S.; et al. *J. Pharmacol. Exp. Ther.* **2011**, *338,* 114–124.
44. Butler, C. R.; Beck, E. M.; Harris, A.; Huang, Z.; McAllister, L. A.; am Ende, C. W.; Fennell, K.; Foley, T. L.; Fonseca, K.; Hawrylik, S. J.; et al. *J. Med. Chem.* **2017**, *60*, 9860–9873.
45. (a) Thomas, J. A.; Koshland, D. E., Jr. *J. Biol. Chem.* **1960**, *235,* 2511–2517. (b) Koshland, D. E., Jr. *Nat. Med.* **1998**, *4,* 1112–1114.
46. Brik, A.; Wong, C.-H. *Org. Biomol. Chem.* **2003**, *1,* 5–14.
47. Lebon, F.; Ledecq, M. *Curr. Med. Chem.* **2000**, *7*, 455–477.
48. Motwani, H. V.; De Rosa, M.; Odell, L. R.; Hallberg, A.; Larhed, M. *Eur. J. Med. Chem.* **2015**, *90*, 462–490.
49. Singh, S.; Sieburth, S. M. N. *Org. Lett.* **2012**, *14,* 4422–4425.
50. Thomson, R.; von Itzstein, M. Discovery and Development of Influenza Virus Sialidase Inhibitor Relenza. In Kazmierski, W. M., ed. *Antiviral Drugs.* Wiley: Hoboken, NJ, **2011**, pp 385–400.
51. (a) Kim, C. U.; Lew, W.; Williams, M. A.; Wu, H.; Zhang, L.; Chen, X.; Escarpe, P. A.; Mendel, D. B.; Laver, W. G.; Stevens, R. C. *J. Med. Chem.* **1998**, *41*, 2451–2460. (b) Lew, W.; Wang, M. Z.; Chen, X.; Rooney, J. F.; Kim, C. Neuraminidase Inhibitors as Anti-Influenza Agents. In De Clercq, E. ed. Methods and Principles in Medicinal Chemistry, 50(Antiviral Drug Strategies). Wiley-VCH: Weinheim, **2011**, pp. 351–376.
52. Babu, Y. S.; Chand, P.; Bantia, S.; Kotian, P.; Dehghani, A.; El-Kattan, Y.; Lin, T.-H.; Hutchison, T. L.; Elliott, A. J.; Parker, C. D.; et al. *J. Med. Chem.* **2000**, *43*, 3482–3486.
53. von Itzstein, M. *Nat. Rev. Drug Discov.* **2007**, *6*, 967–974.
54. Lee, B. *J. Mol. Biol.* **1971**, *61,* 463–469.
55. Kim, K. B.; Crews, C. M. *Nat. Prod. Rep.* **2013**, *30,* 600–604.
56. Groll, M.; Kim, K. B.; Kairies, N.; Huber, R.; Crews, C. M. *J. Am. Chem. Soc.* **2000**, *122*, 1237–1238.
57. Demo, S. D.; Kirk, C. J.; Aujay, M. A; Buchholz, T. J; Dajee, M.; Ho, M. N.; Jiang, J.; Laidig, G. J.; Lewis, E. R.; Parlati, F.; et al. *Cancer Res.* **2007**, *67,* 6383–6391.

58. Zhou, H.-J.; Aujay, M. A.; Bennett, M. K.; Dajee, M.; Demo, S. D.; Fang, Y.; Ho, M. N.; Jiang, J.; Kirk, C. J.; Laidig, G. J.; et al. *J. Med. Chem.* **2009**, *52*, 3028–3038.
59. Ho, H. K.; Chan, J. C. Y.; Hardy, K. D.; Chan, E. C. Y. *Drug Met. Rev.* **2015**, *47,* 21–28.
60. Zhan, P.; Itoh, Y.; Suzuki, T.; Liu, X. *J. Med. Chem.* **2015**, *58*, 7611–7633.
61. Fortier, M. A.; Krishnaswamy, K.; Danyod, G.; Boucher-Kovalik, S.; Chapdalaine, P. *J. Physiol. Pharmacol.* **2008**, *59(Suppl. 1)*, 65–89.
62. (a) Thaler, F.; Mercurio, C. *ChemMedChem* **2014**, *9*, 523–526. (b) Roche, J.; Bertrand, P. *Eur. J. Med. Chem.* **2016**, *121*, 451–483.
63. Lu, X. P.; Li, Z. B.; Ning, Z. Q.; Pan, D. S.; Shan, S.; Guo, X.; Cao, H. X.; Yu, J. D.; Yang, Q. J. *Med. Chem. Rev.* **2017**, *52*, 497–513.
64. Cao, F.; Zwinderman, M. R. H.; Dekker, F. J. *Molecules* **2018,** *23*, 551(1–13).
65. (a) Suzuki, T.; Kouketsu, A.; Itoh, Y.; Hisakawa, S.; Maeda, S.; Yoshida, M.; Nakagawa, H.; Miyata, N. *J. Med. Chem.* **2006,** *49*, 4809–4812. (b) Itoh, Y.; Suzuki, T.; Kouketsu, A.; Suzuki, N.; Maeda, S.; Yoshida, M.; Nakagawa, H.; Miyata, N. *J. Med. Chem.* **2007**, *50*, 5425–5438.
66. Rotella, D. P. *Nat. Rev. Drug Discov.* **2002**, *1*, 674–682.
67. Leftheris, K.; Satoh, Y.; Schafer, P. H.; Man, H.-W. *Med. Chem. Rev.* **2015**, *50*, 171–184.
68. Rogers, K. C.; Oliphant, C. S.; Finks, S. W. *Drugs* **2015,** *75*, 377–395.
69. Ma, T.; Zou, F.; Pusch, St.; Xu, Y.; von Deimling, A.; Zha, X. *J. Med. Chem.* **2018,** *61*, 8981–9003.
70. Kim, E. S. *Drugs* **2017**, *77,* 1705–1711.
71. Popovici-Muller, J.; Lemieux, R. M.; Artin, E.; Saunders, J. O.; Salituro, F. G.; Travins, J.; Cianchetta, G.; Cai, Z.; Zhou, D.; Cui, D.; et al. *ACS Med. Chem. Lett.* **2018**, *9,* 300–305.
72. Parascandola, J. *Trends Pharmacol. Sci.* **1980**, *1*, 189–192.
73. Maehle, A.-H.; Pruell, C.-R.; Halliwell, R. F. *Nat. Rev. Drug Discov.* **2002,** *1*, 637–647.
74. Rasmussen, S. G. F.; Choi, H.-J.; Rosenbaum, D. M.; Kobilka, T. S.; Thian, F. S.; Edwards, P. C.; Burghammer, M.; Ratnala, V. R. P.; Sanishvili, R.; Fischetti, R. F.; et al. *Nature* **2007**, *450*, 383–387.
75. (a) Piscitelli, C. L.; Kean, J.; de Graaf, C.; Deupi, X. *Mol. Pharmacol.* **2015**, *88*, 536–551. (b) Shonberg, J.; Kling, R. C.; Gmeiner, P.; Loeber, S. *Bioorg. Med. Chem.* **2015,** *23,* 3880–3906.
76. Snyder, S. H.; Pasternak, G. W. *Trends Pharmacol. Sci.* **2003**, *24,* 198–205.
77. Corbett, A. D.; Henderson, G.; McKnight, A. T.; Paterson, S. J. *Br. J. Pharmacol.* **2006**, *147(Suppl.1),* S153–S162.
78. Burns, S. M.; Cunningham, C. W.; Mercer, S. L. *ACS Chem. Neurosci.* **2018**, *9,* 2428–2437.
79. Dosa, P. I.; Amin, E. A. *J. Med. Chem.* **2016**, *59,* 810–840.
80. Ganellin, C. R. Discovery of Cimetidine, Ranitidine and Other H_2-Receptor Histamine Antagonists. In Ganellin, C. R.; Roberts, S. M., eds. Medicinal Chemistry: The Role of Organic Chemistry in Drug Research. Academic Press: London, **1994**, pp. 228–254.

81. Rothman, R. B.; Baumann, M. H.; Savage, J. E.; Rauser, L.; McBride, A.; Hufeisen, S. J.; Roth, B. L. *Circulation* **2000**, *102*, 2836–2841.
82. Goernemann, T.; Huebner, H.; Gmeiner, P.; Horowski, R.; Latte, K. P.; Flieger, M.; Pertz, H. H. *J. Pharmacol. Exp. Ther.* **2007**, *324*, 1136–1145.
83. Sriram, K.; Insel, P. A. *Mol. Pharmacol.* **2018**, *93*, 251–258.
84. Wong, D. T.; Bymaster, F. P.; Engleman, E. A. *Life Sci.* **1995**, *57*, 411–441.
85. (a) Marin, P.; Becamel, C.; Dumuis, A.; Bockaert, J. *Curr. Drug Targets* **2012**, *13*, 28–52. (b) Pithadia, A. B.; Jain, S. M. *J. Clin. Med. Res.* **2009**, *1*, 72–82.
86. Fiorino, F.; Severino, B.; Magli, E.; Ciano, A.; Caliendo, G.; Santagada, V.; Frecentese, F.; Perissutti, E. *J. Med. Chem.* **2014**, *57*, 4407–4426.
87. (a) Tepper, S. J.; Rapoport, A. M. *CNS Drugs* **1999**, *12*, 403–417. (b) Tepper, S. J.; Rapoport, A. M.; Sheftell, F. D. *Arch. Neurol.* **2002,** *59*, 1084–1088.
88. (a) Isaac, M.; Slassi, A. *IDrugs*, **2001**, *4*, 189–196. (b) Milson, D. S.; Tepper, S. J.; Rapoport, A. M. *Expert Opin. Pharmacother.* **2000**, *1*, 391–404. (c) Deleu, D.; Hanssens, Y. *J. Clin. Pharmacol.* **2000**, *40*, 687–700.
89. Smith, B. M.; Smith, J. M.; Tsai, J. H.; Schultz, J. A.; Gilson, C. A.; Estrada, S. A.; Chen, R. R.; Park, D. M.; Prieto, E. B.; Gallardo, C. S.; et al. *J. Med. Chem.* **2008**, *51*, 305–313.
90. Chong, Y.; Choo, H. *Expert Opin. Investig. Drugs* **2010**, *19*, 1309–1319.
91. Deeks, E. D. *Drugs* **2015**, *75*, 1815–1822.
92. Sahli, Z. T; Tarazi, F. I. *Expert Opin. Drug Discov.* **2018**, *13*, 103–110.
93. Giger, R.; Mattes, H. *Annu. Rep. Med. Chem.* **2007**, *42*, 195–209.
94. (a) Karila, D.; Freret, T.; Bouet, V.; Boulouard, M.; Dallemagne, P.; Rochais, C. *J. Med. Chem.* **2015**, *58*, 7901–7912. (b) Khoury, R.; Grysman, N.; Gold, J.; Patel, K.; Grossberg, G. T. *Expert Opin. Invest. Drugs* **2018**, *27*, 523–533.
95. Garnock-Jones, K. P. *Drugs* **2016**, *76*, 99–110.
96. Upton, N.; Chuang, T. T.; Hunter, A. J.; Virley, D. J. *Neurotherapeutics* **2008**, *5*, 458–469.
97. (a) Ye, N.; Zhang, A.; Neumeyer, J. L.; Baldessarini, R. J.; Zhen, X.; Zhang, A. *Chem. Rev.* **2013**, *113*, PR123–PR178. (b) Zhang, A.; Neumeyer, J. L.; Baldessarini, R. J. *Chem. Rev.* **2007**, *107*, 274–302.
98. Wu, W.-L.; Burnett, D. A.; Spring, R.; Greenlee, W. J.; Smith, M.; Favreau, L.; Fawzi, A.; Zhang, H.; Lachowicz, J. E. *J. Med. Chem.* **2005**, *48*, 680–693.
99. Xiao, J.; Free, R. B.; Barnaeva, E.; Conroy, J. L.; Doyle, T.; Miller, B.; Bryant-Genevier, M.; Taylor, M. K.; Hu, X.; Dulcey, A. E.; et al. *J. Med. Chem.* **2014**, *57*, 3450–3463.
100. (a) Cheng, J. P.; Leary, J. B.; Edwards, C. M.; Sembhi, A.; Bondi, C. O.; Kline, A. E. *Brain Res.* **2016**, *1640*, 5–14. (b) Leggio, G. M.; Bucolo, C.; Platania, C. B. M.; Drago, F.; Salomone, S. *Pharmacol. Ther.* **2016,** *165*, 164–177.
101. Micheli, F. *ChemMedChem* **2011**, *56*, 1152–1162.
102. Appel, N. M.; Li, S.-H.; Holmes, T. H.; Acri, J. B. *J. Pharmacol. Exp. Ther.* **2015**, *354*, 484–492.
103. Kumar, V.; Bonifazi, A.; Ellenberger, M. P.; Keck, T. M.; Pommier, E.; Rais, R.; Slusher, B. S.; Gardner, E.; You, Z.-B.; Xi, Z.-X.; et al. *J. Med. Chem.* **2016**, *59*, 7634–7650.
104. Lindsley, C. W.; Hopkins, C. R. *J. Med. Chem.* **2017**, *60*, 7233–7243.

105. (a) Ganellin, R. C.; Schwartz, J.-C.; Stark, H. Discovery of Pitolisant, the First Marketed HistamineH3-Receptor Inverse Agonist/Antagonist for Treating Narcolepsy. In Fischer, J.; Rotella, D. P.; Childers, W. E., eds. Successful Drug Discovery. Vol. 3. Wiley-VCH: Weiheim, **2018**, pp. 359–381. (b) Syed, Y. Y. *Drugs* **2016**, *76*, 1313–1318.
106. (a) Whitman, D. B.; Cox, C. D.; Breslin, M. J.; Brashear, K. M.; Schreier, J. D.; Bogusky, M. J.; Bednar, R. A.; Lemaire, W.; Bruno, J. G.; Hartman, G. D.; et al. *ChemMedChem* **2009**, *4*, 1069–1074. (b) Cox, C. D.; Breslin, M. J.; Whitman, D. B.; Schreier, J. D.; McGaughey, G. B.; Bogusky, M. J. et al. *J. Med. Chem.* **2010**, *53,* 5320–5332.
107. (a) Armour, D.; de Groot, M. J.; Edwards, M.; Perros, M.; Price, D. A.; Stammen, B. L.; Wood, A. *ChemMedChem* **2006**, *1*, 706–709. (b) Price, D. A.; Armour, D.; de Groot, M. J.; Leishman, D.; Napier, C.; Perros, M.; Stammen, B. L.; Wood, A. *Bioorg. Med. Chem. Lett.* **2006**, *16*, 4633–4637.
108. (a) Gould, S. E.; Low, J. A.; Marsters, J. C., Jr.; Robarge, K.; Rubin, L. L.; de Sauvage, F. J.; Sutherlin, D. P.; Wong, H.; Yauch, R. L. *Expert Opin. Drug Discov.* **2014,** *9*, 969–984. (b) Burness, C. B. *Drugs* **2016**, *76,* 1559–1566.
109. Kaur, P.; Khatik, G. L. *Mini Rev. Med. Chem.* **2016**, *16,* 531–546.
110. D'Angelo, N. D.; Kim, T.-S.; Andrews, K.; Booker, S. K.; Caenepeel, S.; Chen, K.; D'Amico, D.; Freeman, D.; Jiang, J.; Liu, L.; et al. *J. Med. Chem.* **2011**, *54,* 1789–1811.
111. (a) Santen, R. J.; Yue, W.; Wang, J.-P. *Breast Cancer Res.* **2005**, *7*, S.08. (b) Zahid, M.; Koh, I, E.; Saeed, M.; Rogan, E.; Cavalieri, E. *Chem. Res. Toxicol.* **2006**, *19*, 164–172.
112. Avendano, C.; Menendez, J. C. *Medicinal Chemistry of Anticancer Drugs*. 1st ed. Elsevier Science: Amsterdam, Netherlands. **2008**, pp. 54–91.
113. Piñeiro-Núñez, M. Raloxifene, Evista: A Selective Estrogen Receptor Modulator (SERM). In Li, J. J.; Johnson, D. S., eds. *Modern Drug Synthesis.* Wiley: Hoboken, NJ, **2010**, pp. 309–327.
114. Singh, M. *Med. Res. Rev.* **2001**, *21*, 302–347.
115. Gennari, L.; Merlotti, D.; Martini, G.; Nuti, R. *Expert Opin. Investig. Drugs* **2006**, *15*, 1091–1103.
116. Deeks, E. D. *Drugs* **2018,** *78*, 131–137.
117. Kota, B. P.; Huang, T. H.; Roufogalis, B. D. *Pharmacol. Res.* **2005**, *51*, 85–94.
118. Pirat, C.; Farce, A.; Lebegue, N.; Renault, N.; Furman, C.; Millet, R.; Yous, S.; Speca, S.; Berthelot, P.; Desreumaux, P.; et al. *J. Med. Chem.* **2012**, *55*, 4027–4061.
119. (a) Smith, S. I. *Mol. Cell Biochem.* **2004**, *263*, 189–210. (b) Taygerly, J. P.; Cummins, T. J. *Med. Chem. Rev.* **2016**, *51*, 53–66.
120. (a) Saxenaa, A. K.; Bhuniab, S. S. *Med. Chem. Rev.* **2016**, *51*, 297–310. (b) Malecic, N.; Young, H. S. *Expert Opin. Invest. Drugs* **2016**, *25*, 455–462.
121. Koese, M. *Bioorg. Med. Chem. Lett.* **2017**, *27*, 3611–3620.
122. Lewin, J.; Siu, L. L. *J. Clin. Oncol.* **2015**, *33*, 3372–3374.
123. (a) Stawski, P.; Janovjak, H.; Trauner, D. *Bioorg. Med. Chem.* **2010**, *18*, 7759–7772. (b) Halliwell, R. F. *Trends Pharmacol. Sci.* **2004**, *28*, 214–219.

124. Patel, N. C. Perampanel (Fycompa): AMPA Receptor Antagonist for the Treatment of Seizure. In Li, J. J.; Johnson, D. S., Eds. *Innovative Drug Synthesis.* Wiley: Hoboken, NJ, **2015**, pp. 271–281.
125. Guzzo, P. R. $GABA_A$ Receptor Agonists for Insomnia: Zolpidem (Ambien), Zaleplon (Sonata), Eszopiclone (Estorra, Lunesta), and Indiplon. In Johnson, D.; Li, J. J., Eds. The Art of Drug Synthesis. Wiley: Hoboken, NJ, **2007**, pp. 215–223.
126. Yuen, P. W. Advances in the Development of Methods for The Synthesis of a2δ Ligands [Neurontin (Gabapentin), Lyrica (Pregabalin)]. In Johnson, D.; Li, J. J., Eds. The Art of Drug Synthesis. Wiley: Hoboken, NJ, **2007**, pp 225–240.
127. Zhang, Y.; Ban, H.; Yu, R.; Wang, Z.; Zhang, D. *Future Med. Chem.* **2018**, *10,* 1261–1276.
128. Csuca, O. *Oncology* **2017**, *37(Suppl. 1),* 83–87.
129. Borzilleri, R. M.; Zheng, X.; Schmidt, R. J.; Johnson, J. A.; Kim, S.-H.; DiMarco, J. D.; Fairchild, C. R.; Gougoutas, J. Z.; Lee, F. Y. F.; Long, B. H.; et al. *J. Am. Chem. Soc.* **2000**, *122,* 8890–8897.
130. Sharma, V. K.; Watts, J. K. *Future Med. Chem.* **2015**, *7,* 2221–2242.
131. Wendt, M. D.; Shen, W.; Kunzer, A.; McClellan, W. J.; Bruncko, M.; Oost, T. K.; Ding, H.; Joseph, M. K.; Zhang, H.; Nimmer, P. M.; et al. *J. Med. Chem.* **2006**, *49,* 1165–1181.
132. Park, C.-M.; Bruncko, M.; Adickes, J.; Bauch, J.; Ding, H.; Kunzer, A.; Marsh, K. C.; Nimmer, P.; Shoemaker, A. R.; Song, X.; et al. *J. Med. Chem.* **2008**, *51,* 6902–6915.
133. Souers, A. J.; Leverson, J. D.; Boghaert, E. R.; Ackler, S. L.; Catron, N. D.; Chen, J.; Dayton, B. D.; Ding, H.; Enschede, S. H.; Fairbrother, W. J.; et al. *Nat. Med.* **2013**, *19,* 202–208.

2

Hit/Lead Discovery

To start the medicinal chemistry portion of a drug discovery project, we need a chemical starting point, a hit or a lead. Where do hits or leads come from? We will only briefly review some common practices in this chapter because many dedicated monographs already exist. They include (i). irrational drug design (serendipity); (ii). natural products; (iii). high through-put screening (HTS); (iv). fragment-based lead discovery; and (v). DNA-encoded library (DEL).

2.1 Irrational Drug Design (Serendipity)

Louis Pasteur (1822–1895) famously said, "In the field of experimentation, chance favors the prepared mind." In the early days of drug discovery, nearly all drugs were discovered by serendipity.

The most conspicuous example is probably Alexander Fleming's discovery of penicillin G (**1**). In the summer of 1928, Fleming's discovery of penicillin changed the world. In the context of drug discovery in general, and medicinal chemistry in particular, penicillin G (**1**) is considered a hit/lead compound for all other β-lactamase inhibitors to follow. With appreciation of the β-lactam pharmacophore, numerous analogs have been prepared and evaluated as antibiotics. GSK's amoxicillin (**2**), approved in 1972, is more potent, more stable, and more selective than penicillin G (**1**). More importantly, it is orally bioavailable and may be taken by mouth. When combining with a β-lactamase inhibitor clavulanic acid, the combo drug Augmentin is one of the popular oral antibiotics.

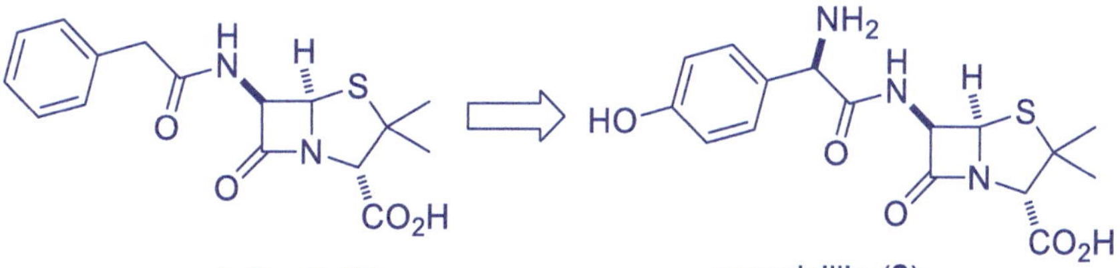

penicillin G (**1**) amoxicillin (**2**)

Medicinal Chemistry for Practitioners, First Edition. Jie Jack Li.

When Barnett Rosenberg discovered cisplatin (Platinol, **3**) as a cancer chemotherapy in 1967, it may be considered as a hit/lead for later platinum-containing chemotherapeutics. Since cisplatin (**3**) is plagued by kidney toxicities, its modification led to carboplatin (Paraplatin, **4**), which has much lower nephrotoxicity. Eventually, oxaliplatin (Eloxatin, **5**) was approved in 1996, and it is devoid of nephrotoxicity.[1]

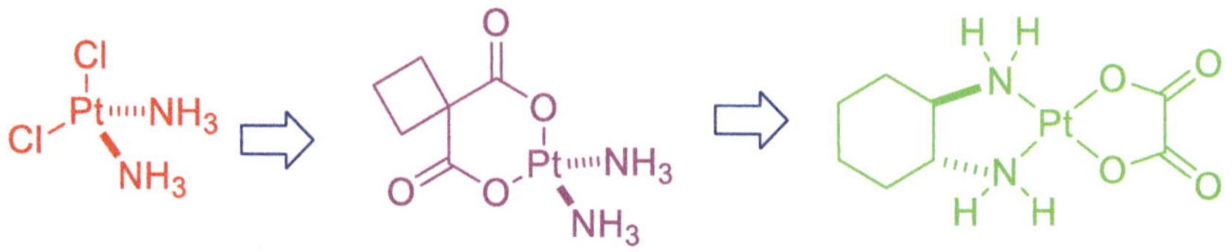

cisplatin (Platinol, **3**) renal damage

carboplatin (Paraplatin, **4**) much lower nephrotoxicity

oxaliplatin (Eloxatin, **5**) devoid of nephrotoxicity

Finding hits, leads, or drugs nowadays is no longer left to the hands of serendipity. We have developed many more rational means to discover drug hits as starting points of drug discovery.

2.2 Natural Products

Mother Nature has bestowed us with a cornucopia of natural products, from where we can find our drug hits.

2.2.1 From Plants

Ancient pharmacopeia was filled with medicines derived from plants. In fact, most traditional Chinese medicines (TCMs) and traditional medicines from other cultures are from seeds, flowers, roots, and stems of plants. In the context of medicinal chemistry, active principles from plants have served as lead compounds to discover new drugs.

Salicylic acid was isolated from willow tree barks. Using it as a starting point to prepare aspirin is one of the early examples of finding a hit/lead from plants. In the same vein, morphine (**6**), isolated from poppy seeds, is the starting material for making heroin. Bayer's Felix Hoffmann was single-handedly responsible for inventing both aspirin and heroin in 1897.

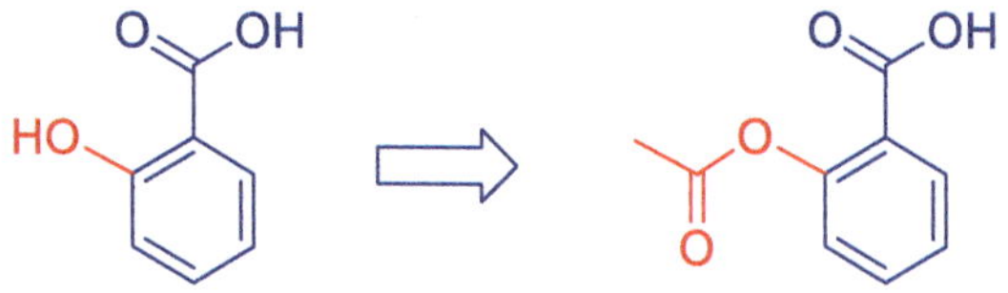

salicylic acid

aspirin (acetylsalicylic acid)

Morphine (**6**) is not only the precursor for heroin, it is also the inspiration for the discovery of nalorphine (Narcan, **7**). While morphine (**6**), a μ receptor agonist, is part of the problem that causes the opioid crisis, nalorphine (**7**), a μ receptor antagonist, is a powerful weapon to combat the scourge. It is remarkable that the difference of merely two carbon atoms can achieve completion interconversion from an agonist to an antagonist.[2]

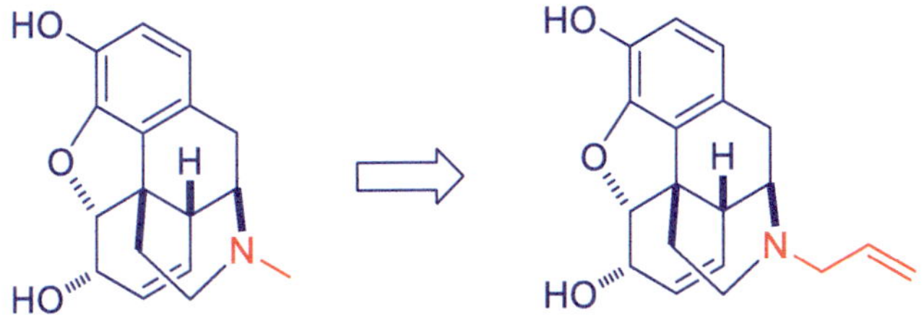

morphine (**6**)
μ receptor **agonist**
K_i = 0.53 nM

nalorphine (Narcan, **7**)
μ receptor **antagonist**
K_i = 0.36 nM

Isolation of camptothecin (**8**) offered an opportunity to investigate its mode of action (MOA). It functions as a DNA topoisomerase I inhibitor to exert its anticancer properties. Camptothecin (**8**) has minimal aqueous solubility with no bioavailablity, making it impossible to dose. However, using it as a hit/lead, installation of solubilizing groups led to irinotecan (Camptosar, **9**), an injection drug, and topotecan (Hycamtin, **10**), an oral drug.[3]

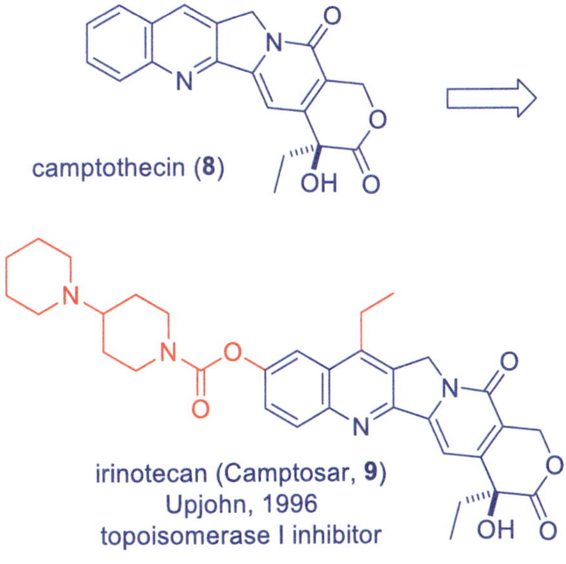

camptothecin (**8**)

irinotecan (Camptosar, **9**)
Upjohn, 1996
topoisomerase I inhibitor

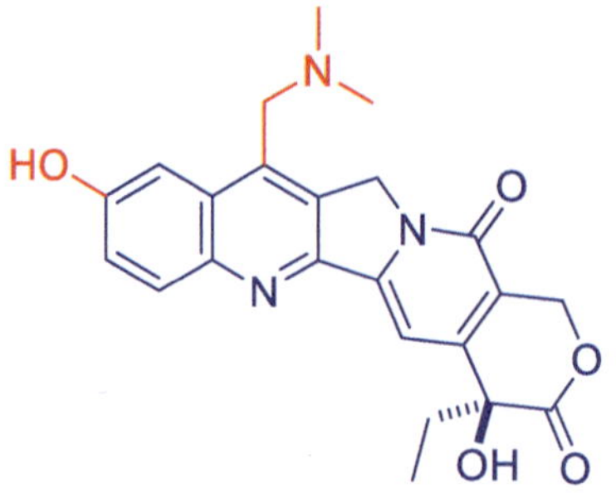

topotecan (Hycamtin, **10**)
GSK, 2007
topoisomerase I inhibitor

BMS's paclitaxel (Taxol, **11**) was initially isolated from the Pacific yew tree barks. If one considers paclitaxel (**11**) as a hit/lead, then its descendent analogs docetaxel (Taxotere, **12**) and cabazitaxel (Jevtana, **13**) are better and more efficacious drugs.[4]

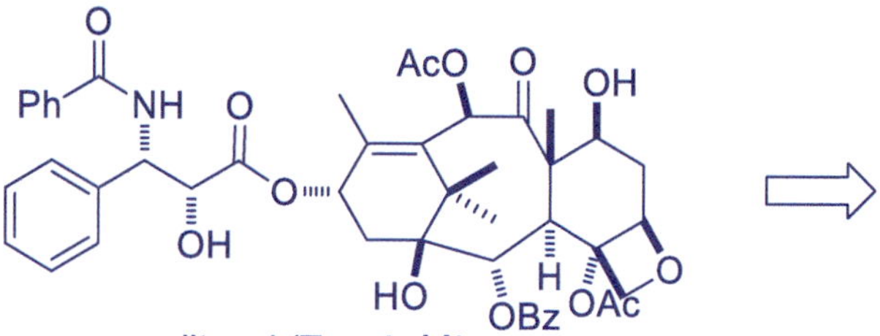

paclitaxel (Taxol, **11**)
BMS, 1993
Disrupt microtubules

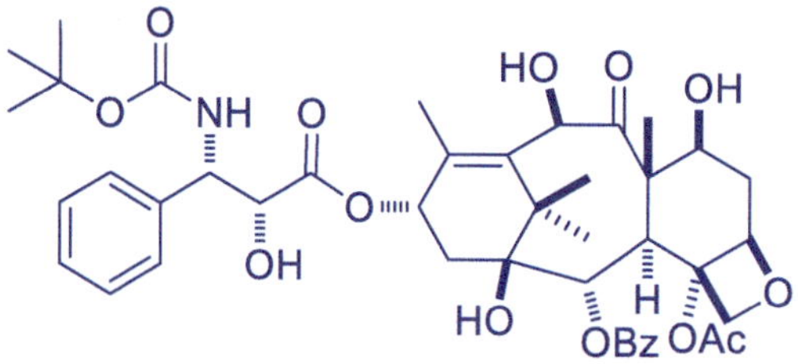

docetaxel (Taxotere, **12**)
Sanofi, 1995
Disrupt microtubule function

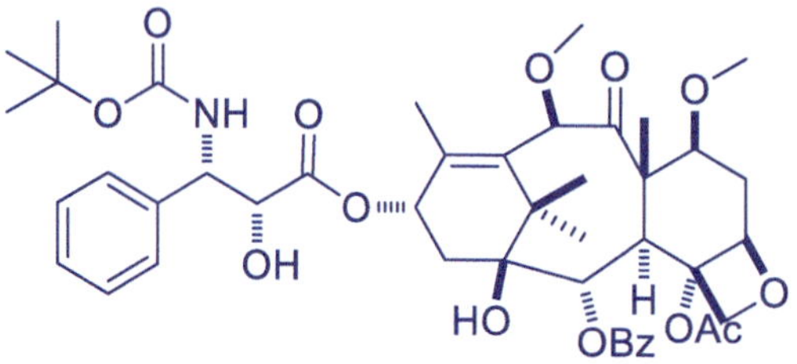

cabazitaxel (Jevtana, **13**)
Sanofi, 2010
Disrupt microtubule function

Warfarin (Coumadin, **14**) owes its genesis to dicumarol (**13**), the active principle isolated from sweet clover (*Melilotus alba* and *Melilotus officinalis*). It turns out that warfarin (**14**) works as an anticoagulant by inhibiting the vitamin K epoxide reductase.

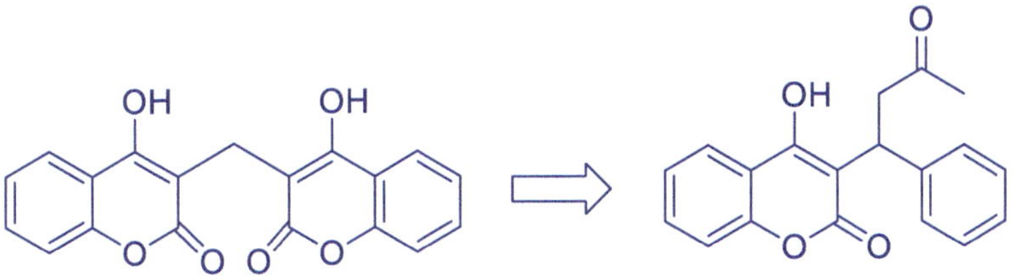

dicumarol (**13**)

warfarin (Coumadin, **14**)
vitamin K antagonist

Although many appreciate metformin (**16**)'s impressive efficacy in treating diabetes and several other diseases, few realize that its genesis traces back to galegine (**15**), a natural product isolated from French Lilac or Goat's Rue (*Galega officinalis*).

galegine (**15**)

metformin (Glucophage, **16**)
Merck, 1972

2.2.2 *From Animals*

Jay McLean isolated heparin from dog's liver in the 1916. Using heparin as the starting point and removing nonpharmacophores, two pentasaccharides fondaparinux (**17**) and idraparinux (**18**) are now popular anticoagulants given via IV.

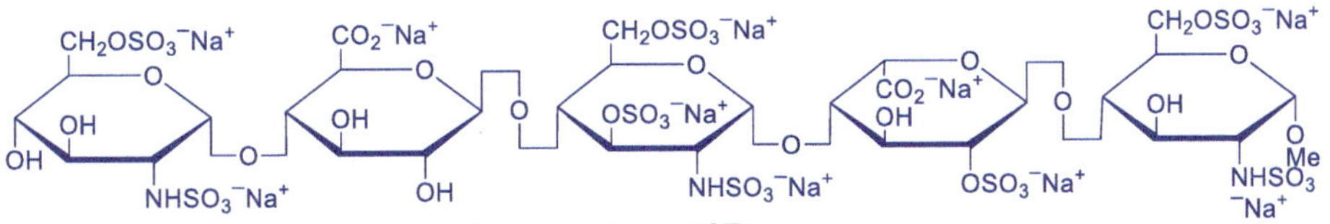

fondaparinux (**17**)

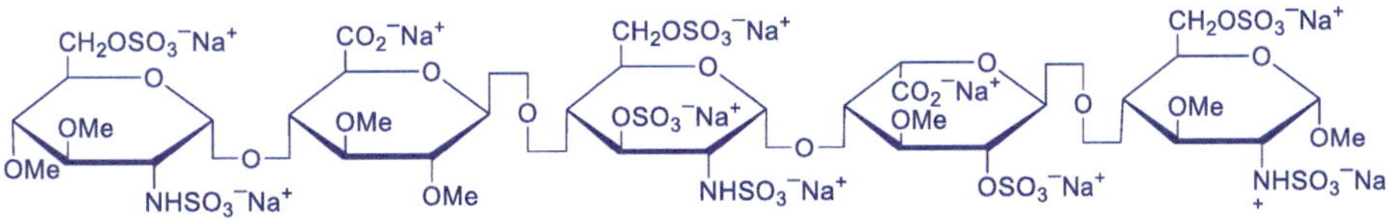

idraparinux (**18**)

Hirudin, a direct thrombin inhibitor, was isolated from European medicinal leech (*Hirudo medicinalis*)'s salivary glands and was the inspiration for the discovery of ximelagatran (Exanta, **20**). AstraZeneca trimmed some inactive segments systemically of hirudin's 65 amino-acid peptide to afford bivalirudin, a 20 amino acid peptide. They further arrived at a pentapeptide, which led to dipeptide melagatran (**19**). An ethyl ester prodrug tactic was combined with conversion of basic amidine to hydroxyamidine, a nearly neutral fragment, delivered ximelagatran (Exanta, **20**). It has an oral bioavailability of 19% in man and can be given by mouth.[5] Regrettably, the drug was withdrawn due to liver toxicity from the market in 2006, only two years after its approval.

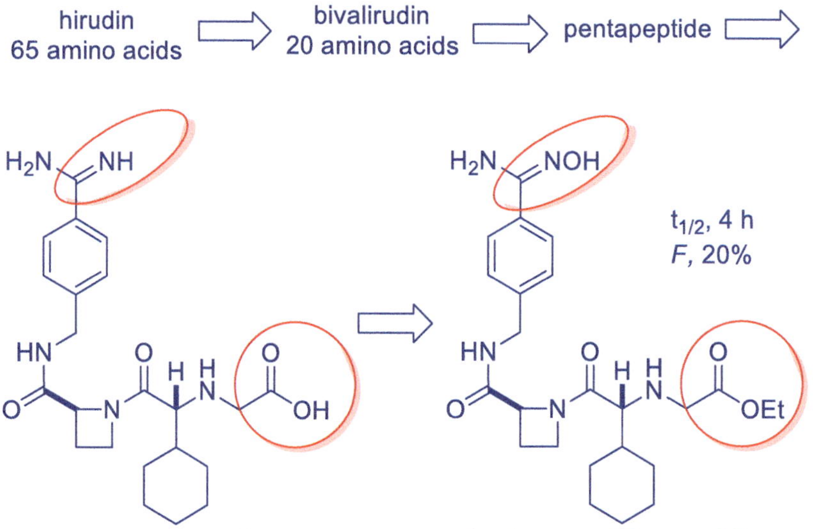

melagatran (**19**) ximelagatran (Exanta, **20**)

The first commercial angiotensin-converting enzyme (ACE) inhibitor was discovered using a snake venom as a starting point. In 1967, John Vane at Oxford University tested a dried extract of the venom of the poisonous Brazilian pit viper, *Bothrops jararaca,* on an *in vitro* preparation of ACE and found it to be a potent inhibitor. In the early 1970s, Squibb isolated teprotide (**21**, a nonapeptide). It was shown to reduced blood pressure in healthy volunteers and confirmed that it was a selective ACE inhibitor in humans. With brilliant insight, David Cushman and Miguel A. Ondetti at Squibb truncated the teprotide molecule and obtained succinoyl-1-proline **22** with an IC_{50} value 330 nM for ACE inhibition. They chose **22** for further modifications because the C-terminal amino acid occurs at the free C-terminus of all the naturally occurring peptidic inhibitors. The activity of **22** demonstrated that a small molecular weight drug could be a potent inhibitor of ACE by occupying only a small fraction of the extended active site cavities because the chemistry of peptide bond hydrolysis is typically dependent on a small number of critical amino acids.

Later on, two important observations were made stemming from succinoyl-1-proline **22**. (i) A moderate boost of activity (1.6-fold) was observed when the carboxylic acid was replaced with a sulfhydryl as in thiol **23**. (ii) A significant increase of the binding activity was obtained when an extra methyl group was installed at the α-position of the amide bond, giving rise to D-(*R*)-2-methylsuccinoyl-1-proline (**24**) with an IC_{50} of 22 nM. The L-(*S*)-enantiomer of **24** was found to be much less active.

A breakthrough came when both features of **23** and **24** were incorporated into one molecule. Thus, they replaced the carboxylate group with a sulfhydryl (–SH) and installed an additional methyl group and achieved a 1,000-fold improvement in potency for ACE inhibition. The drug became the first oral ACE inhibitor, captopril (Capoten, **25**), which was approved by the FDA 1978 and significantly contributed to the management of hypertension and hypertension-related target-organ damage. Squibb arrived at captopril from only 60 compounds logically synthesized and tested.[6] Many more "me-too" ACE inhibitors followed.

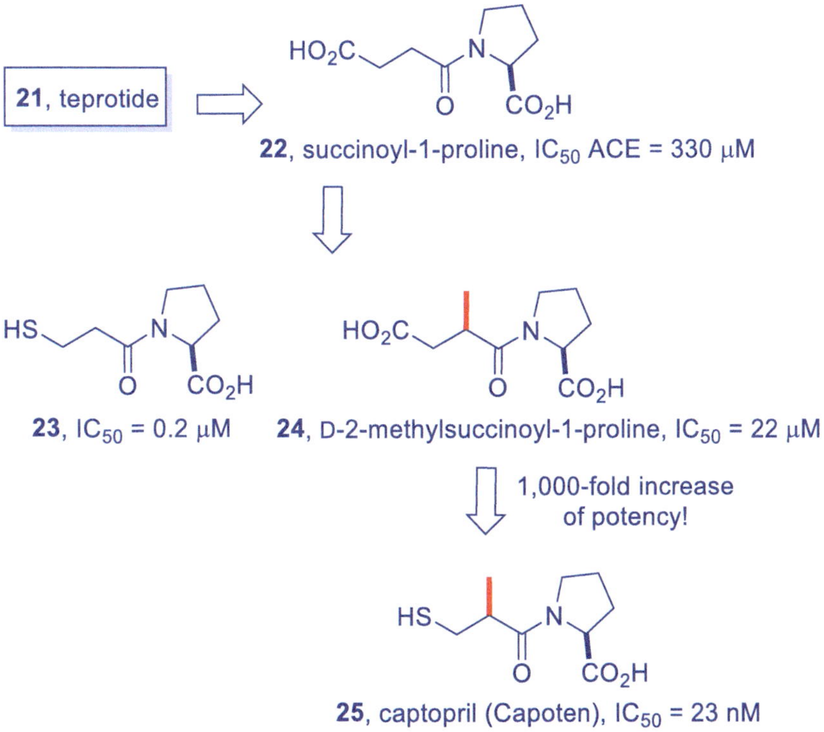

In 1936, Philip Hench and Edward Kendall isolated cortisone from bovine adrenal glands. The emergence of cortisone opened a floodgate of corticosteroids. Today, dozens of them are now on the market. Also in the 1930s, prostaglandins were initially isolated from seminal fluids secreted by prostate glands. It was later found that other organs secret prostataglandins as well. Research on prostaglandins has gone on till this day. One of the recent entries of prostaglandins that garnered the FDA approval is Pfizer's latanoprost (Xalatan, **26**), a prostanoid FP receptor agonist for treating glaucoma.

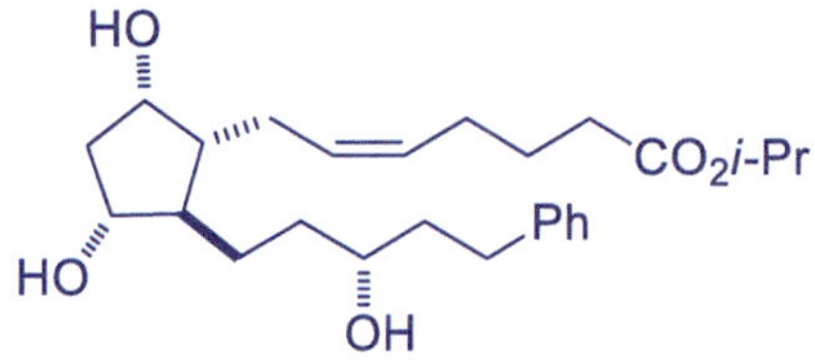

latanoprost (Xalatan, **26**)
Pfizer, 1996
FP receptor agonist

2.2.3 From Microorganisms

Penicillin G (**1**) is actually a secondary metabolite of a fungus *Penicillium notatum*.

Mevastatin (compactin, **27**) was isolated by Aruka Endo in 1973 from the fermentation broth of a fungus *Penicillium citrium*. It was the harbinger of future HMG CoA inhibitors including natural statin such as lovastatin (Mevacor), and totally synthetic statins such as rosuvastatin (Crestor, **28**).

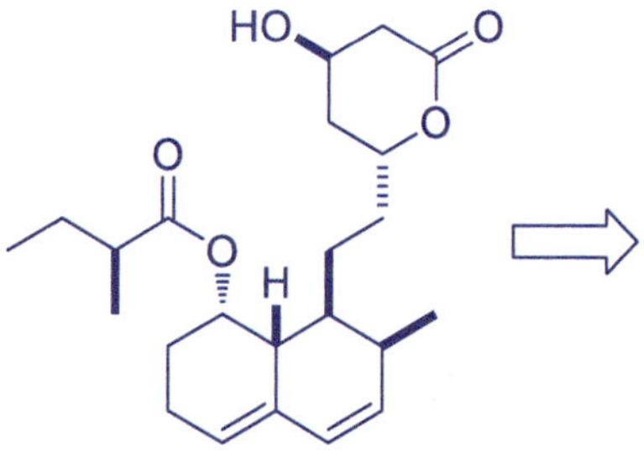

mevastatin (compactin, **27**)

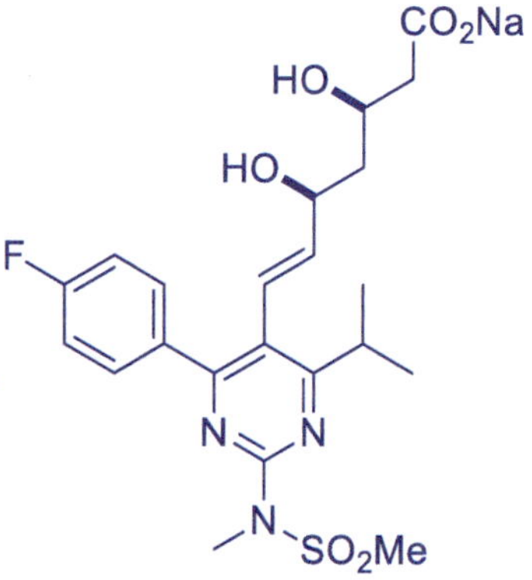

rosuvastatin (Crestor, **28**)

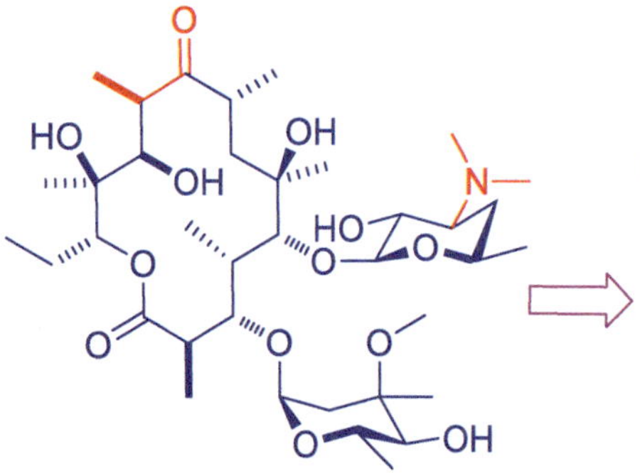

erythromycin (Erythrocin, **29**)

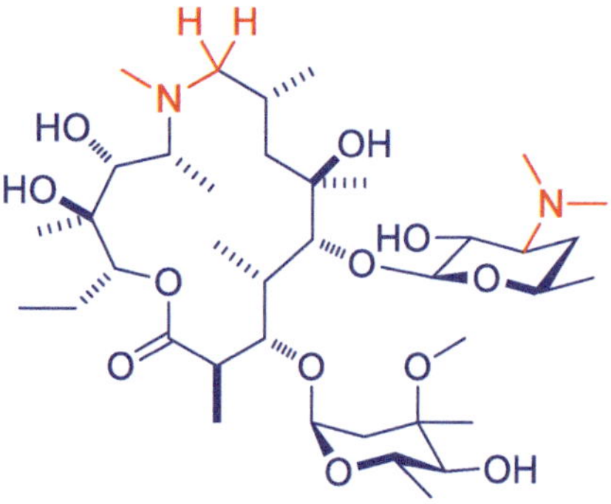

azithromycin (Zithromax, **30**)

Macrolide antibiotic erythromycin (Erythrocin, **29**) was isolated from the metabolic products of a strain of *Streptomyces erythreus* found in soil. Erythromycin (**29**) itself is not a remarkable drug. But using it as a starting point, azithromycin (Zithromax, **30**) was arrived as an excellent aza-macrolide antibiotic.

2.2.4 From Natural Ligands

In order to obtain selective H_2 histamine receptor antagonists as a treatment for peptic ulcer, James Black et al. employed histamine (**31**), a natural ligand though an agonist, as their starting point. The fruit of their labor was cimetidine (Tagamet, **32**) as a selective H_2 receptor antagonist, and the first blockbuster drug ever.[7]

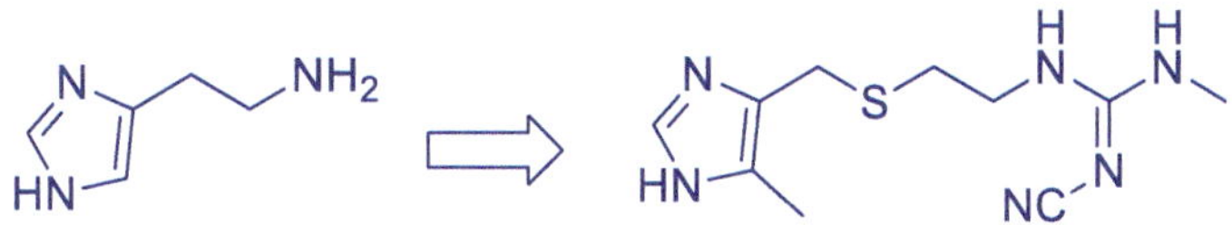

histamine (**31**)
H_2 receptor agonist

cimetidine (Tagamet, **32**)
H_2 receptor antagonist

Triptans, as represented by sumatriptan (Imitrex, **34**), are a class of serotonin $5HT_1$ receptor agonists that have been effective treatments for migraine. They were developed using serotonin (**33**), a natural ligand for the serotonin receptors, as a starting point.[8]

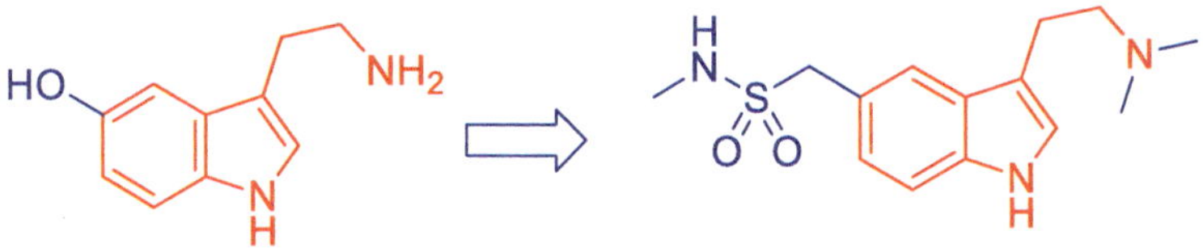

serotonin (**33**)

sumatriptan (Imitrex, **34**)
GlaxoSmithKline, 1995
$5HT_1$ receptor agonist

A natural ligand adrenaline (**35**), an agonist for adrenergic receptors, served as a lead for the discovery of β-blockers such as propranolol (Inderal, **36**). β-blockers are a family of β-adrenergic receptor antagonists to treat hypertension.

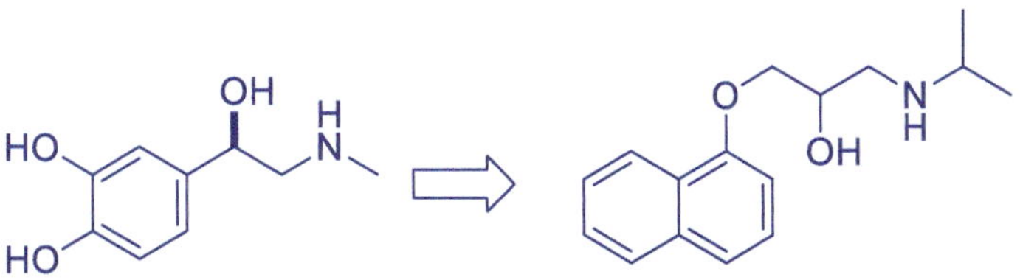

adrenaline (**35**)

propranolol (Inderal, **36**)
ICI, 1962
β-adrenergic receptor antagonist

Neuraminidase (also known as sialidase) is present in all influenza virus types and shares high sequence homology. Sialic acid (*N*-acetylneuraminic acid, or Neu5Ac, **37**) is a key residue that binds to neuraminidase in the virus replication process. Employing sialic acid (**37**) as the lead compound, zanamivir (Relenza, **38**) was discovered. An orally bioavailable influenza virus neuraminidase inhibitor oseltamivir (Tamiflu, **39**) was later discovered using cyclohexene as the core structure.[9]

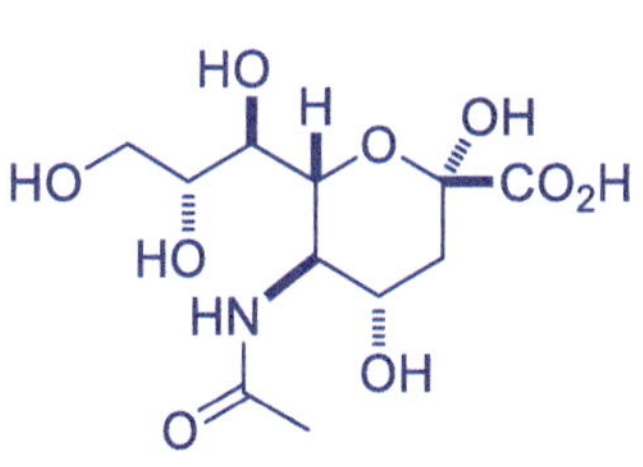

sialic acid (**37**)

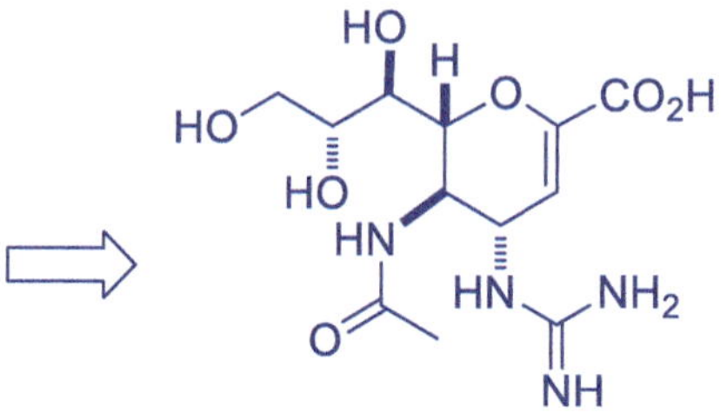

zanamivir (Relenza, **38**)

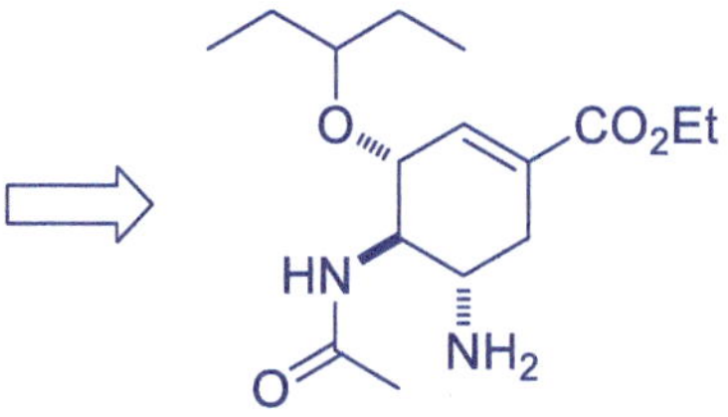

oseltamivir (Tamiflu, **39**)

2.2.5 From Modifying Existing Drugs

On occasions, we can modify an existing drug to amplify its side effects, a process known as selective optimization of side effects (SOSA).

In 1942, Marcel Janbon observed that a sulfa antibiotic isopropylthiadiazole (IPTD, **40**) also had a side effect of hypoglycemia: lowering sugar levels. Scientists took note and developed drugs with more profound hypoglycemia effect. The results were a new class of antidiabetic drugs: sulfonylureas as represented by tolbutamide (Orinase, **41**). They work as potassium channel blockers and stimulate insulin secretion.

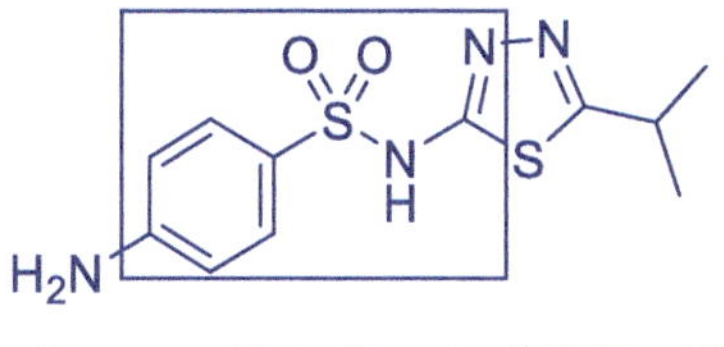

isopropylthiadiazole (IPTD, **40**)

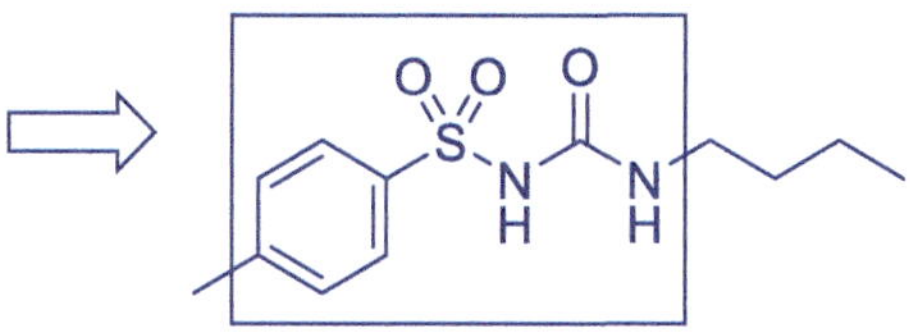

tolbutamide (Orinase, **41**)

In 1987, Steven Brickner became aware of two compounds reported by DuPont: Dup-721 (**43**) and a closely related sulfoxide analog, Dup-105. While reasonably active as antibiotics, both were too toxic. Using those two compounds as leads, Brickner et al.

arrived at linezolid (Zyvox, **44**), a first-in-class antibiotic that inhibits the initial phase of bacterial protein synthesis.[10]

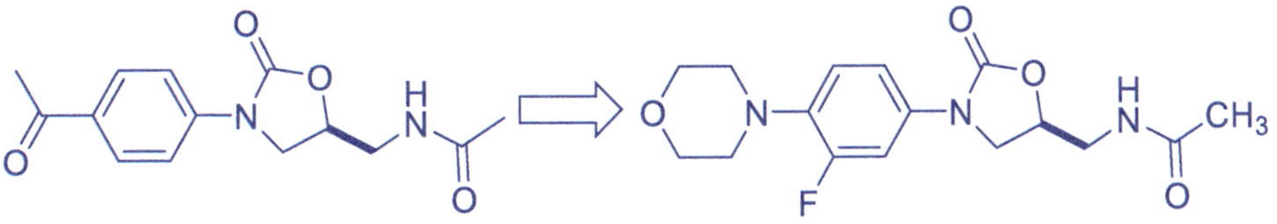

Dup-721 (**43**) linezolid (Zyvox, **44**)

One of the more spectacular successes from the SOSA strategy is probably Sanofi's clopidogrel (Plavix, **47**). Initially, Jean-Pierre Maffrand employed Yoshitomi's tinoridine (Nonflamin, **45**) as a starting point to search for anti-inflammatory drugs. He ended up discovering ticlopidine (Ticlid, **46**) as an anticoagulant. Regrettably, ticlopidine (**46**) was associated with several severe toxicities. Additional efforts as a backup to **46** eventually led to the discovery of clopidogrel (Plavix, **47**), a safe and efficacious anticoagulant with tremendous market success after marketing. Interestingly, its mechanism of action was later elucidated as that one of its active metabolites is a covalent $P2Y_{12}$ inhibitor.[11]

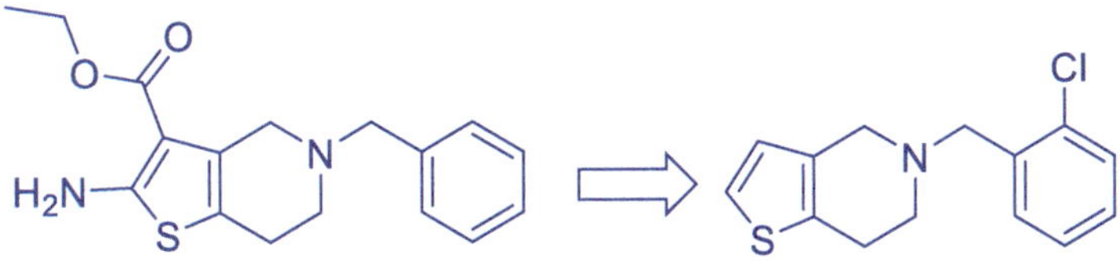

tinoridine (Nonflamin, **45**)
by Yoshitomi

ticlopidine (Ticlid, **46**)
Castaigne, 1979

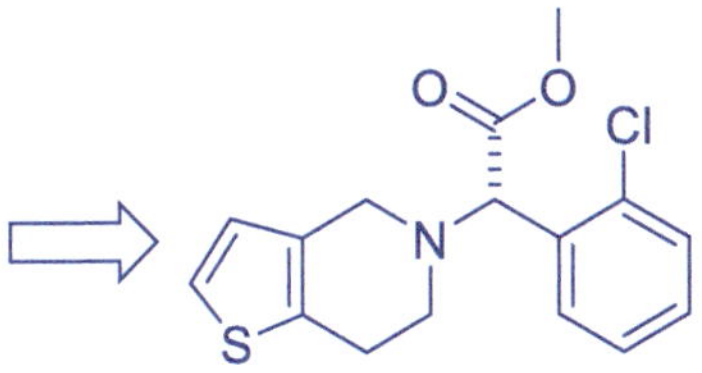

clopidogrel (Plavix, **47**)
Sanofi/BMS, 1993

2.3 High Throughput Screening (HTS)

In the early 1990s, the pharmaceutical companies realized that there was too much at stake to leave to the chances for drug discovery. High throughput screening (HTS) emerged as a consequence of molecular biology revolution that led to the identification of many drug targets. Typically, a big pharma's compound library has 1–3 million compounds. Screening compounds plated on 97 microtiter wells (or 384-well plates or 1536-well plates) against a biological target is monitored by fluorescence spectroscopic profiling.[12] Since a lot of compounds are pan-assay interference compounds (PAINS), it is always advisable to confirm the initial hits using enzymological and biophysical methods (hit validation). Examples of biophysics methods include surface plasmon resonance (SPR), thermal shift assays (TSAs), nuclear magnetic resonance (NMR), calorimetry (isothermal titration and differential scanning), and X-ray crystallography.[13]

HTS plays such an integral part of drug discovery these days that more than 50% approved drugs owe their genesis to a HTS hit(s). GSK's lapatinib (Tykerb), BMS's dasatinib (Sprycel), Bayer/Onyx's sorafenib (Nexavar), Merck's sitagliptin (Januvia), and Pfizer's sunitinib (Sutent) and maraviroc (Selzentry) are just a few examples. A fundamental feature of HTS is that it assumes no *a priori* knowledge of the drug binding site on the target protein.[14]

Nowadays, cell- and organism-based phenotypic assays have been increasingly adopted for HTS. For phenotypic screening, compounds are screened in cell-based assays to test for a change in the activity of specific signaling pathway with no *a priori* knowledge of the potential drug target.[15]

As medicinal chemists, we take part in triaging HTS hits. This is when all our learnings and experience come to play. Here instead of going through the minutia of HTS, a few examples are given to illustrate how HTS hits yielded marketed drugs. We start with the discovery of nevirapine.

Boehringer Ingelheim's nevirapine (Viramune, **52**) owes its origin to an HTS hit. To discover nonnucleoside inhibitors of HIV-1 reverse transcriptase, they screened their worldwide compound repositories against the enzyme. In 1988, a hit **48** emerged with an IC_{50} of 6 μM. It was originally made for the anti-muscarinic receptor analogs as antiulcer drugs. Three months and thousands of compounds later, LS (**49**) was identified with an IC_{50} of 350 nM. Bis-pyridyl analog **50** proved potent (IC_{50}, 125 nM) yet N-demethylation and N-deethylation readily took place. Rearranging the methyl to the left pyridine ring led to a very potent analog **51** (IC_{50}, 35 nM). However, nevirapine (**52**, IC_{50}, 84 nM) was chosen as the development candidate because the cyclopropyl group is more resistant to N-dealkylation even though it was not as potent as **51**.[16]

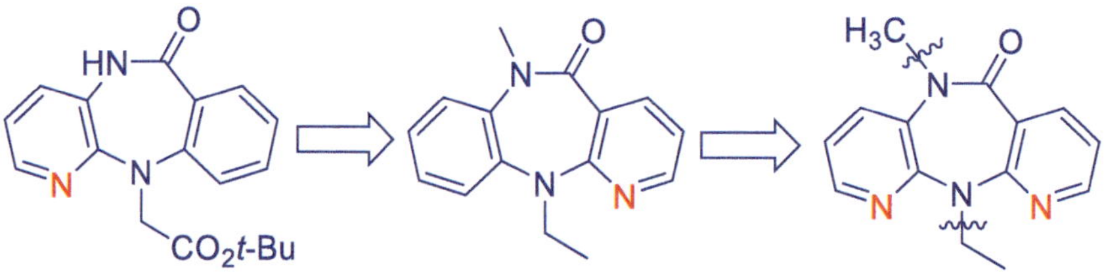

48, IC_{50} = 6 μM **49**, LS, IC_{50} = 350 nM **50**, IC_{50} = 125 nM

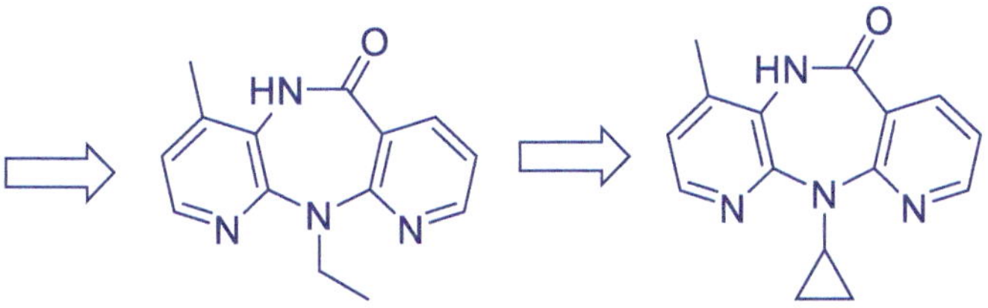

51, IC_{50} = 35 nM

nevirapine (Viramune, **52**)
IC_{50} = 84 nM

One of the early kinase inhibitors on the market, Sugen's sunitinib (Sutent, **57**) was discovered based on HTS hits. In 1994, Sugen initiated an HTS targeting tumor angiogenesis by inhibiting vascular endothelial growth factor receptor (VEGFR) catalytic activity. One of the initial hits indolin-2-one SU4312 (**53**) inhibited cellular tyrosine activity of another kinase, platelet-derived growth factor receptor (PDGFR, IC_{50}, 12 μM). Structure–activity relationship (SAR) investigation around indolin-2-ones led to the identification of highly selective VEGFR2 inhibitors. For instance, SU5416 (semaxanib, **54**) was 20-fold selective for VEGFR2 against PDGFR and it was moved to clinical trials. While the trials provided proof-of-concept for this mechanism of action, its poor pharmacokinetic properties (poor solubility only allowed IV dosing) prevented moving it further forward.

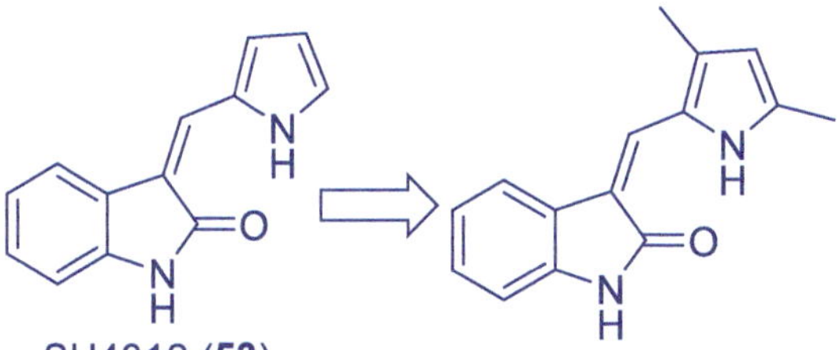

SU4312 (**53**)
PDGFR IC_{50},12 μM (cell)

SU5416 (semaxanib, **54**)
PDGFR IC_{50}, 20 μM (cell)
VEGFR2 IC_{50}, 1.0 μM (cell)
Discontinued, poor PK (IV)
POC, for MOA of inhibiting angiogenesis

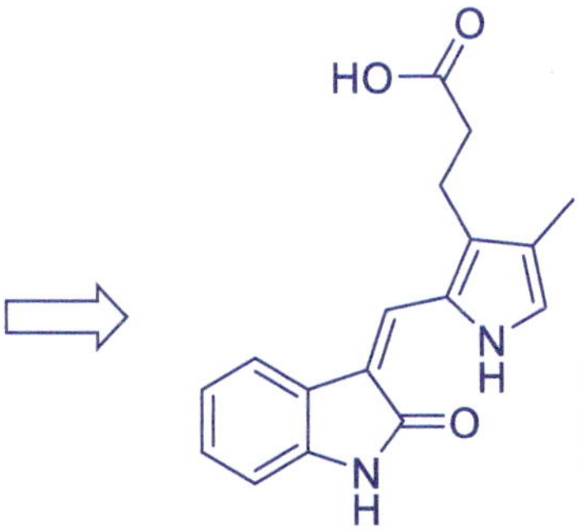

SU5402 (**55**)
Improved activity vs. VEGFR1,2,3

SBDD

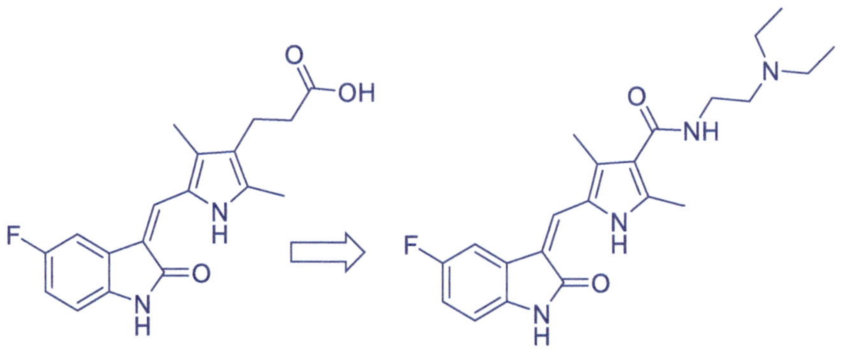

SU6668 (**56**)
PDGFR IC_{50}, 0.008 μM (cell)
VEGFR2 IC_{50}, 2.1 μM (cell)

sunitinib (Sutent, **57**)
VEGFR and PDGFR inhibitor
PDGFR IC_{50}, 0.008 μM (cell)
VEGFR2 IC_{50}, 0.008 μM (cell)

Co-crystallization of SU5402 (**55**) with FGFR1 shed light to its binding conformation. The propionic acid chain was incorporated to interact with the basic arginine residue in the ATP-binding pocket. More soluble, potent, and selective (against PDGFR) SU6668 (**56**) was brought to clinical trials as an oral compound. Eventually, sunitinib (Sutent, **57**) was identified as a dual PDGFR and VEGFR2 inhibitor with optimal pharmacokinetic and safety profiles. Interestingly, its major metabolite, the N-deethylation product had comparable biological activity as sunitinib (**57**).[17]

Merck discovered sitagliptin (Januvia, **61**) starting from HTS hits. From their HTS campaign against the dipeptidyl peptidase-4 (DPP-4) enzyme, two leads were identified. One was the β-amino acid proline amide **58** and the other was β-amino piperazine **59**. Extensive SAR investigations around the two leads eventually coalesced to give triazole **60**, although it had poor pharmacokinetic properties. Additional fluorine on the difluorophenyl and replacing the ethyl group with trifluoromethyl on **60** delivered sitagliptin (Januvia, **61**), that is potent, selective, and orally bioavailable (F = 44% in rat).[18]

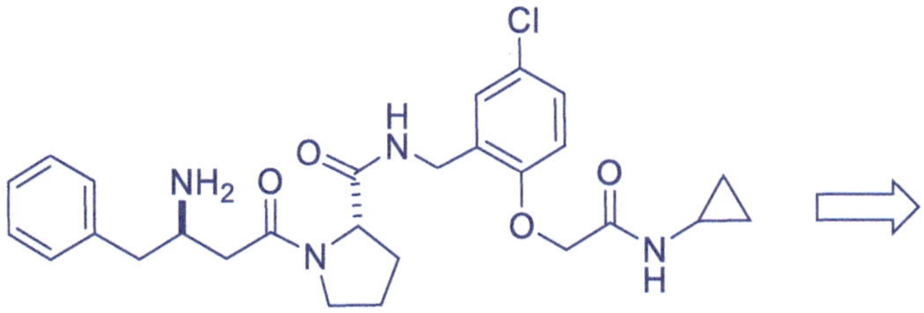

58, DPP-4 IC_{50} = 1,900 nM

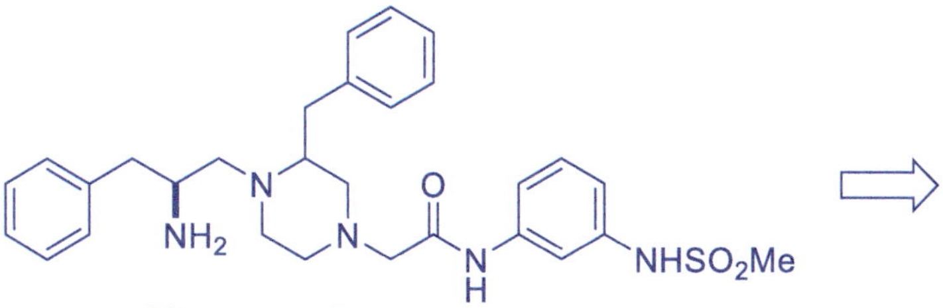

59, DPP-4 IC_{50} = 11,000 nM

60, DPP-4 IC_{50} = 231 nM

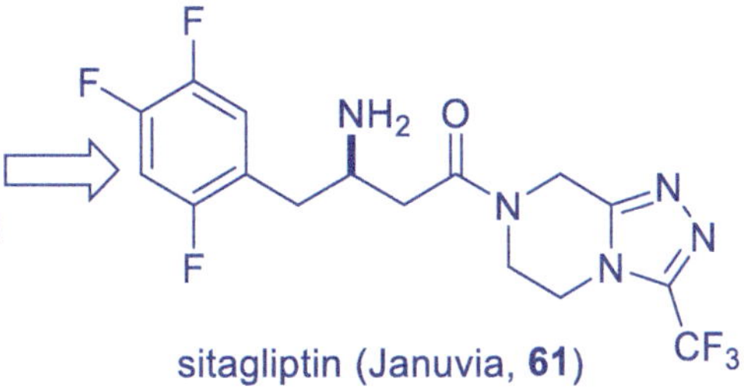

sitagliptin (Januvia, **61**)
DPP-4 IC_{50} = 18 nM,
Rat bioavailability = 76%

A review was published to argue for the merits of HTS in 2011.[19]

2.4 Fragment-based Lead Discovery

In comparison to HTS that looks for hits by screening compounds randomly, structure-based drug design (SBDD) is more rational. Much has been written about SBDD in the literature. Here we move on to fragment-based lead discovery, or fragment-based drug discovery (FBDD).

During the last decade, FBDD has firmly been established as a productive approach to find viable fragment leads. At least two marketed drugs owe their genesis to FBDD: Plexxikon's vemurafenib (Zelboraf, **66**) and Abbvie's venetoclax (Venclexta, **73**).

Unlike HTS library collections, fragment libraries are normally not large, numbering from dozens to thousands. As the name fragment implies, they are small molecules with less than 20 heavy atoms (molecular weights 100–250). They generally have low number of functional groups and are just big enough to have enough interactions with targets, yet small enough to minimize unfavorable interactions. They are exposed to the targets at high concentrations (μM to mM) so to detect even weak interactions. Highly sensitive screening approach is required to detect fragment binding. NMR and SPR are frequently used to detect direct binding. Protein–ligand X-ray crystallography by soaking has found many applications, too. Since these biophysical methods only detect binding, it is useful to have an orthogonal assay to remove non-specific binders.

Once fragments are found, optimization is carried out through careful structure-based growth or through linking different fragments to grow affinity. There are several strategies to "grow" fragments. Fragment-linking or tethering is a popular method. Since they bind to proximal part of the active site, the two fragments joined by a linker with appropriate length will give a larger and higher affinity binding molecule, although identifying suitable linkers is not trivial and may take many reiterations. Fragment evolution is another popular approach where an initial fragment is optimized by adding functionality to bind adjacent regions of the active site.

The rules for *fragments* in the context of FBDD are somewhat different from the rule of five (Ro5) for drugs. Congreve et al. at Astex proposed a "rule of 3" (Ro3) in 2003. It is also known as "Astex Rule of Three" (Table 2.1).[20]

Table 2.1 Ro3 and Ro5.

Variable	Rule of 3	Rule of 5
Clog *P*	≤ 3	≤ 5
No. of N	yes	yes
No. of O	<9	<10
H-bond donors	≤ 3	≤ 5
MW	≤ 300	≤ 500
No. rotatable bonds	≤ 3	≤ 10
Polar surface area (PSA, Å)	≤ 60	≤ 140

The caveat of FBDD is that the fragment hits will be weak compared to traditional screening hits. Therefore, they need to be tested at high concentrations, which bring the solubility issue. High concentrations also exacerbate the colloidal aggregates. Even low-level impurities (such as Zn, EDTA) can cause problems at high concentrations.

An excellent book published in 2016 collects all aspects of FBDD.[21] Here are the two marketed drugs originated from FBDD approach.

Plexxikon's vemurafenib (Zelboraf, **66**) was the first marketed drug discovered employing the FBDD (or scaffold-based drug design) strategy under the guidance of co-crystallography. No sooner than the BRAF V600E mutant allele as a cancer target became known in 2002, Plexxikon began pursuing this target because $BRAF^{V600E}$ is the most frequent oncogenic protein kinase mutation known and exists only in tumors that are dependent on the well-known RAF/MEK/ERK pathway. A library of 20,000 fragment compounds with molecular weights ranging from 150 to 350 (fewer than eight hydrogen bond donors and acceptors and few rotatable bonds) was screened at a concentration of 200 μM. One of the 238 high throughput screening (HTS) hits, 7-azaindole (**62**), bound to the ATP site, co-crystallized with a kinase called proviral integration site of moloney murine leukemia virus-1 (PIM1) enzyme. Meanwhile, 3-anilinyl-7-azaindole **63** also co-crystallized with PIM1 with an IC_{50} value of approximately 100 μM for PIM1. The 7-azaindole scaffold **63** represented a general framework capable of presenting two hydrogen bonding interactions with the kinase hinge region. Minor variations afforded benzyl-7-azaindole **64**, which co-crystallized with another kinase, fibroblast growth factor receptor-1 (FGFR1), with an IC_{50} value of 1.9 μM for FGFR1. Structure–activity relationship (SAR) investigations led to PLX4720 (**5**),[22] which was a potent and selective (including wide-type B-Raf and many other kinases) $BRAF^{V600E}$ inhibitor with an IC_{50} value of 13 nM. Installation of a chlorophenyl fragment to replace the 5-chlorine atom on the 7-azaindole core of **65** led to vemurafenib (**66**),[23] which displayed similar potency for BRAF (31 nM) and c-RAF-1 (48 nM) and selectivity against other kinases, including wide-type B-Raf (100 nM). It was chosen for development over **65** because its pharmacokinetic properties scaled more favorably in beagle dogs and cynomolgus monkeys. The FDA approved vemurafenib (Zelboraf, **66**) for the treatment of BRAF-mutant metastatic melanoma in 2011.[24]

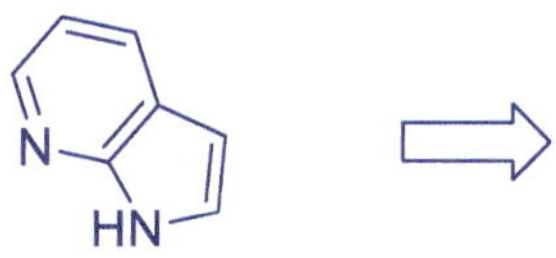

7-azaindole (**62**)
crystallized with PIM1
PIM1, IC_{50} > 200 μM, LE < 0.56

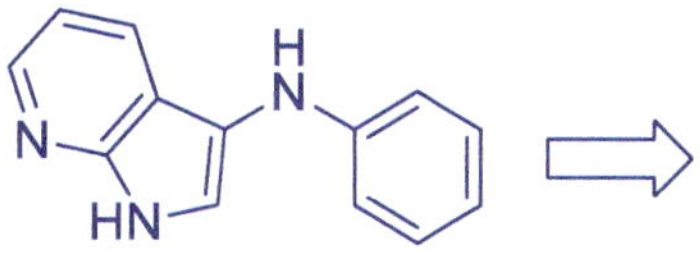

anilinyl-7-azaindole **63**
crystallized with PIM1
PIM1, IC_{50}, ~100 μM, LE = 0.34

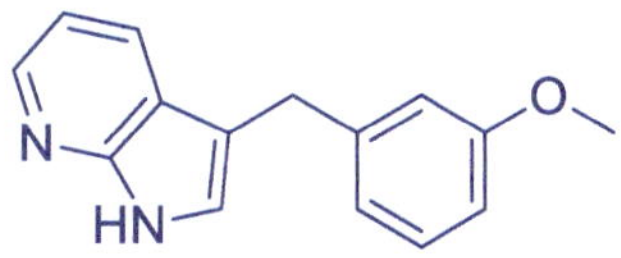

benzyl-7-azaindole **64**
crystallized with FGFR1
FGFR1, IC_{50}, 1.9 μM, LE = 0.43

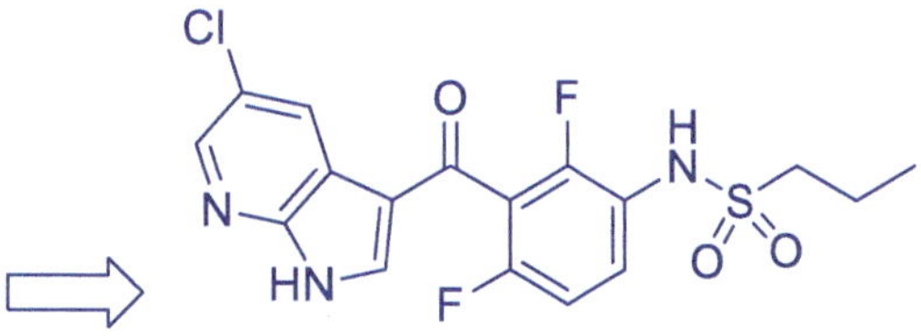

PLX4720 (**65**)
$BRAF^{V660E}$, IC_{50}, 13 nM, LE = 0.40

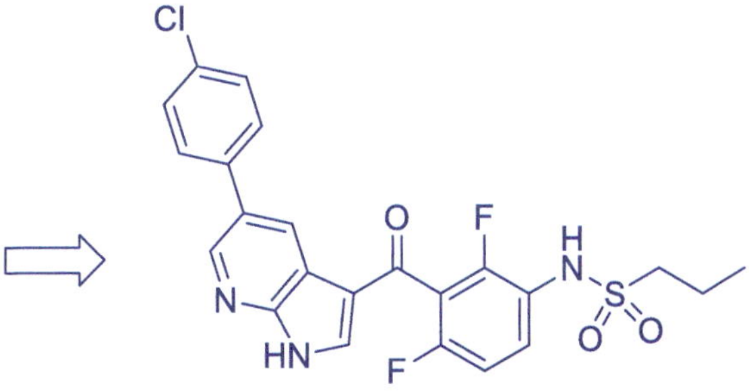

vemurafenib (Zelboraf, **66**)
$BRAF^{V660E}$, IC_{50}, 31 nM, LE = 0.31

Abbvie's venetoclax (Venclexta, **73**) is a B-cell lymphoma 2 (Bcl-2) inhibitor discovered by employing the FBDD strategy. Instead of the co-crystallography tactic, "SAR by NMR" method was key to generate their fragment hits. From initial screening a 10,000 compound library with MW <215 at 1 mM concentration, *p*-fluorophenyl-benzoic acid (**67**, LE = ligand efficiency) emerged as one of the first-site (P1) ligands. Later on, screening a 3,500 compound library with MW ~150 at 5 mM concentration identified

5,6,7,8-tetrahydro-naphthalen-1-ol (**68**) as the second site (P2) ligand.[25] The choice of using acyl-sulfonamide **69** as the carboxylic acid isostere enabled elongation of right-hand portion to provide with **70**, which now occupied third binding site (P3). Regrettably, **70**, experienced serum deactivation from domain II of human serum albumin (HSA-III) binding. Careful structure-based optimization resulted in ABT-737 (**71**), which had more polar amines and had reduced protein binding.

Replacing one of the phenyl rings with a cyclohexene structure boosted oral bioavailability and subsequent removal of the nitrophenyl structural alert led to the discovery of navitoclax (**72**) as a potent and orally bioavailable Bcl-2 inhibitor, although it was not selective against Bcl-x_L. Thus, navitoclax (**72**) was inflicted by dose-limiting platelet depletion side effect.[26] Eventually, the fourth binding site (P4) was replaced with a 7-azaindole ether and its N atom captured an additional hydrogen bond with Arg104 on the target, giving rise to venetoclax (Venclexta, **73**) as a potent, selective (against Bcl-x_L, Bcl-w, and Bcl-1), and orally bioavailable Bcl-2 inhibitor. In 2016, it was approved by the FDA for treating chronic lymphocytic leukemia (CLL) with the 17p deletion.[27,28]

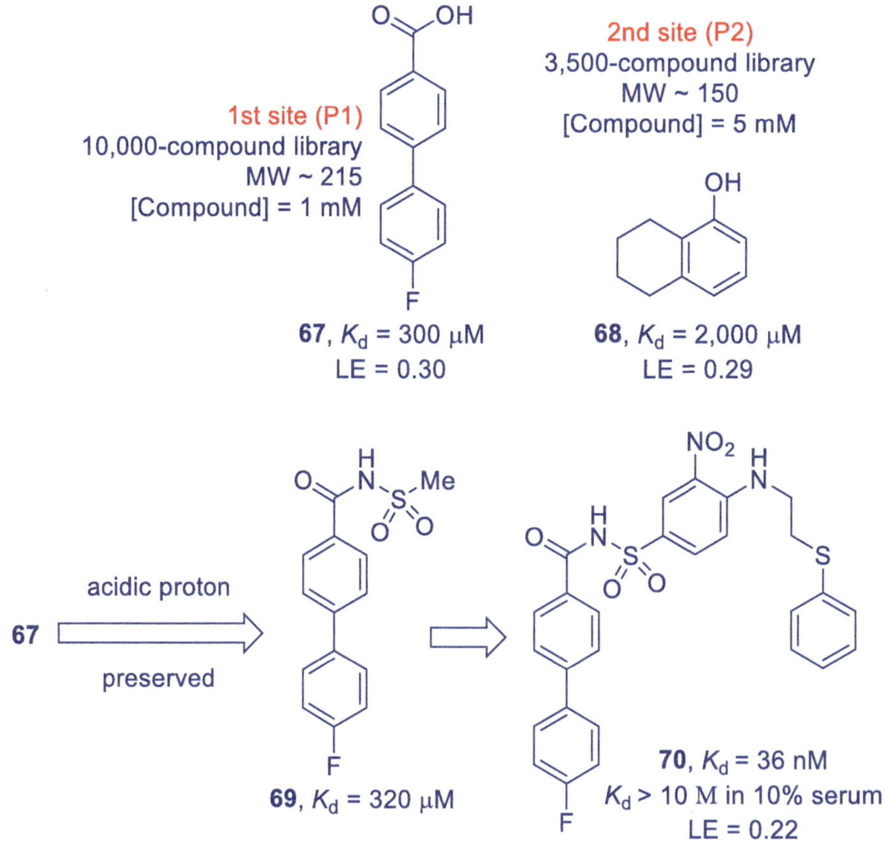

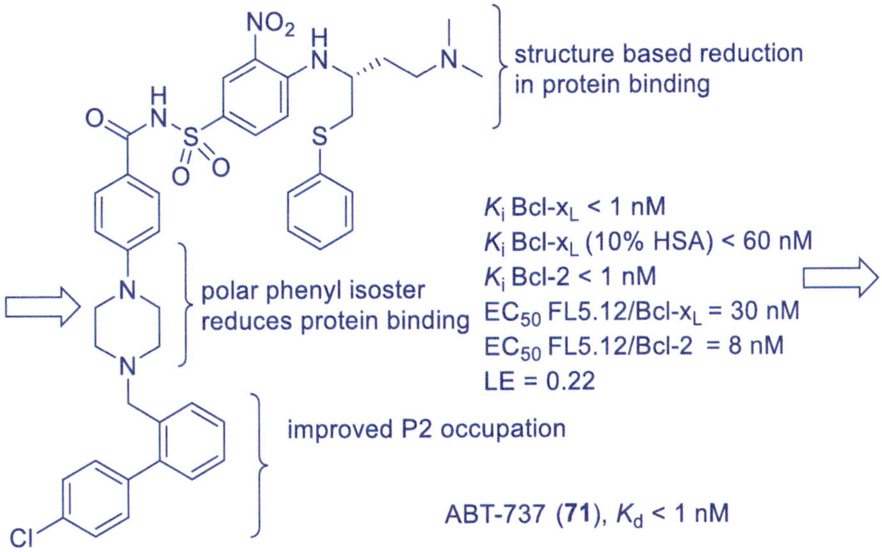

ABT-737 (**71**), K_d < 1 nM

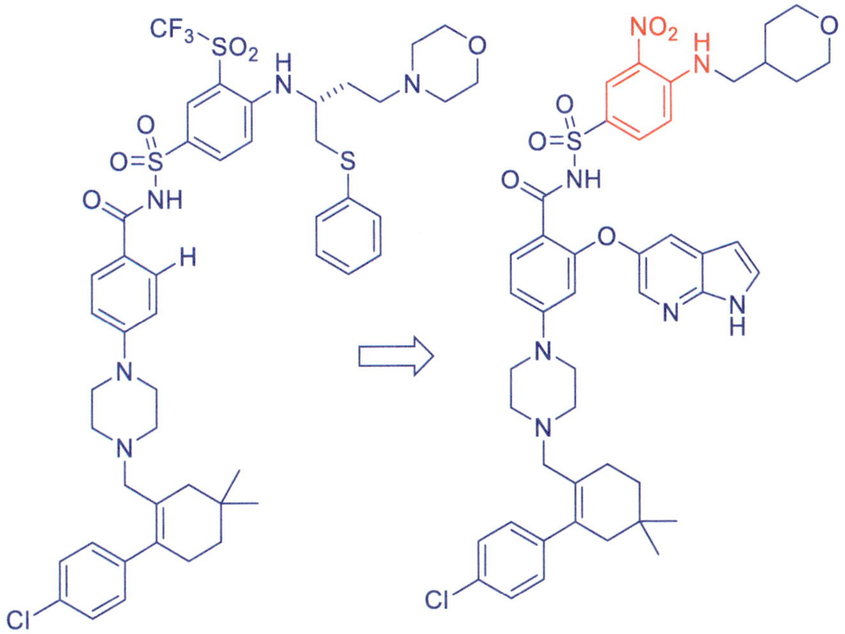

navitoclax (**72**)
K_i = 0.04 nM, F ~ 30%, LE > 0.20

venetoclax (Venclexta, **73**)
K_i = 0.01 nM, F ~ 29%, LE > 0.25

2.5 DNA-encoded Library (DEL)

DNA-encoded libraries (DELs) are libraries of small molecules tagged with unique identifier DNA sequences which can be efficiently screened against biological targets. Amplifying and sequencing the DNA identifier tag for a bound compound effectively identifies the compound. As DEL pioneers Sydney Brenner and Richard Lerner explained: "By coupling genetics and the versatility of organic chemical synthesis, we have extended the range of analysis to chemicals that are not themselves part of biological systems."[29]

DEL is a powerful tool to generate large number of compound libraries by sequentially recording chemistry information using unique DNA sequences. Simple split-pool mix strategy can produce more than 100 million compounds per library. It uses a small fraction of the amount of target protein (~0.3 nmol) used for a typical HTS. Since the requirement is of each attomoles of each mole per screen, it is relatively inexpensive to setup. DEL is considered as "poor man's HTS."

Practical operation for generating a DEL library follows a general sequence shown below:[30]

1. Tag chemical building blocks with short DNA sequence tags.
2. React tagged building blocks with each other using split-and-pool synthesis to yield compounds with longer unique DNA tags. It has been observed that libraries limited to 2–3 cycles of chemistry provide an optimum balance between structural diversity, synthetic yield, and ligand molecular properties.
3. Screen tagged compounds as mixtures against immobilized protein target using affinity chromatography.
4. Wash/elute to retain only bound compounds.
5. Amplify and sequence DNA tag to identify enriched binders.
6. Confirm binding through re-synthesis of DNA-free compounds.

DEL approach has been successful, especially in the kinase field. By 2018, two clinical candidates originated from DEL hits. Both projects took place at GSK.

Receptor interacting protein-1 (RIP1) kinase plays an important role in tumor necrosis factor (TNF)-mediated inflammation. GSK's initial HTS of seven million compounds failed to deliver drug-like compounds. Screening three-cycle amino acid core DEL library with 7.7 billion compounds yielded a remarkably potent and selective unoptimized 1.6 nM hit GSK'481 (**74**). It was prepared using two building blocks (BBs) joined by an amide bond. The benzoxazepinone core structure is unique and atypical of the purported chemotype of the traditional kinase space. For an unoptimized hit, GSK'481 (**74**) already exhibited good oral systemic exposure in rat. Limited optimization yielded GSK2982772 (**75**) with good potency, selectivity (against >450 off-target kinases), good exposure, and good PK profile (85% bioavailability in cynomolgus). GSK2982772 (**75**) was advanced to PIIa trials in 2017 for treating inflammatory disorders such as ulcerative psoriasis.[31]

biochemical
IC_{50} = 1.6 nM
cellular
IC_{50} = 10 nM

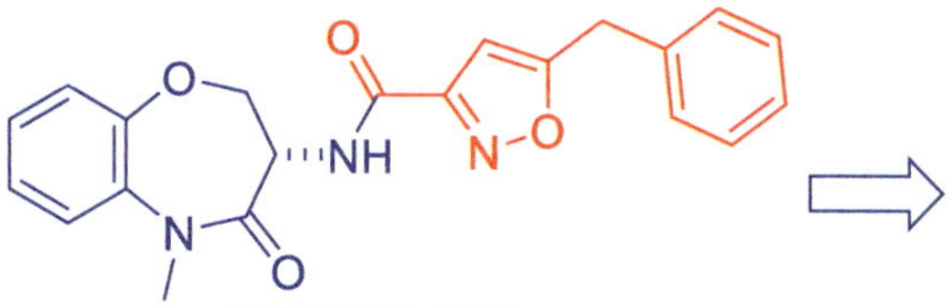

GSK'481 (**74**)

biochemical
IC_{50} = 1.0 nM
cellular
IC_{50} = 6.3 nM

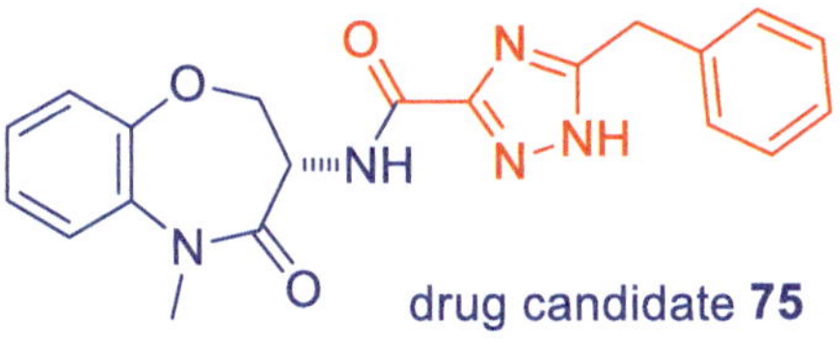

drug candidate **75**

Different from the majority of kinase targets interrogated by DEL technique, soluble epoxide hydrolase (sHE) is, as the name implies, a hydrolase. It converts lipid epoxides to their corresponding diols and is a target for cardioprotection and inflammation such as chronic obstructive pulmonary disease (COPD). Previous HTS and FBDD campaigns at GSK failed to deliver any tractable hits. From 100 million compounds, their initial DEL off-DNA hit triazine **76** was assembled using three building blocks (BBs). It was already reasonably potent (IC_{50}, 40 nM). Subsequent hit-to-lead (H2L) efforts arrived at piperazine **77** with enhanced potency and superior molecular properties. Further optimization on "developability" parameters such as aqueous solubility and oral bioavailability led to clinical candidate **78**. It has an excellent selectivity against other targets, displaying a relatively high free fraction in human blood (12.8%) and good oral bioavailability in rat and dog (94 and 100%, respectively). In Phase I clinical trials, it demonstrated target inhibition in obese smokers with no serious adverse events.[32,33]

DEL hit **76**
MW = 517
IC_{50} 40 nM
pIC_{50} = 8.1
LE = 0.29, BEI = 16
Clog *P* = 5.4

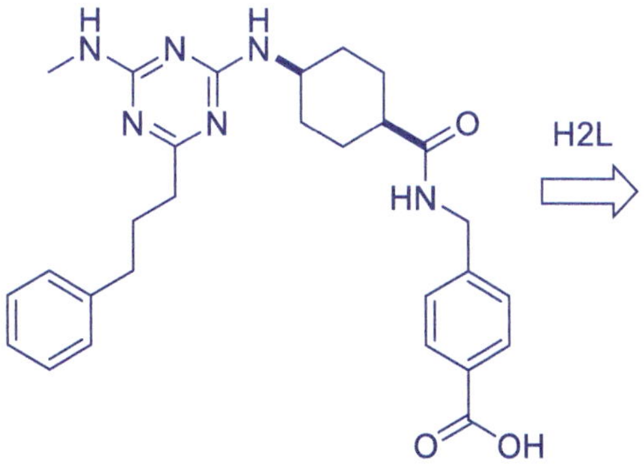

H2L compound **77**
MW = 506
pIC_{50} = 8.5
LE = 0.32, BEI = 17
Clog P = 4.5

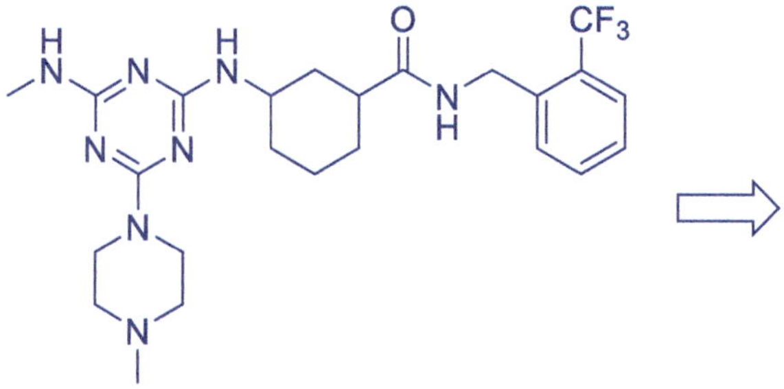

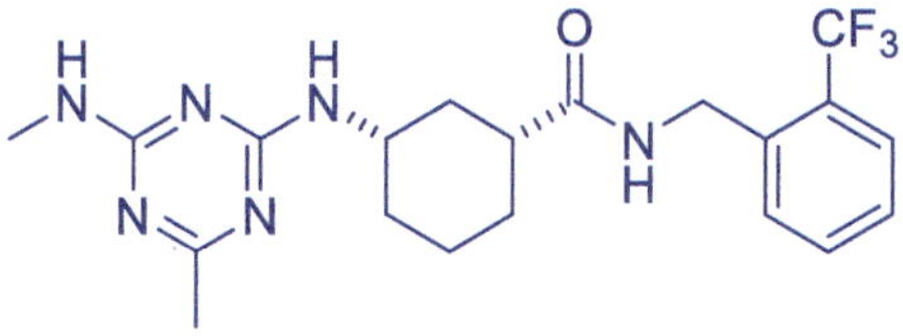

GSK2256294 (**78**)
MW = 422
sEH, IC_{50} = 27 pM
cell, IC_{50} = 6.83 nM

2.6 PROTAC

PROTAC stands for proteolysis targeting chimera which functions as a protein degrader. It uses hetero-bifunctional small molecules to remove specific proteins from cells to achieve targeted protein degradation.

Protein synthesis and degradation are highly regulated cellular processes that are essential for normal cell division and cell survival. Many of the cellular processes underlying carcinogenesis and cancer progression are due to an imbalance in proteins that control cell division (e.g., cyclins), apoptosis (e.g., pro-apoptotic protein Bax), tumor suppression (e.g., p53), and stress response (e.g., NF-κB). In the normal healthy cell, the majority of intracellular proteins that require degradation, due to damage, misfolding, or transient signaling molecules, are labeled through polyubiquitination, targeting them for proteolysis within the multicatalytic 26S proteasome. On the other hand, the 26S proteasome is a cylindrical structure comprised of a 20S catalytic core with caspase-like, trypsin-like, and chymotrypsin-like activities. The catalytic core is capped by two 19S regulatory subunits that are involved in directing the entry of polyubiquitin-tagged proteins into the enzyme complex.

We are not strangers to targeted protein degradation and encountered a protein degrader in Chapter 1 with fulvestrant (Faslodex, **79**), the first-in-class selective estrogen receptor degrader (SERD) approved by the FDA in 2002 used in breast cancer therapy.[34] The next generation SERD, GDC-08010 (brilanestrant, **80**), is an orally bioavailable drug being evaluated in clinical trials to treat breast cancer patients resistant to standard endocrine treatments.[35]

Other known protein degraders including thalidomide (**81**), lenalidomide (**82**), and pomalidomide (**83**) are immunomodulatory drugs (IMiDs). They bind to cereblon (CRBN), a substrate receptor of cullin-4 RING E3 ligase (CRL) complex, which results in polyubiquitination and degradation of transcription factors (TFs) such as Ikaros (IKZF1) and Aiolos (IKZF3).[36] Crystallographic studies now establish that IMiDs bind CRBN to form a cryptic interface that promotes recruitment of IKZF1 and IKZF3.

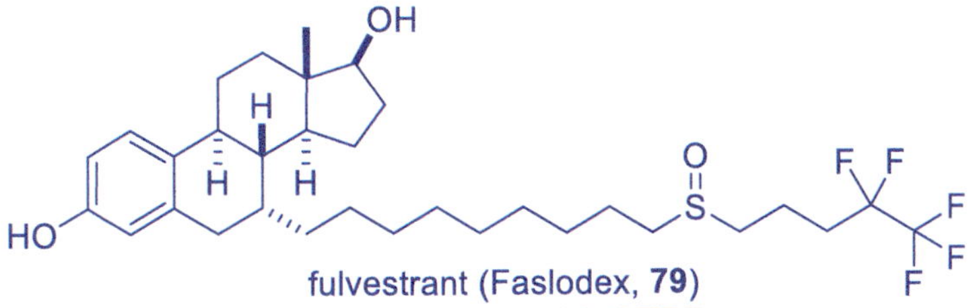

fulvestrant (Faslodex, **79**)
AstraZeneca, 2002, SERD

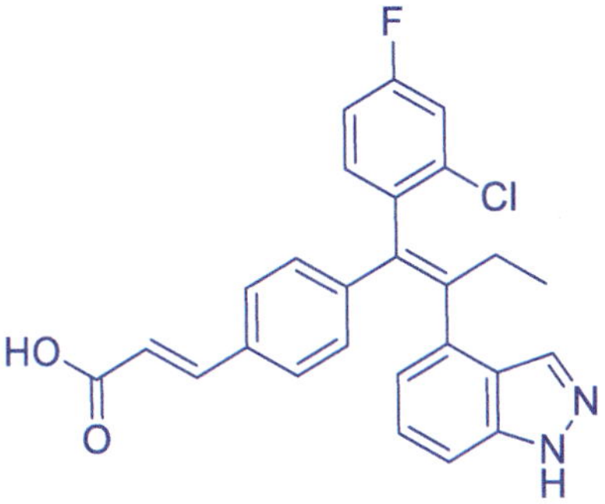

GDC-08010
(brilanestrant, **80**)
SERD

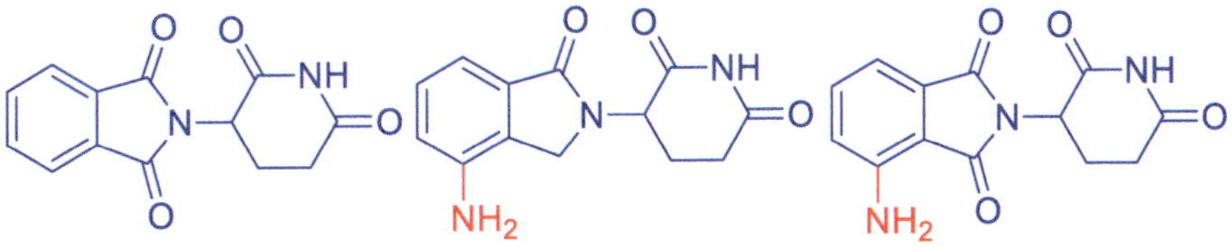

thalidomide (**81**) lenalidomide (**82**) pomalidomide (**83**)

PROTAC takes advantage of proteome's ability to degrade proteins by linking protein of interest (POI) with an E3 ligase recognition domain. So PROTAC consist of a ligand for the POI, a flexible linker, and E3 ubiquitin ligase ligand. We can consider PROTAC as a "glue" to facilitate ternary complex formation of the POI with an E3 ligase. Once that accomplished, ubiquitin can be transferred from E2 to the target protein, which is eventually degraded by 26S proteasome. PROTACs appear to be highly modular, enabling degradation of different targets by different ubiquitin ligases through simple ligand exchange. They operate independently of protein–protein interactions (PPIs).

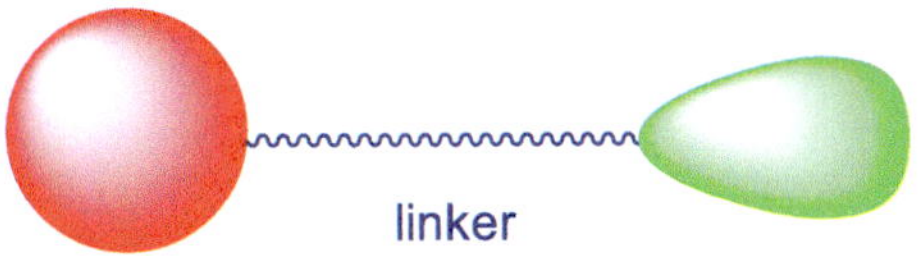

There are several advantages of PROTAC over reversible or covalent inhibition. PROTAC could achieve pico-molar activity from modestly active ligands because the cooperative complex formation and the catalytic nature of the bispecific molecules. Since there are hundreds of proteins of interest and human genome encodes more than 600 E3 ligases,[37] PROTAC offers seemingly limitless targets for drug discovery. According to Churcher, there are at least five advantages for PROTAC over traditional small molecule antagonists:[38]

1. High cellular potency driven by the catalytic MOA,
2. Highly selective degradation,
3. Wide applicability across cells and *in vivo* systems,
4. Potential for extended pharmacodynamic duration of action, and
5. The opportunity to mediate novel pharmacology.

The first PROTAC was described by Deshaies and Crews in 2001 concerning cellular degradation of the aminopeptidase MetAP2 using a hybrid of ovalicin, a small covalent inhibitor of MetAP2, linking an IκBα phosphotide epitope known to bind ubiquitin E3 ligase SCF$^{\beta TrCP}$. But ovalicin is a covalent inhibitor and the ligase ligand is a peptide with six glycine residues. It suffers from low potency (μM), poor cellular permeability, and metabolically unstablity.[39]

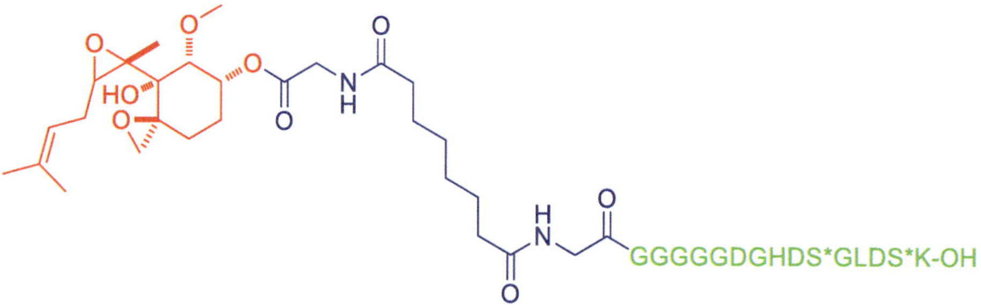

first PROTAC **84**, astericks indicate phosphorylated residues

In 2008, efforts to search for non-peptide drug-like E3 ligase binding moieties led to the discovery of nutlin, a binder to the E3 ligase murine double minute 2 (MDM2), a 90 kDa protein whose natural substrate is p53. Crews et al. chose to recruit the androgen receptor (AR) via the PROTACs to the E3 ligase MDM2. PROTAC **85** has pharmacophore of selective androgen receptor modulator (SARM) bicalutamide (Casodex) on the left and nutlin as the E3 ligase ligand at the right. It led to partial degradation of AR within the cells at a concentration of 10 μM and the degradation was proven to be proteasomally dependent.[40]

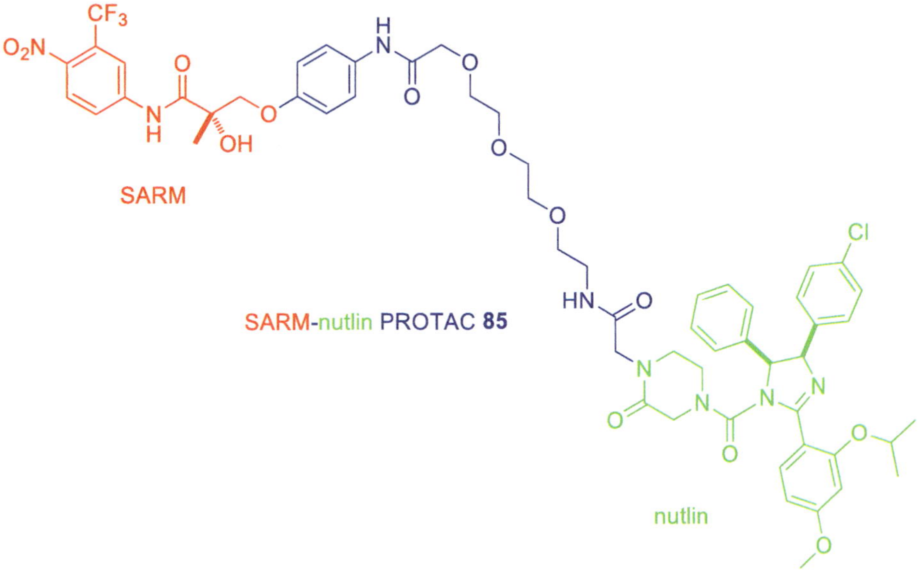

SARM-nutlin PROTAC **85**

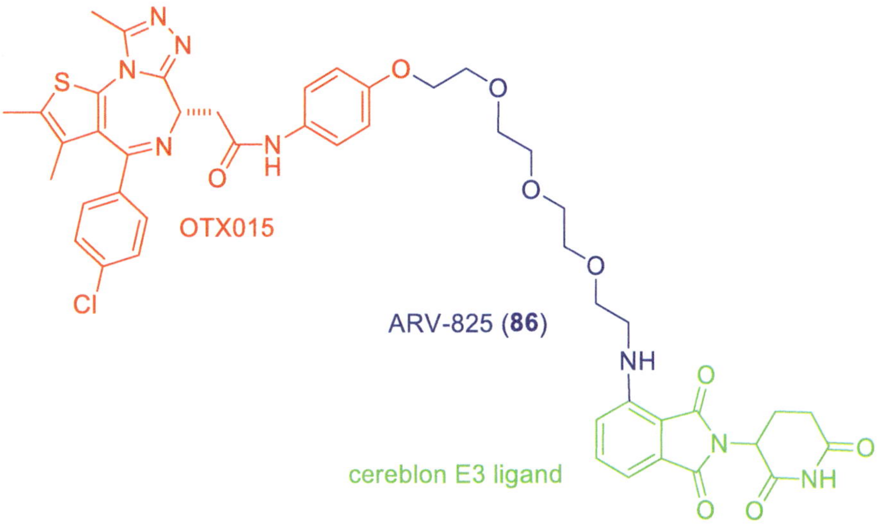

ARV-825 (**86**)

The bromodomain and extraterminal (BET) proteins include the ubiquitously expressed BRD2, BRD3, and BRD4. They recruit transcriptional regulatory complexes to acetylated chromatin, thereby controlling specific networks of genes involved in cellular proliferation and cell cycle progression. They play important roles in translational regulation, epigenetics, and cancer. Arvinas scientists and Crews' group succeeded in hijacking the E3 ubiquitin ligase CRBN to effectively target BRD4. Tethering a pan-BET selective bromodomain inhibitor OTX015 with thalidomide-based CRBN ligand, they arrived at ARV-825 (**86**). It was shown to recruit BRD4 to the E3 ubiquitin ligase CRBN, leading to fast, efficient, and prolonged degradation of BRD4 in all Burkitt's lymphoma (BL) cell lines tested. Therefore, ARV-825 (**86**) more effectively suppressed c-MYC levels and downstream signaling than small molecule BRD4 inhibitors OTX015 and JQ1, resulting in more effective cell proliferation inhibition and apoptosis induction in BL.[41]

Back in 2014, guided by X-ray crystal structures, Ciulli's group identified the von Hippel–Lindau (VHL) ligand **87**. It is a nanomolar ligand for the PPI between the VHL E3 ubiquitin ligase and the hypoxia inducible factor-alpha (HIF-1α). It is of the best-in-class ligand for pVHL's LHS2 region. It has a good affinity for CRL2^{VHL} (CRL = cullin-RING ligase) with a K_d value of 185 nM. VHL ligand **87** has found widespread applications since then. Apparently, the *t*-butyl group is the best substituent to offer the best affinity.[42]

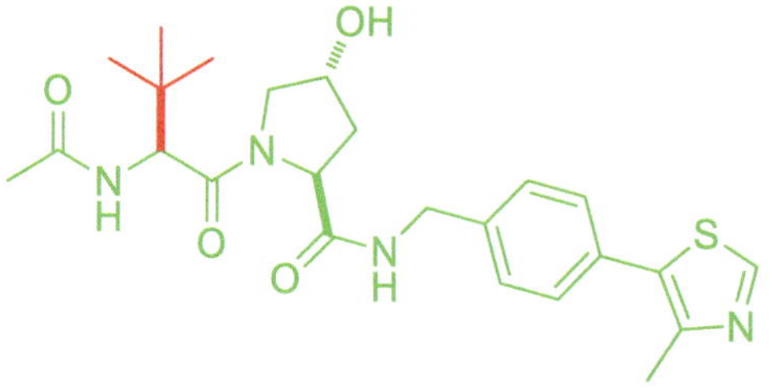

VHL ligand **87**
K_d = 185 nM
LE = 0.28

In 2015, Ciulli's group coupled the pan-BET selective bromodomain inhibitor JQ1 and a VHL E3 ubiquitin ligase ligand similar to **87** and prepared several PROTACs including MZ1 (**88**). Compound MZ1 (**88**) potently and rapidly induced reversible, long-lasting, and unexpectedly selective removal of BRD4 over BRD2 and BRD3, although JQ1 itself is not selective against BRD2–4. The activity of MZ1 (**88**) is dependent on binding to VHL but is achieved at a sufficiently low concentration not to induce stabilization of HIF-1α. It showed degradation of the BET family of epigenetic bromodomain-containing proteins that is consistent with selective targeting of BRD4.[43]

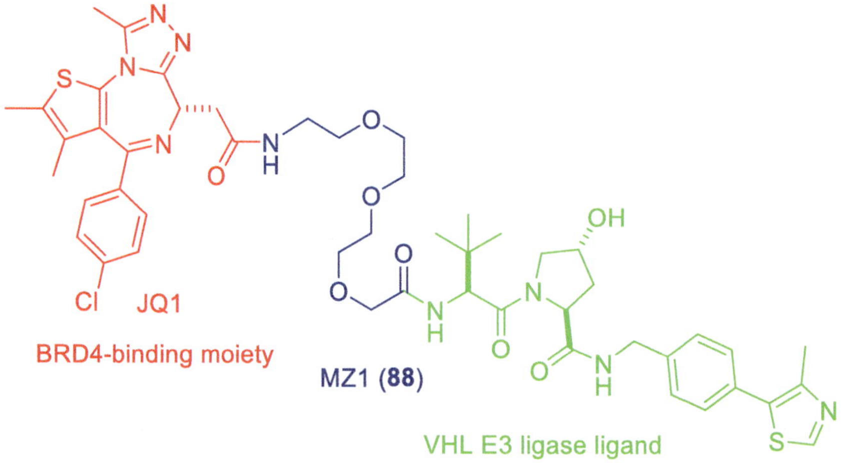

MZ1 (**88**)

In 2016, Arvinas and Crews described their PROTAC-induced pan-BET protein degrader ARV-771 (**89**). Along with a shorter linker, its BRD4-binding moiety was JQ1, it employed a VHL-binding moiety with an extra (*S*)-methyl substituent. In cellular levels, ARV-771 (**89**) was highly active, achieving a DC_{50} (the drug concentration that results in 50% protein degradation) values <1 nM. Superior to BET inhibitors such as JQ1 and OTX015, subcutaneous delivery of ARV-771 (**89**) resulted in suppression of both AR signaling and AR levels and led to tumor regression in a castration-resistant prostate cancer (CRPC) mouse xenograft model. This study was the first to demonstrate efficacy with a small-molecule BET degrader in a solid-tumor malignancy and potentially represented an important therapeutic advance in the treatment of CRPC. Interesting, the E3 ligase dependence of the degradation was confirmed using a VHL-nonbinding diastereomeric control ARV-766, a diastereomer of ARV-771 (**89**) with configuration inversion at the *t*-butyl group was completely inactive. As expected, ARV-766 was unable to recruit the ligase function and gave no degradation.[44]

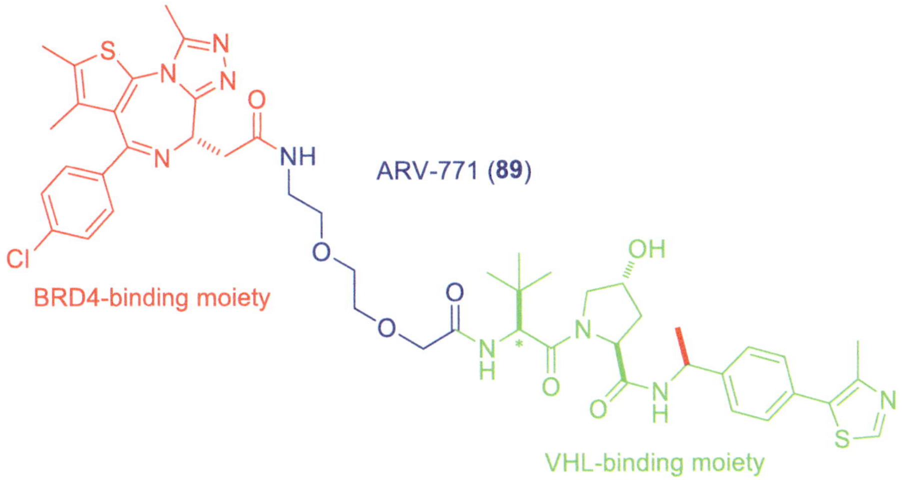

ARV-771 (**89**)

In 2015, Bradner's group reported their discovery of dBET1 (**90**) after having shown that the carboxyl group on JQ1 and the aryl ring of thalidomide can tolerate chemical substitution, therefore good egress positions. They capitalized on the fact that IMiDs are E3 $CRL4^{CRBN}$ ligands and created PROTACs containing bromodomain ligands such as JQ1 attached to an IMiD. dBET1 (**90**) resulted in rapid, efficient, and highly specific degradation of BRD2, BRD3, and BRD4 in cultured cells, although JQ1 is a ligand with no intrinsic binding preference. It showed increased apopototic response of primary acute myeloid leukemia (AML) cells in comparison to efficacious tool compound JQ1. Regrettably, because of poor pharmacokinetics (PK), the drug's impressive antitumor activity required daily injection of compound (50 mg/kg body weight daily, ip, intraperitoneal) that has a terminal half-life of only 40 min.[45]

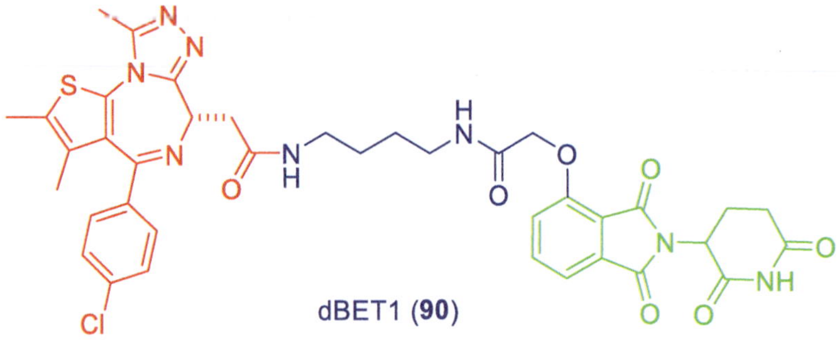

dBET1 (**90**)

In addition, they also demonstrated that PROTACs [dFKBP-1 and dFKBP-2 (**91**)] based on an FK506 binding protein (FKBP12) ligand can be used to degrade FKBP fusion proteins, which may prove to be extremely useful for controllable elimination of specific proteins in engineered cell lines or animals. Both dFKBP-1 and dFKBP-2 (**91**) may also serve as useful tool compounds in control of fusion protein stability.[45]

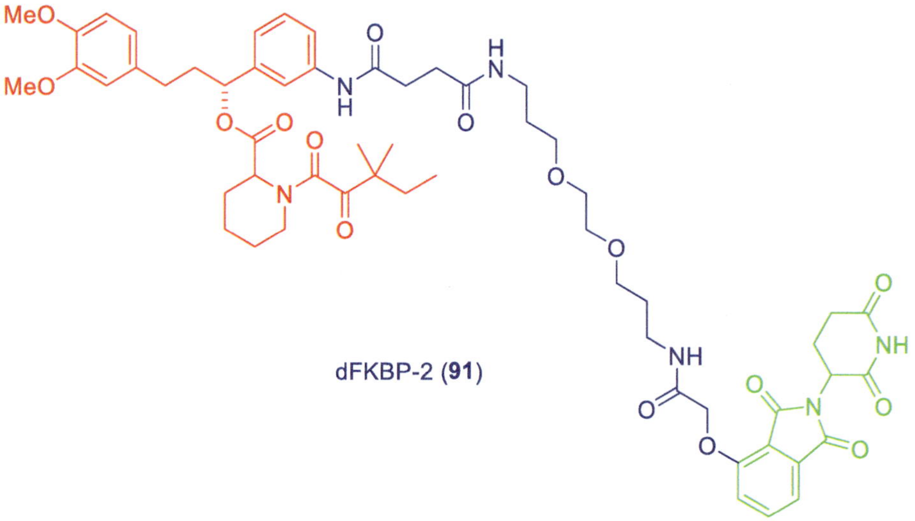

dFKBP-2 (**91**)

Year 2015 was truly a banner year for PROTACs. Crews' group reported their major advances in PROTACs. They employed the newly discovered high-affinity, small molecule ligand for VHL, which retained the hydroxyproline moiety and introduced a *t*-butyl group. The VHL ligand had a K_d of 320 nM. On the other hand, estrogen-related receptor-α (EERα) is an orphan nuclear hormone receptor implicated as a master regulator of several biological processes. Connecting the VHL ligand with an EERα selective (over other ERR isoforms) inhibitor, PROTAC **92** was assembled. It had a DC_{50} of ~100 nM and a D_{max} (maximal level of degradation) of 86%. Again, the epimer (inversion of configuration point asterisked) was inactive, confirming the role of VHL ligand in ubiquitination.[46]

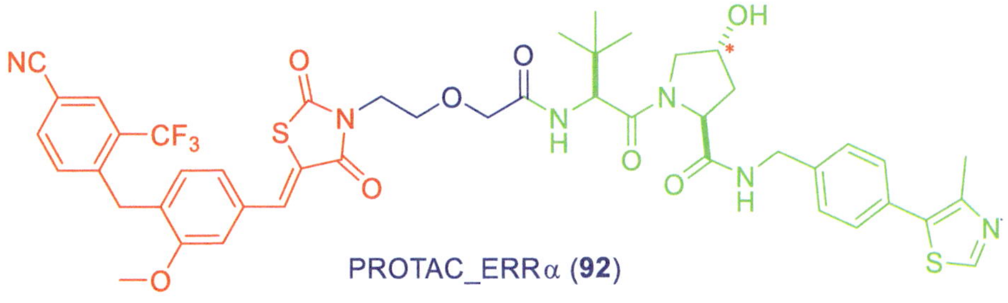

PROTAC_ERRα (**92**)

Another RPTOAC, compound **93**, linking an inhibitor of the serine-threonine kinase RIPK2 with the VHL ligand, had a 12-atom linker, which was tested to be optimal. It had a D_{max} >95% at concentrations >10 nM and a DC_{50} of 1.4 nM. By pretreating the cells with the proteasome inhibitor epoxomicin, they confirmed the proteasomal dependence of the degradation. Both **92** and **93** mediated catalytic ubiquitination and were highly specific for their targets. More encouragingly, both of them were efficient in *in vivo* knockdown in mice.[46]

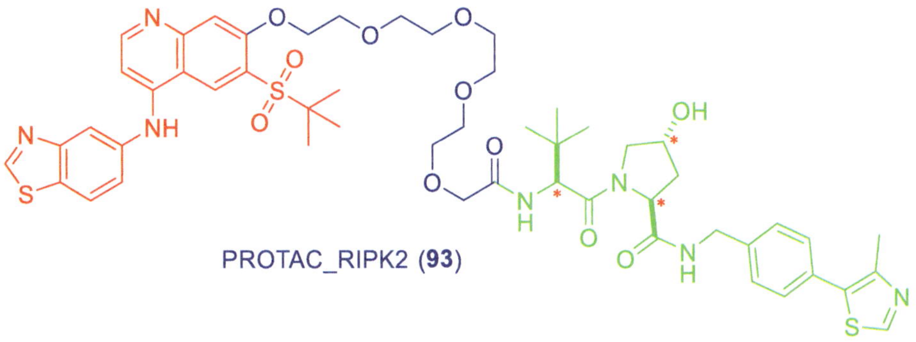

PROTAC_RIPK2 (**93**)

In 2016, Crews' group delved into the degradation of oncogenic Bcr-Abl employing modular PROTAC design. Surprisingly, despite many linkers attempted and regardless either Cereblon or Von Hippel Lindau E3 ligase ligands used, imatinib-containing PROTACs lost affinity for the phosphorylated and nonphosphorylated form of Abl compare to the parent compound imatinib. In the same vein, bosutinib (Bosulif, **94**)-VHL did not induce degradation of Bcr-Abl or c-Abl. Thankfully, dasatinib (Sprycel, **95**)-VHL induced a clear (>65%) decrease of c-Abl at 1 μM PROTAC concentration. Therefore, independent of simple target binding, the inhibitor warhead [imatinib, bosutinib (**94**), or dasatinib (**95**)] largely determines the capability of a PROTAC to induce c-ABL degradation.[47]

Switching E3 ligase ligand from VHL to CRBN, dasatinib (**95**)–CRBN PROTAC not only induced degradation of c-Abl (>85% at 1 μM) but also induced Bcr-Abl degradation. By optimizing the linker, they achieved an EC_{50} of 4.4 nM for dasatinib (**95**)-6-2-2-6-CRBN PROTAC. Therefore, the capacity of a PROTAC to induce degradation involves more than just target binding: the identity of the inhibitor warhead and the recruited E3 ligase largely determine the degradation profiles of the compounds;

thus, as a starting point for PROTAC development, both the target ligand and the recruited E3 ligase should be varied to rapidly generate a PROTAC with the desired degradation profile.[47]

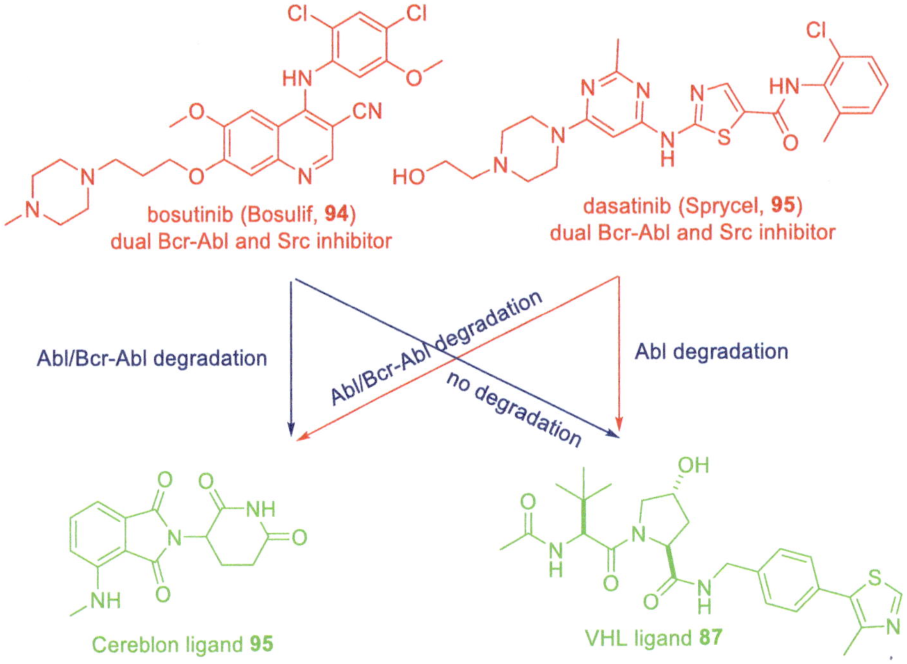

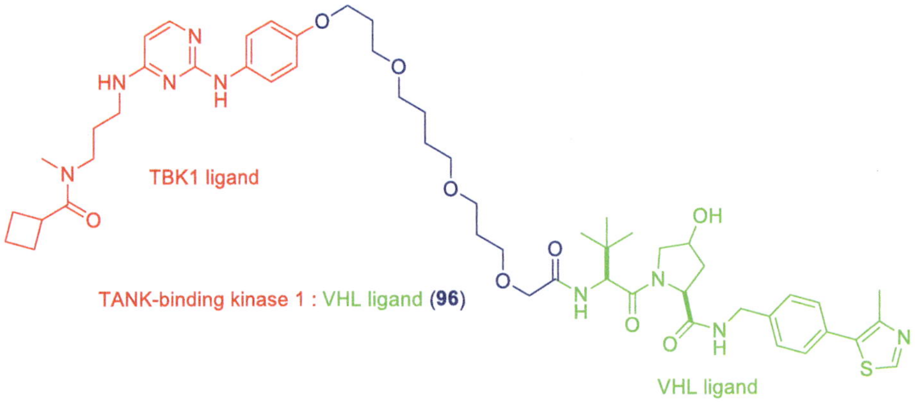

TANK-binding kinase 1 (TBK1) is a serine/threonine kinase and a noncanonical member of the IKK family implicated in diverse cellular functions, including innate immune response as well as tumorigenesis and development. By carefully scrutinizing the SAR of a series of TBK1 inhibitors, Arvinas and Crews found appropriate egress positions for linkage of the POI and VHL ligand. Among the PROTACs that they prepared, compound **96** had the optimal linker and provided greater potency and selectivity. In the paper published in 2018, they revealed that they began to optimize leads into orally active drug candidates.[48]

Also in 2018, Wang's group prepared a BET:CRBN PROTAC **97** (BETd-260) with a short linker. They took advantage of the potent and selective azacarbazole-based BET inhibitors and tethers them onto thalidomide/lenalidomide as ligand for cereblon/Cullin4A. Since their modeled structure showed that the 2-carboxamide group attached to the [6,5,6] tricyclic system in their BET inhibitor is exposed to solvent, they chose that position as their egress point to attached the linker. Among over a dozen linkers of different lengths that they extensively explored, $-(CH_2)_{4-7}NH-$ such as the one on **97** proved to be optimal. Not only did it effectively induce degradation of BRD2–4 at 30–100 pM concentration in a 3 h treatment of the RS4;11 leukemia cells, but also achieved an IC_{50} value of 51 pM in inhibition of the RS4;11 xenograft tumors with no signs of toxicity in mice.[49]

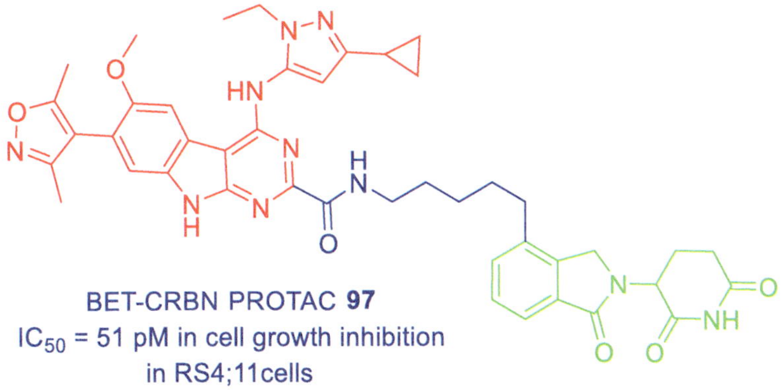

BET-CRBN PROTAC **97**
IC_{50} = 51 pM in cell growth inhibition
in RS4;11cells

The transcription factor p53 plays a pivotal role in apoptosis of cells, and its inactivation is a major contributing factor in tumorigenesis. The numerous functions of p53 are regulated by MDM2. MDM2 and MDMX proteins provide the inhibition of p53 tumor suppressor, thus allowing for accelerated mutation-driven cancer microevolution. During the last two decades, many MDM2/X-p53 inhibitors have been invented to reactivate p53 in p53^{wt} cells. Several small molecule inhibitors of the MDM2–p53 PPI have been brought forward to clinical trials with mixed results.[50]

Wang's group prepared PROTAC MD-224 (**98**) linking a MDM2–p53 inhibitor MI-1061 on the left and IMiDs as E3 ligase binders after careful linker optimization. The CRL4–CRBN E3 ubiquitin ligase on MD-224 (**98**) degraded MDM2, but co-treatment of **98** with lenalidomide (**82**), a CRBN binder, effectively blocked MDM2 degradation via competitive displacement of CRBN from the ternary complex, confirming the drug was on target. PROTAC **98** was tested as a nanomolar drug in cell and efficacious in RS4;11 xenograft animal models when given multiple IV-dosing at 25 mg/kg every second day (Q2D).[50]

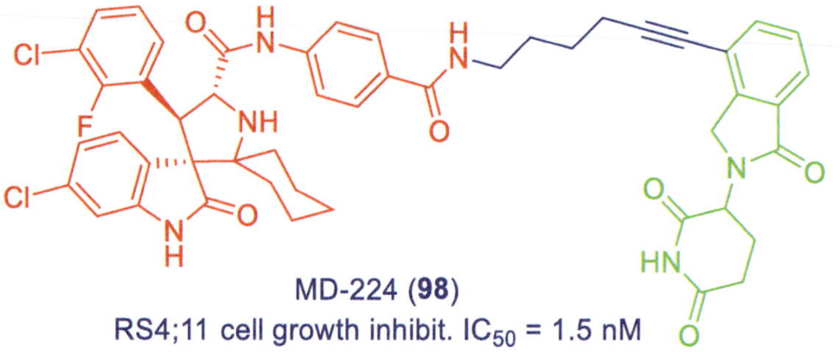

MD-224 (**98**)
RS4;11 cell growth inhibit. IC_{50} = 1.5 nM
RS4;11 xenograft = 50% regression
at 25 mpk, iv Q2D dosing

Arvinas took advantage of not only MDM2 protein as a therapeutic molecular target but also as its ubiquitinating properties to target additional proteins of therapeutic potential, androgen receptor as represented by PROTAC compound **99**, BRD4, c-Jun N-terminal kinases (JNK), enhancer of Zeste homolog 2 (EZH2), ER, or rapidly accelerated fibrosarcoma (RAF) proteins. For PROTAC A2435 (**99**), its left-hand portion is the pharmacophore of androgen receptor antagonist enzalutamide (Xtandi). Its right-hand portion is a known MDM2 inhibitor as the ubiquitinating protein.[51]

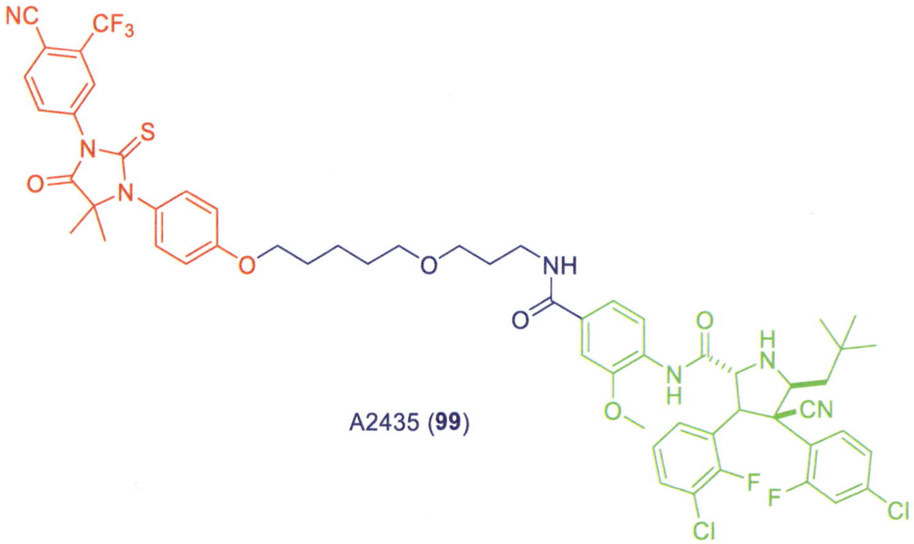

A2435 (**99**)

In March 2019, Arvinas's ARV-110, an AR degrader, entered Phase I clinical trials, which was followed by initiation of Phase I clinical trials for ARV-470, an ER degrader, in the fall. The utility of PROTACs may be revealed in the near future. Like Deshaies prophesied: "The Gold rush is on!"[45(b)]

2.7 Further Reading

Holenz, J.; ed. Lead Generations, Methods, Strategies, and Case Studies. Wiley-VCH: Weinheim, **2016**.

2.8 References

1. Cheff, D. M.; Hall, M. D. *J. Med. Chem.* **2017**, *60,* 4517–4532.
2. Ulukan, H.; Swaan, P. W. *Drugs* **2002**, *62,* 2039–2057.
3. Dosa, P. I.; Amin, E. A. *J. Med. Chem.* **2016**, *59,* 810–840.
4. Ojima, I.; Lichtenthal, B.; Lee, S.; Wang, C.; Wang, X. *Exp. Opin. Ther. Pat.* **2016**, *26,* 1–20.
5. Gustafsson D. *Semin. Vasc. Med.* **2005**, *5*, 227–234.
6. (a) Ondetti, M. A.; Williams, N. J.; Sabo, E. F.; Pluscec, J.; Weaver, E. R.; Kocy, O. *Biochemistry* **1971**, *10*, 4033–4039. (b) Cushman, D. W.; Ondetti, M. A. *Nature Med.* **1999**, *5*, 1110–1112.
7. Ganellin, C. R. Discovery of Cimetidine, Ranitidine and Other H_2-Receptor Histamine Antagonists. In Ganellin, C. R.; Roberts, S. M., eds. *Medicinal Chemistry: The Role of Organic Chemistry in Drug Research.* Academic Press: London, **1994**, pp. 228–254.
8. Link, A.; Link, B. *Pharmazie* **2002**, *31*, 486–493.
9. (a) McClellan, K.; Perry, C. M. *Drugs* **2001**, *61,* 263–283. (b) Lew, W.; Wang, M. Z.; Chen, X.; Rooney, J. F.; Kim, C. Neuraminidase Inhibitors as Anti-Influenza Agents. In De Clercq, E., ed. *Antiviral Drug Strategies*. Wiley-VCH: Weinheim, **2001**.
10. Brickner, S. J.; Hutchinson, D. K.; Barbachyn, M. R.; Manninen, P. R.; Ulanowicz, D. A.; Garmon, S. A.; Grega, K. C.; Hendges, S. K.; Toops, D. S.; Ford, C, W.; Zurenko, G. E. *J. Med. Chem.* **1996**, *39*, 673–679.
11. (a) Kupka, D.; Sibbing, D. *Exp. Opin. Drug Metab. Toxicol.* **2018**, *14*, 303–315. (b) Zetterberg, F.; Svensson, P. *Bioorg. Med. Chem. Lett.* **2016**, *26*, 2739–2754.
12. Simeonov, A.; Jadhav, A.; Thomas, C. J.; Wang, Y.; Huang, R.; Southall, N. T.; Shinn, P.; Smith, J.; Austin, C. P.; Auld, D. S.; et al. *J. Med. Chem.* **2008**, *51*, 2363–2371.
13. Genick, C. C.; Wright, S. K. *Exp. Opin. Drug Discov.* **2017**, *12*, 897–907.
14. Jhoti, H.; Rees, S.; Solari, R. *Exp. Opin. Drug Discov.* **2013**, *8*, 1449–1453.
15. Clemons, P. A. *Curr. Opin. Chem. Biol.* **2004**, *8*, 334–338.
16. Adams, J.; Merluzzi, V. J. Discovery of Nevirapine a Non-nucleoside Inhibitor of HIV-1 Reverse Transcriptase. In Adams, J.; Merluzzi, V. J., eds. *The Search for Antiviral Drugs, Case Hisotories from Concept to Clinic.* Birkhäuser: Boston, MA, **1993**, pp. 45–70.
17. Sun, C. L.; Christensen, J. G.; McMahon, G. Discovery and Development of Sunitinib (SU11248): A Multitarget Tyrosine Kinase Inhibitor of Tumor Growth, Survival, and Angiogenesis. In Li, R.; Stafford, J. A., eds. *Kinase Inhibitor Drugs.* Wiley: Hoboken, NJ.; **2009**, pp. 3–39.
18. Parmee, E. R.; SinhaRoy, R.; Xu, F.; Givand, J. C.; Rosen, L. A. Discovery and Development of DPP-4 Inhibitor Januvia (Sitagliptin). In *Case Studies in Modern*

Drug Discovery and Development. Huang, X.; Aslanian, R. G., eds.; Wiley: Hoboken, NJ.; **2012**, pp. 10–44.

19. Macarron, R.; Banks, M. N.; Bojanic, D.; Burns, D. J.; Cirovic, D. A.; Garyantes, T.; Green, D. V. S.; Hertzberg, R. P.; Janzen, W. P.; Paslay, J. W.; et al. *Nat. Rev. Drug Discov.* **2011**, *10,* 188–195.
20. (a) Congreve, M.; Carr, R.; Murray, C.; Jhoti, H. *Drug Discov. Today* **2003**, *8*, 876–877. (b) Jhoti, H.; Williams, G.; Rees, D. C.; Murray, C. W. *Nat. Rev. Drug Discov.* **2013**, *12,* 644–645.
21. Erlanson, D. A.; Jahnke, W., eds. *Fragment-based Drug Discovery.* Wiley-VCH: Weinheim, **2016**.
22. Tsai, J.; Lee, J. T.; Wang, W.; Zhang, J.; Cho, H.; Mamo, S.; Bremer, R.; Gillette, S.; Kong, J.; Haass, N. K.; et al. *Proc. Natl. Acad. Sci. USA* **2008**, *105*, 3041–3046.
23. Bollag, G.; Hirth, P.; Tsai, J.; Zhang, J.; Ibrahim, P. N.; Cho, H.; Spevak, W.; Zhang, C.; Zhang, Y.; Habets, G.; et al. *Nature* **2010**, *467*, 596–599.
24. Bollag, G.; Tsai, J.; Zhang, J.; Zhang, C.; Ibrahim, P.; Nolop, K.; Hirth, P. *Nat. Rev. Drug Discov.* **2012**, *11,* 873–886.
25. Wendt, M. D.; Shen, W.; Kunzer, A.; McClellan, W. J.; Bruncko, M.; Oost, T. K.; Ding, H.; Joseph, M. K.; Zhang, H.; Nimmer, P. M.; et al. *J. Med. Chem.* **2006**, *49*, 1165–1181.
26. Park, C.-M.; Bruncko, M.; Adickes, J.; Bauch, J.; Ding, H.; Kunzer, A.; Marsh, K. C.; Nimmer, P.; Shoemaker, A. R.; Song, X.; et al. *J. Med. Chem.* **2008**, *51*, 6902–6915.
27. Souers, A. J.; Leverson, J. D.; Boghaert, E. R.; Ackler, S. L.; Catron, N. D.; Chen, J.; Dayton, B. D.; Ding, H.; Enschede, S. H.; Fairbrother, W. J.; et al. *Nat. Med.* **2013**, *19*, 202–208.
28. Valenti, D.; Hristeva, S.; Tzalis, D.; Ottmann, C. *Eur. J. Med. Chem.* **2019**, *167*, 76–95.
29. Brenner, S.; Lenner, R. A. *Proc. Natl. Acad. Sci. USA* **1992**, *89*, 5381–5383.
30. Goodnow, R. A. Jr. (ed.) *A Handbook for DNA-Encoded Chemistry: Theory and Applications for Exploring Chemical Space and Drug Discovery.* Wiley: Hoboken, NJ, **2014**.
31. Harris, P. A.; King, B. W.; Bandyopadhyay, D.; Berger, S. B.; Campobasso, N.; Capriotti, C. A.; Cox, J. A.; Dare, L.; Dong, X.; Finger, J. N.; et al. *J. Med. Chem.* **2016**, *59*, 2163–2178.
32. Belyanskaya, S. L.; Ding, Y.; Callahan, J. F.; Lazaar, A. L.; Israel, D. I. *ChemBioChem* **2017**, *18*, 837–842.
33. Satz, A. *ACS Med. Chem. Lett.* **2018**, *9*, 408–410.
34. Deeks, E. D. *Drugs* **2018**, *78*, 131–137.
35. Lai, A.; Kahraman, M.; Govek, S.; Nagasawa, J.; Bonnefous, C.; Julien, J.; Douglas, K.; Sensintaffar, J.; Lu, N.; Lee, K.-J.; et al. *J. Med. Chem.* **2015**, *58*, 4888–4904.
36. Lu, G.; Middleton, R. E.; Sun, H.; Naniong, M.; Ott, C. J.; Mitsiades, C. S.; Wong, K.-K.; Bradner, J. E.; Kaelin, W. G., Jr. *Science* **2014**, *343*, 305–309.
37. Zhao, Y.; Sun, Y. *Curr. Pharm. Des.* **2013**, *19,* 3215–3225.
38. Churcher, I. *J. Med. Chem.* **2018**, *61*, 444–452.
39. Sakamoto, K. M.; Kim, K. B.; Kumagai, A.; Mercurio, F.; Crews, C. M.; Deshaies, R. J. *Proc. Natl. Acad. Sci. USA* **2001**, *98*, 8554–8559.

40. Schneekloth, A. R.; Pucheault, M.; Tae, H. S.; Crews, C. M. *Bioorg. Med. Chem. Lett.* **2008**, *18*, 5904–5908.
41. Lu, J.; Qian, Y.; Altieri, M.; Dong, H.; Wang, J.; Raina, K.; Hines, J.; Winkler, J. D.; Crews, A. P.; Coleman, K. G.; Crews, C. M. *Chem. Biol.* **2015**, *22*, 755–763.
42. Galdeano, C.; Gadd, M. S.; Soares, P.; Scaffidi, S.; van Molle, I.; Birced, I.; Hewitt, S.; Dias, D. M.; Ciulli, A. *J. Med. Chem.* **2014**, *57*, 8657–8663.
43. Zengerle, M.; Chan, K. H.; Ciulli, A. *ACS Chem. Biol.* **2015**, *10*, 1770–1777.
44. Raina, K.; Lu, J.; Qian, Y.; Altieri, M.; Gordon, D.; Rossi, A. M. K.; Wang, J.; Chen, X.; Dong, H.; Siu, K.; Winkler, J. D.; Crew, A. P.; Crews, C. M.; Coleman, K. G. *Proc. Natl. Acad. Sci. USA* **2016**, *113*, 7124–7129.
45. (a) Winter, G. E.; Buckley, D. L.; Paulk, J.; Roberts, J. M.; Souza, A.; Dhe-Paganon, S.; Bradner, J. E. *Science* **2015**, *348*, 1376–1381. (b) Deshaies, R. J. *Nat. Chem. Biol.* **2015**, *11*, 634–635.
46. Bondeson, D. P.; Mares, A.; Smith, I. E.; Ko, E.; Campos, S.; Miah, A. H.; Mulholland, K. E.; Routly, N.; Buckley, D. L.; Gustafson, J. L.; et al. *Nat. Chem. Biol.* **2015**, *11*, 611–617.
47. Lai, A. C.; Toure, M.; Hellerschmied, D.; Salami, J.; Jaime-Figueroa, S.; Ko, E.; Hines, J.; Crews, C. M. *Angew. Chem. Int. Ed.* **2016**, *55*, 807–810.
48. Crew, A. P.; Raina, K.; Dong, H.; Qian, Y.; Wang, J.; Vigil, D.; Serebrenik, Y. V.; Hamman, B. D.; Morgan, A.; Ferraro, C.; Winkler, J. D.; Crews, C. M. *J. Med. Chem.* **2018**, *61*, 583–598.
49. Zhou, B.; Hu, J.; Xu, F.; Chen, Z.; Bai, L.; Fernandez-Salas, E.; Lin, M.; Liu, L.; Yang, C.-Y.; Zhao, Y.; et al. *J. Med. Chem.* **2018**, *61*, 462–481.
50. (a) Wurz, R. P.; Cee, V. J. *J. Med. Chem.* **2019**, *62*, 445–447. (b) Li, Y.; Yang, J.; Aguilar, A.; McEachern, D.; Przybranowski, S.; Liu, L.; Yang, C.; Wang, M.; Han, X.; Wang, S. *J. Med. Chem.* **2019**, *62*, 448–466.
51. Crew, A. P.; Crews, C. M.; Dong, H.; et al. US Patent 2017/0008904 (**2017**).

3

Pharmacokinetics (ADME)

ADME stands for absorption, distribution, metabolism, and excretion. They are the four pillars that govern the pharmacokinetics of drugs.

When the patient takes a pill, the patient and the drug interact with each other. What the drug does to the body is called *pharmacodynamics* (PD). The PD of a drug could be, at the end, relieving pain, lowering cholesterol level, shrinking tumors, or killing bacteria or viruses. On the other hand, what the patient's body does to the drug is known as *pharmacokinetics* (PK). PK is of abundant importance for drug discovery because no matter how efficacious a drug is, it is still useless if it does not reach the target(s) that causes the disease. For an oral drug, it goes through four stages in human body: absorption, distribution, metabolism and excretion, which are the focus of this chapter on pharmacokinetics.

Before 1991, potency was the major thrust of medicinal chemistry. The medicinal chemist's mentality was to make the drug as potent as possible and then handed it over to the "formulation people" to make it bioavailable. Regrettably, one cannot make a silk purse out of a sow's ear. In fact, in 1991, 40% of drugs in clinical trials failed due to poor PK/bioavailability. That untenable attrition rate sent the message home. The industry took notice and started paying closer attention to drugs′ PK profiles. As a consequence, 10 years later in 2000, the attrition rate from poor PK in clinical trials was reduced five-fold to merely 8%.[1]

3.1 Physicochemical Properties

A drug′s ADME is largely influenced by its physicochemical properties. Here we start by discussing lipophilicity, hydrogen bonding, polar surface area (PSA), and number of rotatable bonds, then the famous rule of 5 (Ro5), followed by ligand efficiency (LE) and lipophilic ligand efficiency (LLE or LipE).

3.1.1 Lipophilicity

Lipophilicity is a measure of how greasy a molecule is. It has a profound impact on a drug's ADME because it is closely associated with drug's solubility, plasma protein binding (PPB), metabolic clearance, volume of distribution, enzyme/receptor binding,

Medicinal Chemistry for Practitioners, First Edition. Jie Jack Li.

and more.[2] A quantitative measure of a molecule's lipophilicity is its *partition coefficient*, *P*, which is the ratio of the equilibrium concentrations of a dissolved solute in a two-phase system containing two largely immiscible solvents. Traditionally, the two immiscible solvents are 1-octanol (o) and water (w). As shown in Figure 3.1, the partition coefficient *P* is defined in Eq. (3.1):

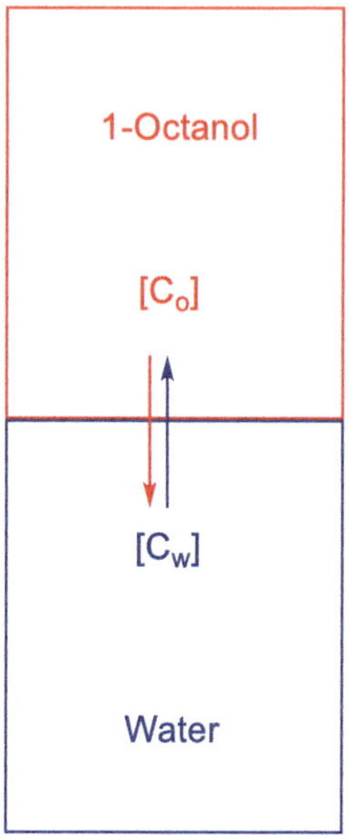

Figure 3.1 Partition of a neutral molecule between 1-octanol (o) and water (w).

$$P = P_{ow} = [C_o]/[C_w] \quad (3.1)$$

Since the value of partition coefficient *P* could be unwieldy with a wide range of numbers, log *P*, defined as Eq. (3.2), is regularly used instead as a more manageable measure of lipophilicity:

$$\text{Log } P = \log[C_o]/[C_w] \quad (3.2)$$

Log *P* impacts nearly all aspects of a drug's pharmacokinetics. As the value of log *P* increases, binding to targets such as receptors and enzymes is increased. In the past, medicinal chemists kept making larger and larger molecules and were happy to see the potency grow. Indeed, lipophilicity enhances a drug's binding as a nonspecific driving force for the partition of the drug into the binding site by raising its free energy in water. Unfortunately, molecular inflation,[3] or molecular obesity and obsession with potency,[4] is harmful to drug design because increased lipophilicity makes the molecules less drug-like with lower bioavailability. As the log *P* value increases, the aqueous solubility decreases, although absorption through the membrane increases. A molecule with a larger log *P* also tends to have higher binding to CYP450-metabolizing enzymes, thus a higher chance for drug–drug interactions (DDIs). A molecule with a larger log *P* value also is inclined to have tighter binding to human ether-a-go-go (hERG, Kv11.1) potassium ion channel and elevate the cardiovascular toxicity risk. Last, but not the least, high log *P* value correlates to high PPB, which has an impact on efficacious concentration of the drug. It is not all surprising that some argued that log *P* is the most consequential property with regard to drugs′ attrition. Historical data tend to support this notion as well.[5]

Nowadays, computational chemistry is so advanced that calculated log *P* value of a given molecule is readily acquired. Clog *P*, for calculated log *P*, is a daily vernacular of drug discovery. Even the simple, ubiquitous Chemdraw® program can calculate Clog *P* and topological polar surface area (tPSA) with a click of button at *Show Chemical Properties Window* under *View*.

P and Clog *P* are adequate in quantifying a drug's lipophilicity for *neutral* molecules. However, it is more complicated when it comes to ionizable acids or bases because their concentrations in octanol and water vary depending upon the degree of ionization (Figure 3.2). The significance of acid–base properties in drug discovery has been well documented.[6] For acids and bases, *distribution coefficient D* is a more appropriate measurement of lipophilicity at a given pH. It is a function of both lipophilicity of the un-ionized compound and degree of ionization.

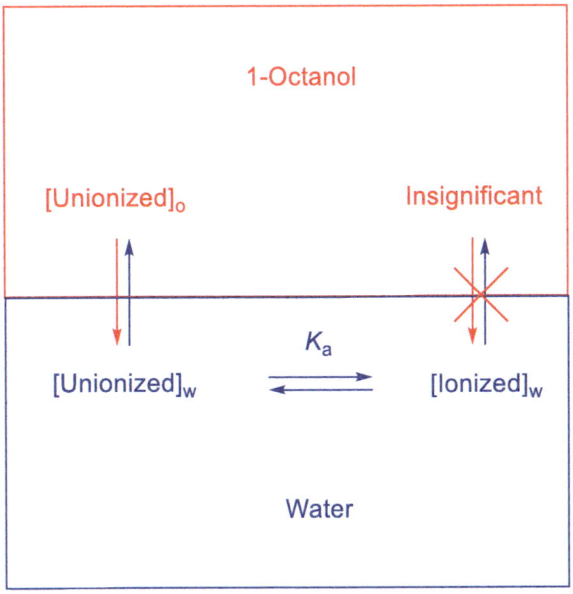

Figure 3.2 Partition of an acid or a base between 1-octanol (o) and water (w).

For an acid,

$$HA_{aq} \rightleftharpoons H^{+}_{w} + A^{-}_{w} \tag{3.3}$$

$$D = [HA]_o/\{[HA]_w + [A^-]_w\} \tag{3.4}$$

For a base,

$$BH^+_w \rightleftharpoons B_w + H^+_w \qquad (3.5)$$

$$D = [B]_o/\{[BH^+]_w + [B]_w\} \qquad (3.6)$$

Like partition coefficient *P*, distribution coefficient *D* is usually expressed as log *D* so that the values are more manageable and easier to make comparisons. The most popular log *D* is log $D_{7.4}$: 7.4 is the pH value of human blood. Why does human blood have a pH of 7.4? The short answer is evolution. Different organs have different pH values, stomach, for instance, is very acidic (pH ~ 1.5–3.5 due to the presence of gastric acid, i.e., HCl), enzymes, receptors, and many other biological entities in human body function optimally at pH 7.4 as a consequence of evolution. Large deviation from pH 7.4 is likely an indication of a disease. If the pH value of one's blood is below 7.0, one is likely suffering from acidosis, which could be the result a number of diseases.

Many marketed drugs contain the acid functional group(s). Indomethacin (Indocin, **1**) is a non-steroidal anti-inflammatory drug (NSAID) and atorvastatin calcium (Lipitor, **2**) is a 3-hydroxy-3-methylglutaryl-CoA (HMG-CoA) inhibitor for lowering cholesterol levels. Furthermore, montelukast sodium (Singulair, **3**) is a leukotriene receptor antagonist for the treatment of asthma, and ciprofibrate (**4**) lowers triglyceride level and boost high-density lipoprotein (HDL)-cholesterol level. They all contain one carboxylic acid. For singly ionizing acids such as **1–4**, calculation of their log *D* is straightforward:[7]

$$\log D = \log P - \log [1 + 10^{(pH - pK_a)}] \qquad (3.7)$$

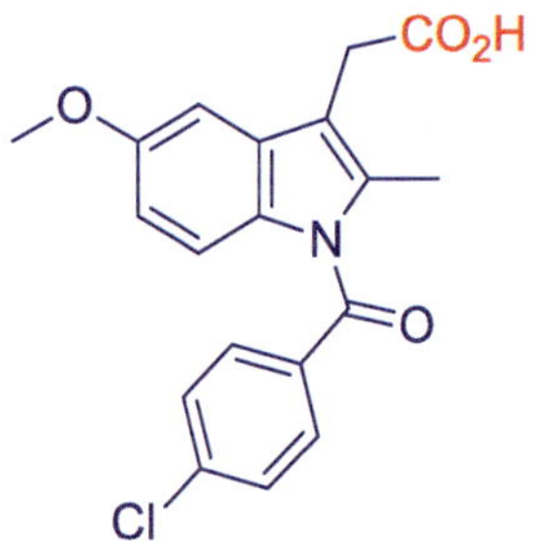

indomethacin (Indocin, **1**)

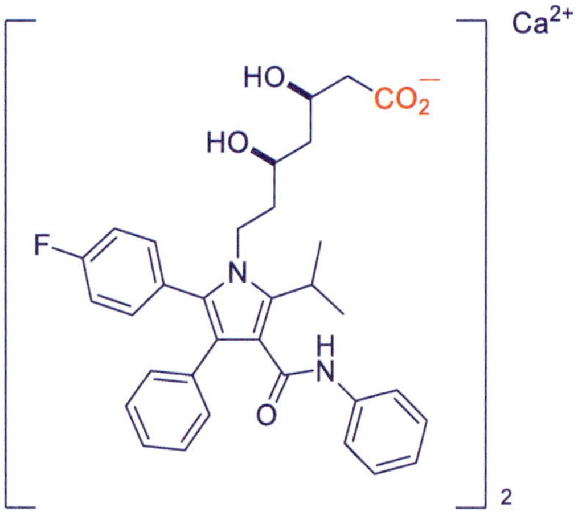

atorvastatin calcium (Lipitor, **2**)

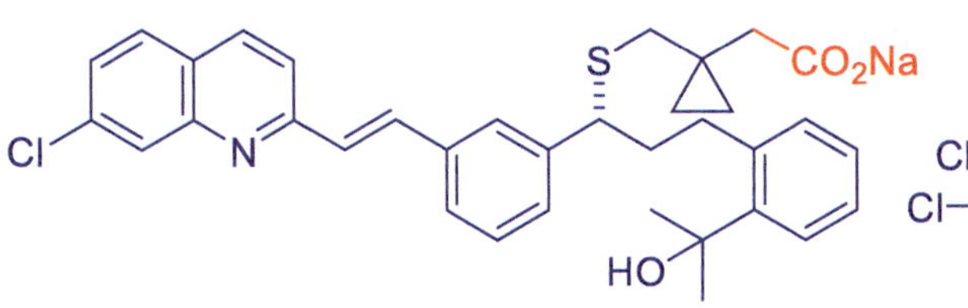

montelukast sodium (Singulair, **3**)

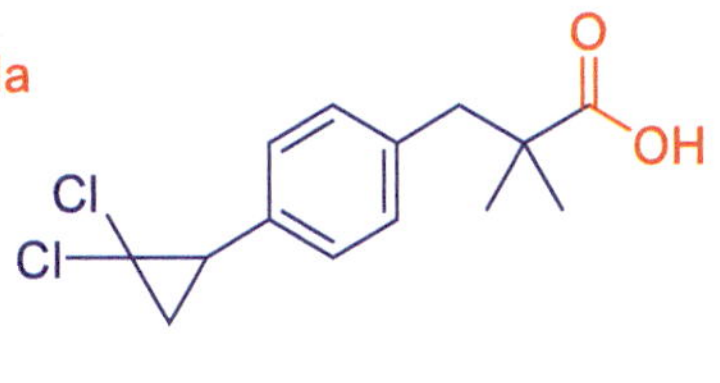

ciprofibrate (**4**)

Normally, for an acid with a pK_a 5 at pH 7.0, its log D is approximately (log P – 2). In case of indomethacin (Indocin, **1**), it has a pK_a value of 4.5. In a very acidic environment, pH 2.0 for instance, the log D is the same as log P: 4.25 since 100% of the molecules are unionized. At pH 4.5, 50% of the drug remains unionized and its log D is 3.8. Under very basic conditions, pH 9.5 for example, merely 0.001% of the drug remains un-ionized since essentially all drug molecules are ionized and its log D is –0.8.

There are also drugs that contain more than one carboxylic acid such as pemetrexed (Alimta, **5**), a folate analog metabolic inhibitor for treating pleural mesothelioma. There are yet other drugs with both acidic and basic functional groups such as pregabalin (Lyrica, **6**, likely exists as a zwitterion) as a modulator of the $\alpha_2\delta$ unit of the calcium channel for treating epilepsy and neuropathic pain. Calculation of their log D values is more complicated and is beyond the scope of this book.

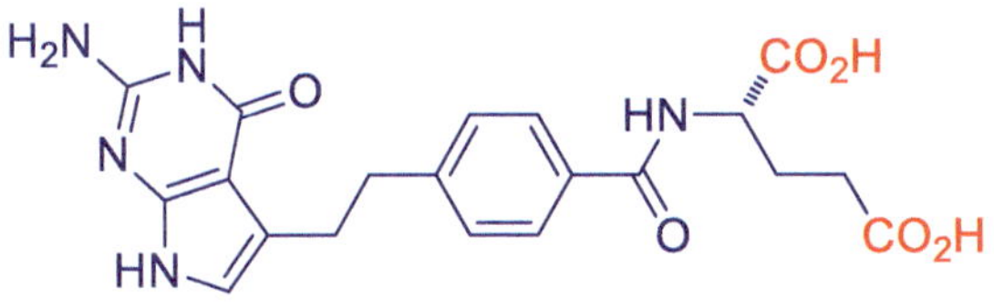

pemetrexed (Alimta, **5**)

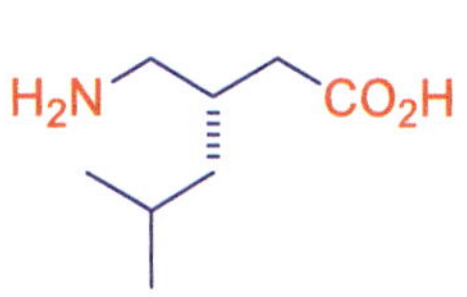

pregabalin (Lyrica, **6**)

Inordinate amount of marketed drugs contain basic amines. Sertraline hydrochloride (Zoloft, **7**), a selective serotonin reuptake inhibitor (SSRI) for treating depression, contains one basic nitrogen atom. Clopidogrel sulfate (Plavix, **8**), an anticoagulant, also contains one basic nitrogen atom. For mono-bases:

$$\log D = \log\,[1 + 10^{(pK_a - pH)}] \tag{3.8}$$

At pH 7.0, log D is approximately (log P – 2) for a base with pK_a of 9.

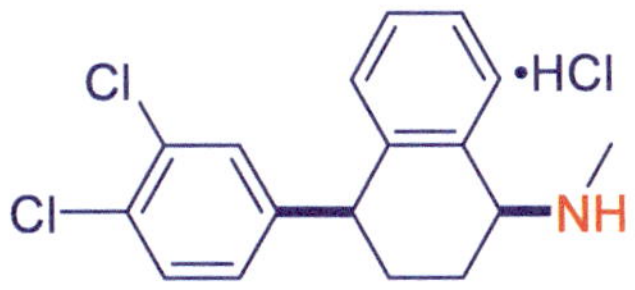

sertraline hydrochloride (Zoloft, **7**)

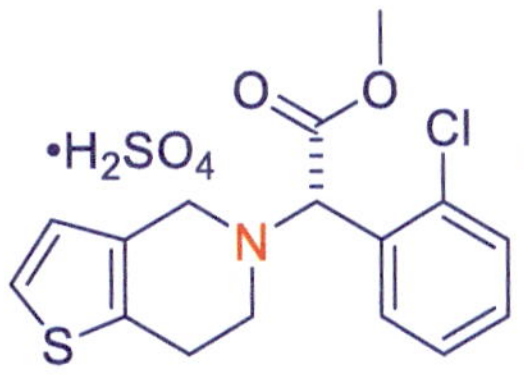

clopidogrel sulfate (Plavix, **8**)

There are drugs with two or more basic nitrogen atoms. Cetirizine (Zyrtec, **9**), an anti-histamine for treating allergy, possesses two nitrogen atoms. Sitagliptin phosphate (Januvia, **10**), for the treatment of diabetes mellitus type 2 (DMT2), has five nitrogen atoms. Why the prevalence of nitrogen atoms in many drugs?

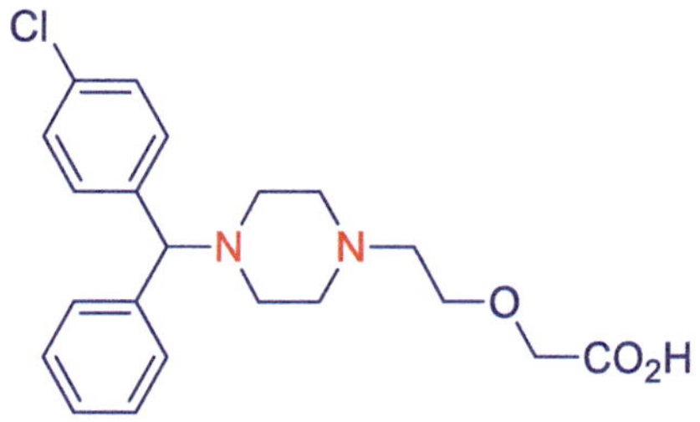

cetirizine (Zyrtec, **9**)

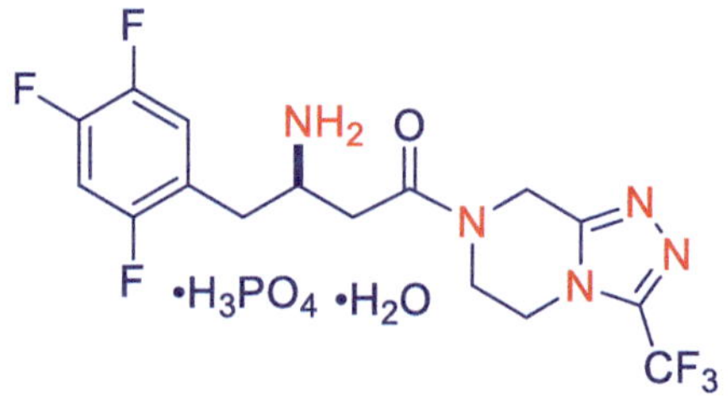

sitagliptin phosphate (Januvia, **10**)

In order for a drug to pass through cell membranes, a dichotomy is at play. On the one hand, the drug should be slightly hydrophilic so that it can dissolve in water. On the other hand, it should be somewhat lipophilic so that it may cross the cell membranes. Amines fit the bill well. Amines′ pK_a values are in the range of 6 to 8, thus they are partially ionized at blood pH 7.45. They can easily equilibrate between their ionized and nonionized forms with a good balance of the dual requirements of water and fat solubility. They can cross cell membrane in the nonionized form, while the ionized form gives good water solubility and permits good binding interactions with its target's binding sites. Striking a balance of lipophilicity is one of the drug design conundrums.

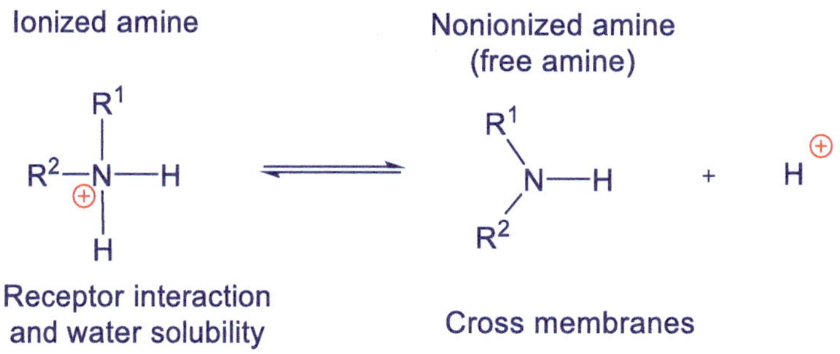

3.1.2 Hydrogen Bonding

Hydrogen bonding influences interactions between a drug and its target such as a receptor or an enzyme. The oxygen and nitrogen atoms on the drug serve as hydrogen bond acceptors, while the OH, SH, and NH groups act as hydrogen bond donors.

Not only is hydrogen bonding crucial to a drug's potency but it also contributes significantly to its physicochemical properties as well. For a drug dissolved in water, intermolecular hydrogen bonds with each other are virtually nonexistent between drug molecules themselves, which are overwhelmingly surrounded by water molecules. To form a hydrogen bond between a donor and an acceptor, both must first break their hydrogen bonds with surrounding water molecules. Because most oral drugs are absorbed by transcellular absorption, neutral molecules are favored over solvated molecules. However, desolvation and formation of a "naked" molecule is not favored thermodynamically if the compound forms many hydrogen and/or ionic bonds with water. As a consequence, drugs with too many hydrogen bond donors and/or acceptors experience difficulty getting from the gut into the blood. In 1988, Young et al. investigated the role hydrogen bonding played in the penetration of antihistamines into the central nervous system (CNS).[8] They concluded that excessive hydrogen bonding prevented access to the CNS.

With regard to penetrating the cell membrane, carbohydrates, metal ions, neurotransmitters, and insulin are exceptions to the rule because they are absorbed with the aid of active transports. See Section 3.2.4 for more details.

Intramolecular hydrogen bonds on drugs are more readily formed in water since they are much more favorable entropically. Intramolecular hydrogen bonding frequently boosts cell membrane penetration. For instance, amido-carbamates **11** and **12** have identical PSAs, yet compound **12** [$P_{app(A\to B)}$ = 43 nm/s) is four-times more cell-permeable (Caco-2) than **11** [$P_{app(A\to B)}$ = 117 nm/s] by virtue of the intramolecular hydrogen bond.[9] Caco-2 permeability assay is a popular cellular method. The Madin–Darby canine kidney (MDCK) cell permeability assay is also frequently used.

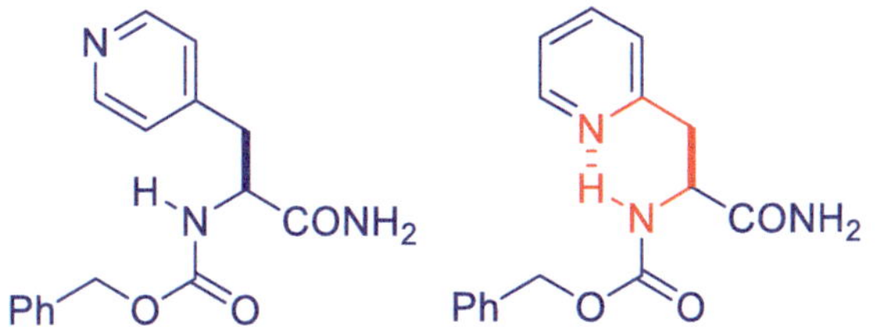

11 $P_{app(A\ to\ B)}$ = 43 nm/s

11′ $P_{app(A\ to\ B)}$ = 177 nm/s
4x more cell-permeable (Caco-2)

Another well-known case involving intramolecular hydrogen bonding is cyclosporine A (CsA, **13**, molecular weight 1206). Measurements of partition coefficient *P* indicated that the hydrogen bonding capacity of CsA (**13**) changed dramatically from in an apolar solvent (where it is internally hydrogen bonded) to in a polar solvent (where it exposes its hydrogen bonding groups to the solvent).[10a]

Cyclosporine A (**13**) is one of the very few macrocycles (including avermection, midecamycin, and rapamycin) that possess good oral bioavailability. Most macrocycles with many polar groups do not cross cell membranes because they are too polar, but cyclosporine A (**13′**) does due to the existence of its four intramolecular hydrogen bonds, which lock its conformation and raise its log *P*. This phenomenon is dubbed as chemical chameleon or cyclosporine A chameleon, insinuating that CsA (**13**), normally a polar compound, "disguises" itself as a greasier molecule CsA (**13′**) by forming intramolecular bonds in order to cross the cell membrane.

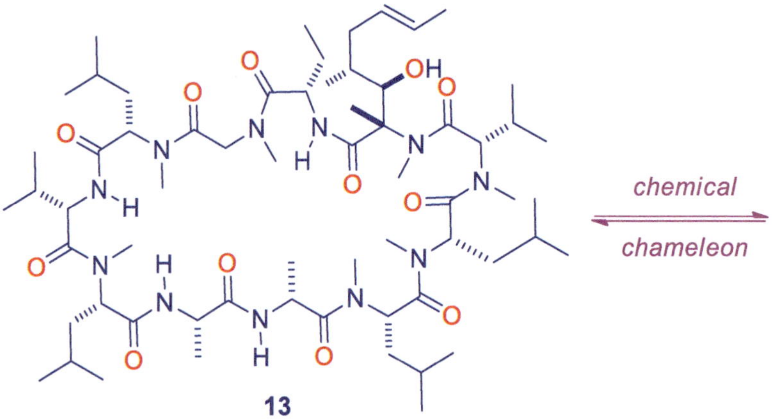

13

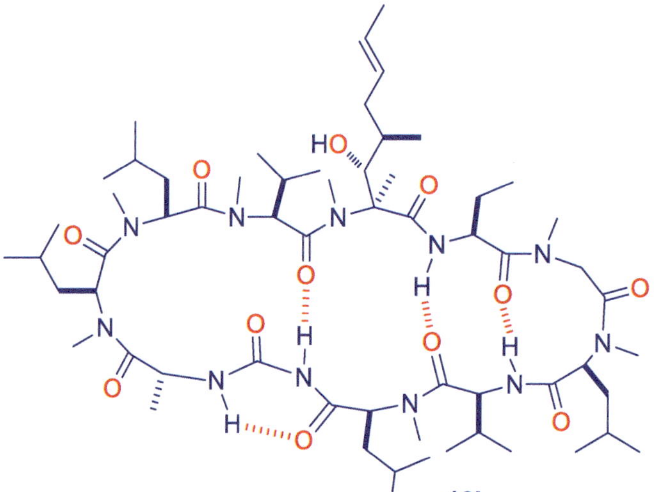

13'

Very recently, Gilead chemists made a Herculean effort to discover bioavailable macrocycles stemming from sanglifehrin A.[10b] *En route* to their best macrocycle, they encounter a situation where an intramolecular hydrogen bond also made tremendous difference in bioavailability. Isoquinoline **14**, with no potential for an internal H-bond, has a log *D* of 2.0 at pH 7.4, very poor apical to basolateral permeability, and high efflux (ratio of *BA*/*AB* = 79-fold). Not unexpectedly, quinoline **14′**, with an intramolecular H-bond, has a measured log *D* value of 3.2 that is 1.2 unit higher than that of **14**. This is a case in point to highlight the importance of employing measured log *D* data during the structure-activity relationship (SAR) optimization process, especially for constrained molecules where dynamic hydrogen bonds can impact lipophilicity. Somewhat unexpectedly, quinoline **14′** has a solubility of 55 μM/mL in aqueous media that is nine-fold more soluble than that of the parent compound **14**. This phenomenon is presumed to be due to the flexibility of the macrocycle or "chameleonic effect" that allowed it to readily adopt alternate conformations in more polar environments. The chameleonic behavior of macrocyclic compounds adapting to different environments through making or breaking of intramolecular hydrogen bonds has been noted as drug delivery tactic for compounds in chemical space *beyond rule of five* (bRo5).[10b]

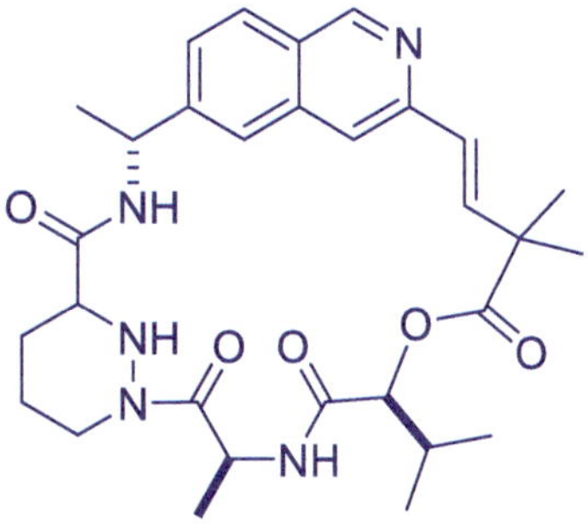

isoquinoline **14**
without intramolecular H-bond
log *D* = 2.0 (pH 7.4)
Caco-2, AB/BA = 0.1/7.9 10^{-6} cm^{-1}
Solubility, 6 μg/mL

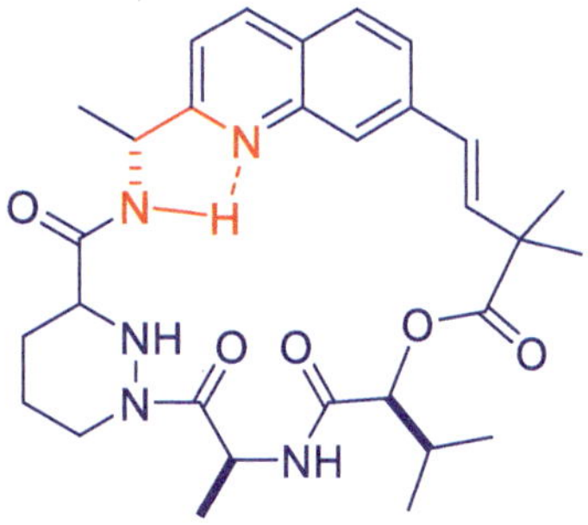

quinoline **14'**
with intramolecular H-bond
log *D* = 3.2 (pH 7.4)
Caco-2, AB/BA = 17/47 10^{-6} cm^{-1}
Solubility, 55 μg/mL

Extensive effort has been made to improve membrane permeability and absorption by taking advantage of intramolecular hydrogen bonding in bRo5 chemical space.[11] It is hypothesized that formation of intramolecular hydrogen bonds in drug molecules shields polarity, thus offering improved membrane permeability and intestinal absorption. Application of hydrogen bonding calculations in property-based drug design has been reviewed.[12] As a side, statistically, the chance that intramolecular hydrogen bonding improves biological activities is 50%, virtually a coin toss.[13]

3.1.3. Polar Surface Area

Molecular size is an important factor affecting biological activities, but it is also very difficult to measure. There are various ways of gauging molecular size: molecular weight (probably most significant); electron density; polar surface area (PSA); van der Waals surface; and molar refractivity. Among these, PSA is a simple measure of total hydrogen bonding capacity. It is defined as a sum of surface of polar atoms (usually oxygen and nitrogen atoms).

Having investigated the impact of PSA on over 2,000 drugs in Phase II or later stages of clinical trials including 45 drugs on the market, Kelder and colleagues determined that PSA is the dominating determinant for oral drug absorption and brain penetration of drugs.[14] They concluded that orally active drugs transported passively by the transcellular route should not exceed a PSA of 120 Å. For CNS drugs, their PSA values should not exceed 70 Å.

From studying the oral bioavailability of over 1,100 drugs in rats, Veber et al. arrived at two criteria for drugs to be orally bioavailable[15] (i). 10 or fewer rotatable bonds; and (ii). polar surface area equals to or less than 140 Å (or 12 or fewer H-bond donors and acceptors).

3.1.4 Rotatable Bonds

A rotatable bond is defined as any single bond, not in a ring, bound to a nonterminal heavy (nonhydrogen) atom. Amide C–N bonds are not rotatable because of their high barrier to rotation, thus possessing a partial double bond character. The number of rotatable bonds (n_{rot}) influences both bioavailability and binding potency. Generally speaking, when all is equal or similar for two drugs, the one with fewer rotatable bonds has higher absorption. For instance, propranolol (Inderal, **15**) was the first-in-class beta-blocker for the treatment of hypertension marketed since 1962 by ICI and atenolol (Tenormin, **15′**) was out in 1976, also by ICI. The rotatable bond count for propranolol (**15**) is 6 and that of atenolol (**15′**) is 8 since the C–N bond does not count as one. The *absorption* for propranolol (**15**) is 90% and that of atenolol (**15′**) is 50%.[16] Their bioavailability is more complex since it involves metabolism and clearance. In fact, propranolol (**15**) suffers from first-pass metabolism, therefore, has a lower bioavailability of 30% than that of atenolol (**15′**)'s, 50%.

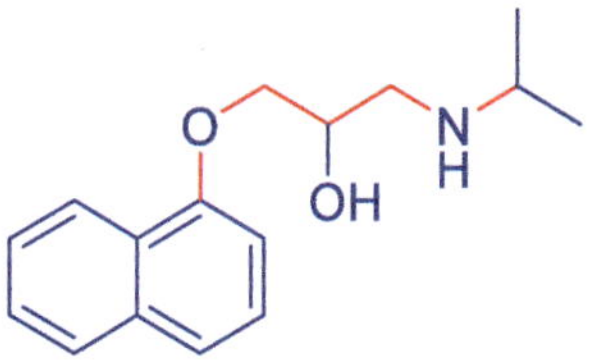

Propranolol (Inderal, **15**)
n_{rot}, 6
Absorption > 90%
Bioavailability, 30%
Clearance, 1,000 mL/min
Protein binding, 93%

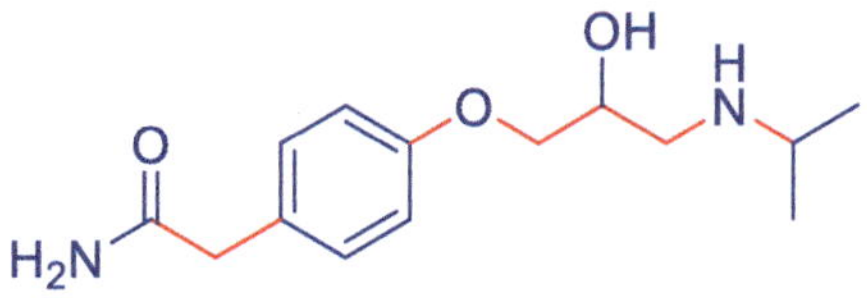

Atenolol (Tenormin, **15′**)
n_{rot}, 8
Absorption 50%
Bioavailability, 50%
Clearance, 130 mL/min
Protein binding, 5%

Veber et al. investigated factors influencing oral bioavailability in drug candidates.[15] Somewhat unexpectedly, they discovered that the effect of molecular rigidity, as represented by number of rotatable bond count, is independent of molecular weight. They tallied fractions of compounds with a rat oral bioavailability of 20% or greater as a function of molecular weight and rotatable bond count (n_{rot}). They concluded that, regardless of molecular weight:

1. ~65% of drug candidates (DCs) have a rat bioavailability greater than 20% if their $n_{rot} \leq 7$;
2. ~33% of DCs have a rat bioavailability greater than 20% if $7 < n_{rot} \leq 10$; and
3. ~25% of DCs have a rat bioavailability greater than 20% if $n_{rot} > 10$.

3.1.5 Rule of 5

From their extensive analysis of pharmacokinetics parameters Lipinski and colleagues concluded that marketed drugs are mostly small and moderately lipophilic molecules.[17,18] Lipinski's Ro5 predicts that a drug may have poor solubility and permeability (marked as an "*Alert*") if the compound exceeds two or more of the following four limits:

1. Molecular weight (*M*) >500.0
2. Clog *P* (*C*) >5.0
3. Hydrogen bond acceptors (*A*) >10
4. Hydrogen bond donor (*D*) >5

Otherwise, the compound is marked "OK."

Einstein said: "Everything in science should be made as simple as possible, but not simpler." Thanks to its simplicity, Lipinski's Ro5 has had a profound and, mostly positive, impact on drug discovery. A chemist immediately becomes *alert* if the molecule is outside Lipinski's space. Therefore, recent medicinal chemistry has produced drug candidates that are more drug-like.

But every coin has two sides. No doubt, Lipinski's Ro5 has been one of the most influential, if not *the* most influential, rules with regard to solubility and permeability of drugs. However, approximately 6% of oral drugs on the market are bRo5.[19] It is even more in the last few years, in the span of three years from 2014 to 2017, 21% new drug approvals (12 drugs) are bRo5, especially in the fields of oncology and hepatitis C virus (HCV).

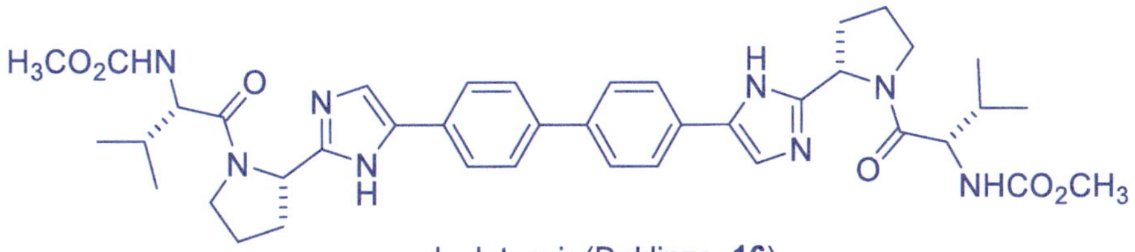

daclatasvir (Daklinza, **16**)
BMS, 2015, HCV NS5A Inhibitor
MW, 738; Clog *P*, 1.3; HBD, 4; N+O, 14, *F*%, 67%!

For example, daclatasvir (Daklinza, **16**) is an HCV NS5A inhibitor. Both of its molecular weight (738) and number of hydrogen bond acceptor (14) are outside Lipinski space, but it has an oral bioavailability of 67% for humans. Meanwhile, venetoclax (Venclexta, **17**), a BCL-2 inhibitor for the treatment of chronic lymphocytic leukemia (CLL), is a posterchild for bRo5. Yet, it is orally bioavailable with an *F*% value of 65%. Even more remarkably, it is the first marketed drug with the mechanism of action (MOA) of modulating intramolecular protein–protein interaction (PPI). DeGoey et al. speculated that those large and highly lipophilic molecules succeed as drugs via a fundamentally different route from most marketed drugs.[19] Strictly following the Ro5 would have overlooked those life-saving medicines.

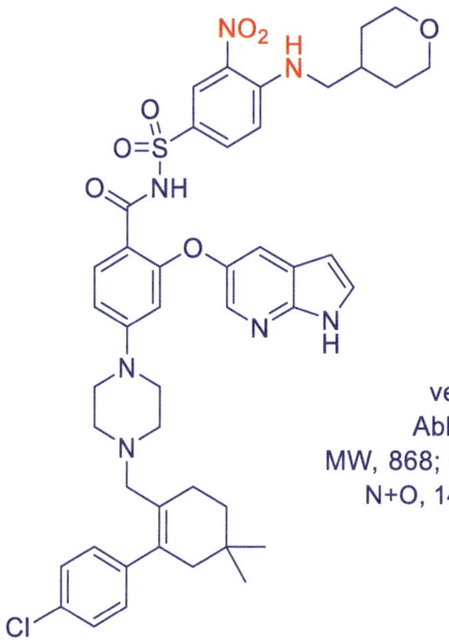

venetoclax (Venclexta, **16**)
Abbvie, 2016, BCL-2 inhibitor
MW, 868; Clog *P*, 10.4; Clog *D*, 6.25; HBD, 3;
N+O, 14, n_{rot}, 13; tPSA, 172; *F*%, 65%!

Ro5 has been successfully employed for drugs. However, the rules for *fragments* in the context of fragment-based drug discovery (FBDD) are somewhat different. Congreve et al. at Astex proposed a "rule of 3" (Ro3) in 2003.[20] It is also known as "Astex Rule of Three."

Table 3.1 Ro3 and Ro5

Variable	Rule of 3	Rule of 5
Clog *P*	≤3	≤5
No. of N	yes	yes
No. of O	<9	<10
H-bond donors	≤3	≤5
MW	≤300	≤500
No. of rotatable bonds	≤3	≤10
PSA (Å)	≤60	≤140

3.1.6 Ligand Efficiency

Medicinal chemistry has become more sophisticated nowadays. Gone are the days when a medicinal chemist focused solely on potency. Much attention is now paid to drug-likeness. To combat the exorable rise of molecular weight as a false prophet, Hopkins et al. in 2004 proposed a concept *ligand efficiency* that attempts to "normalize" the potency of a lead with respect to molecular weight as a useful metric for *lead selection*.[21]

During the process of triaging hits from high throughput screen (HTS) and selection of lead compounds, low molecular weight compounds that achieve their full binding potential have low overall activity (K_i >10 μM) and are frequently ignored as leads. In contrast, high molecular weight compounds that are biologically active (K_i <10 μM), often inefficient ligands, are frequently pursued as leads. However, active efficient ligands have a better than average binding energy per atom, and are biologically active (K_i <10 μM). They are excellent lead material as a starting point to work with.

To that end, Hopkins proposed ligand efficiency as a measurement of the binding energy per atom of a ligand to its binding partner, a protein such as a receptor or an enzyme. It is used to assist in narrowing focus on lead compounds with optimal combinations of physicochemical properties and pharmacological properties. Mathematically, ligand efficiency (LE, or Δg) can be defined as the ratio of Gibbs free energy (ΔG) to the number of nonhydrogen atoms of the compound (N):[21]

$$\mathrm{LE} = \Delta g = \Delta G/N \qquad (3.9)$$

where $\Delta G = -RT \bullet \ln K_d$ and N is the number of nonhydrogen atoms (i.e., heavy atoms, sometimes it is expressed as heavy atom count: HAC = N). LE can also be expressed as:

$$\mathrm{LE\ (kcal/mol/atom)} = (1.37 * \log \mathrm{IC}_{50})/N = [1.37 * p(\mathrm{activity})]/N \qquad (3.10)$$

It was observed that LE of 0.3 kcal/mol/nonhydrogen atom added would be a realistic minimum gain as the compounds evolve from "lead-like" to "drug-like."[22] A drug candidate normally has an LE value greater than 0.35 kcal/mol/atom.

Ironically, last decade saw proliferation of ligand efficiency metrics (LEMs).[23] PEI, stands for percent efficiency index; BEI, binding efficiency index; SEI, surface efficiency index; SILE, size-independent ligand efficiency; GE, group efficiency, and so on. Like an arms race, everyone has to have his own "personal ligand efficiency." To keep things manageable, only Hopkins's LE (Δg) will be employed most of the times in this book.

Not everyone is a big fan of LEMs. Kenny and Mantanari initially accused LEMs guilty of "inflation of correlation in the pursuit of drug-likeness."[24] Later on, they opined that "ligand efficiency metrics considered harmful."[25] They argued that LEMs distort our perception of the relationship between activity and the risk factor(s) with which we choose to normalize it. Moreover, neither the scaling nor the offsetting transformations used for normalizing activity has any physiochemical basis and excessive reliance on metrics inhibits more thorough examination of data. In contrast, they favor ligand-efficiency dependent lipophilicity (LELP) as a better measure of compound quality (*vide infra*).

3.1.7 Lipophilic Ligand Efficiency

Lipophilic efficiency (LipE), also known as lipophilic ligand efficiency (LLE), was introduced by Leeson and Springthorpe in 2007.[26] It is a parameter linking potency and lipophilicity in an attempt to estimate druglikeness.

$$\mathrm{LipE} = \mathrm{p(activity)} - \log D_{7.4} = \mathrm{pIC}_{50} - \log D_{7.4} \qquad (3.11)$$

In addition to pIC_{50}, p(activity) may be $\mathrm{p}K_d$ or $\mathrm{p}K_i$ as well.

The compound with the highest LipE is the most efficient expression of potency for lipophilicity. This will drive low *in vivo* dose because of low free drug level requirement (potency) and low clearance (driven by lower log D for the same series). For oral drug candidates, even though compounds have high LipE, however, their high molecular weight and high PSA or basic amine will likely have high risks in terms of absorption and selectivity.

Keserű in 2009 proposed the concept of ligand-efficiency dependent lipophilicity (LELP),[27] which is defined as the ratio of log P and LE, thus depicting the price of ligand efficiency paid in log P:

$$\text{LELP} = (\log P)/\text{LE} \tag{3.12}$$

By comparing how both LipE and LELP behave for compounds at different stages of drug discovery, Keserű concluded that LipE is more sensible for the *development stages* and does not prefer fragment-type hits that are otherwise considered to be promising starting points for lead discovery.[28] In contrast, LELP incorporates molecular size and penalizes the increase in log P more than LipE; therefore, it has the advantage for *ADMET-related issues* over LipE. Here T stands for toxicity.

In 2013, Shultz compared several composite parameters, or efficiency indices, including LE, LipE, and LELP via a matched molecular pair (MMP) analysis.[29] Despite multiple attempts to correct LE for size with modifications such as SILE, LEMs cannot normalize for potency under ideal conditions. LELP also fell short upon closer scrutiny since the assumption of potency per heavy atom is not valid for almost all drug-like compounds that Shultz examined. In contrast to other empirically derived composite parameters based on HAC, LipE sets consistent expectations regardless of molecular weight or relative potency, and can be used to generate consistent expectations for any matched molecular pair (MMP). Shultz further demonstrated that LipE most strongly correlates with compound quality as defined by enthalpy-driven binding, thus providing the basis of LipE over other metrics in enthalpy optimization.[30] In short, LipE is the lipophilic efficiency of choice.

Meanwell and Johnson published two scholastic and informative reviews on LEMs in 2016 and 2018, respectively.[31]

3.2 Absorption

3.2.1 Definition of Pharmacokinetics Parameters

A drug's pharmacokinetics is a number's game. In order to understand the rules of the engagement, we need to become acclimated to common PK parameters. As shown in Figure 3.3, a drug's journey through human body goes through three stages: absorption phase, absorption rate is higher than elimination rate; post-absorption phase, elimination rate is higher than absorption rate; and elimination phase, no significant absorption occurs (only elimination process).

Among conventional PK parameters, t_{max} is the time that it takes for the drug to reach its *maximum concentration*, which is denoted as C_{max}. The t_{max} is independent of dose and is dependent on the rate constants for absorption and elimination. The phase from the time when the drug is given to the time reaching C_{max} is the absorption phase because the concentration of the drug increases. At C_{max}, sometimes called *peak concentration*, the rate of drug absorbed is equal to the rate of drug eliminated. Therefore, the net rate of concentration change is zero. Once C_{max} is reached, the phase becomes the elimination phase until C_{min} (*minimal concentration*) is reached because the concentration becomes lower and lower until it is beyond the limit of detection. AUC stands for area-under-curve, which is a good measurement of the body's quantitative exposure to a drug (i.e., bioavailability).

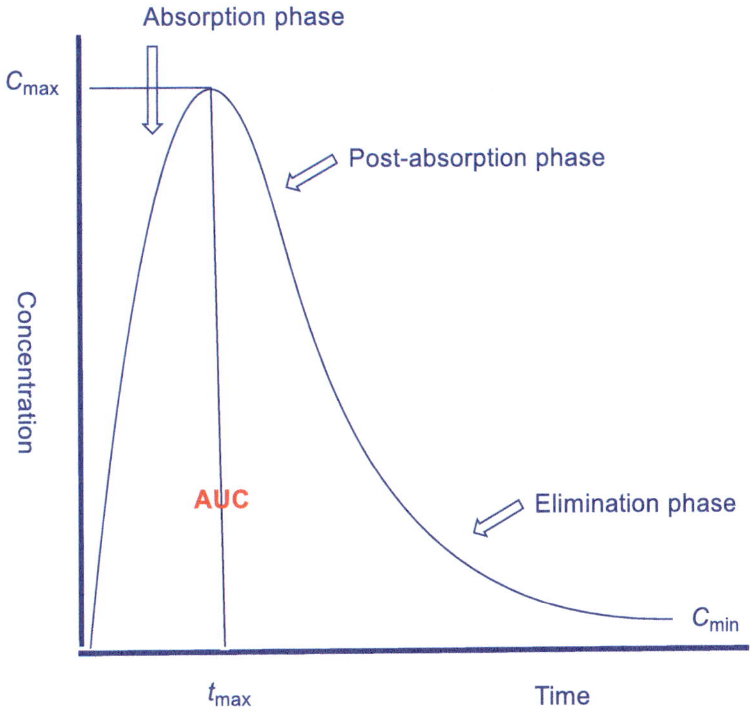

Figure 3.3 Oral pharmacokinetics parameters.

Volume of distribution (V_d) is an imaginary parameter. Often V_{ss} is employed because it is preferred to calculate the volume of distribution at the steady-state (SS).

Table 3.2. Estimation of volume distribution, V_{ss} (L/kg)

Species	Low	Moderate	High	Very high
All	<0.6	0.6–5	5–100	>100
	<body water	body water–5	5–100	>100

Also known as apparent volume of distribution, V_d represents the instantaneous drug concentration (t = 0) immediately after drug equilibration in the body. Mathematically, volume of distribution is defined as Eq. (3.13):

$$V_d = \text{Dose/Concentration} = D/C \quad (3.13)$$

D, dose, is the total amount of drug in body and C is the plasma drug concentration.

V_d, an abstract term, often uses "L/kg" as its unit. But sometimes "L" is employed as the unit, assuming the drug is dosed to a 70-kg adult. Normally, it is assumed that 70 L is the average body volume and the average blood volume is 5 L. The values of V_d vary drastically. It could be quite small. For instance, a drug confined to the blood only, its V_d is merely 0.08 L/kg. It could be moderate. Antipyrine is readily diffused through cell membrane and readily distributed throughout extra- and intracellular water volume, with a V_d value is 0.6 L/kg. For basic drugs, its volume could be very high because they have higher affinity for the tissue than the plasma protein. It could be as high as 10,000 L, unquestionably an artificial number since nobody has that much body fluid. It only has mathematical significance.

There are nominal trends for volume of distribution (V_d) with regard to the types of compounds. For acidic compounds, their $V_d < 0.4$ L/kg. Acidic drugs are mainly confined to plasma with limited tissue distribution. The V_d values for neutral compounds range from 0.4 L/kg to 1.0 L/kg. Normally, neutral drugs are uniformly distributed between plasma and tissues. Finally, V_d of basic compounds are > 1.0 L/kg since they may be more concentrated in particular tissue. An estimation for V_{ss} is listed in Table 3.2.

More details on volume of distribution will be discussed in Section 3.3.

Clearance is the total volume of blood from which a drug completely cleared per unit time. The units are thus volume per time, usually mL/min/kg:

$$\text{Cl (L/h)} = \text{dose/AUC [mg/(mg*h/L]} \quad (3.14)$$

Alternatively,

$$\text{Cl} = \text{Dr}/C_{ss} \quad (3.15)$$

Dr is the dose rate and C_{ss} stands for steady-state plasma concentration. A rough estimation for Cl cross the species is listed in Table 3.3.

Table 3.3 Clearance, Cl (mL/min/kg).

Species	Low	Moderate	High	Very High
Rat	<7	7–53	53–70	>70
Dog	<4	4–26	26–35	>35
Monkey	<4	4–33	33–44	>44
Human	<2	2–15	15–20	>20
	$1/10Q_h$	$1/10Q_h$–$3/4Q_h$	$3/4Q_h$–Q_h	$>Q_h$

And half-life: $t_{1/2} = (0.693 * V_d)/\text{Cl}$ (3.16)

Normally, the higher is the clearance, the lower the half-life ($t_{1/2}$), which in turn is the time taken for the amount of drug in the body (or the plasma concentration) to fall by half. The relationship between half-life and volume distribution and clearance may be expressed as Eq. (3.16). An estimation of half-life for four species is listed in Table 3.4 and the half-life values in human for 15 common drugs are collected in Table 3.5.

Table 3.4 Half-life, $t_{1/2}$ (h).

Species	Low	Moderate	High	Very high
Rat	<1	1–4	4–10	>10
Dog	<2	2–6	6–12	>12
Monkey	<2	2–6	6–12	>12
Human	<3	3–8	8–14	>14

Table 3.5 Half-life values of 15 common drugs.

Drugs	$t_{1/2}$ (h)
Tubocurarine	0.2
Penicillin	0.5
Insulin	0.7
Erythromycin	1.5
Hydrocortisone	1.7
Ethyl biscoumacetate	2.4
Prednisolone	3.4
Imipramine	3.5
Aspirin	6.0
Sulfadimidine	7.0
Tetracycline	9.0
Glutethimide	10.0
Sulfadimethoxine	30.0
Dicumarol	32.0
Vitamin D	40.0

Half-life values have real-life impact. For instance, if a drug has a half-life of 12 h or longer, a once-daily (qd in Latin) regimen is suitable. But if a drug has a short half-life, 2 h or 3 h, for example, it has to be taken twice (bid) or even thrice (tid) a day in order to maintain efficacious bioavailability as measured by AUC.

On the TV commercial for tadalafil (Cialis), a phosphodiesterase type 5 (PDE5) inhibitor for the treatment of erectile dysfunction (ED), it touts "......daily use helps you ready when the moment is right......" insinuating that it is superior to sildenafil (Viagra) and vardenafil (Levitra). This advertisement actually has some scientific veracity. The half-life for tadalafil is 17.5 h, thus it only needs to be taken once daily.[32] This may explain why Cialis is favored by the Europeans as "the weekend pill." In contrast, half-lives of sildenafil and vardenafil are 3.8 h and 4.7 h, respectively. They are taken one hour prior.

A key PK parameter is bioavailability denoted as $F\%$. It is defined as in Eq. (3.17), and estimation of $F\%$ is listed in Table 3.6. Factors that influence bioavailability of a drug include: first pass hepatic metabolism; solubility of a drug; chemical stability; and nature of drug formulation.

$$F\% = \frac{AUC_{Oral}}{AUC_{IV}} \times \frac{Dose_{IV}}{Dose_{Oral}} \tag{3.17}$$

If the dosages are the same for both oral and IV dosing, then,

$$F\% = AUC_{Oral}/AUC_{IV} \tag{3.18}$$

Bioavailability, *F*%.

Species	Very poor	Poor	Moderate	Excellent
All	<5%	5–20%	20–75%	>75%

3.2.2 Improving Solubility

The Food and Drug Administration (FDA)'s Biopharmaceutical Classification System (BCS) divides drugs into four classes as shown in Figure 3.4.[33]

Class I High solubility High permeability	**Class II** Low solubility High permeability
Class III High solubility Low permeability	**Class IV** Low solubility Low permeability

Figure 3.4 FDA's Biopharmaceutical Classification System (BCS).

For a drug to be absorbed, it has to be dissolved first. Not surprisingly, aqueous solubility is a key factor to influence a drug's bioavailability. A superb review by Walker on improving solubility via structural modification was published in 2015.[34] Tactics to improve a compound's solubility include (i). attaching a basic side-chain; (ii). disruption of aromaticity; (iii). disrupting hydrogen bonding; and (iv). certain subtle changes.

3.2.2.1 Attach a Basic Side-Chain

When Zimmermann et al. at Novartis discovered the first receptor tyrosine kinase (RTK) inhibitor imatinib (Gleevec, **19**), a basic side-chain containing piperazine was attached to phenylaminopyrimidine **18** to improve its solubility.[35] The fact that the piperazine also enhanced ligand binding of **19** to the Bcr–Abl kinase was a pleasant, unexpected surprise. The excellent aqueous solubility catapulted imatinib (**19**) to achieve a remarkable bioavailability of 98% in human. The piperazine ring is a privileged motif in drugs. A quick glance of FDA-approved drugs on the market revealed that quite a few drugs contain this "enchanted" ring in general and kinase inhibitors in particular. Piperazine-containing kinase inhibitors include dasatinib (Sprycel), bosutinib (Bosulif), ponatinib (Iclusig), palbociclib (Ibrance), ribociclib (Kisqali), abemaciclib (Verzanio), nintedanib (Ofev), and brigatinib (Alunbrig).

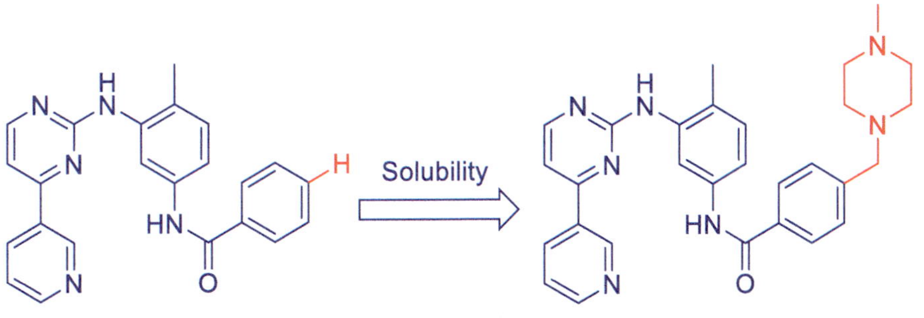

Phenylaminopyrimidine **18**

imatinib (Gleevec, **19**)
$t_{1/2}$, 19 h; *F*, 98%; sol, 250 mg/mL

4-Aminoquinazoline **20** is a potent kinase insert domain receptor (KDR) inhibitor. In an effort to boost its aqueous solubility, a basic piperidine ring was installed on the side chain to replace the triazole, which resulted in **21** with up to a 500-fold improvement of solubility at pH 7.4.[36]

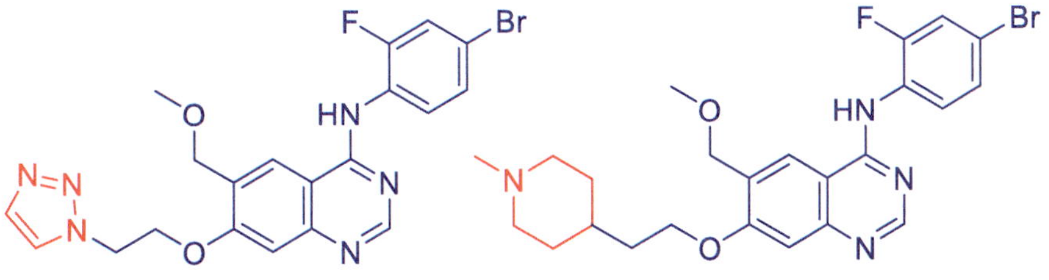

20, pK_a = 5.3
sol, 0.7 μM

21, pK_a = 9.4, 5.3
sol, 330 μM

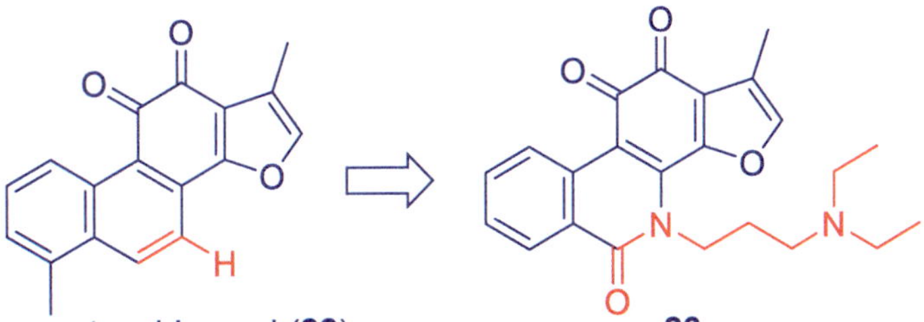

tanshinone I (**22**)
sol, <10^{-4} mg/mL
F%, ~ 0%

23
15.7 mg/mL
21%

Isolated from traditional Chinese medicine *Danshen*, natural product tanshinone I (**22**) showed moderate anti-cancer activities, but it has an abysmal aqueous solubility hence poor pharmacokinetics.[37] Systemic modifications of the parent structure led to **23** with a lactam core and a diethylamino-propyl side-chain. It displayed potent anti-proliferative activities with an aqueous solubility of 15.7 mg/mL.

3.2.2.2 Disruption of Aromaticity

In order for a drug to dissolve, its crystalline lattices must be broken first. The more crystalline the material is, the more difficult for it to dissolve. As a compound progresses through the pipeline, one often finds that its solubility keeps "diminishing." This is the result of an artificial artifact. When the compound is prepared for the first time, the purity requirement is not that stringent, greater than 95–98% purity is more than sufficient for the purposes of biochemical and cellular assays. As the compound progresses from a hit to a lead, then to a drug candidate, the criteria for its purity grow. Many of them are crystalline. In fact, crystalline forms are vital intellectual properties associated with innovative medicines. Empirically, π–π stacking enhances crystallinity of compounds containing aromatic rings. Therefore, disruption of planarity thus aromaticity results in improvement of solubility.[38]

Sulfonamides **24** and **25** are γ-secretase inhibitors with similar potencies.[39] Employing bicyclo[1.1.1]pentane (BCP) as a bioisostere for the fluorophenyl moiety, the F_{sp3} count more than doubled from 0.25 to 0.52. F_{sp3} is the fraction of sp^3 hybridized heavy atoms, an alternative to number of aromatic rings (#Ar).[40,41] This disruption of aromaticity translated to a higher LipE value of 5.95 for **25** from 4.95 for **24**. More important, this maneuver also translated into the practical advantage of improved kinetic and thermodynamic aqueous solubility, and increased membrane permeability, probably brought about by a reduction in lipophilicity since the log D is reduced by 0.9.

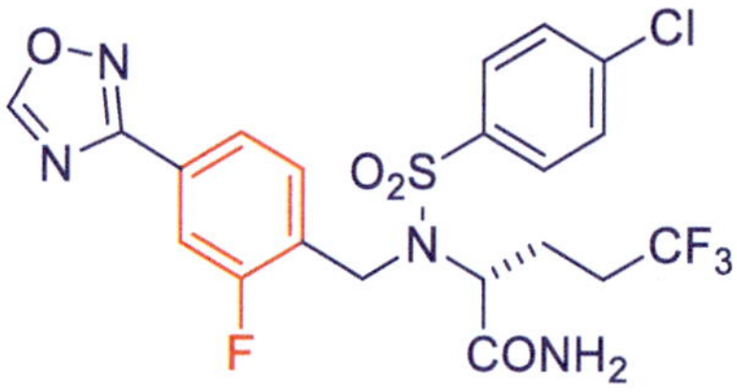

24, pIC_{50} = 9.65
log D = 4.7
F_{sp3} = 0.25
LE = 0.40
LLE = 4.95
LELP = 11.75
sol pH 7.4 = 0.9 μM

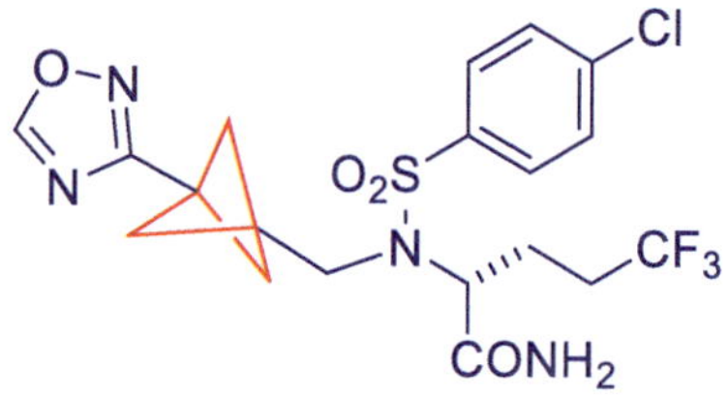

25, pIC_{50} = 9.65
log D = 3.8
F_{sp3} = 0.52
LE = 0.43
LLE = 5.95
LELP = 8.83
sol pH 7.4 = 29.4 μM

Vanilloid receptor-1 (transient receptor potential channel-1, or TRPV1) antagonist 4-oxopyrimidine **26** behaves like "brick dust" with no solubility to speak of. Disruption of the aromaticity of the trifluorophenyl motif resulted in **27**, which is bestowed with a 13-fold boost of aqueous thermodynamic solubility over **26**.[42,43] The fact that partially saturated **27** has a significantly lower melting point is a good indication that reducing the planarity disrupted the crystal-stacking capacity. cPFI stands for calculated Property Forecast Index.

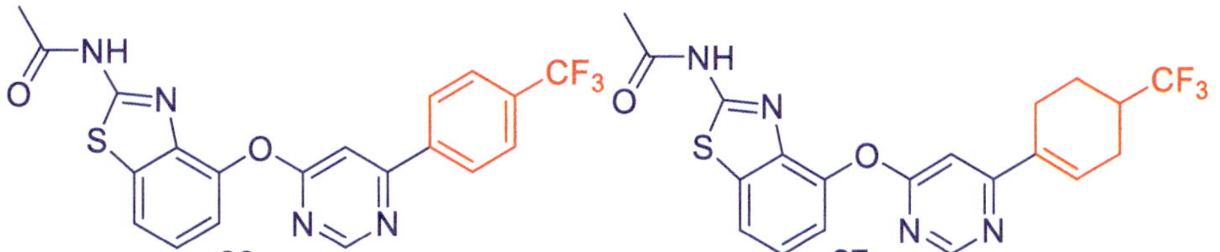

	26	27
solubility (0.01 M HCl)	<1 μg/mL	13 μg/mL
Clog *P*	4.6	3.7
mp	219–221 °C	139–131 °C
F_{sp3}	0.1	0.3
cPFI	10.0	8.5

Pyrazolopyridine inhibitor of B-Raf^{V600E} **28** has a poor aqueous solubility because the molecule is flat with extensive π–π stacking. Two maneuvers were taken to disrupt the planarity. One of the two fluorine atoms was replaced with a chlorine atom, which might increase the energy barrier for the phenyl–amide single bond rotation and the methoxyl group was replaced with the bulkier cyclopropyl group to afford analog **29**, whose melting point is 65 °C lower than that of **28**. The solubility of **29** at pH 7.4 is 14 times higher than that of **28**.[44]

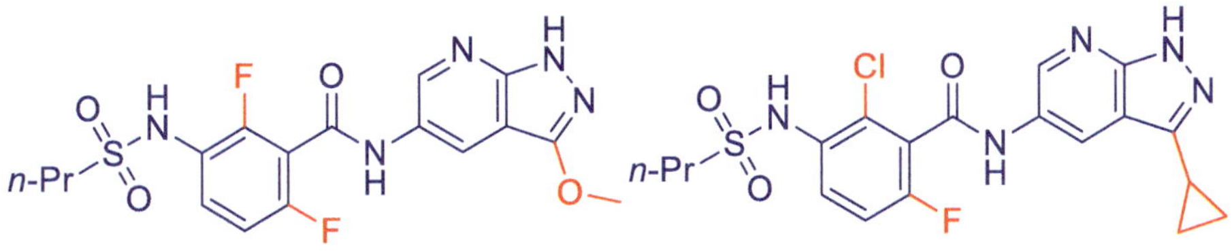

28	29
mp = 229 °C	164 °C
Clog *P* = 1.61	2.19
sol pH 7.4 = 9 μg/mL	127 μg/mL

3.2.2.3 Disrupting Hydrogen Bonding

For a solvate molecule to dissolve in water, the existing intermolecular hydrogen bonding must be broken so that the solvate molecule may form intermolecular hydrogen bonding with water molecules. Thalidomide (**30**) is probably one of the most infamous drugs because of its flaming teratogenicity. It has a high melting point (275 °C) and low lipophilicity. It is highly crystalline partially because of the presence of a hydrogen bond donor on the imide ring. Not surprisingly, thalidomide (**30**) has an abysmal solubility of 52.1 μg/L. *N*-Methyl-thalidomide (**31**), with the hydrogen bond donor eliminated, has a melting point (159 °C) that is a drastic 116 °C plummet from **30**'s 275 °C.[45]

Concurrently, **31** has an aqueous solubility of 275.9 μg/L, a more than five-fold boost over that of **30**. Apparently, the loss of the imido-hydrogen's ability to form hydrogen bond with water is more than compensated for the reduced crystallinity of the compound. However, *N*-propyl-thalidomide (**32**) and *N*-pentyl-thalidomide (**33**) have even lower melting points as a consequence of further reduced crystallinity, which is insufficient to compensate the damage that increased lipophilicity does to their solubility. As a result, *N*-propyl-thalidomide (**32**) has a similar solubility to that of thalidomide (**30**), and *N*-pentyl-thalidomide (**33**) is even less soluble than thalidomide (**30**).

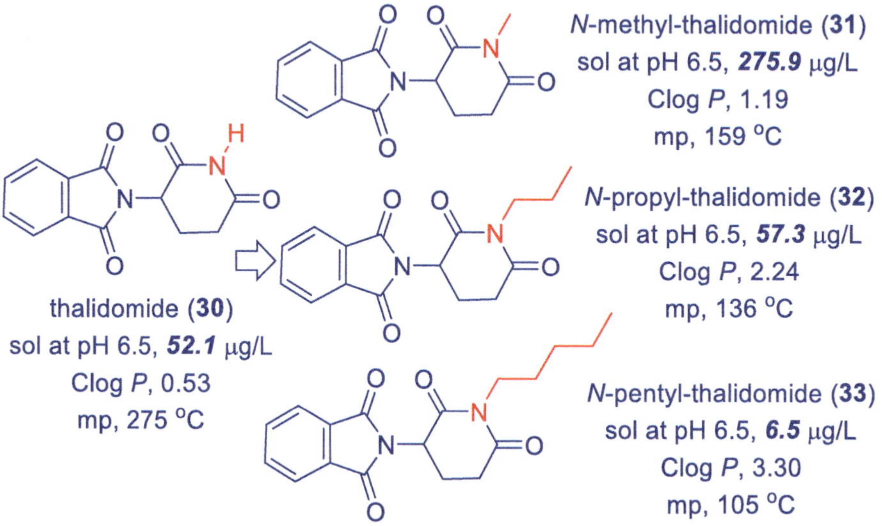

3.2.2.4 Subtle Changes

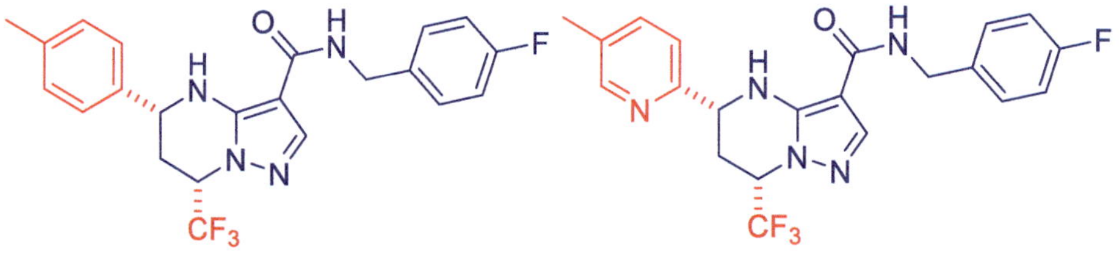

34, log *P* = 6.3
sol pH 6.8 <4 mM

35, log *P* = 4.4
sol pH 6.8 = 9 mM

Tetrahydropyrazolopyrimidine carboxamide **34** is a potent anti-tubercular agent, but it suffers from low aqueous solubility (<4 mM at pH 6.8).[46] In an effort to boost solubility while maintaining potency, a 2-pyridyl group was employed to replace the *p*-tolyl substituent to give **35**, which has a lower log *P* and a slightly higher water solubility (9 mM). For an unknown reason, replacement of the core 7-trifluororomethyl substituent in **35** with difluoromethyl afforded compound **36**, which interestingly also significantly increased its aqueous solubility to 212 mM. The most soluble compound **37** was arrived when 7-trifluororomethyl substituent was combined with two 2-pyridyl group to give the lowest log *P* of 3.2 and a solubility of 347 mM.

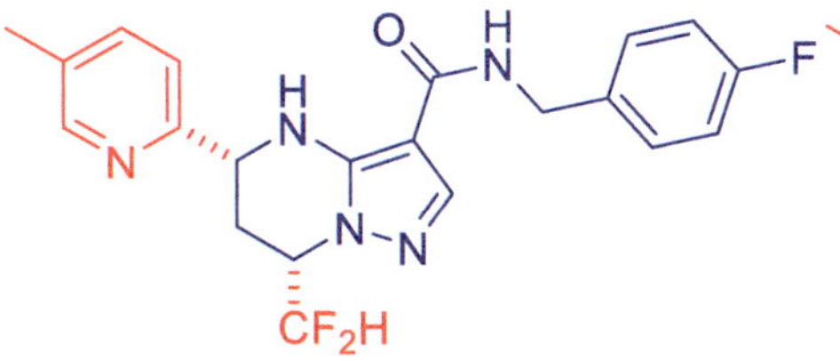

36, log *P* = 4.6
sol pH 6.8 = 212 mM

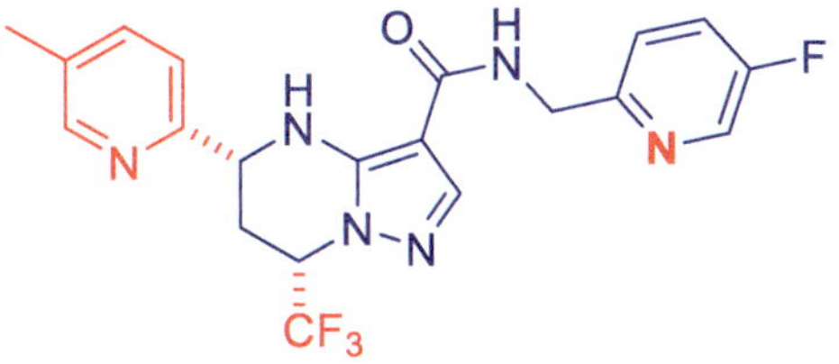

37, log *P* = 3.2
sol pH 6.8 = 347 mM

The impact of a "magic methyl" on the potency of a compound has been widely known.[47] Less known is that it could have a profound impact on a compound's solubility as well. Tetrahydropyrazolo[1,5-*a*]pyrazine **38** is a potent and selective as *a*taxia *t*elangiectasia and *R*ad-3-related protein (ATR) inhibitor with a poor solubility of 3 μM.[48] A simple addition of a methyl substituent on the piperazine ring led to **39**, which has an excellent solubility of 188 μM, an impressive 63-fold increase. This may be explained by the fact that the F_{sp3} value increased five-fold from the replacement of hydrogen with a methyl group. Again, F_{sp3} is the fraction of sp^3 hybridized heavy atoms. The presence of the methyl moiety probably exerted enough steric hindrance to prevent free rotation of the azaindole ring.

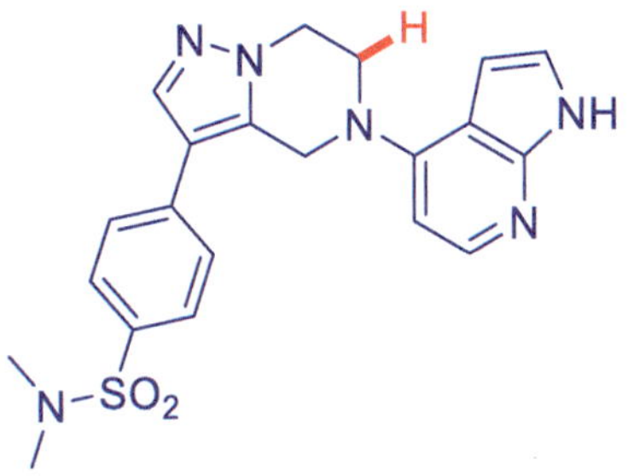

38, ATR IC_{50} = 1 nM
LLE = 4.5
sol, 3 μM
F_{sp3}, 0.1

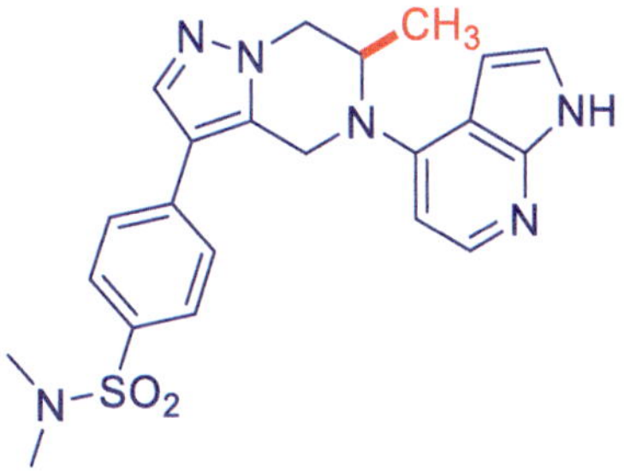

39, ATR IC_{50} = 0.5 nM, cell, 37 nM
LLE = 5.4
sol, 188 μM
F_{sp3}, 0.5

Chemistry is magical and never ceases to amaze. Sometimes, a subtle difference could make a striking impact. A matched pair of melanin concentrating hormone receptor 1 (MCHR1) agonists **40** and **41** have comparable potency. Yet **41** is more than 200,000-fold more soluble than **40** in DMSO at pH 7.4.[49] The profound difference is apparently the consequence of different dipole moments of the two regioisomeric oxadiazoles.

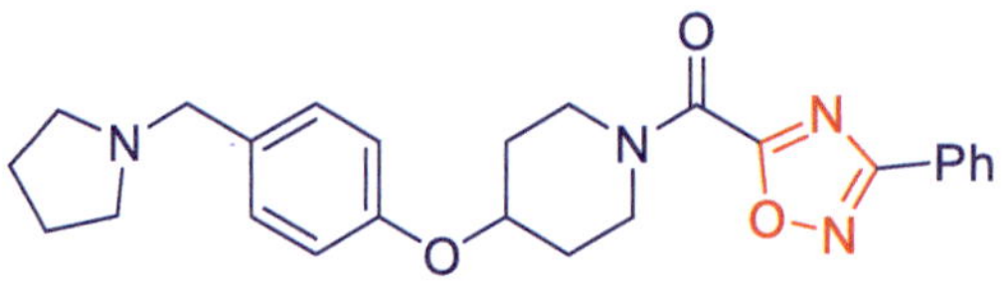

40
IC_{50}, 43 nM
log *D*, 3.2
LLE, 3.9
sol, <0.005 μM

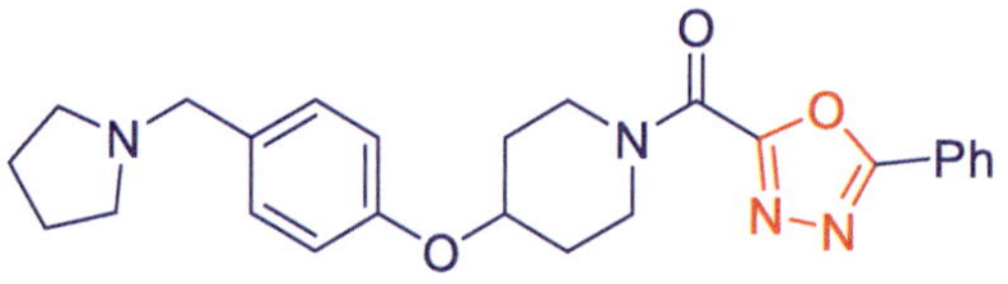

41
IC_{50}, 75 nM
log *D*, 2.2
LLE, 5.2
sol, >100 μM

3.2.3 Absorption by Diffusion

Experimentally, many methods exist to measure a drug's permeability. They may be divided into cellular methods and noncellular methods.

Caco-2 permeability assay is the most popular *cellular* methods. Developed at Sloan–Kettering in the 1970s, the Caco-2 cell line is a continuous cell of heterogeneous human epithelial colorectal adenocarcinoma cells. The cells, when cultured on semi-porous filters, form confluent monolayers that model the intestinal epithelial barrier for permeability assays.[50]

One of the most popular *noncellular* methods is the parallel artificial membrane permeability assay (PAMPA). It has been employed to determine the permeability of substances from a donor compartment through a lipid-infused artificial membrane into an acceptor compartment.[51]

When the solution of a drug in plasma reaches the exterior of a cell, the concentration of the drug inside the cell (cytoplasm) is zero. Shear physics demands diffusion of the drug from outside the cell membrane to cytoplasm until the concentrations reach equilibrium. There are two means of diffusion: transcellular absorption and paracellular absorption.

Transcellular absorption, as depicted in Figure 3.5, is the main route of absorption for most oral drugs. As many as 90% drugs get absorbed via this route of absorption. Needless to say, the drug must be in the solution at cell surface. Since the drug has to be neutral to pass through the cell membrane, its pK_a value is very consequential. Meanwhile, lipophilicity is also important, ideally with log *D* at 1–4 range. The lipid-soluble unionized drug diffuses across the lipid bio-membrane in the direction of their concentration gradient. It does not need energy. Molecules within Lipinski's Ro5 have better chances of permeating the cell membrane. Generally speaking, diffusion through lipid of cell membrane depends on area, diffusion gradient, diffusion coefficient, lipid solubility, etc. Compounds with molecular weights less than 200 are absorbed transcellularly with ease. Compounds with molecular weights between 200 and 300 may, or may not, permeate the cell membrane via transcellular absorption. Drugs with molecular weights greater than 300 would have difficulty crossing membrane via transcellular absorption.[52] As far as macrocycles are concerned, a small portion of them *may* cross cell membranes as well via extensive intramolecular bonding (see Section 3.1.1 and reference 19 for details).

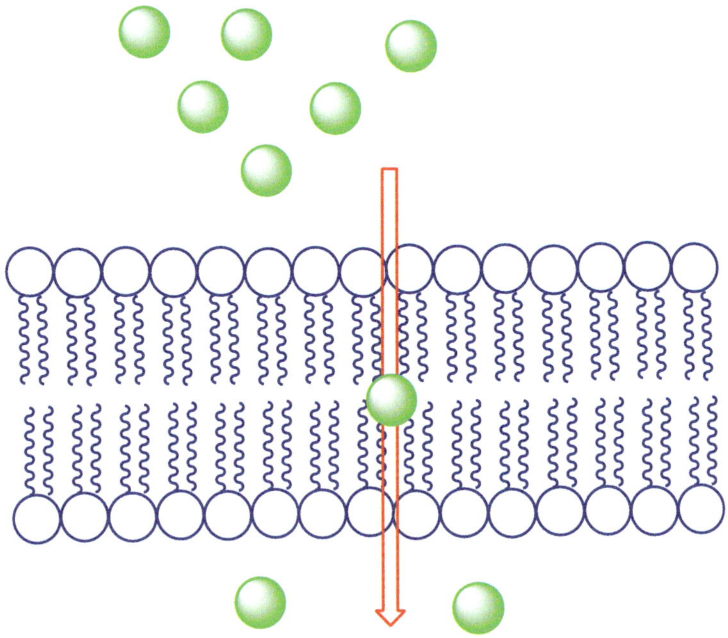

Figure 3.5 Transcellular absorption.

Approximately 5–10% drugs permeate through the cell membrane via *paracellular absorption* where a drug passes through gaps between cells. Passage of a drug through aqueous pores in the membrane through paracelullar spaces is also known as filtration. The moving force is hydrostatic or osmotic pressure. Lipid insoluble drugs cross the biomembrane by filtration only if their molecular size is smaller than the diameter of the enlarged aqueous pores. The filtration has importance mainly at the level of renal glomerulus, where the size of capillaries have large pores (40 Å) and most drugs (even albumin) can filtrate. The brain capillary pores have small sizes, therefore, allowing molecules with only smaller sizes to penetrate, whereas larger molecules have harder time crossing the blood–brain barrier (BBB).

3.2.4 Absorption by Active Transports

Some molecules get across the cell membrane not via diffusion, either transcellular or paracellular. Indeed, glucose, ions (proton, sodium, potassium, calcium, etc.), and neurotransmitters are absorbed by active transports.

An active transport is also known as a carrier transport or a carrier-mediated transport,[53a] or a membrane transport protein. As depicted in Figure 3.6, it acts as a *ferry-boat* to transport the molecule across the lipid region of the membrane. Active transports are specific, and only shuttle particular molecules across. For instance, levodopa (L-DOPA, **66**) and methyldopa are actively absorbed from the gut by aromatic amino acid transport. Examples also include iron in gut; L-DOPA (**66**) at BBB; and anion/cation transport in kidney. Active transports are also saturable and competitively inhibited by analogs that utilize the same carrier.[53b]

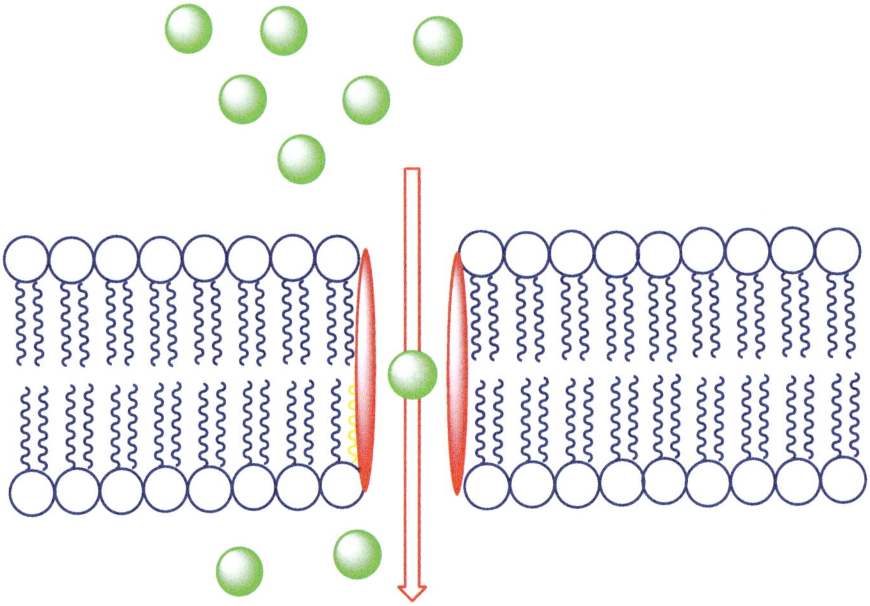

Figure 3.6 Absorption by active transport.

Among many active transports, ion transporters (ion channels) are responsible for shuttling ions across the cell membrane. The ions include proton, sodium, potassium, calcium, etc. One particular ion channel, hydrogen potassium (H^+, K^+) ATPase is colloquially known as proton pump. Inhibition of proton pump is the MOA for popular ulcer drugs omeprazole (Prilosec) and esomeprazole (Nexium).[54] Calcium (Ca^{++}) channel blockers include well-known antihypertensive drugs nifedipine (Adalat, **42**) and amlodipine besylate (Norvasc, **43**).

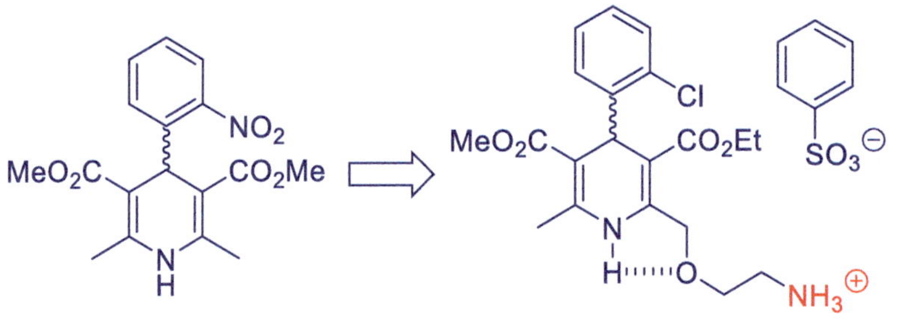

nifedipine (Adalat, **42**)
$t_{1/2}$ ~ 1 h

amlodipine besylate (Norvasc, **43**)
$t_{1/2}$ ~ 34 h, F 74%, V_d 16 L/kg

A more menacing ion channel is a potassium ion (K^+) channel known as human ether-a-go-go (hERG, Kv11.1).[55] hERG channel, promiscuous by design, is large with fourfold symmetry with multiple binding sites. It plays a central role in cardiac repolarization. Drugs that are hERG substrates tend to have QTc prolongation and consequently cardiotoxicities.[56] Some drugs have been withdrawn due to hERG channel inhibition. The poster child for hERG issue is Pfizer's dofetilide (Tikosyn, **44**), a class III antiarrhythmic agent. Since it is a potent hERG channel substrate, it has been implicated with a serious side effect called torsades de pointes (TdP), a rare but serious condition manifested as QT prolongation on ECG. Another two well-known hERG substrates are Janssen's antipsychotic haloperidol (Haldol, **45**) and Pfizer's antibiotic trovafloxacin (Trovan, **46**), respectively.

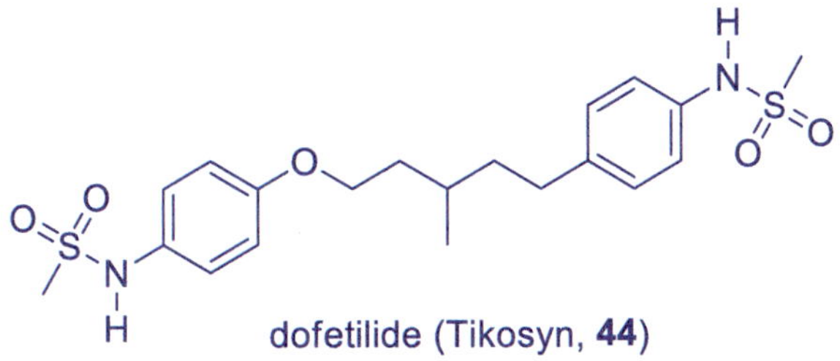

dofetilide (Tikosyn, **44**)

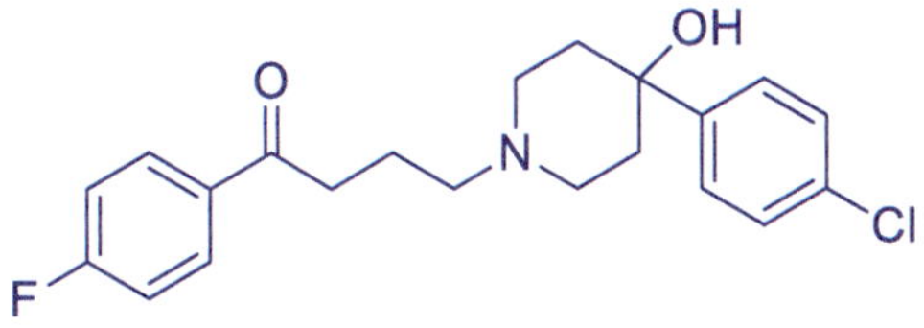

haloperidol (Haldol, **45**)

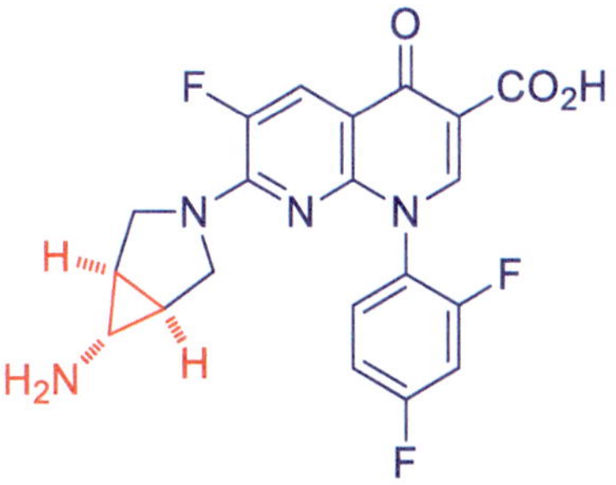

trovafloxacin (Trovan, **46**)

Several tactics exist to mitigate the hERG issue including:

(i) Structural modifications by disrupting interactions with Tyr652 and Phe656 amino acids on the protein. Adding peripheral aryl rings, introducing constraints, and varying stereochemistry have been successful.

(ii) Replacing phenyl with heteroaryl ring. Merck's lead compound **47** as an antagonist of adenosine receptor subtype 2A (A_{2A}) was potent enough with a K_i value of 5.5 nM in an assay against human A_{2A} receptor.[57] Unfortunately, its Clog *P* is relatively high (4.0) and its distal fluorophenyl ring forms favorable interactions with *four Phe656* in the tetrameric hERG channel, which leads to high hERG activity (IC_{50} = 1.55 μM). Taking a page from their past successes, Merck chemists replace the fluorophenyl moiety with a five-membered heterocycles and all resulting derivatives had their hERG activity mitigated. Most impressively, dimethylthiazole analog **48** saw both boost of potency for A_{2A} and diminishing of the hERG activity.

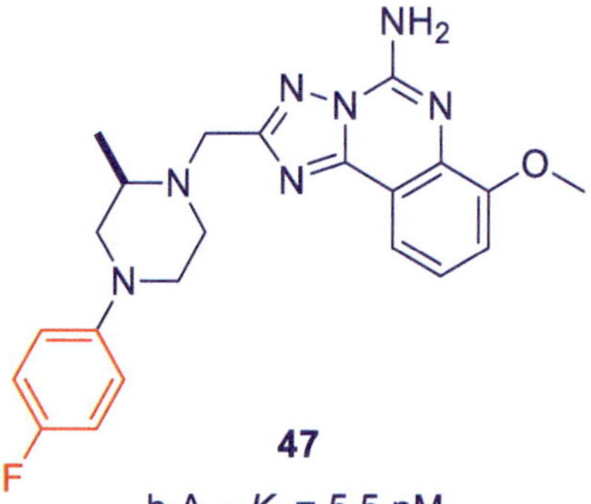

47

h-A_{2A} K_i = 5.5 nM
Clog P = 4.0
hERG K_i = 1.55 μM

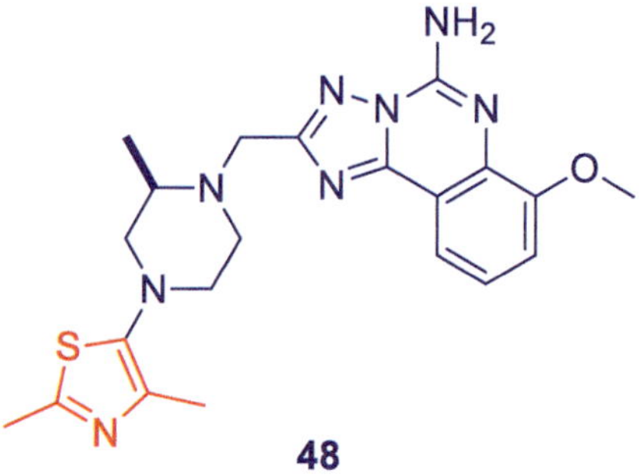

48

h-A_{2A} K_i = 1.4 nM
Clog P = 3.6
hERG K_i > 60 μM

(iii) Control lipophilicity. Often, there is a relationship between blocking the hERG and the measure of lipophilicity. If so, there is a lipophilic binding site that is accessed from the intracellular domain. Therefore, increasing polarity will interrupt lipophilic interactions. As a rule of thumb, molecules with a Clog P value smaller than 3.0 tend to have reduced tendency to bind to the hERG channel.

Compound **49**, also an A_{2A} receptor antagonist, was prepared by Merck chemists where the fluorophenyl motif on **47** was replaced with a pyrimidyl substituent.[57] It was found to be essentially inactive for hERG, probably due to the higher polarity of the molecule with a calculated Clog P as low as 2.0.

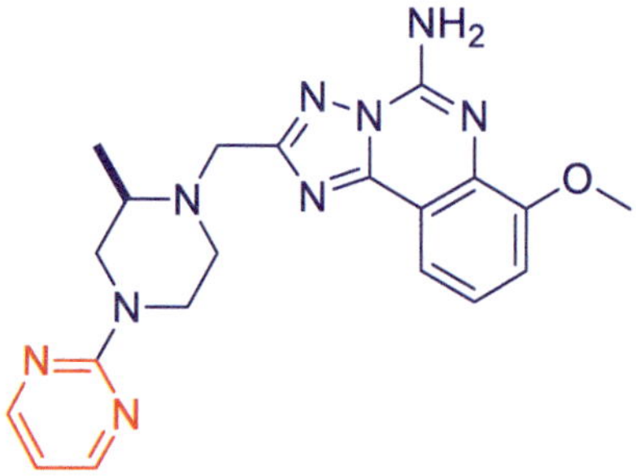

49

h-A_{2A} K_i = 26.5 nM
Clog P = 2.0
hERG K_i = > 60 μM

The renal outer medullary potassium channel (ROMK) inhibitors are potentially diuretics/natriuretics. Merck's ROMK inhibitor **50** had a very active hERG activity.[58] Many reiterations of SAR optimizations resulted in MK-7145 (**51**) whose hERG activity was significantly mitigated, probably thanks to the exponential decrease of lipophilicity with Clog P value of –0.51. MK-7145 (**51**) was the first small molecular ROMK inhibitor in clinical trials.

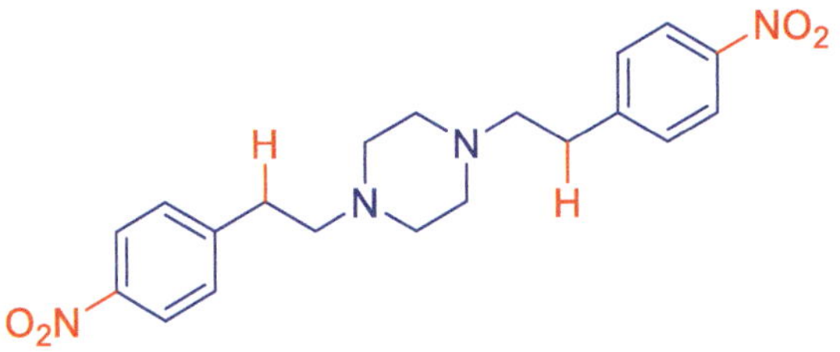

50
ROMK IC_{50} = 52 nM
hERG IC_{50} = 7 nM
Clog *P*: 3.3

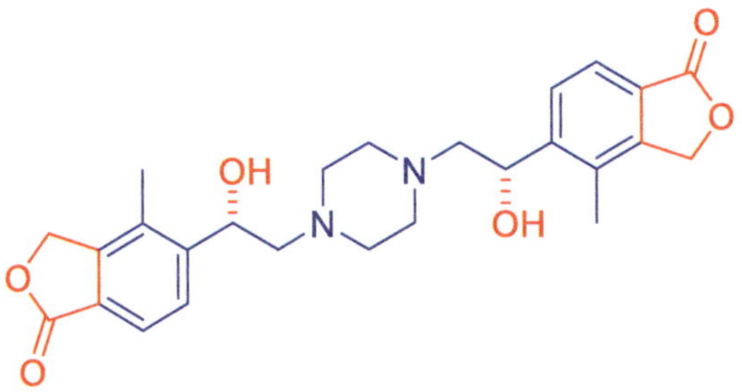

MK-7145 (**51**)
ROMK IC_{50} = 6.8 nM
hERG IC_{50} = 23 μM
Clog *P*: –0.51

(iv) Control of pK_a. The basic amine is not required for the hERG blockage and aromatic or lipophilic groups could have similar functions. Therefore, lower the pK_a of the basic nitrogen often reduces the hERG activity by disrupting cation–π interaction. Furthermore, shielding the basic center by bulky groups or constraints also retards binding to the hERG channel.

Building upon Merck's success of selective non-peptide neurokinin (NK_1) antagonist aprepitant (Emend) for the treatment of chemotherapy-induced nausea and vomiting, Huscroft and coworkers at Merck interrogated a series of 1-phenyl-8-azabicyclo[3.2.2]octane ethers. While compound **52** had good potency and efficacy in animal models, its interactions with hERG channel was less favorable with a K_i value of 100 nM.[59] A seemingly minute change of installing a fluorine atom at the α-position of tetrazole altered the pK_a value from **52**'s 7.3 to the fluoro-analog **53**'s 5.0. As a consequence, the selectivity over the hERG channel was soundly improved.

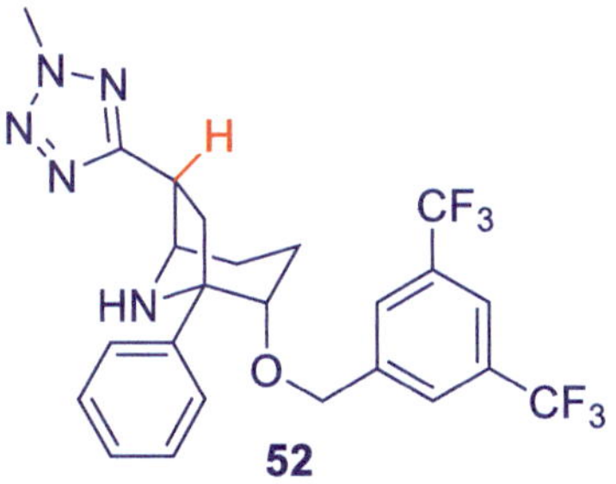

52
hNK1 IC_{50} = 1 nM
hERG K_i = 100 nM
Clog *P* = 4.3
pK_a = 7.3

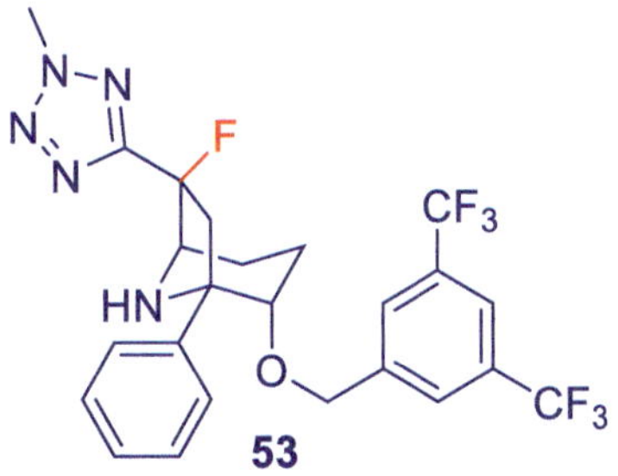

53
hNK1 IC_{50} = 1.5 nM
hERG K_i > 10,000 nM
Clog *P* = 5.1
pK_a = 5

(v) Formation of zwitterions limits the membrane permeability of a compound. Therefore, forming a zwitterions of a drug prevents access to the transmembrane binding site and minimizes potential interaction with hERG.

Terfenadine (Seldane, **54**), a lipophilic second-generation histamine H_1 receptor antagonist (antihistamine), is not a very good drug. Its log D is 2.11 with a pK_a of 8.6. Like many lipophilic and basic drugs, it binds to the potassium ion channel hERG and causes QT elongation and consequently cardiotoxicities. It was withdrawn in the 1980s. In contrast, its major metabolite, fexofenadine (Allegra, **55**), is a good drug. Being a hydrophilic zwitterion (log D = ~0.4), it is a class III substance according to BCS: high solubility, low permeability. While lipophilic terfenadine (Seldane, **54**) is metabolized extensively, the hydrophilic fexofenadine (Allegra, **55**) has negligible metabolism. It is selective over hERG (and many other biological targets) and is devoid of cardiotoxicity. The transformation from **54** to **55** may offer an invaluable lesson on how to overcome hERG issues that have plagued many lipophilic and basic compounds. Finally, protein binding for **54** to **55** is quite different as well. While lipophilic **54** is 97% plasma protein-bound in humans (high); hydrophilic **55** is ~65% protein-bound (moderate).[60]

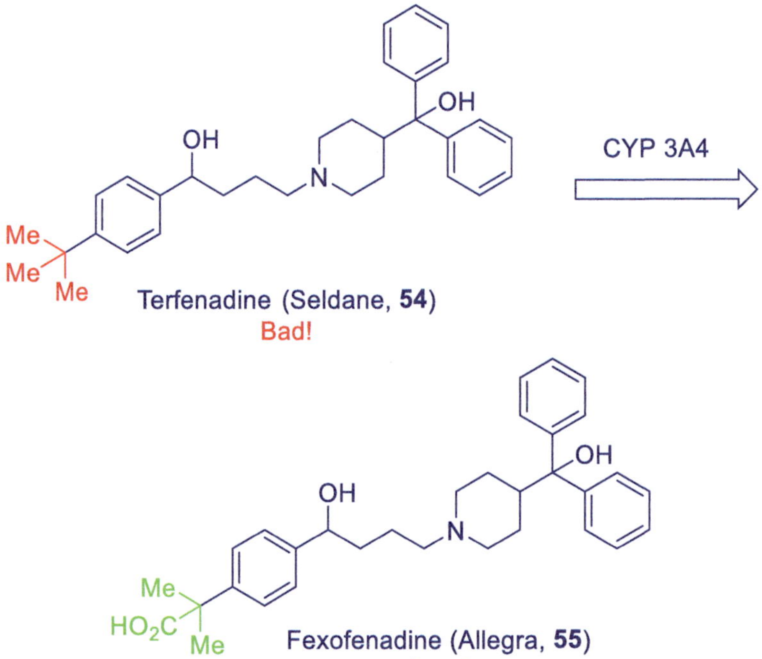

Terfenadine (Seldane, **54**)
Bad!

Fexofenadine (Allegra, **55**)
Good!

In the practice of medicinal chemistry, several tactics have been employed individually or altogether to mitigate or eliminate interactions with hERG.

Membrane glucose transporters (GLUT1–4) facilitate transporting glucose, so do another family of active transports, sodium–glucose linked transporter (SGLT) proteins. One particular isoform SGLT-2 has proven to be a viable target for treating type-2 diabetes. Canagliflozin (Invokana, **56**) and apagliflozin (Farxiga, **57**) are the first two SGLT-2 inhibitors approved by the FDA in 2013 and 2014, respectively.[61]

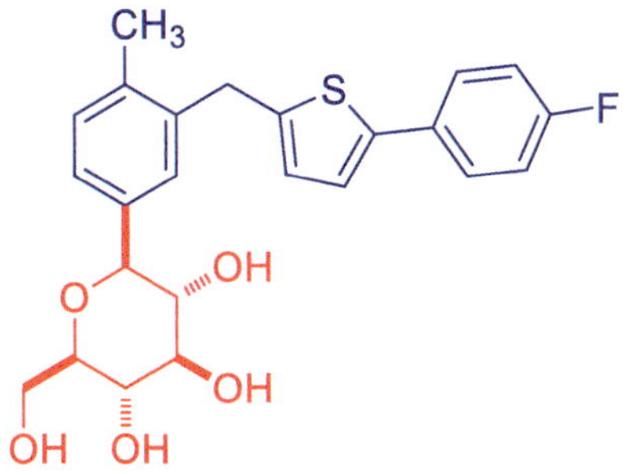

canagliflozin (Invokana, **56**)

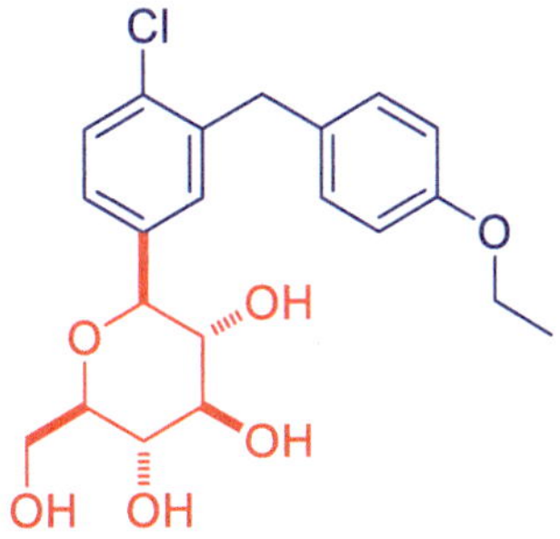

dapagliflozin (Farxiga, **57**)

Another essential class of active transports are neurotransmitter transporters, responsible for the reuptake of neurotransmitters such as serotonin, dopamine, glycine, and γ-aminobutyric acid (GABA). One prominent family of modern antidepressants are selective serotonin reuptake inhibitors (SSRIs). They include household names such as fluoxetine (Prozac, **58**), paroxetine hydrochloride (Paxil, **59**), and sertraline (Zoloft, 7).

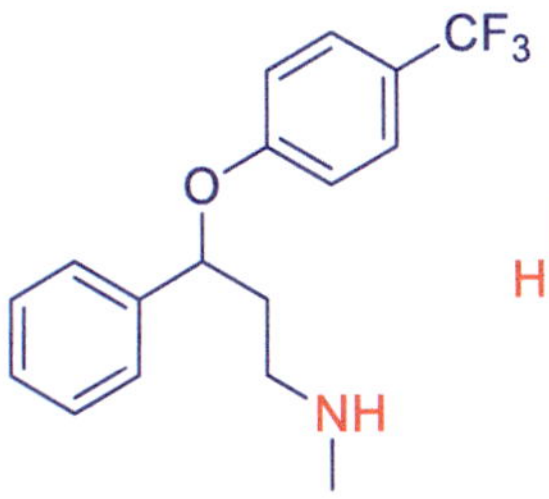

fluoxetine (Prozac, **58**)

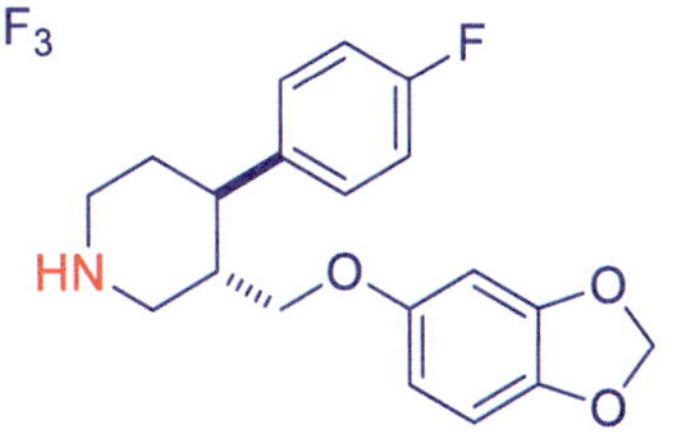

paroxetine hydrochloride (Paxil, **59**)

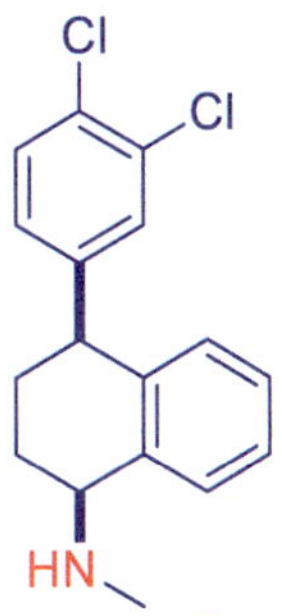

sertraline (Zoloft, **7**)

3.2.5 Absorption by Pinocytosis

Pinocytosis involves the invagination of a part of the cell membrane and trapping within the cell of a small vesicle containing extracellular constituents. The vesicle contents can then be released within the cell, or extruded from the other side of the cell. Pinocytosis is pivotal for the transport of some macromolecules, e.g., insulin through BBB and botulinum toxin in gut.

3.3 Distribution

3.3.1 Around the Blood Supply, to Tissues, and to Cells

Once a drug is absorbed, it is subsequently distributed around the blood supply and to tissues and cells. Distribution is the process by which a drug reversibly leaves the blood stream and enters the interstitial or cellular fluid of the body. Intestinal fluid, intracellular fluid, and transcellular fluid are 16%, 35%, and 2% of the body mass, respectively. Meanwhile, plasma is 5% of body mass and fat is 20%. Details on physiologic volumes of five common species of interest are compiled in Table 3.6.

Table 3.6 Physiologic volumes of body fluids across nonclinical species and human.

	Mouse (0.02 kg)	Rat (0.25 kg)	Dog (10 kg)	Monkey (5 kg)	Human (70 kg)
Total body water (mL)	14.5	167	6,036	3,465	42,000
Intracellular fluid (mL)	–	92.8	3,276	2,425	23,800
Extracellular fluid (mL)	–	74.2	2,760	1,040	18,200
Plasma volume (mL)	1.0	7.8	515	224	3,000

All of the fluid in the body (total body water) in which a drug can be dissolved may be roughly divided into three compartments: intravascular (blood plasma found within blood vessels); interstitial/tissue (fluid surrounding cells), and intracellular (fluid within cells, i.e., cytosol). The distribution of a drug into these compartments is dictated by its physical and chemical properties. Compounds distribute differentially within body and PPB may limit distribution. Drugs may accumulate in specific organs or become bound to specific tissue constituents (tissue storage). Not surprisingly, liver, kidneys, and other excretory organs often show high concentrations of compounds. Most conspicuously, lipophilic compounds may accumulate in fatty tissues. For instance, thiopental, ether, and minocycline tend to collect in adipose tissues. Additional examples of tissue storage include iodine in thyroid gland; calcium, tetracyclines in bones and teeth; digoxin (to muscle proteins) in heart and skeletal muscles; chloroquine, tetracyclines, and digoxin in liver; tetracyclines and digoxin in kidney; chloropromazine, isoniazid, and acetazolamide in the brain; chloroquine (to nucleoproteins and causes retinopathy) to retina; and finally, ephedrine and atropine (to melanin) in iris.

Overall, volume of distribution (V_d) of a drug is determined by its partitioning across various membranes; binding to tissue components; binding to blood components; and physiological volumes. Apparent volume of distribution (V_d) is a *primary PK parameter* and could be greater than 10,000 L. Such an astronomical number of V_d means that the majority of the drug is in the tissue and very little is in the plasma circulating. The larger the volume of distribution, the more likely that the drug is found in the tissues of the body. In contrast, the smaller is the volume of distribution, the more likely is the drug confined to the circulatory system. As shown in Table 3.7, the volume of distribution (V_d) for acidic, neutral, and basic compounds are different. Generally speaking, acidic drugs have the lowest value of V_d (<0.4 L/kg), whereas basic drugs have the highest value of V_d (>1.0 L/kg). Neutral drugs have moderate values of V_d (0.4–1.0 L/kg). The units L/Kg and L may be easily interconverted by assuming a "generic" person's bodyweight is 70 kg.

Table 3.7 Volume of distribution (V_d) for acidic, neutral, and basic compounds.

Compounds	V_d (L/kg)	V_d (L)
Acidic	<0.4	<28
Neutral	0.4–1.0	28–70
Basic	>1.0	>70

A wide range of V_d values are shown in drugs **60**–**63**. The V_d value for Warfarin (**60**) is low and that of theophylline (**61**) is moderate. However, the V_d values for basic drugs quinidine (**62**) and imipramine (**63**) are considered high. In fact, 2,100 L for the V_d of imipramine (**63**) is ranked "very high." One explanation of for the high values of V_d for lipophilic amine-containing drugs is lysosomotropism (also known as lysosomal trapping or lysosomal sequestration) the phenomenon that lipophilic amines (log P >1) and amphiphilic drugs (cationic amphiphilic drugs) with ionizable amines (pK_a >6) can accumulate in lysosomes.[62] Therefore, the presence of basic amines normally leads to increase of tissue affinity, thus boosts the V_d value.

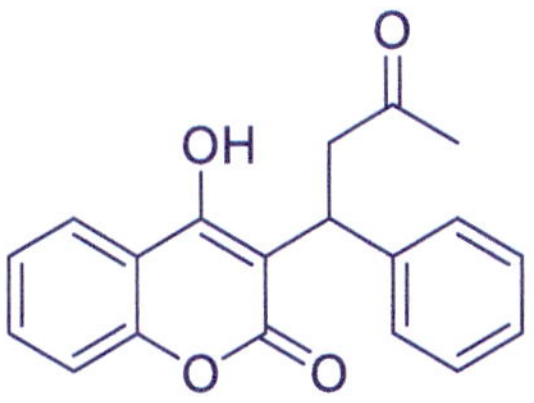

Warfarin (**60**), V_d = 8 L

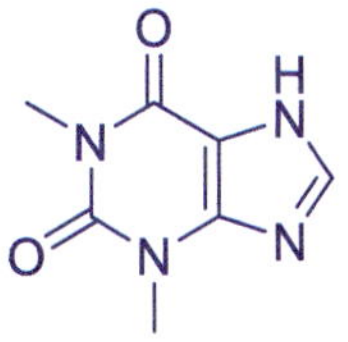

theophylline (**61**), V_d = 35 L

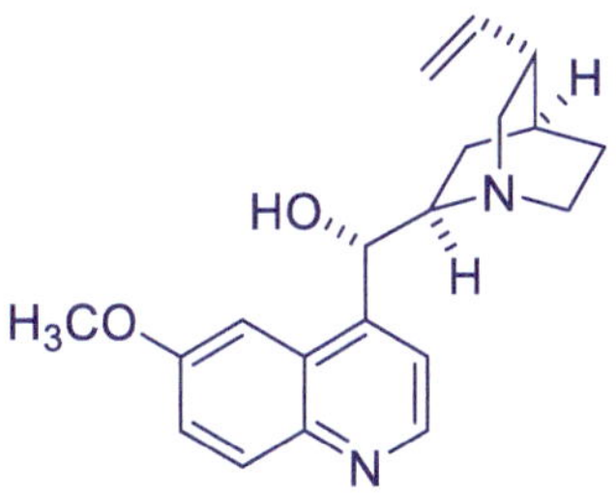

quinidine (**62**), V_d = 150 L

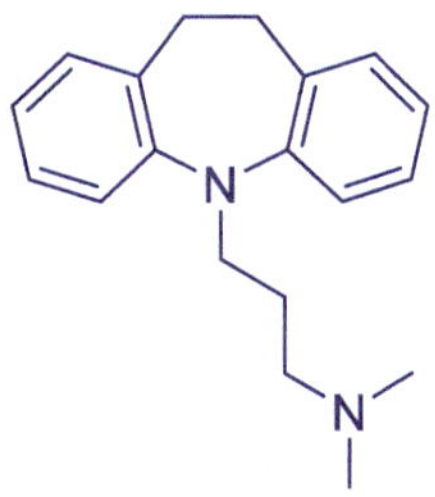

imipramine (**63**), V_d = 2,100 L

A well-known example is progression of the first-generation to second- and third-generation calcium channel blockers. Nifedipine (Adalat, **42**), a first-generation calcium channel blocker, is a neutral drug with a moderate V_d of 0.75 L/kg. It has a short half-life of 2 h, thus has to be taken three times a day. In stark contrast, amlodipine

(Norvasc, **43**), a third-generation calcium channel blocker, has a basic primary amine sidechain. Thanks to, presumably, its lysosomotropism, it has a very high V_d of 21 L/kg, which translates to a half-life of 35 h, enabling a qd (once daily) regimen.[63]

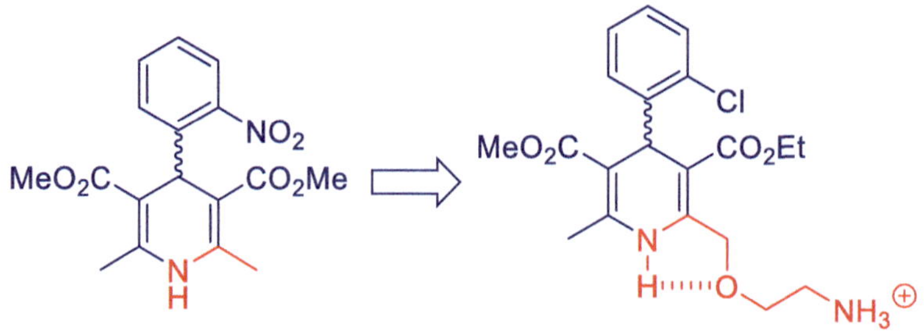

nifedipine (Adalat, **42**)
V_d, 0.75 L/kg
$t_{1/2}$, 2 h

amlodipine (Norvasc, **43**)
21 L/kg
35 h!

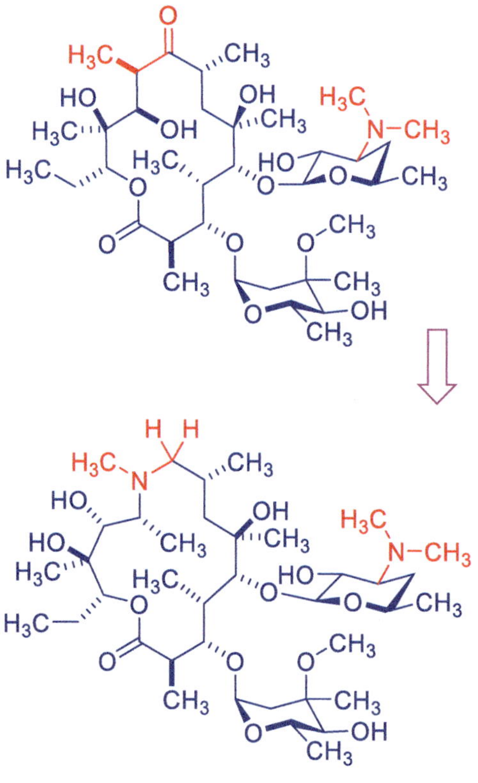

erythromycin (Erythrocin, **64**)
V_d, **4.8 L/kg**
Cl, 55 mL/kg
$t_{1/2}$, 3 h
tissue/serum ratio, 0.5–5x
F%, 25%
4x daily

azithromycin (Zithromax, **65**)
V_d, **62 L/kg**
Cl, 15 mL/kg
$t_{1/2}$, 18 h
tissue/serum ratio, 10–100x
F%, 37%
qd

Many macrolide antibiotics are inferior drugs in terms of PK. Some are Pgp inhibitors (*vide infra*) in addition to other flaws. Erythromycin (Erythrocin, **64**), for example, has only *one* basic nitrogen atom, a V_d value of 4.8 L/kg, and a half-life of 3 h. It has to be taken four times a day. Remarkably, azithromycin (Zithromax, **65**), which bears striking resemblance to erythromycin (**64**) but with *two* basic nitrogen atoms, is an excellent azamicrolide with a V_d value of 62 L/kg and a half-life of 18 h. This allows a qd regimen and makes it an exceptional antibiotic.[64]

3.3.2 Blood–Brain Barrier

Among over 7,000 drugs on the market, less than 5% of them are for treating CNS disorders. CNS drugs must cross the *BBB* to enter the brain first and then exert their pharmacological effects. Endogenous influx transporters resided at the BBB shuttle nutrients such as carbohydrates, monocarboxylic acid, neutral/basic amino acids, and purine nucleosides to the brain, but the BBB blocks xenobiotic from entering the brain. Neurotransmitter dopamine (DA, **67**) plays an essential role in Parkinson's disease, psychosis, attention deficit hyperactivity disorder (ADHD), and many other mental disorders. However, armed with three polar groups, dopamine (**67**) is too polar to enter the brain via BBB. In contrast, its precursor levodopa (**66**) does thanks to the amino acid transport located on BBB. Once in the brain, levodopa (**66**) is metabolized to dopamine (**67**).

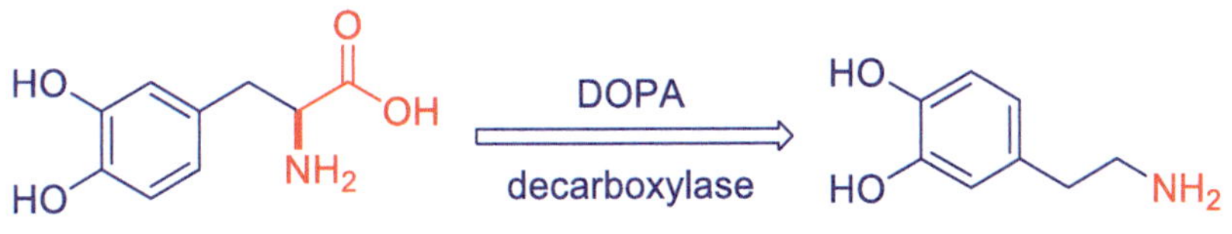

levodopa (**66**) dopamine (**67**)

Barring benefiting from active transporters, the requirements for CNS drugs' physicochemical properties are more stringent than other drugs in order to cross the BBB via passive diffusion (see Figure 3.7 for a comparison of their physicochemical property spaces).[65]

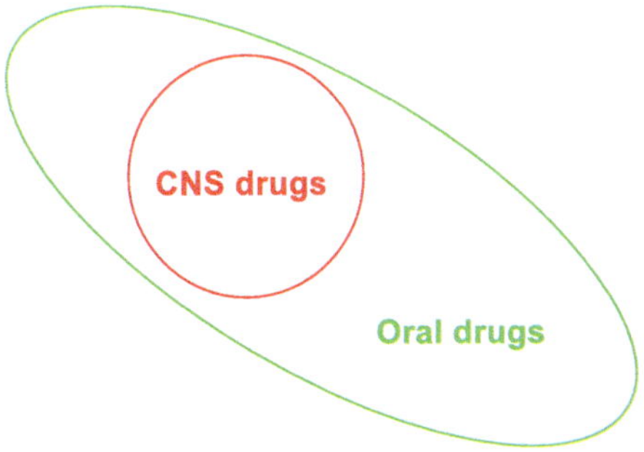

Figure 3.7 Physicochemical property space.

Optimization of physicochemical and pharmacokinetic properties may lead to the drug getting through the membrane and avoiding efflux transporters along the apical side of the BBB. Normally, high potential for hydrogen bonding generally results in decreased BBB permeability, thus highly polar molecules with strong hydrogen bonding capacity do not traverse BBB readily. After canvassing 1,500 drugs filtered from United States Adopted Names (USAN) or International Non-proprietary Names (INN) for good CNS penetration, Lipinski arrived at a rule for CNS penetration.[66] It states that a drug is likely to have good CNS penetration if its:

1. Molecular weight ≤400
2. log *P* ≤5
3. Hydrogen bond donor ≤3
4. Hydrogen bond acceptor ≤7

Otherwise, the drug is unlikely to have good BBB penetration.

BBB is largely comprised of lipids thus only lipid-soluble, unionized drugs penetrate to act on the CNS. Conversion of morphine (**68**, log *P* <–0.22) to the bis-acetylated derivative Heroin (**69**) significantly elevated the lipophilicity as reflected by the boost of the Clog *P* value (0.68). The fact that Heroin (**69**)'s brain penetration is 100-fold greater than that of morphine (**68**) speaks volumes for the fact that lipophilic drugs are more likely to penetrate BBB.

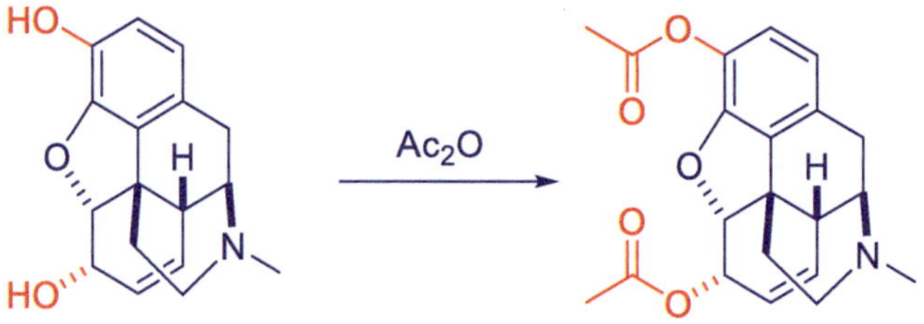

morphine (**68**)
Clog *P*: –0.22

Heroin (**69**)
Clog *P*: 0.68

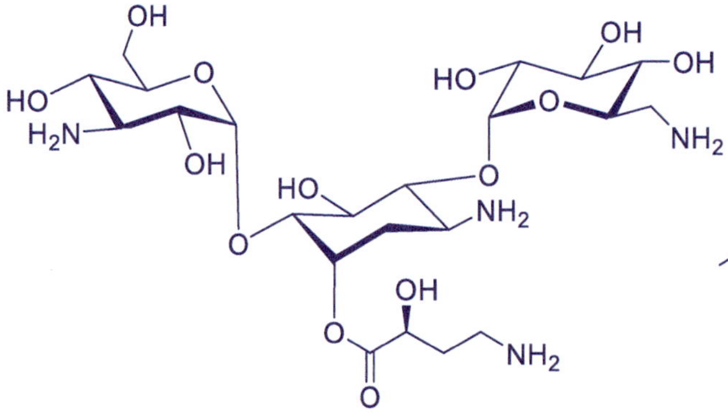

amikacin (**70**)
Clog *P*: –6.15

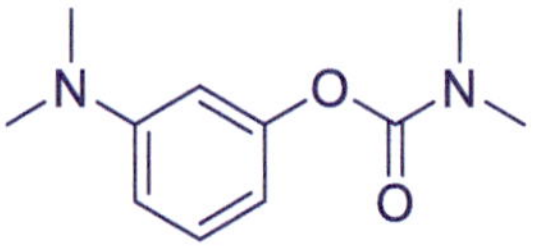

neostigmine (**71**)
Clog *P*: 1.70

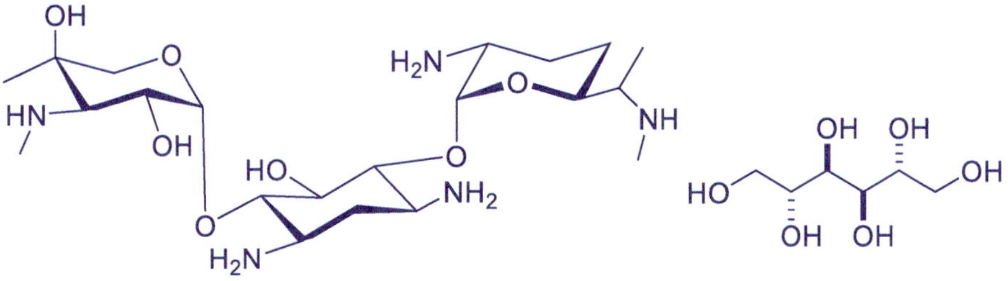

gentamicin C1 (**72**)
Clog *P*: −1.80

mannitol (**73**)
Clog *P*: −2.05

In contrast, BBB severely limits the entry of nonlipid soluble, hydrophilic drugs (nominally defined as drugs with log $P < 0$) as exemplified by amikacin (**70**), neostigmine (**71**), gentamycin (**72**), mannitol (**73**), etc. Interestingly, inflammation of the meninges of the brain actually increases drugs' permeability of the BBB.

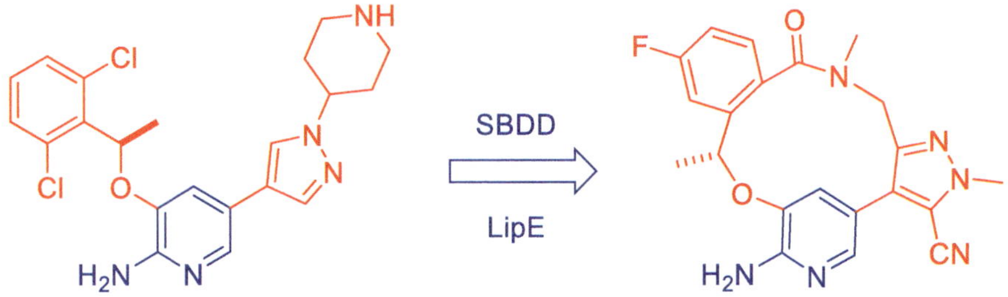

	crizotinib (Xalkori, **74**)	lorlatinib (Lorbrena, **75**)
ALK cell IC_{50}	80 nM	1.3 nM
ALK-L1196M cell IC_{50}	843 nM	21 nM
MDR BA/AB	45.0	1.5

Pfizer's crizotinib (Xalkori, **74**) is an anaplastic lymphoma kinase (ALK) inhibitor. It was the first ALK inhibitor approved by the FDA in 2011 to treat ALK-positive lung cancer patients. Regrettably, cancers treated by crizotinib (**74**) tend to develop resistance and metathesize in the brain. Resistant patient samples revealed a variety of point mutations in the kinase domain of ALK including the L1196M gatekeeping mutation. Furthermore, the drug itself has a cerebrospinal fluid (CSF) to free plasma ratio of 0.03, indicating a low probability of distributing to the brain. Therefore, it is advantageous if a drug is developed to penetrate the BBB. Johnson and his coworkers arrived at lorlatinib (Lobrena, **75**), a macrocyclic analog of crizotinib (**74**) after an extensive SAR campaign. Macrocycle **75** has a combination of broad-spectrum potency for both ALK kinase and its L1196M gatekeeper mutation, favorable central nervous system ADME, a high degree of kinase selectivity, and a multiple-drug resistance (MDR) value of 1.5 as opposed to 45.0 for the linear **74**.[67] The value of $MDR_{B\rightarrow A/A\rightarrow B}$ is a measure of *P-glycoprotein (Pgp)* 1-mediated efflux. The smaller is the number, the less likely is the drug pumped out of BBB. The numerical values of $MDR_{B\rightarrow A/A\rightarrow B}$ as a measure of *P-glycoprotein (Pgp)* 1-mediated efflux. The bidirectional assay for Pgp is often carried out employing the MDCK cell permeability assay.[50]

3.3.3 Efflux Transporters

The transformation of **74**→**75** touched upon the concept of Pgp (permeability glycoprotein), the most prevalent drug efflux transporter. Since Pgp is encoded by the multidrug resistance-1 (MDR-1) gene, Pgp and MDR-1 are sometimes interchangeable in colloquial settings. Pgp belongs to a class of ATP-binding cassette (ABC) transporters. In humans, as many as a dozen efflux transporters (transport proteins) exist in various tissues such as liver, intestine, kidney, and BBB. Contrary to active drug transports, which ferry drugs across the cell membrane from outside the cell to cytoplasm, *efflux transporters* shuttle drug *outside* the cell membranes. In addition to Pgp, other transport proteins include the organic anion transporter (OAT) family; multidrug resistance-associated protein (MRP); breast cancer resistance protein (BCRP), etc. Specific transporters may aid influx, or alternatively, promote efflux of a drug.

In the context of this book, we only focus on the most important efflux transporter: Pgp, which can transport drugs back out of the gut wall and into the gut lumen, thus reducing absorption. It transports drugs out of the kidney and into the urine. Pgp is mainly expressed in cells of large/small intestines, liver, kidney, pancreas, and the BBB and plays an important role in pumping foreign substance/toxins out of the cells in the gut and/or the brain, etc. Pgp is often a problem with CNS drugs and it is widely assumed to be a major determinant of brain penetration. Human body has evolved this way so that xenobiotics would not invade the brain so easily.

Pgp has been implicated as a cause of multidrug resistance in tumor cells since it is often overexpressed in tumor cells. In the 1970s, it was reported that some chemotherapy drugs were not bioavailable and Pgp was established to be the culprit. Juliano and Ling at University of Toronto in 1976 observed that Chinese hamster ovary (CHO) cells selected for resistance to colchicine (**76**) displayed a pleiotropic cross-resistance to a plethora of amphiphilic drugs.[68] They discovered Pgp, a surface glycoprotein modulating drug permeability in CHO cell mutants. As one of the efflux pumps, Pgp removes hydrophobic substrates directly from the plasma membrane. A 170 kDa protein, it is an ATP-dependent, multidrug efflux pump, effluxing cytotoxic drugs out of the cells.

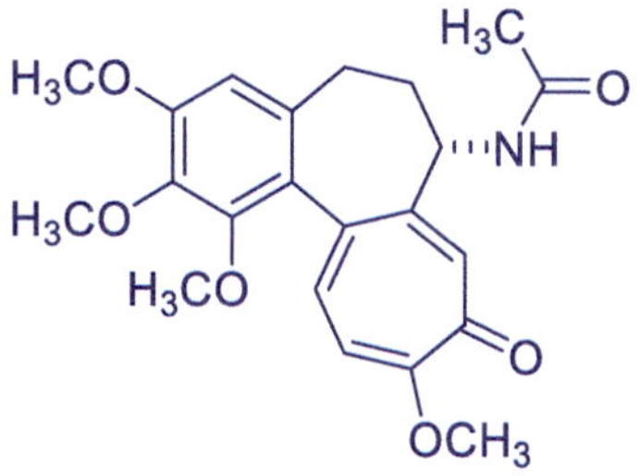

colchicine (**76**)

Half of the marketed drugs are Pgp substrates. *Pgp substrates* are defined as compounds transported by the Pgp, whereas *Pgp inhibitors* are compounds that have been shown to inhibit Pgp.[69] Pgp is characterized by having a promiscuous binding pocket that allows for hydrophobic and aromatic interactions which allow for a variety of structurally diverse drugs to be transported out of the cell from the plasma membrane, resulting in low intracellular drug levels. Most Pgp substrates tend to be amphipathic in nature, containing both hydrophobic and hydrophilic moieties that are spatially separated (think detergent). Examples of Pgp substrates include macrolide antibiotic erythromycin (**64**), DNA intercalator doxorubicin (Adriamycin, **77**), and human immunodeficiency virus (HIV) protease inhibitor indinavir (Crixivan, **78**), etc. On the other hand, representative Pgp inhibitors are quinidine (**62**), calcium channel blocker verapamil (**79**), antifungal ketoconazole (**80**), etc.

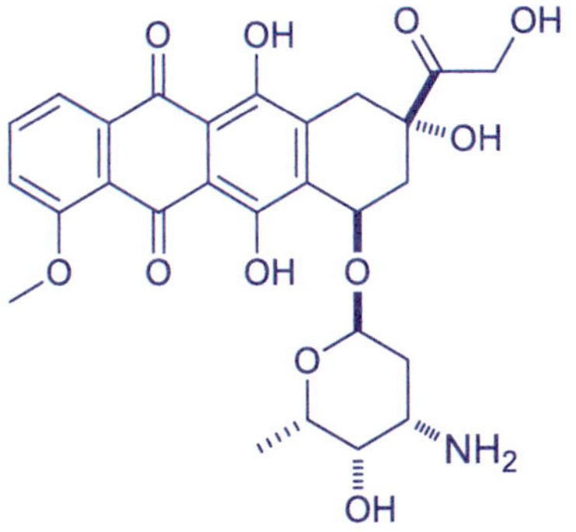

doxorubicin (**77**)
Pgp substrate

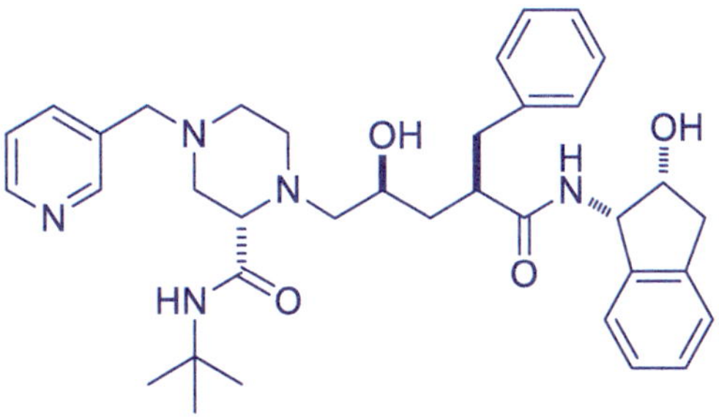

indinavir (Crixivan, **78**)
Pgp substrate

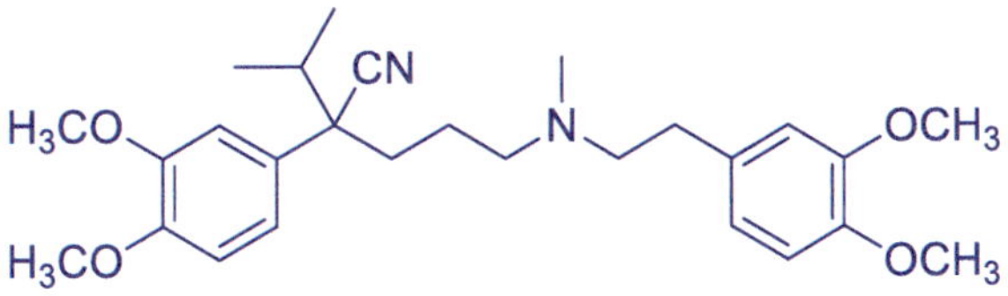

verapamil (**79**)
Pgp inhibitor

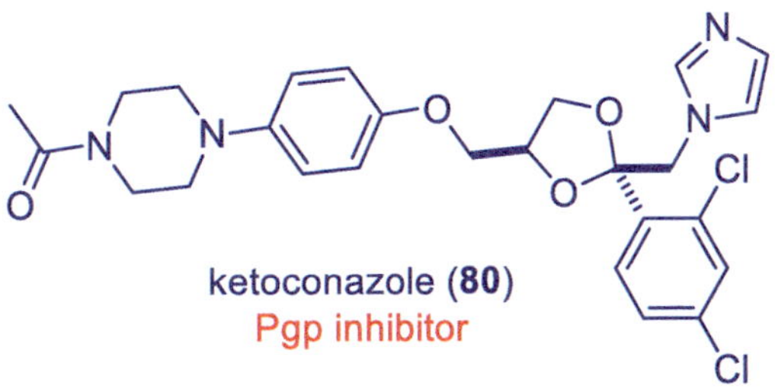

ketoconazole (**80**)
Pgp inhibitor

Pgp-mediated efflux is a potential source of peculiarities of drug pharmacokinetics, such as nonlinearity including dose-dependent absorption, drug–drug interactions (DDIs), intestinal secretion, and limited access to the brain. Strategies to mitigate the Pgp issue include (i). Co-administer an effective/selective Pgp inhibitor that does not cause cytotoxic effects and is reversible with the drug; and (ii). Evade Pgp by optimizing physicochemical properties to make the drug's permeability higher going into the cell than going out.

Petrauskas and colleagues proposed a rule of 4 (Ro4) regarding Pgp substrates from their extensive survey of existing drugs.[69] It states that a compound is more likely to be a Pgp substrate if its:

$N + O \geq 8$;
MW > 400; and
pK_a > 4.

In contrast, a compound is more likely to be a non-Pgp substrate if its:

$N + O \leq 8$;
MW > 400; and
pK_a < 8 (acids and neutrals).

Many tactics exist to abrogate the Pgp issue:[70,71] introduce steric hindrance to the hydrogen bond-donating atoms by attaching a bulky group; methylate the nitrogen atom; decrease hydrogen bond acceptor potential by adding an adjacent electron-withdrawing group; replace or removing the hydrogen bonding group, e.g., amide; modify structural features to interfere with Pgp binding, e.g., adding a strong acid; And modify log P to reduce penetration into the lipid bilayer where binding to Pgp occurs.

Tetracyclic compound **81** is a chemotherapy plagued with cytotoxic drug resistance as a consequence of being a Pgp substrate.[72] A Mannich reaction of **81** offered the corresponding 3-aminomethyl derivatives **82** and **83**, respectively. The maneuver conferred a salient feature to the resulting two compounds, namely, the potency for tumor cells otherwise resistant to a variety of anticancer drugs. It is likely that the steric hindrance of cyclic amines piperazine and quinuclidine on **82** and **83** minimized the hydrogen bonding-donating potential of the adjacent phenol group.

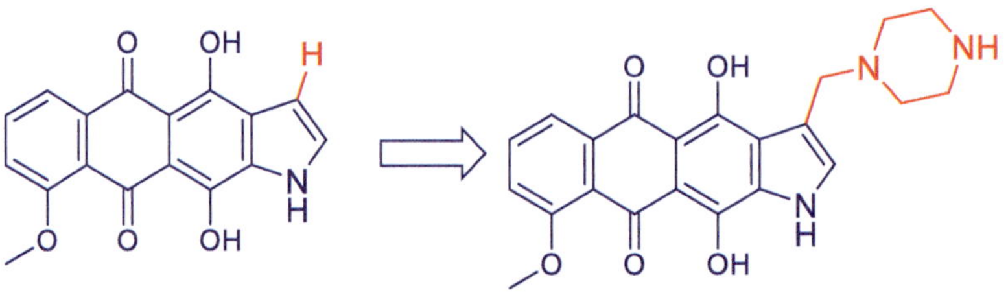

81, Pgp substrate **82**, **NOT** a Pgp substrate

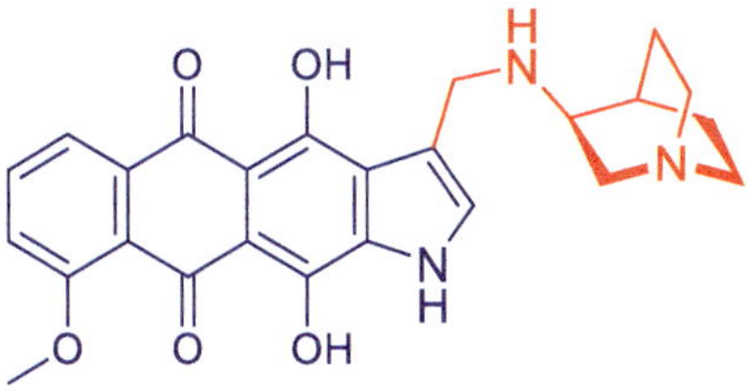

83, **NOT** a Pgp substrate

Vinca alkaloids, initially isolated from Madagascar periwinkles in the 1950s, are superb drugs for treating breast cancer, leukemia, and Hodgkin's disease even by today's standards. Their MOA is through depolymerization of tubulins, coincidentally, the same MOA of colchicine (**76**)'s as well. Inconveniently, they develop clinical resistance mediated by overexpression of the drug efflux pump phosphoglyco-protein Pgp. Vinblastine (Velban, **84**, R = OH), for instance, has a $P_{app(B\rightarrow A)}$ value of 38.2 × 10^6 cm/s, along with an efflux ratio of 16.2 and an 87% Pgp ATPase activity. Boger's group undertook a Herculean effort to tackle the Pgp issue and succeeded in maintaining, or even improving, the potency and simultaneously overcoming Pgp-derived efflux and resistance.[73] Their hard work was rewarded with the success for variations at the C20′ position. Among 180 amides that they made to replace the C20′–OH group on vinblastine (**84**), many showed similar potency but without the Pgp liability. When R is the bicyclic benzamide group, the amide derivative **85** is 11-fold more potent than vinblastine (**84**) in the HCT116 assay. Furthermore, **85** has a $P_{app(B\rightarrow A)}$ value of 1.5 × 10^6 cm/s, along with an efflux ratio of 2.2 and no detectable Pgp ATPase activity.

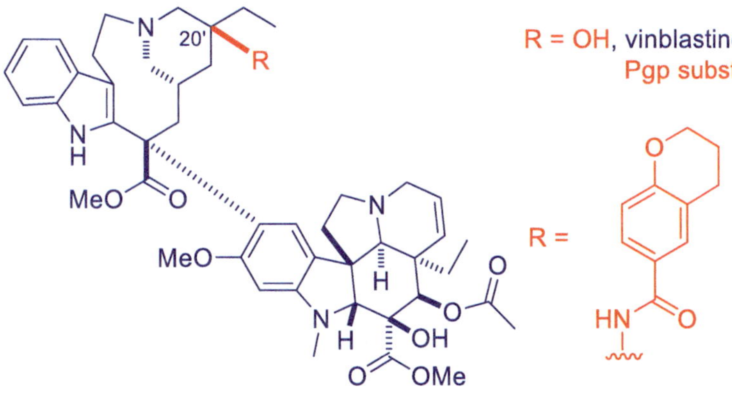

R = OH, vinblastine (Velban, **84**)
Pgp substrate

R = (bicyclic benzamide) **85**
NOT a Pgp substrate

Taxoids such as paclitaxel (Taxol, **86**), docetaxel (Taxotere), and cabazitaxel (Jevtana) are also remarkable cancer drugs, especially effective in treating breast cancer (BRCA) and castrate-resistant prostate cancer (CRPC).[74] Their MOA is stabilization of microtubules (whose function is the formation of the mitotic spindle responsible for separation of chromosomes). All taxoids are Pgp substrates and develop clinical resistance in due course. The Ojima group carried out extensive SAR investigations around paclitaxel (**86**). Among them, a new taxoid SB-T-1214 (**87**) is 25-fold more potent than paclitaxel (**86**). The exceptional activity may be ascribed to an effective inhibition of Pgp binding by the modified C-10 moiety (replacing acetate with the cyclopropane carboxylate). The magic of success stems from the astute observation that modifications at C-10 are tolerated for the activity against normal cancer cell-lines, but the activity against a drug-resistant human breast cancer cell-line expressing multidrug resistance (MDR) phenotype MCF7-R is highly dependent on the structure of the C-10 modifier. The other variation on **87** is replacement of the original C-3′ phenyl group with the isobutenyl substituent.[75]

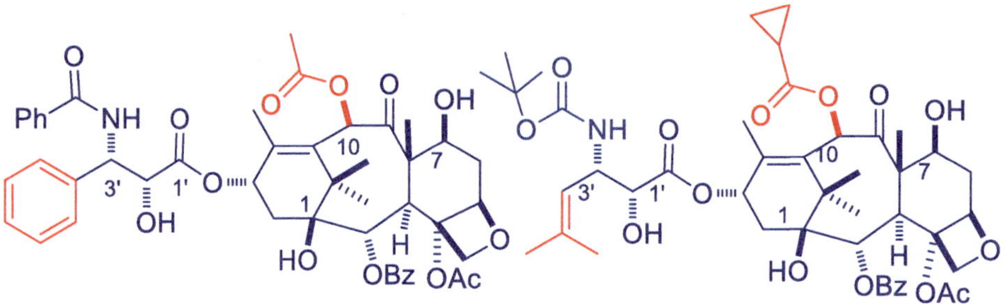

paclitaxel (Taxol, **86**)
potent
Pgp substrate

SB-T-1214 (**87**)
25X more potent
Not a Pgp substrate

Remarkably, a *bona fide* Pgp substrate is devoid of other Pgp substrates′ pitfall of multidrug resistance (MDR). Dp44mT (**88**) is highly effective and selective against a variety of belligerent solid human tumors *in vivo* by the intravenous and/or oral routes. It is transported into the lysosome by Pgp, causing lysosomal targeting of Dp44mT (**88**), and resulting in enhanced cytotoxicity. It overcomes MDR via utilization of lysosomal Pgp transport activity. The fact that the beneficial effect does not happen to more cancer drugs is because Dp44mT (**88**) is unique with three characteristics. (i), it is a Pgp substrate; (ii), it becomes charged at acidic pH to enable accumulation in lysosomes; and (iii), the agent causes marked redox stress in the acidic lysosome, leading to cytotoxic reactive oxygen species (ROS) that induces lysosomal membrane permeabilization (LMP) and apoptosis (cell death).[76]

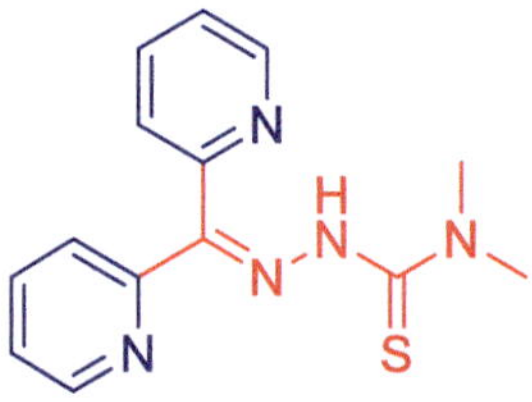

Dp44mT (**88**)
Hijacking lysosomal Pgp

3.3.4 Plasma Protein Binding

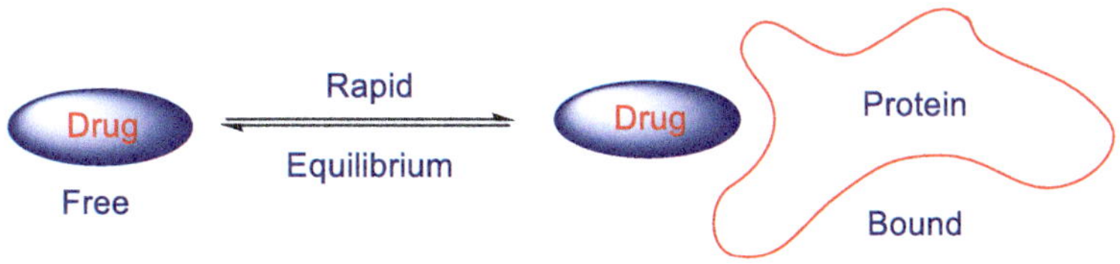

Figure 3.8 A drug's plasma protein binding.

Drugs can bind to protein macromolecules in the blood, a phenomenon known as *plasma–protein binding* (PPB).[77] As shown in Figure 3.8, the protein-bound form of the drug must dissociate from the protein in order to be useful because only unbound compound is available for distribution into tissues. There are three types of plasma proteins: human serum albumin (HSA) and α-1 acid glycoprotein (AAG) are the two more abundant proteins; whereas the third plasma protein, lipoprotein, is of less importance for PPB. The characteristics of the three types of plasma proteins are listed in Table 3.8.

Table 3.8 Common drug-binding proteins.

	Albumin	α-1 Acid glycoprotein	Lipoproteins
MW (g/mol)	67,000	42,000	200,000–2,400,000
Concentration (g/dL)	3.5–5	0.04–0.1	Variable
Half-life (days)	19	5	Up to 6

Human serum albumin (HSA), the most abundant protein in human blood plasma, has more than six distinctive binding sites including two for long-chain fatty acids; one for bilirubin; and two for acidic drugs (Site I for warfarin and phenylbutazone, etc. and Site II for diazepam, ibuprofen, etc.). On the other hand, AAG has only one selective site for basic drugs such as disopyramide and ligocaine. Acidic drugs, in particular, bind to serum albumin and tend to have higher PPB than neutral/basic drugs. Meanwhile, bases bind to AAG. For bases and neutrals, PPB is proportional to log *D*.

Drugs extensively bound to plasma proteins are largely restricted to the vascular compartment and have low V_d (e.g., warfarin is 99% protein-bound and its V_d is 0.1 L/kg). Drugs sequestrated in tissues may have V_d values much higher than the total body water or even body mass (e.g., propranolol, 3.5 L/kg and digoxin, 6 L/kg) because most of the drugs is present in other tissues, and the plasma concentration is low. In case of poisoning, drugs with large V_d are not easily removed by hemodialysis. Approximation of PPB is tabulated in Table 3.9.

Table 3.9 Approximation of plasma protein binding.

0–50% bound	Negligible
50–90%	Moderate
90–99%	High
>99%	Very high

Clinical implications of drugs′ PPB are summarized below:[78]

1. There is an equilibration between the PPB fraction of the drug and the free molecules of the drug. The PPB fraction is not available for action.
2. The drugs with high physicochemical affinity for plasma proteins (e.g., aspirin, sulfonamides, chloramphenicol) can replace the other drugs (e.g., acenocoumarol, warfarin) or endogenous compounds (bilirubin) with lower affinity.
3. High degree of protein binding makes the drug long-acting, because bound fraction is not available for metabolism, unless it is actively excreted by the liver or kidney tubules.
4. Generally expressed plasma concentrations of the drug refer to bound as well as free drug.
5. In hypoalbuminemia, binding may be reduced and high concentration of free drug may be attained (e.g., phenytoin).

Some representative drugs′ PPB data are listed in Table 3.10.[79] An excellent review on PPB was published in 2013.[80]

Table 3.10 Some representative drugs′ plasma protein binding.[79]

Drug	MW	f_b%
Acetaminophen (**109**)	151	0
Aspirin (**118**)	180	55
Camptothecin (**183**)	348	98
Captopril	217	30
Clozapine	327	95
Fluorouracil (5-FU)	130	11
Fluoxetine (**58**)	309	94
Haloperidol (**45**)	376	92
Heroin (**70**)	369	35
Ibuprofen	206	99
Ketoconazole (**80**)	531	99
Methotrexate	454	46
Morphine (**68**)	285	35
Omeprazole	345	95
Quinidine (**63**)	324	87
Salicylic acid (**119**)	138	95
Sertraline (**7**)	306	99
Streptomycin (**177**)	582	48
Tamoxifen	564	99
Tetracycline	444	18
Tetrahydrocannabinol	314	95
Tancomycin	1,449	30
Warfarin (**61**)	308	99

MW = molecular weight; f_b% = percentage of plasma binding.

Below are some examples where PPB issues were solved.

En route to the discovery of venetoclax (Venclexta, **16**), acylsulfonamide **89** was identified as a Bcl-X_L inhibitor with a K_i of 36 nM. However, its Bcl-X_L inhibition was nearly obliterated when **89** was tested in the presence of 10% human serum. It was also deactivated by 69-fold in the presence of 1% human serum, indicating strong protein binding.[81] Further scrutiny revealed that binding to site 3 of human serum albumin (HSA-III) was the main driver for compound **89**'s serum deactivation. Insight into the structural difference between Bcl-X_L and HSA-III led to analogs **90** and **91** after installation of the (*R*)-dimethylaminoethyl motif at site 3 region and replacements of the fluorobenzene moiety. Both **90** and **91** emerged as potent Bcl-X_L inhibitors with approximately 1 nM affinity. Furthermore, deactivation from serum binding was greatly reduced, in particular, the targeted HSA-III affinity was reduced by over 2 orders of magnitude as measured in the presence of 3% fetal bovine serum (FBS).

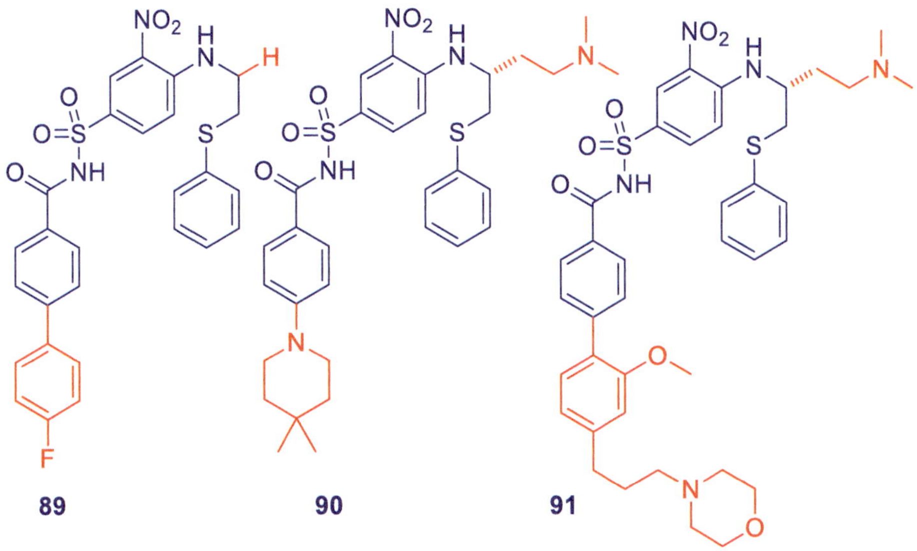

89 **90** **91**

Phenylalanine derivative **92** is an intrinsically potent dipeptidyl peptidase-4 (DPP-IV) inhibitor with an IC_{50} value of 12 nM. Disappointingly, it experienced a high serum shift of 32-fold due to high PPB. Replacing the fluorophenyl group in the homophenylalanine series with polar heterocycles such as a methylpyridone in **93** led to reduced serum shift. But it is metabolically labile, forming the demthylated metabolite. Introduction of bicyclic fused heterocycle as in compound **94** improved the PPB with 11-fold *in vivo* potency shift in the presence of serum. Eventually, installation of an additional fluorine atom, along with a small permutation of the fused bi-heterocycle, resulted in compound **95** as a potent and orally active DPP-4 inhibitor with much reduced PPB.[82]

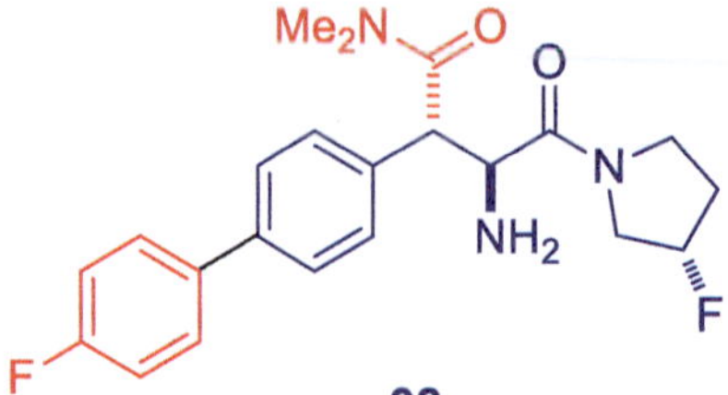

92

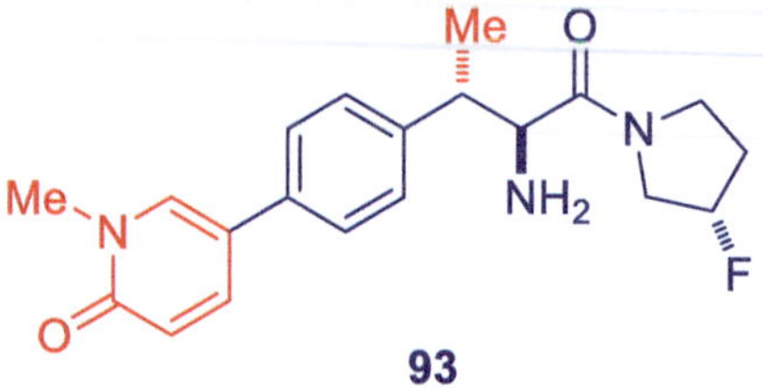

93

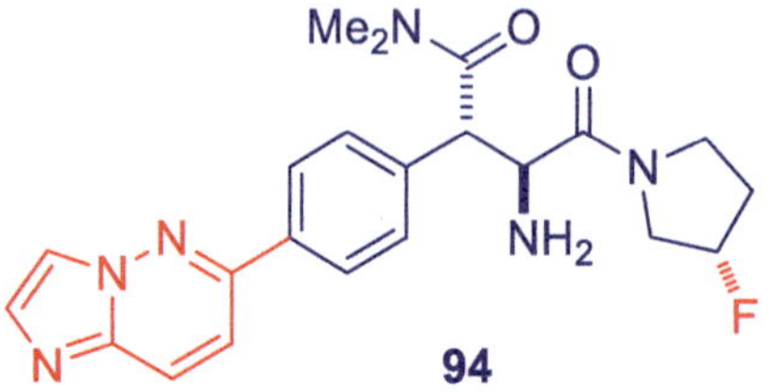

94

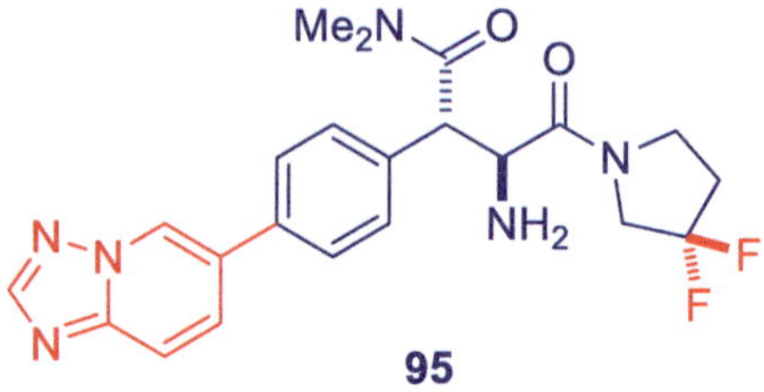

95

2-Aminopyrazole PNU-2922137 (**96**) was a cyclin-dependent kinase-2 (CDK-2) inhibitor (IC_{50}, 37 nM) endowed with *in vivo* antitumor activity in a mouse xenograft model. But it suffered from two major drawbacks. One is low aqueous solubility, 20–50 μM at pH 7.0 depending on its crystalline forms. The other is being a strong binder to HSA (99%) in a preclinical assay. It was correctly speculated that the greasy naphthyl moiety played a critical role in imparting the unfavorable properties. An effort to make PNU-2922137 (**96**) more drug-like led to the corresponding *para*-δ-lactam derivative **97** to reduce the lipophilicity. The resulting five-membered *para*-δ-lactam derivative **97** retained the CDK-2 inhibitory activity (IC_{50}, 37 nM, equipotent to that of **96**) and scored a >10-fold increase in buffer solubility over PNU-2922137 (**96**). More importantly, PPB was consistently reduced from 99% to 74%.[83]

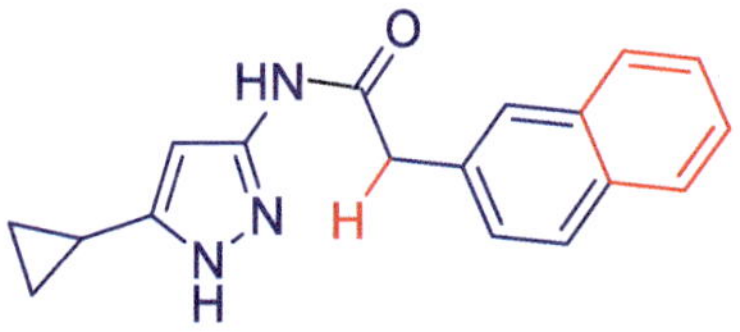

PNU-2922137 (**96**)

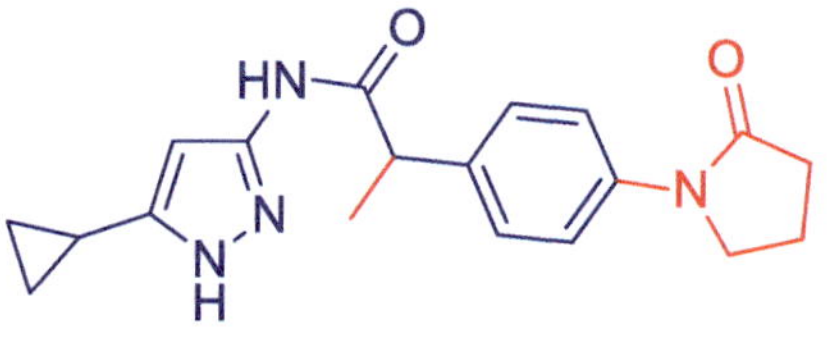

97

In 2002, Benet and Hoener published a landmark article, claiming that changes in PPB have little clinical relevance.[84] They argued that there are very limited cases (~25 out of 456 drugs examined) when protein binding changes may be relevant clinically. Greater than 95% clinically successful drugs are protein-bound, and the nominal value of the dose is more of a function of drug potency and less driven by the degree of protein binding. From calculation of PK parameters, they concluded the belief that the effective concentration of all drugs depends on protein binding is not correct. Rather, exposure is more relevant. In fact, for all low extraction ratio (hepatic clearance over hepatic blood flow) drugs, regardless of route of administration, the total exposure is independent of protein binding. Their argument was that drug distribution and drug elimination will change to compensate for the increased free drug clearance when there are changes in PPB. Therefore, changes in protein binding caused by DDIs or disease–drug interactions will usually not influence the clinical exposure of a patient to a therapeutic agent. Nonetheless, the authors cautioned rare cases of a drug with high extraction ratio and narrow therapeutic index that is given parentally. In those cases, changes in PPB will have clinical significance.

Benet and Horner's assertion resonated with Liu and co-workers. In 2015, Liu et al. also reported that low PPB does not necessarily lead to high *in vivo* unbound plasma concentration and optimize the compound structure to increase *in vivo* unbound drug concentration.[85]

One case of PPB's clinical relevance may be highlighted by the Pentothol's demise in Pearl Harbor in 1943.[86] Thiopentone (Pentothol, **99**) is an analog of pentobarbital (Nembutal, **98**). The sulfur derivative **99** as an anesthetic has a rapid action but with a short duration. Due to its lipophilicity, it is highly plasma protein bound. On the Day of Infamy, wounded GIs were given Pentothol injections. Since the free drug needed time to dissociate from plasma protein to impart its PD effects, repetitive multiple injections were given and caused massive overdose. The morale of the Pentothol story is that PPB could mean life-and-death on occasions.

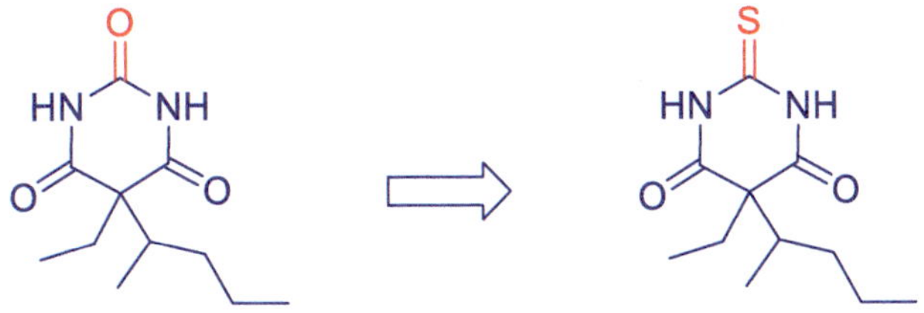

pentobarbital (Nembutal, **98**) thiopental (Pentothol, **99**)

3.4. Metabolism

3.4.1. Overview of Drug Metabolism

Aside from water and most hydrophilic dugs, all other molecules/drugs are metabolized. This is actually essential because lipophilic drugs would circulate in the body for a long time, causing untoward side effects if not eliminated in due course. In most cases, metabolism converts lipophilic compounds to hydrophilic metabolites, which are then eliminated/excreted from the body. Metabolism is chemical alteration of the drugs in the body. The primary site for drug metabolism is the liver, which is of the uttermost importance with regard to a drug's biotransformations. Other sites of metabolism are the kidney, intestine, lungs, and plasma.

Most hydrophilic drugs, such as amikacin (**70**), neostigmine (**71**), gentamycin (**72**), and mannitol (**73**), are not biotransformed, and they are excreted unaltered. Renal excretion is the major route of elimination for hydrophilic drugs.

As far as potency is concerned during drug metabolism, the resultant metabolites could be either active or inactive. Most drugs and their active metabolites are converted to less active or inactive metabolites, e.g., sertraline hydrochloride (Zoloft, **7**), propranolol (Inderal, **14**), morphine (**68**), phenobarbital (Luminal, **100**), etc. One of the major metabolite of phenobarbital (**100**) is the ring-opening product, which is completely inactive. Meanwhile, some drugs are converted to one or more active metabolites, e.g., diazepam (Valium, **101**) and amitriptyline (Elavil, **102**).

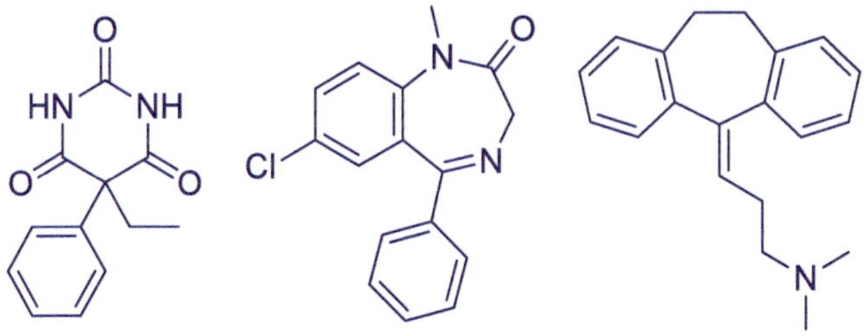

phenobarbital (Luminal, **100**) — diazepam (Valium, **101**) — amitriptyline (Elavil, **102**)

Another good example of active metabolites is atorvastatin (Lipitor, **2**). Its two major metabolites are 2-hydroxy-atorvastatin (**103**) and 4-hydroxy-atorvastatin (**104**), respectively.[87] They are active metabolites that are equipotent as the parent drug **2**. It was a pleasant surprise not envisioned even by the inventor, Bruce Roth, while designing the drug.

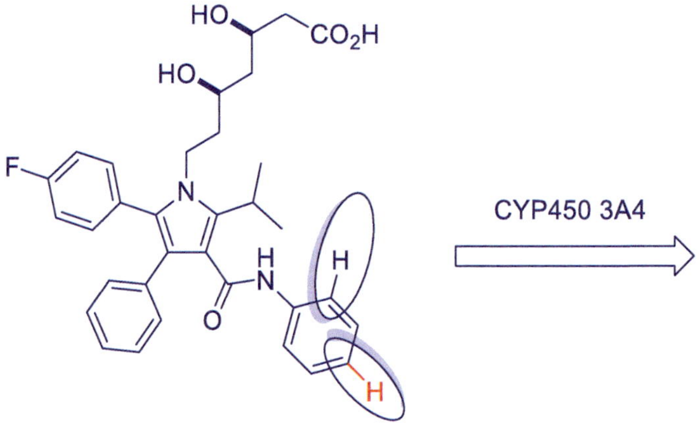

atorvastatin (Lipitor, **2**)

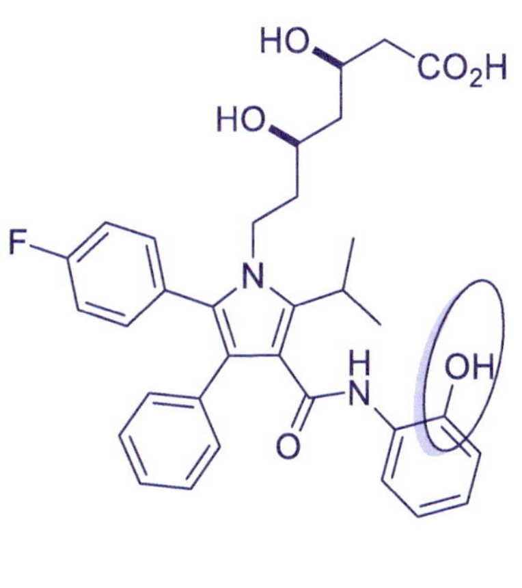

2-hydroxy-atorvastatin (**103**)

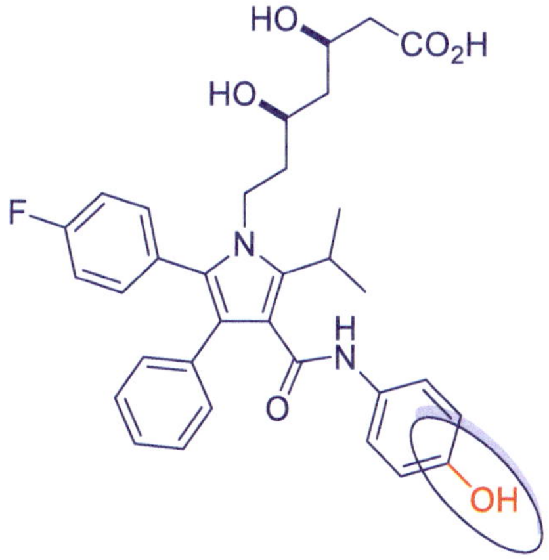

4-hydroxy-atorvastatin (**104**)

Metabolism is key for bioactivation of prodrugs, which are inactive as such and need conversion in the body (i.e., metabolism) to one or more active metabolites. The prodrug may offer certain advantages. Their active forms may be more stable; they can have better bioavailability, or other desirable pharmacokinetic properties or less side effects and toxicity. For instance, during Merck's search of their angiotensin-converting enzyme (ACE) inhibitors, enalaprilat (**105**) was potent enough, yet had low bioavailability because the molecule is too polar to permeate the cell's membrane. The corresponding ethyl ester, enalapril (Vasotec, **106**) was lipophilic enough to diffuse through cell membrane. The prodrug **106** itself is not active in biochemical *in vitro* assays although enalaprilat (**105**) is.[88] More on prodrugs at the end of this chapter.

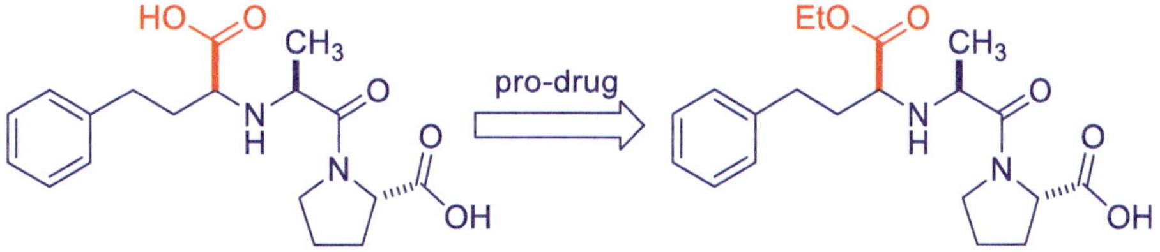

enalaprilat (**105**)

enalapril (Vasotec, **106**)

With regard to toxicity of the metabolites, drugs may be converted to less toxic/effective materials; more toxic/effective materials; or materials with different type of effect or toxicity. For example, an effective antidepressant venlafaxine (Effexor, **107**) is a dual inhibitor of serotonin (SERT) and norepinephrine (NE), although it is five-fold more potent inhibiting SERT than NE. Its major metabolite (56%) performed by CYP 2D6 is O-desmethyl-venlafaxine (desvenlafaxine, Pristiq, **108**).[89] While venlafaxine (**107**) is a good antidepressant, desvenlafaxine (Pristiq, **108**) is an even better one because of its simpler metabolism, lower risk of DDI, and lack of the need for extensive titration to achieve therapeutic efficacy.[90] Desvenlafaxine (Pristiq, **108**) itself has been on the market for the treatment of major depressive disorder (MDD) since 2008.

venlafaxine (Effexor, **107**)
Good

desvenlafaxine (Pristiq, **108**)
Better!

Acetaminophen (Tylenol, **109**) is not an NSAID (nonsteroid anti-inflammatory drug). It relieves fever and headaches by inhibiting the synthesis of prostaglandins. One of its major metabolites by CYP3A4 and 2E1 is *N*-acetyl-*p*-benzoquinone imine (NAPBQI, **110**), a reactive metabolite, which is responsible for the heptatoxicities.[91] In this case, the drug is good, but its metabolite is bad, just like many drugs with reactive metabolites. More on reactive metabolites in Chapter 5.

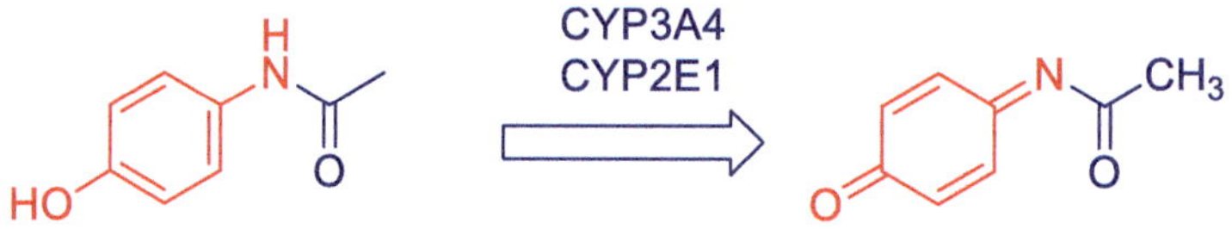

acetaminophen (Tylenol, **109**)
OK

NAPBQI (**110**)
Bad!

We already described that lipophilic terfenadine (Seldane, **54**) is metabolized extensively and has hERG issues. But its major metabolite, the hydrophilic fexofenadine (Allegra, **55**) has negligible metabolism and is selective over hERG and devoid of cardiotoxicity.[92]

3.4.2 Cytochrome P450 Enzymes

As shown in Figure 3.9, drugs are metabolized mostly by a class of enzymes called cytochrome P450 (CYP450) enzymes. They are so named because they are bound to membranes within a cell (cyto) and contain a heme pigment (chrome and P) that absorbs light at a wavelength of 450 nm when exposed to carbon monoxide. CYP450 enzymes are a superfamily of 18 heme-containing enzyme families, which may be further divided into 43 subfamilies and more than 200 CYP450 isoforms. Chief among them are CYP450 3A4 and 2D6. In fact, CYP 3A4 carries out biotransformations of the largest number (~50%) of drugs. Other important CYPs are 1A2, 2C9, 2C19, and 3A5. In all, these six CYP enzymes are responsible for metabolizing 90% of drugs.[93] In addition to the liver, these isoforms are expressed in the intestine and the kidney too. The intestine is responsible for the first-pass metabolism of drugs.

Inhibition of CYP450 3A4 by erythromycin (**64**), clarithromycin, verapamil (**79**), ketoconazole (**80**), itraconazole, diltiazem, and a constituent of grapefruit juice is responsible for unwanted interactions with terfenadine (Seldane, **54**) and many other drugs. The phenomenon is an example of DDI.

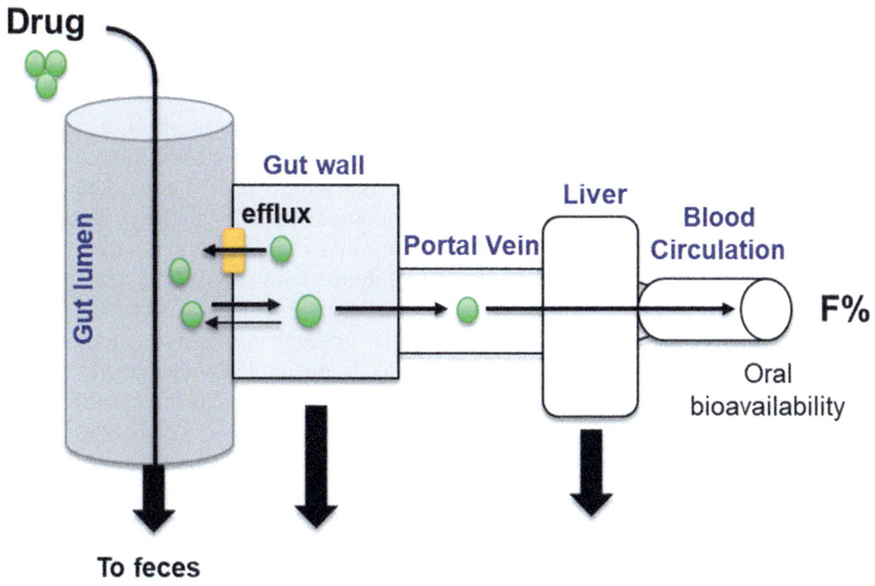

Figure 3.9 Route of oral drugs.

3.4.3. Drug–Drug Interactions

The drug–drug interaction (DDI), in a broad sense, is a modification of the effect of a drug when administered with another drug. But in a narrower sense with regard to drug metabolism, DDI refers to the fact that toxicity often ensues when two co-administered drugs are metabolized by the same isoform of CYP450 enzymes. For example, if drugs A and B are both metabolized by CYP450 3A4, as it so often happens, the enzyme is pre-occupied by metabolizing drug A, it no longer possesses the capacity to metabolize drug B. Without the benefit being biotransformed, untoward toxicities often manifest.

With regard to drugs as ligands for CYP450 enzymes, they may be divided into three categories: substrates, inducers, and inhibitors. CYP450 substrates are ligands that are metabolized by the enzymes. Examples of CYP3A4 substrates include amiodarone (Cordadrone), cimetidine (Tagamet), fluoxetine (Prozac, **58**), antifungal ketoconazole (Nizoral, **80**), HIV protease inhibitors [e.g., indinavir (Crixivan, **78**)], macrolide antibiotics (with an exception for azithromycin **65**), nefazodone (Serzone), and grapefruit juice.

CYP450 inducers will increase the enzyme activity by increasing enzyme synthesis. Examples of inducers include rifampicin, phenytoin (**182**), carbamazepine (**125**), and phenobarbital (Luminal, **100**). On the other hand, CYP450 inhibitors will decreases the activity of the enzyme and may slow down the metabolism of substrates, generally leading to an increased drug effect. CYP450 inhibitors may be divided to reversible inhibitors and irreversible inhibitors.

Reversible inhibitors are the most common mechanism of DDI. Ketoconazole (**80**)′s major drug interactions with CYP3A4 could result in 100-fold changes in pharmacokinetics. Another class of reversible inhibitors including fluoxetine (Prozac, **58**), paroxetine (Paxil, **59**), and quinidine (**62**) have major drug interactions with CYP2D6.

Irreversible inhibitors form stable complexes with the CYP450 enzymes. This type of metabolism may be detrimental to drugs, making them "undesirable." For example, calcium channel blocker mibefradil (Posicor) saw two to ten-fold changes in pharmacokinetics as a consequence of being a potent CYP3A4 irreversible inhibitor. The drug was removed from the market due to its interactions with CYP3A4 and potential for DDIs, some of them deadly. Clarithromycin, troleandomycin, and erythromycin (**64**) are also irreversible inhibitors of CYP3A4 with two to six-fold changes in pharmacokinetics. Ritonavir (Norvir, **116**), also an irreversible inhibitor with 2–50-fold changes in pharmacokinetics, has a black-box warning due to drug interactions with CYP3A4. Cisapride (Propulsid) and astemizole (Hismanal) undergo extensive metabolism, primarily by CYP3A4. They were both withdrawn due to metabolism-related side effects such as QT prolongation. Finally, troglitazone (Rezulin), a peroxisome proliferator-activated receptor-γ (PPAR-γ) agonist for the treatment of type-II diabetes, was hepatotoxic due to reactive intermediates formed by CYP450 metabolism.

One of the more conspicuous examples to showcase the demise of DDIs is cerivastatin sodium (**112**), an HMG-CoA inhibitor for lowering the low-density lipoprotein (LDL)-cholesterol levels. It was brought to the market in 1998 by Bayer with the trade name of Baycol. Since it is a potent CYP2C8 inhibitor,[94] rhabdomyolysis (muscle weakness) emerged as the consequence of DDI when taken together with other cholesterol-lowering drugs such as gemfibrozil (Lopid, **113**). A member of the fibrate family of drugs, gemfibrozil (**113**) is also a CYP2C8 substrate. Baycol (**112**) was withdrawn in 2001 since it caused severe rhabdomyolysis when taken with other fibrates. A list of marketed statins′ metabolism by CYP3A4 is compiled in Table 3.11.

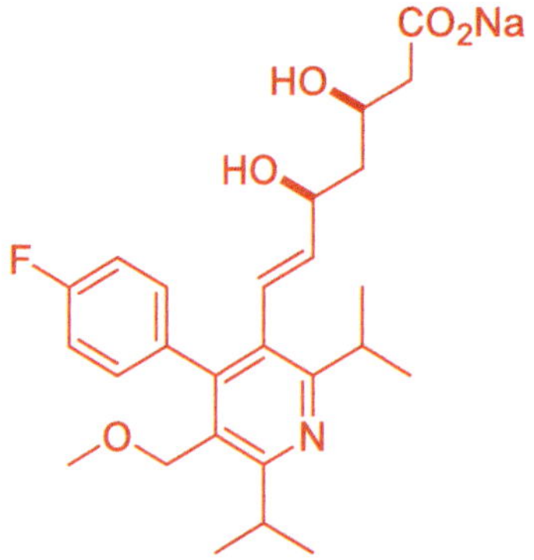

Cerivastatin sodium (Baycol, **112**)

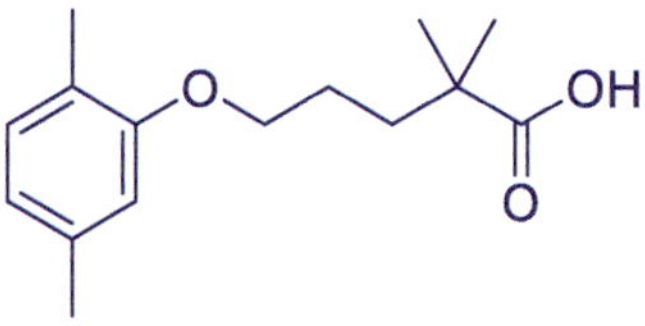

Gemfibrozil (Lopid, **113**)

Table 3.11 CYP450 3A4 and statin metabolism.

CYP450 3A4	Metabolism
Rosuvastatin	No
Pravastatin	No
Atorvastatin	Yes
Simvastatin	Yes
Simvastatin/ezetimibe	Yes

Drinking a glass of orange juice in the morning is good for you. But if you take your medicine with it, you should be aware of *the grapefruit juice effect*.

Grapefruits contain furanocoumarin derivatives that are rapid, potent, mechanism-based inhibitors (MBIs) of intestinal CYP3A4. Its major ingredient bergamottin (**114**) is oxidized by CYP3A4 to bergamottin epoxide (**115**).[95] Bergamottin (**114**) inhibits CYP3A4 via protein modification with a K_i value of 7 μM. It also inhibits CYP1A2, 2A6, 2C9, 2C19, 2D6, and 2E1 in human liver microsome (HLM). Therefore, when a drug is taken with grapefruit juice, its bioavailability is frequently boosted. For instance, grapefruit juice is found to increase felodipine's bioavailability to 164–469% and nifedipine (**42**)'s *F*% to 134%. Drugs that undergo high pre-systemic (enteric) metabolism may exhibit pharmacological effects in high dose/high plasma levels. They should not be taken with grapefruit juice. Those drugs include amiodarone, astemizole, buspirone, cilostazol, cyclosporine A (**13**), etoposide, indinavir (Crixivan, **78**), midazolam, nifedipine (**42**), pimozide, saquinavir, sildenafil, some statins, terfenadine (**54**), etc.[96,97]

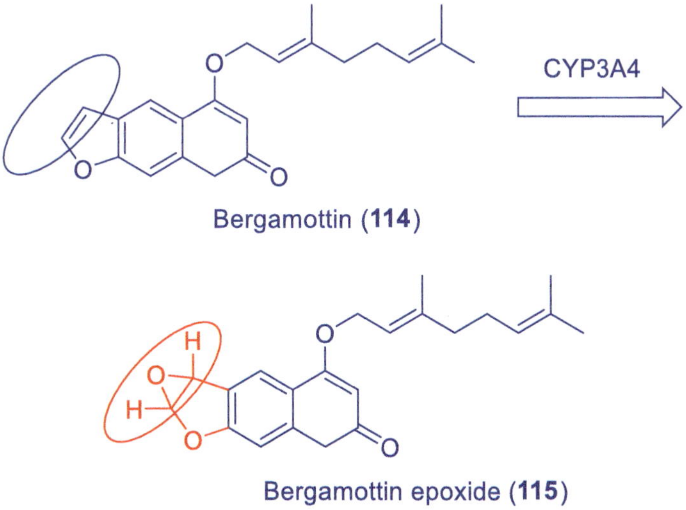

As the cases exemplified above, metabolism-related issues have been recognized by regulatory agencies and the pharmaceutical industry. Much effort has been devoted to the understanding of metabolism. Early prediction and elimination of such metabolism "problematical" compounds may avoid safety issues, regulatory obstacles, and market pressures.

Not all DDIs are pernicious. Sometimes, advantages may be taken to boost drugs′ bioavailability. For example, cyclosporine A (**13**) is an immunosuppressant with an oral bioavailability of approximately 25%. Since it is a CYP3A4 substrate and so is antifungal drug ketoconazole (**80**), the use of a cyclosporine–ketoconazole (**13–80**) combination boosts cyclosporine A (**13**)′s bioavailability and makes renal transplantation affordable in developing countries.[98] In the field of HIV protease inhibitors, both saquinavir and ritonavir (Norvir, **116**) have low bioavailabilities, which is why they have to be taken in large doses individually. Since both of them are CYP3A4 substrates, co-administering them together would elevate their oral bioavailability. In fact, ritonavir (Norvir, **116**) has been employed in another pharmacokinetic enhancement of protease inhibitor therapy with the ritonavir–lopinavir combination.

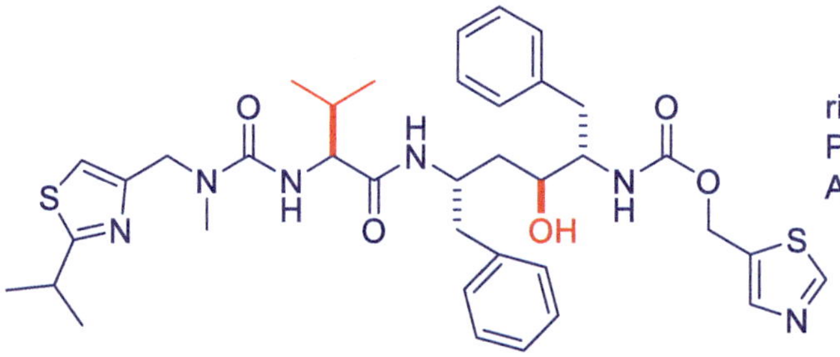

ritonavir (Norvir, **116**)
Protease inhibitor
Abbott, 1996

Furthermore, Gilead carried out an SAR investigation around ritonavir (Norvir, **116**) and arrived at cobicistat (Tybost, **117**) as a potent and selective CYP3A4 inhibitor with no anti-HIV protease activities.[99] Cobicistat (**117**) is now employed to combine with other HIV drugs to boost their bioavailability. For example, it is combined with elvitegravir, an HIV integrase inhibitor, to achieve higher concentrations of elvitegravir in the body with lower dosing, theoretically enhancing elvitegravir's viral suppression while diminishing its adverse side effects.

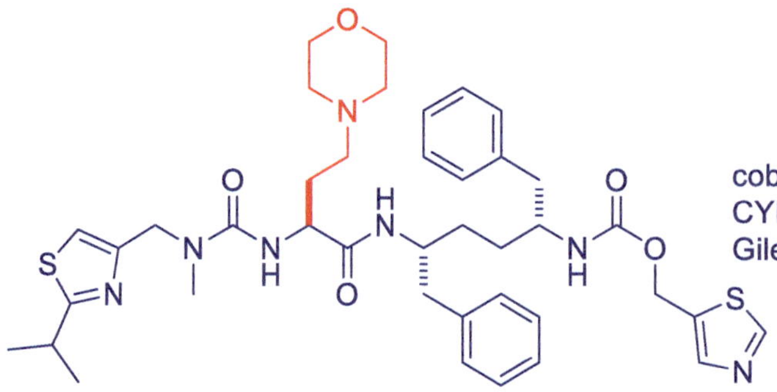

cobicistat (Tybost, **117**)
CYP450 3A4 Inhibitor
Gilead, 2012

3.4.4. *Phase I Metabolism*

Drug metabolism may be divided to two phases: Phase I metabolism and Phase II metabolism. *Phase I metabolism* refers to functional group transformations of the original drug, converting it to a more polar molecule(s). Metabolic hydrolysis of Aspirin (**118**) to salicylic acid (**119**) is a good example of Phase I metabolism. *Phase II metabolism*, also known as conjugation, is the process of appending a very polar and highly hydrophilic molecule (glucose or sulfate, for example) to appropriately functionalized parent compound or Phase I metabolite. It may be exemplified by conversion of salicylic acid (**119**) to its corresponding glucuronide **120**.

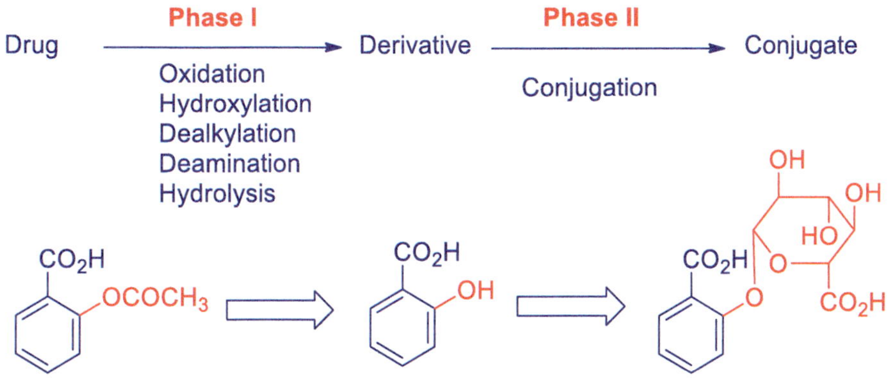

Aspirin (**118**) salicyclic acid (**119**) glucuronide **120**

As shown below, the types of reactions for Phase I metabolism are oxidation, reduction, hydrolysis (from **118** to **119**, for instance), cyclization, and de-cyclization.

Oxidation (alcohols and aldehydes)

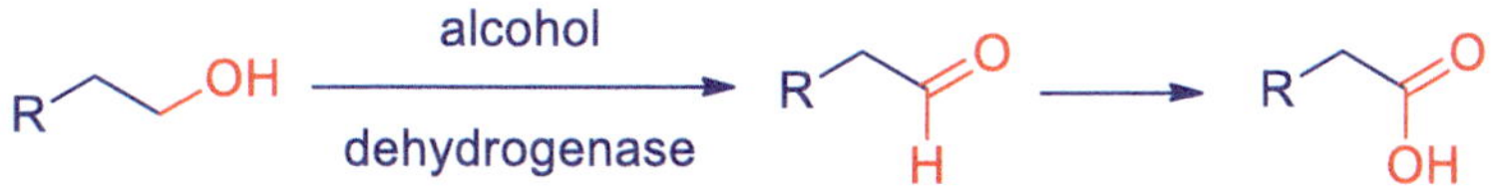

Reductions: (ketones, double bonds, nitro and azo compounds, sulfoxides and *N*-oxides, disulfides, quinone, dehalogenation)

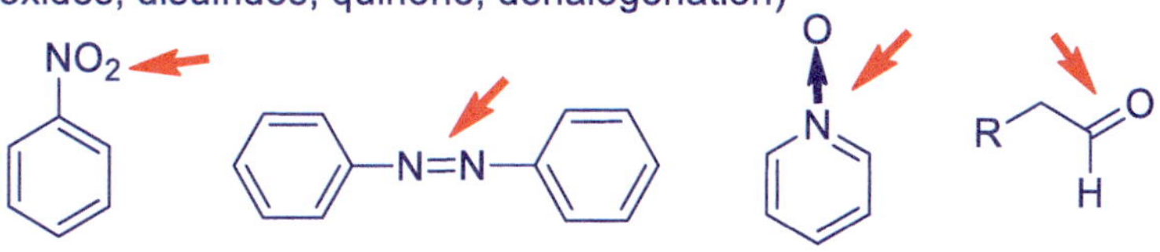

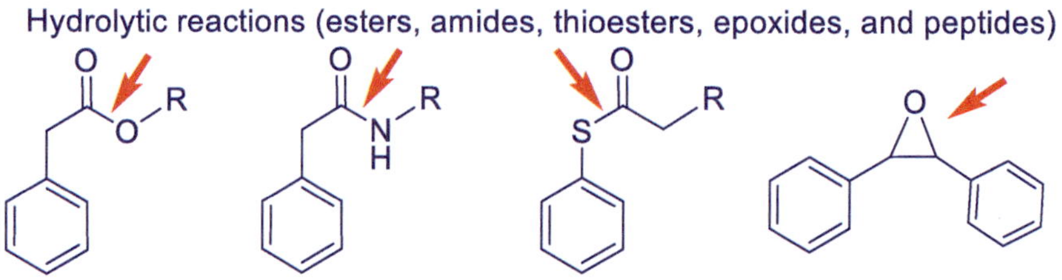

3.4.4.1 Oxidation

Oxidation is the most important drug-metabolizing reaction. Some consider that, for a good reason, Phase I metabolism is largely an oxidative process. Various oxidative metabolisms are hydroxylation; oxygenation at carbon, nitrogen, or sulfur atoms; N-dealkylation or O-dealkylation, oxidative deamination, etc. They are mostly carried out by a group of monooxygenases in the liver, and the final step involves CYP450 and O_2.

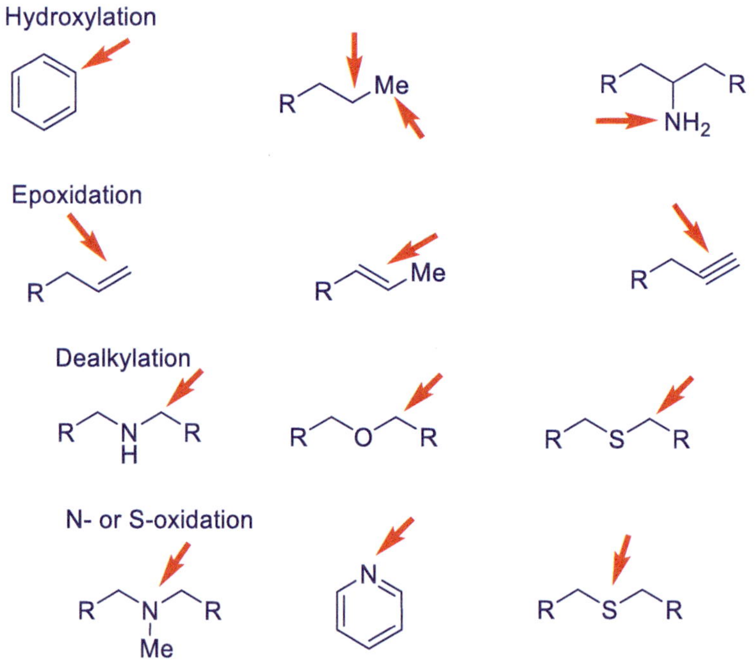

Hydroxylation is a prevalent oxidation process for Phase I metabolism. CYP3A4 oxidation of atorvastatin (**2**) to the corresponding 2-hydroxy-atorvastatin (**103**) and 4-hydroxy-atorvastatin (**104**), respectively, is a quintessential example.[87] The major metabolite of celecoxib (Celebrex, **121**) is the corresponding benzylic hydroxylation product **122** carried out by mixed-function oxidases (MFO).[100] Alcohol **122** is further oxidized to the corresponding carboxylic acid, which then undergoes glucuronidation and other Phase II metabolisms before being eliminated.

The benzylic proton is more vulnerable to metabolic hydroxylation. SCH-48461 (**123**) is readily oxidized to the corresponding benzyl alcohol *in vivo*. Concurrent demethylation provided the major metabolite SCH-57215 (**124**).[101] Addressing these two points of metabolic vulnerability and two additional "soft spots" led to the discovery of eztimibe (Zetia).

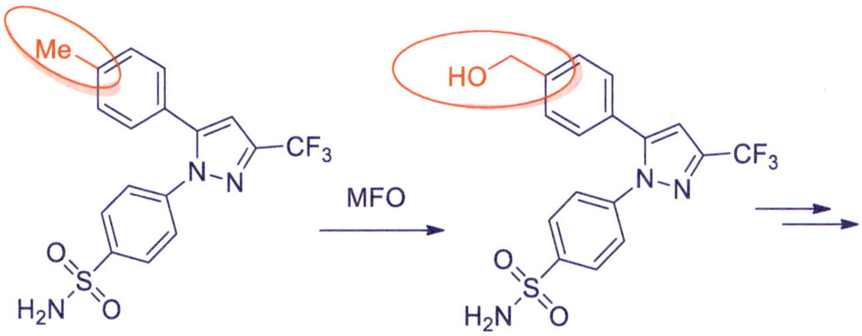

celecoxib (Celebrex, **121**) **122**

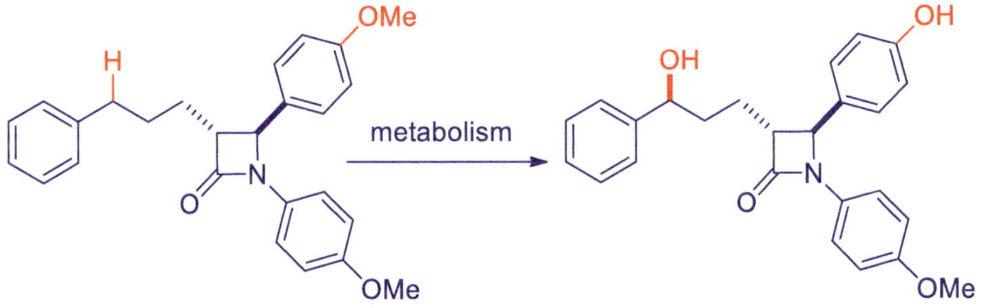

SCH-48461 (**123**) SCH-57215 (**124**)

Epoxidation during the Phase I metabolism is well represented. Carbamazepine (Tegretol, **125**), an anticonvulsant, is first oxidized to carbamazepine 10,11-epoxide (**126**) by CYP3A4 during its Phase I metabolism. Epoxide **126**, in turn, is subsequently converted to carbamazepine 10,11-diol (**127**) by epoxide hydrolase, which may be partially responsible for some of **125**'s idiosyncratic adverse drug reactions (IADRs). Meanwhile, epoxide **126** may be opened by glutathione (GSH) to afford the more innocuous alcohol **128**.[102]

carbamazepine (Tegretol, **125**) carbamazepine 10,11-epoxide (**126**)

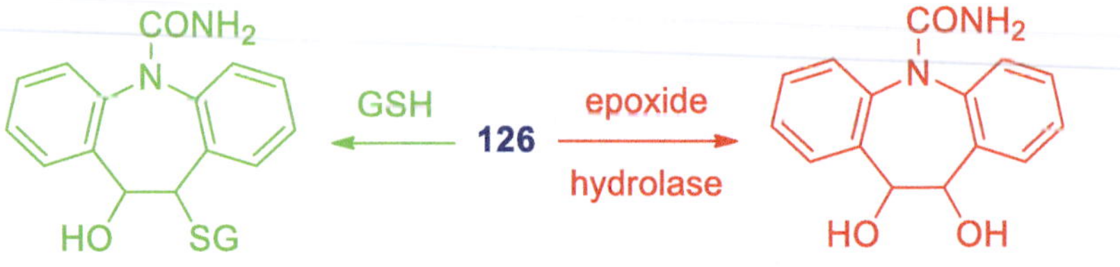

128 carbamazepine 10,11-diol (**127**)

Some drugs' toxicity may be attributed to its metabolic epoxidation. Aflatoxin B_1 (AFB_1, **129**) is a potent hepatocarcinogen. It is activated by both CYP3A4 and CYP2A6 to afford aflatoxin B1 *exo*-8,9-epoxide (**130**), which is unstable in water and reacts with deoxynucleic acid (DNA) to give adducts in high yield (>98%). Epoxide **130** is largely responsible for aflatoxin B_1 (**129**)'s hepatocarcinogenicity. In contrast, aflatoxin B_1 (**129**) is also metabolized by CYP1A2 to give rise to a less dangerous product aflatoxin B1 *endo*-8,9 epoxide (**131**), which is nongenotoxic.[103]

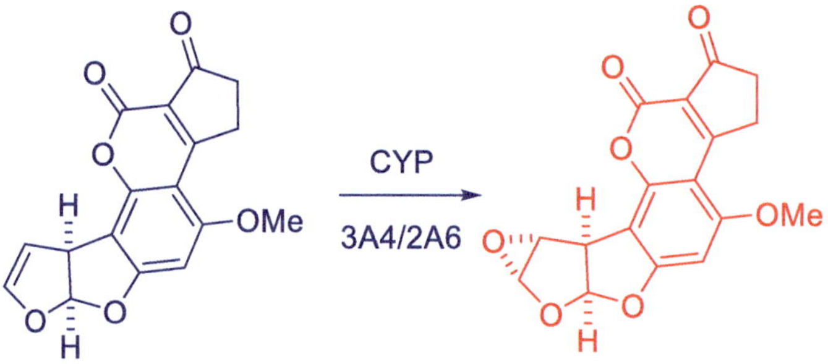

aflatoxin B1 (**129**) aflatoxin B1 *exo*-8,9 epoxide (**130**) **Hepatocarcinogen**

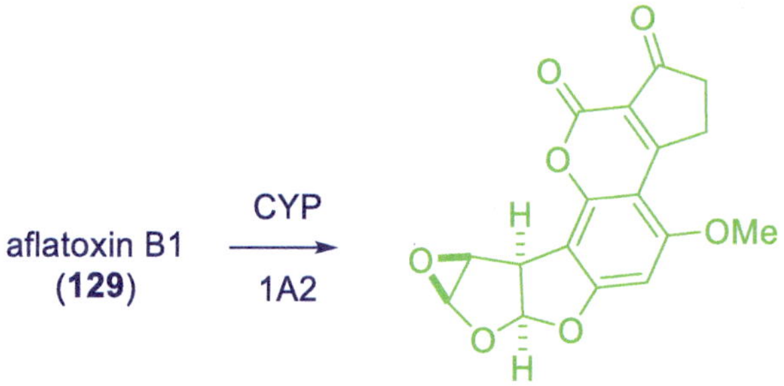

aflatoxin B1 *endo*-8,9 epoxide (**131**) **Non-genotoxic**

As far as dealkylation is concerned during Phase I metabolism, there are O-dealkylation, N-dealkylation, and S-dealkylation:

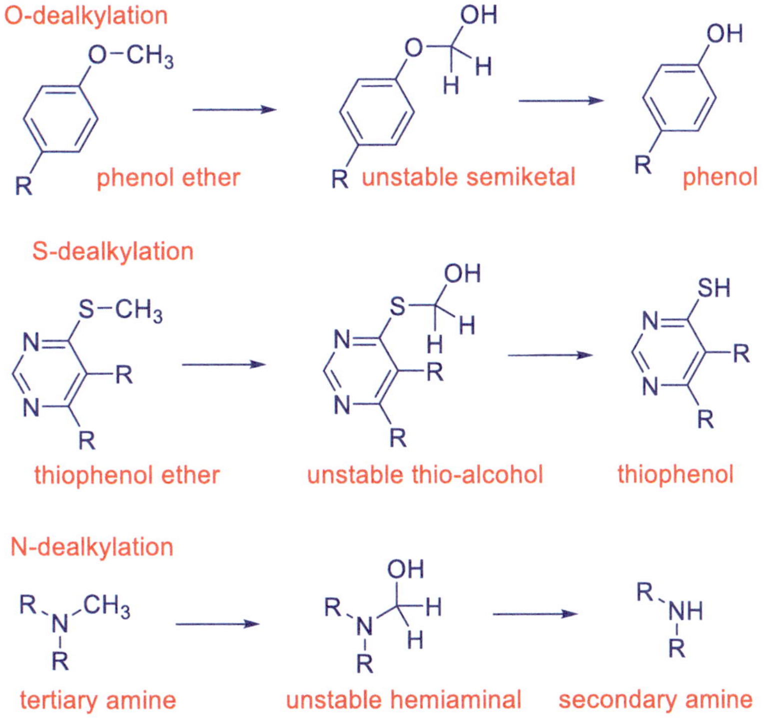

Metabolic conversion from SCH-48461 (**123**) to SCH-57215 (**124**) touched on the subject of O-de-alkylation in general and O-de-methylation in particular. Another example of O-de-methylation may be traced back to the metabolism of venlafaxine (Effexor, **107**) by CYP2D6 to produce O-desmethyl-venlafaxine (desvenlafaxine, **108**).[89]

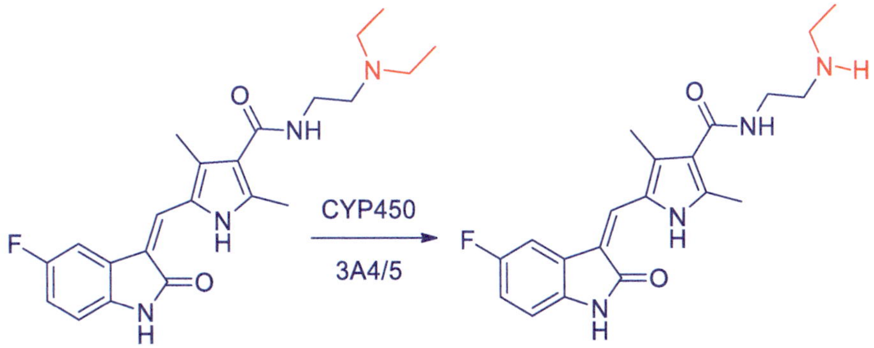

sunitinib (Sutent, **132**) SU-12662 (**133**)

An example of N-dealkylation is metabolism of sunitinib (Sutent, **132**) to produce SU-12662 (**133**).[104] The de-ethylated metabolite **133** has the same potency as the parent drug **132** for PDGFR-α, PDGFR-β, VEGF2, and Kit. Furthermore, the de-ethylated metabolite **133** accumulates 7–10-fold whereas the parent drug **132** accumulates three to four-fold.

Metabolic S-dealkylation is rare simply because short-chain alkyl-sulfide-containing drugs are not common. 6-Mercaptopurine (**136**) was found to be the major metabolite of 6-methylthiopurine (**134**).[105] Presumably, 6-methylthiopurine (**134**) is first oxidized to the unstable hydroxyl intermediate **135**, which then spontaneously collapses to deliver 6-mercaptopurine (**136**).

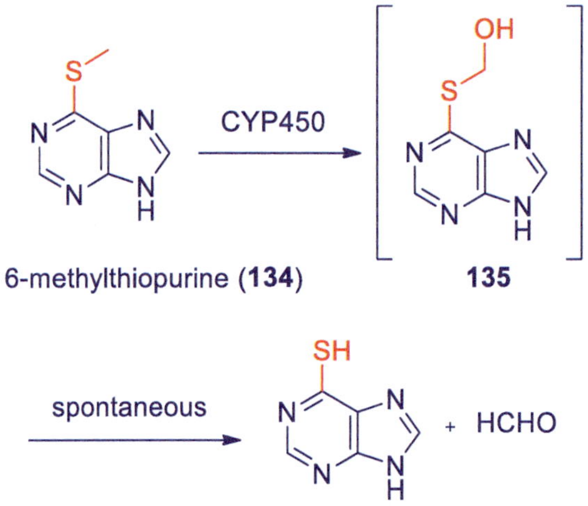

6-methylthiopurine (**134**) **135**

6-mercaptoopurine (**136**)

3.4.4.2 Reduction

Reduction is the reverse of oxidation and involves CYP450 enzymes working in the opposite direction. Metabolic reduction is responsible for the metabolism of one of the earliest synthetic drugs in history. Prontosil (**137**), discovered by Gerhard Domagk in 1930, is actually a prodrug. It is reduced by bacterial nitro-reductase in the intestines to sulfanilamide (**138**, the actual active drug). It has been speculated that an azo-anion free radical intermediate is involved.[106]

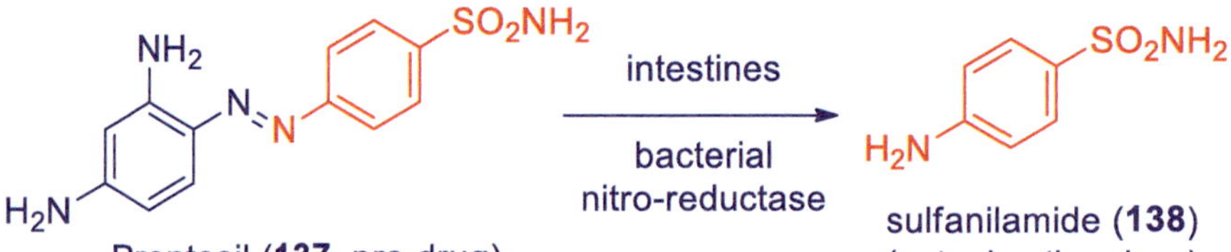

Prontosil (**137**, pro-drug) sulfanilamide (**138**) (actual active drug)

A notorious example of metabolic reduction is probably the reduction of nitroaromatic drugs to the corresponding aniline metabolites.[107,108] Certain drugs containing a nitroaromatic moiety (e.g., tolcapone, nimesulide, nilutamide, flutamide, and nitrofurantoin) have been associated with organ-selective toxicity including rare cases of idiosyncratic liver injury. They display a broad spectrum of mutagenic, genotoxic, and carcinogenic properties. Although the transformation from nitrobenzene to aniline *in vivo* is a complex process involving electron (single electron or two electrons) and proton transfers, key intermediates invariably consist of highly carcinogenic nitrosobenzene and phenylhydroxylamine.

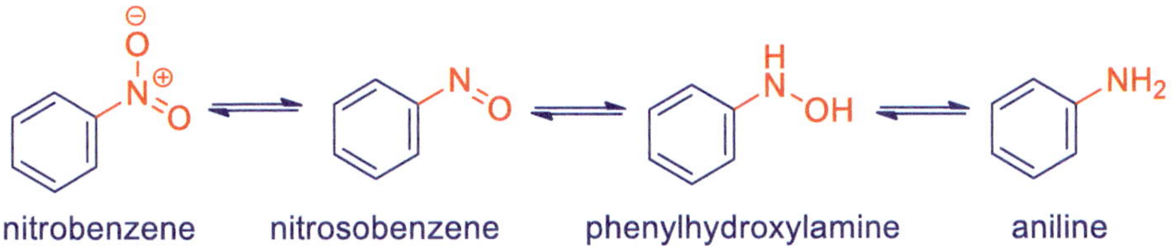

The skeletal muscle relaxant dantrolene (Dantrium, **139**) is metabolized to its major metabolite, aniline **140**, by flavin-dependent NADPH-CYP450 reductase. Formation of the hydroxylamine intermediate may be associated with dantrolene (**139**)'s liver injury.[109]

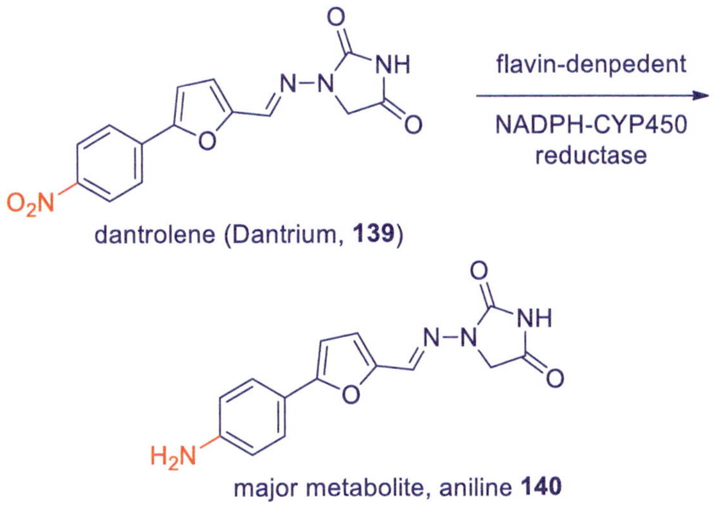

3.4.4.3 Hydrolysis

Hydrolysis means adding water. For an ester-containing drug, hydrolysis is cleavage of the ester by taking up a molecule of water employing esterase. Similarly, amides and polypeptides are hydrolyzed by amidases and peptidases, respectively. Hydrolysis occurs in the liver, intestines, plasma, and other tissues. Examples are choline-esters, procaine, lidocaine, pethidine, and oxytocin.

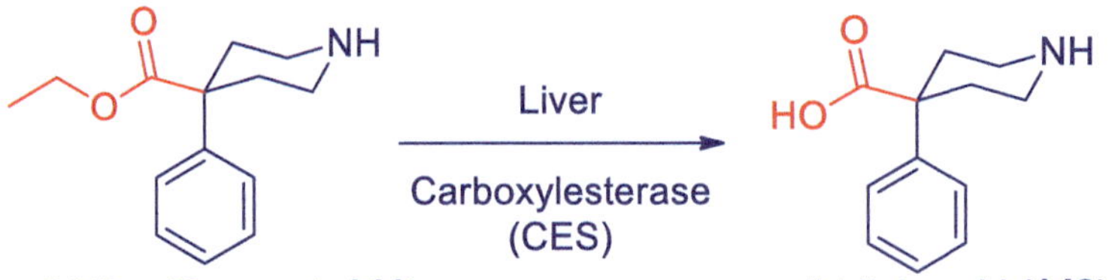

pethidine (Demerol, **141**) pethidinic acid (**142**)

Pethidine (Demerol, **141**) is an old synthetic opioid pain medication of the phenylpiperidine class. It is quickly hydrolyzed in the liver to pethidinic acid (**142**, which is inactive) by a prevalent esterase called liver carboxylesterase (CES).[110]

3.4.4.4 Cyclization

Metabolic cyclization is formation of a ring structure from a straight-chain compound. An old treatment for malaria, proguanil (**143**) is a dihydrofolate reductase (DHFR) inhibitor. A biguanide-type drug, the linear proguanil (**143**) is oxidatively metabolized to a cyclic active metabolite cycloguanil (**144**) by CYP2C19.[111]

proguanil (**143**) cycloguanil (**144**)

3.4.4.5 Decyclization

Metabolic decyclization is ring-opening of a cyclic molecule such as phenytoin (**182**) and barbiturates. One of the barbiturates primidone (**145**) is first metabolized by CYP2C9/C19 to phenobarbital (Luminal, **100**), which is further metabolized to the ring-opening product phenylethylmalonamide (PEMA, **146**) as the major metabolite.[112] PEMA (**146**) may be further metabolized to the corresponding inactive bisacid.

primidone (**145**) phenobarbital (Luminal, **100**)

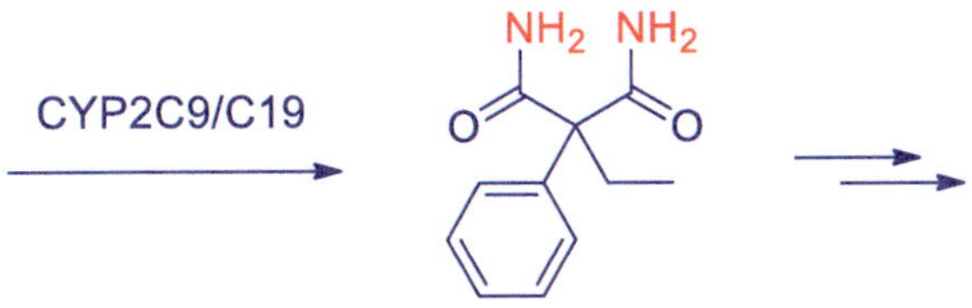

phenylethylmalonamide (PEMA, **146**)

Metabolism from the active parent drug to inactive metabolite such as **145**→**146** takes place frequently. The opposite takes place as well when an inactive prodrug is metabolized to the active metabolite. The success of clopidogrel (Plavix, **8**) is an excellent example. Remarkably, despite its enormous commercial success, the identity of its active metabolite **148** was not elucidated until 1999, two years after it was on the market. It was isolated after exposure of clopidogrel (**8**) or 2-oxo-clopidogrel (**147**) to human hepatic microsomes. Metabolite **148** was determined to be an antagonist of the $P2Y_{12}$ purinergic receptor and prevents binding of adenosine diphosphate (ADP) to the $P2Y_{12}$ receptor. However, clopidogrel (**8**) itself is not active *in vitro*, but is activated *in vivo* by CYP450-mediated hepatic metabolism to give the active metabolite **148**. This particular metabolic pathway involves a decyclization, ring-opening process.[113–115]

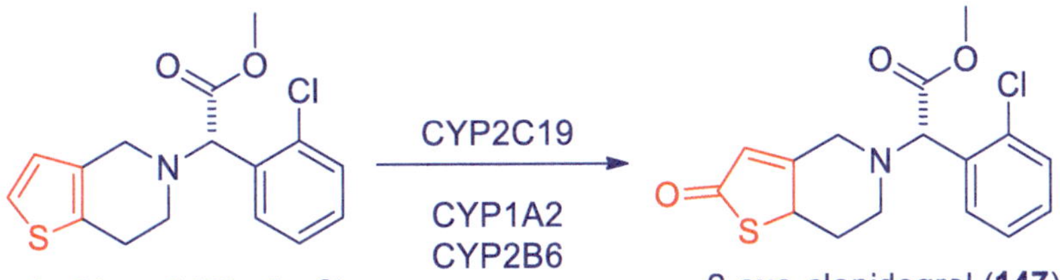

clopidogrel (Plavix, **8**)
inactive *in vitro*

2-oxo-clopidogrel (**147**)
inactive *in vitro*

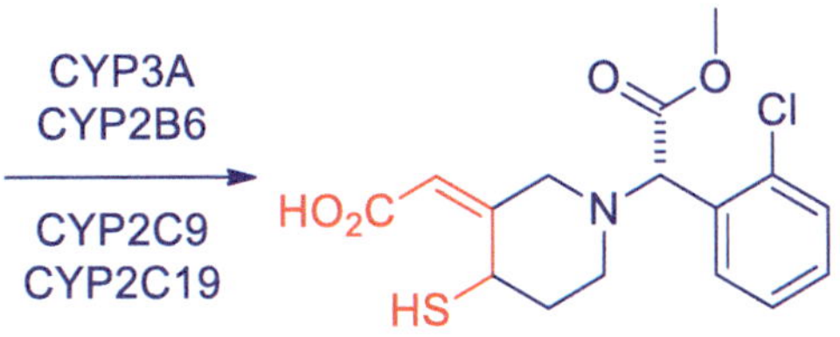

the active metabolite (**148**)
active *in vitro*
as a $P2Y_{12}$ inhibitor

There are also cases where both parent drug and the de-cyclization metabolite are active. Leflunomide (Arava, **149**) was approved by FDA in 1998 as a disease-modifying rheumatoid arthritis (RA) treatment. The drug works through a complex poly-pharmacology. It inhibits dihydroorotate dehydrogenase (DHODH), the signal transducer and activator of transcription 6 (STAT6), Janus kinase-3 (JAK3) and platelet-derived growth factor receptor (PDGFR) and is plagued with many side effects. *In vivo,* it is de-cylized to the isoxazole ring-opening metabolites, teriflunomide (**150**). With a better safety profile, teriflunomide (Aubagio, **150**) was approved by the FDA in 2012 for treating multiple sclerosis (MS).[116]

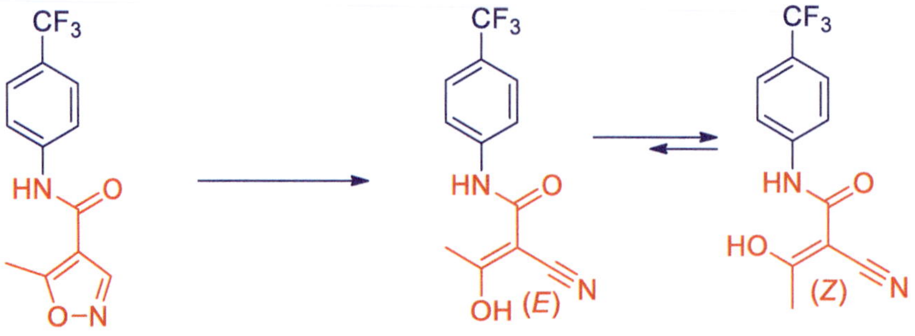

leflunomide (Arava, **149**) teriflunomide (Aubagio, **150**)

Phase I metabolism is often problematic if the compound undergoes extensive metabolism to afford inactive metabolites, or worth still, reactive metabolites. There are many approaches to address Phase I metabolism issues:[117] (i) Reducing the lipophilicity of the drug; (ii) Blocking a site of hydroxylation by replacing the hydrogen(s) with fluorine(s); (iii) Blocking a site of metabolism through cyclization; (iv) Eliminating or replacing a functional group with an isostere less susceptible to metabolism; or (v) Changing the chirality near or at the site of metabolism. This makes sense because the CYP enzymes are chiral, therefore metabolize different chirality differently. If the (*R*)-stereochemical center is metabolized, chances are the corresponding (*S*)-stereochemical center may be resistant to the metabolism. The following case studies employ some of the tactics to overcome Phase I metabolism issues.

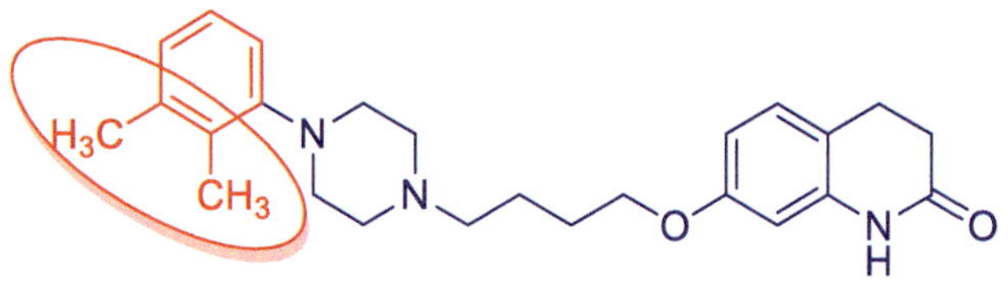

OPC-4392 (**151**), ED_{50} = 41 μmol/kg, p.o.

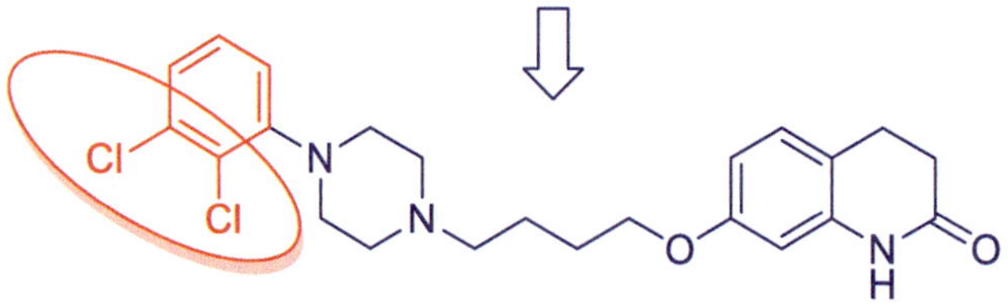

aripiprazole (Abilify, **152**), ED_{50} = 0.6 μmol/kg, p.o.
Otsuka/BMS, 2002

OPC-4392 (**151**) is a unique D_2 antagonist discovered by Otsuka. It suffered a low efficacy with an ED_{50} of 41 mmol/kg, po (taken orally). Scrutiny of its metabolism revealed that the two methyl groups readily underwent hydroxylation and the diols were further oxidized to the corresponding inactive carboxylic acids. Switching the two methyl groups to two chlorine atoms led to a molecule (**152**) that is more resistant to the metabolism. The resulting compound OPC-14597 (aripiprazole, Abilify, **152**) is more efficacious with an ED_{50} of 0.6 mmol/kg, p.o. It was approved by the FDA in 2002 as an effective and unique antipsychotic.[118]

Switching methyl group on **151** to chlorine substituent on **152** retarded drug metabolism. Sometimes occasions arise when metabolism needs to be hastened. While the chlorophenyl derivative S236 (**153**) was a reasonable cyclooxygenase-2 (COX-2) inhibitor, it had a long half-life in rats (117 h). Having a drug staying in the body too long could result in safety issues. Simply switching the chlorine atom to a methyl group afforded celecoxib (Celebrex, **154**), which has a more reasonable half-life of 3.5 h in rats and 12 h in humans.[119]

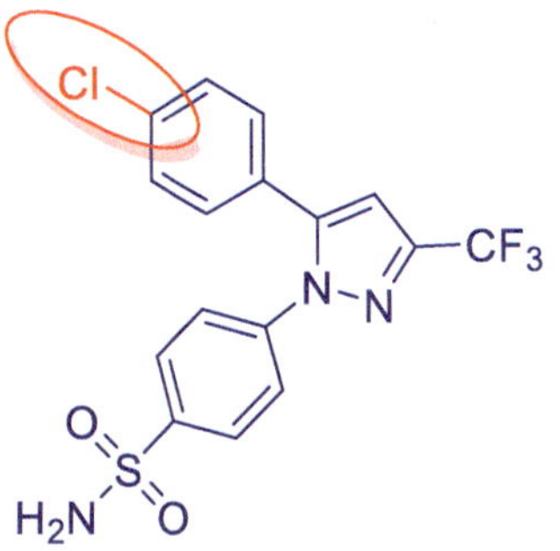

(S236, **153**), $t_{1/2}$ = 117 h in rats

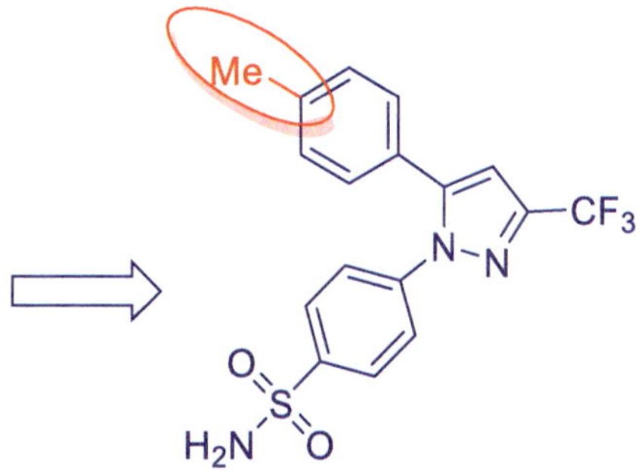

celecoxib (Celebrex, **154**)
$t_{1/2}$ = 3.5 h in rats
$t_{1/2}$ = 12 h in humans

Reducing lipophilicity of a drug may minimize its time-dependent inhibition (TDI) of CYP450 stemmed from mechanism-based inhibition (MBI) of the CYP450 enzymes, although the maneuver is not always a guarantee.[120] A successful example may be found in addressing the issue associated with STAT6 inhibitor **155**. With a Clog *P* value of 4.53, STAT6 inhibitor **155** showed clear MBI by CYP3A4 and is TDI-positive. Replacing morpholine on **155** with *N*-hydroxylethyl-piperazine in **156** reduced the Clog *P* value to 3.21. No TDI of CY3A4 could be detected for **156**, most likely an outcome from the lipophilicity reduction.[121]

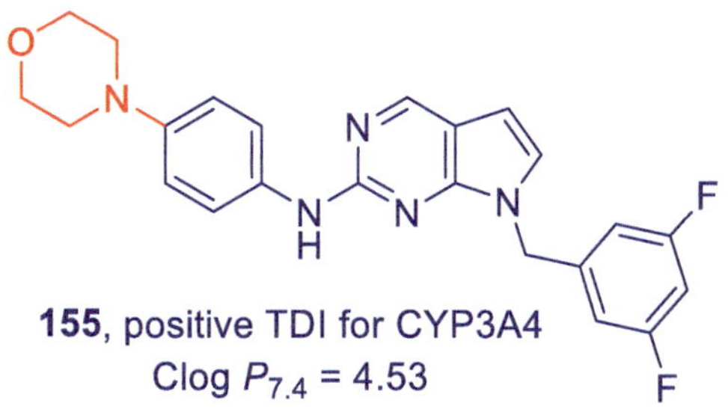

155, positive TDI for CYP3A4
Clog $P_{7.4}$ = 4.53

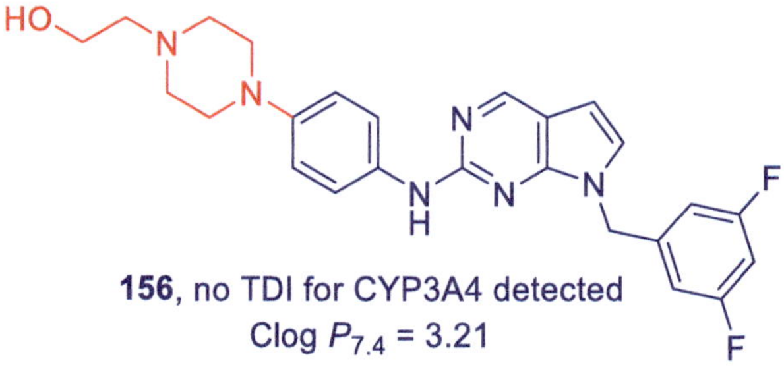

156, no TDI for CYP3A4 detected
Clog $P_{7.4}$ = 3.21

3.4.5. Phase II Metabolism

Phase II metabolism involves conjugation of the parent drug or its Phase I metabolite with an endogenous substrate to form a polar and highly-ionized compound in the majority of the cases, which is more readily excreted in urine or bile. The six most important endogenous substrates are (i). Glucuronic acid; (ii). Sulfate; (iii). Glutathione; (iv). Glycine; (v). Acetylation; (vi). Methylation. Operations (i)–(iv) generate more polar metabolites whereas operations (v) and (vi) afford less polar metabolites. The less polar metabolites often serve to minimize the biological activities of the parent drug or Phase I metabolites. The six conjugation reactions are discussed below.

3.4.5.1 Glucuronidation

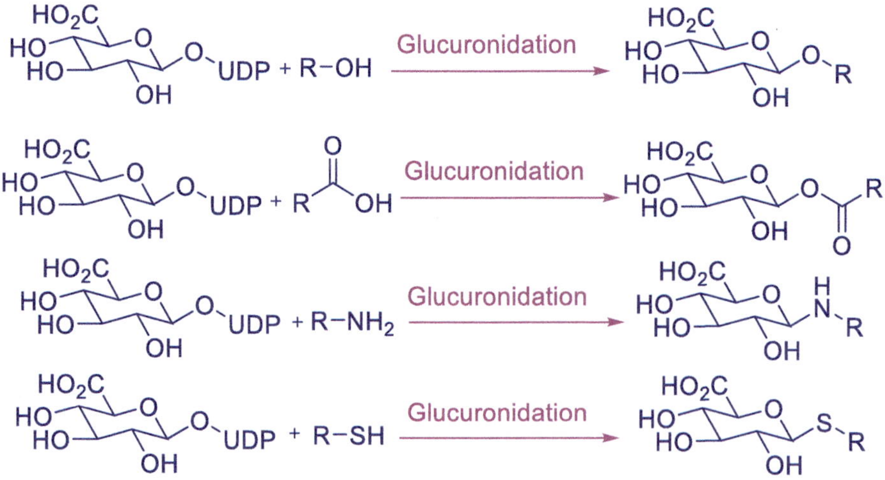

Glucuronide conjugation is the most important Phase II metabolism. Uridine diphosphate (UDP)-glucuronic acid forms glucuronides with primary, secondary, and tertiary alcohols (OH), carboxylic acids (CO_2H), amines (NH_2), amides, thiols (SH), phenols; hydroxylamines; aromatic and aliphatic carboxylic acids; carbamic acids, amino and sulfhydryl groups. Drugs prone to glucuronidation are morphine (**68**), chloramphenicol (**157**), aspirin (**118**), metronidazole, bilirubin, thyroxine, etc. Drug glucuronides, excreted in bile, can be hydrolyzed in the gut by bacteria, producing β-glucuronidase. The liberated drug is reabsorbed and undergoes the same fate. This enterohepatic recirculation of some drugs prolongs their action.

Chloramphenicol (Chloromycetin, **157**) is an example of a parent drug to form a glucuronide. It is an old broad-spectrum antibiotic discovered by Parke–Davis at the end of the 1940s. While an efficacious drug, it is no longer used in developed countries due to its side effects (the most serious of which is blood dyscrasias). The parent drug **157**, bearing two hydroxyl groups, forms two O-glucuronides with glucuronic acid to afford chloramphenicol 1-O-glucuronide (**158**) and chloramphenicol 3-O-glucuronide (**159**), respectively. The glucuronidation process was found to be catalyzed by the UDP-glucuronosyltransferase-2B7 (UGT2B7).[122]

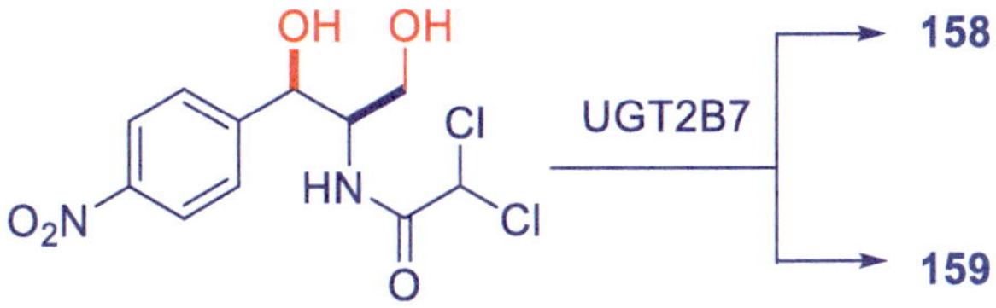

chloramphenicol (Chloromycetin, **157**)

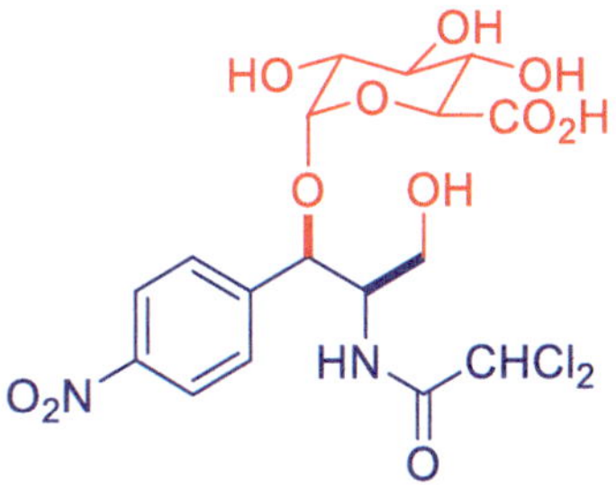

chloramphenicol
1-O-glucuronide (**158**)

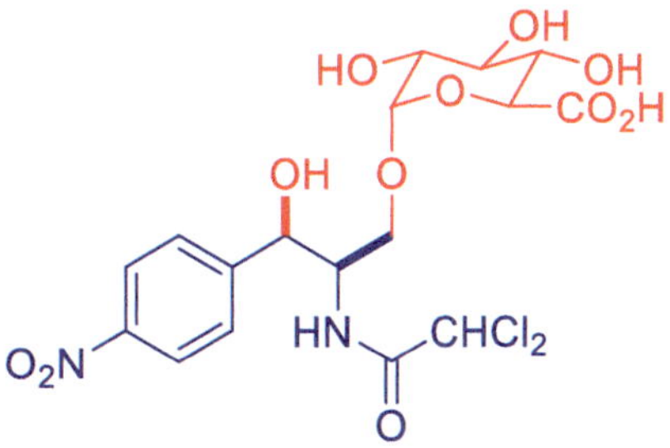

chloramphenicol
3-O-glucuronide (**159**)

3.4.5.2 *Sulfate Conjugation*

Alcohols, phenols, and hydroxylamines [e.g., chloramphenicol (**157**), adrenal, and sex steroids] are sulfated with the aid of sulfotransferase. Sulfation may lead to reactive carbonium or nitrenium ions (in case of hydroxylamine, for example) with loss of sulfate. As shown below, an alcohol is sulfated by 3′-phosphoadenosine-5′-phosphosulfate (PAPS).

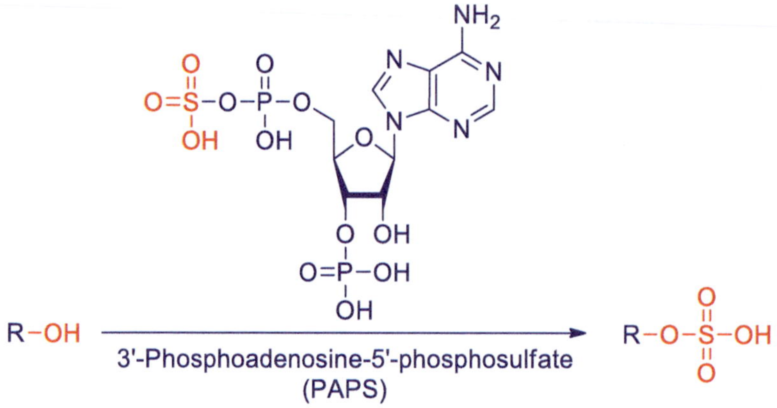

Amines, just like alcohols, are prone to sulfation as well. Moxifloxacin (Avelox, **160**) is a second-generation fluoroquinolone antibiotic. Its MOA is inhibiting DNA gyrase (specifically, type II/IV topoisomerases, enzymes needed to separate bacterial DNA), thereby inhibiting cell replication. Since it has two polar functional groups, a carboxylic acid and a secondary amine, respectively, the parent drug undergoes extensive (52%) Phase II metabolism involving both glucuronidation and sulfation. As shown below, moxifloxacin (**160**) is sulfated by PAPS at the secondary amine site to form moxifloxacin sulfate (**161**).[123]

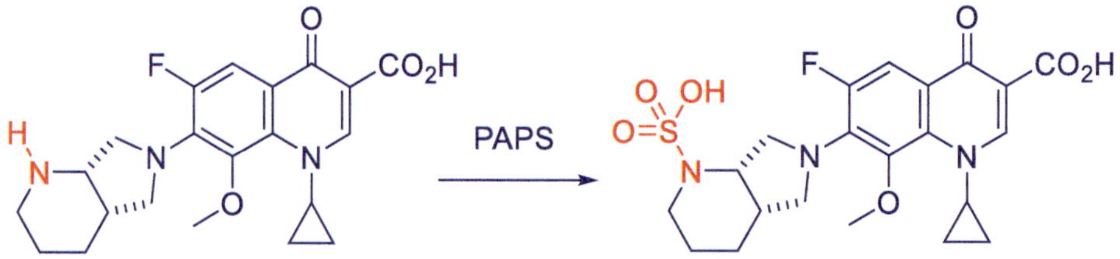

moxifloxacin, free base (Avelox, **160**)   moxifloxacin sulfate (**161**)

3.4.5.3 Glutathione Conjugation

Glutathione (GSH, **162**) is "one of the good guys" in drug metabolism. Its functions often manifest via its thiol group, a reducing agent, which prevents damage caused by reactive oxygen species (ROS) such as radicals, peroxides, and heavy metals. In the context of Phase II metabolism, glutathione conjugation may be (i) Displacement to an electron-withdrawing group such as halogen or nitro group; or (ii) Attack of an arene oxide intermediate; or (iii) Addition to an electron-deficient double bond such as Michael acceptors as in the case of the major metabolite of acetaminophen (**109**).

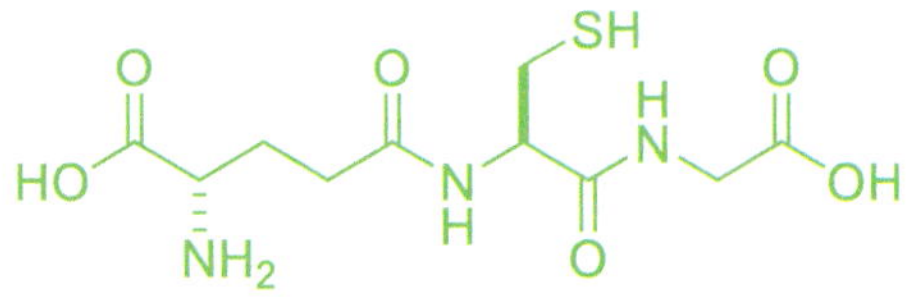

glutathione (GSH, **162**)

Acetaminophen (Tylenol, **109**)'s major metabolite from metabolism by CYP3A4 and 2E1 is *N*-acetyl-*p*-benzoquinone imine (NAPBQI, **110**), a reactive metabolite, which is responsible for the hepatotoxicity. Toxicity ensues when NAPBQI (**110**) reacts with proteins and nucleic acids and forms covalent bonds. Conversely, detoxification takes place when NAPBQI (**110**) reacts with glutathione (GSH, **162**) and forms covalent bonds to produce acetaminophen–glutathione conjugate (**163**).[124]

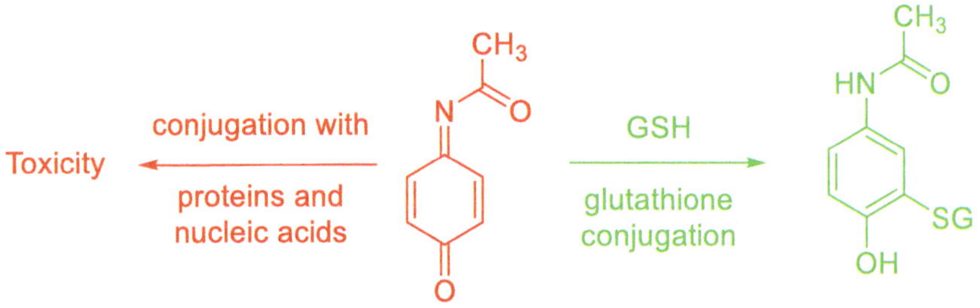

NAPBQI (**110**) acetaminophen–glutathione (**163**)

3.4.5.4 Amino Acid Conjugation

Carboxylic acids, especially aromatic acids and arylacetic acids, tend to form conjugates with amino acids including glycine, glutamine, and taurine.

Salicylic acid (**119**) is an NSAID containing a phenol and a carboxylic acid. It is phosphorylated with adenosine 5′-triphosphate (ATP) to form acyl monophosphate (Step 1). This activated carbonyl then reacts with coenzyme A (CoA) that contains a terminal thiol (Step 2). The resultant thioester reacts with glycine to deliver the salicylic acid–glycine conjugate as amide **164** (Step 3).

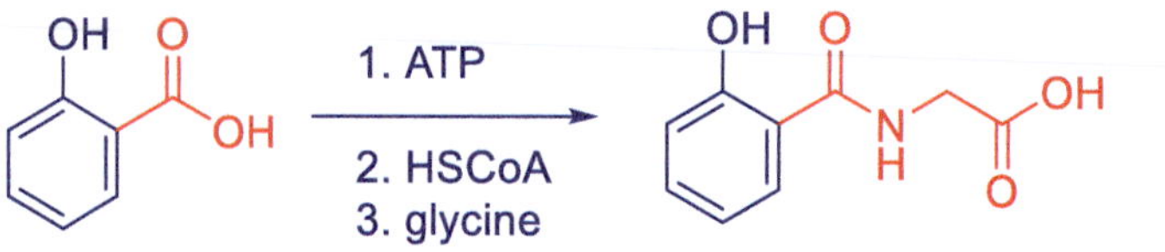

salicylic acid (**119**)
Clog *P* = 2.10

salicylic acid-glycine conjugate (**164**)
Clog *P* = 1.28

3.4.5.5 Acetylation

Compounds (e.g., sulfonamides, isoniazid) having amino or hydrazine residues are acetylated with the help of acetyl-CoA (**165**). Secondary and tertiary amines are not acetylated presumably due to steric hindrance. Multiple genes control the acetyl transferases and rate of acetylation shows genetic polymorphism for slow and fast acetylators.

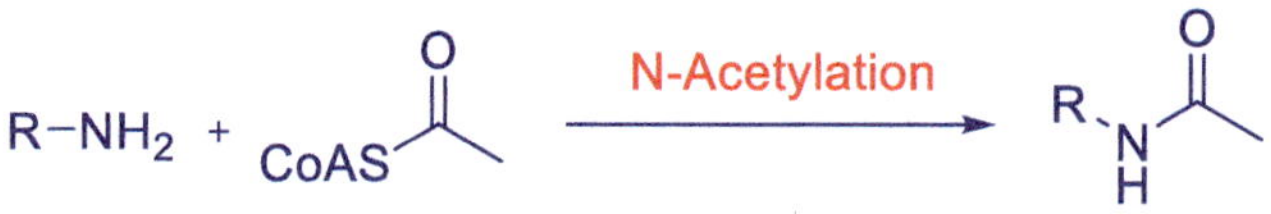

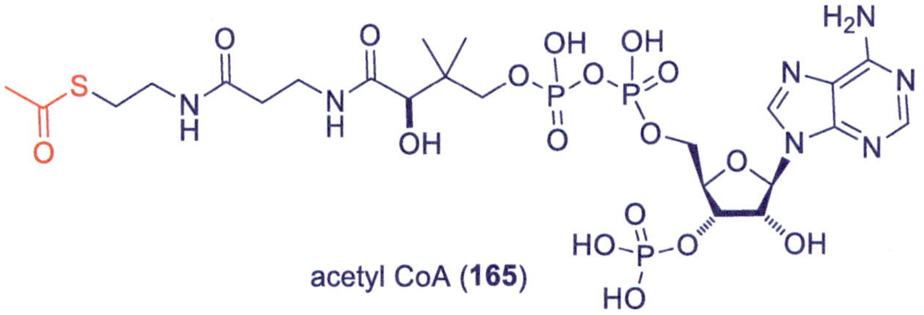

acetyl CoA (**165**)

The aniline group on procainamide (Pronestyl, **166**) undergoes acetylation during its Phase II metabolism to produce acetamide **167**.[125] Acetylation products acetyl-procainamide (**167**, Clog *P*: 1.64) is more lipophilic than procainamide (**166**, Clog *P*: 1.42). The purpose of acetylation here is to deactivate the biological activity.

acetyl CoA (**165**)

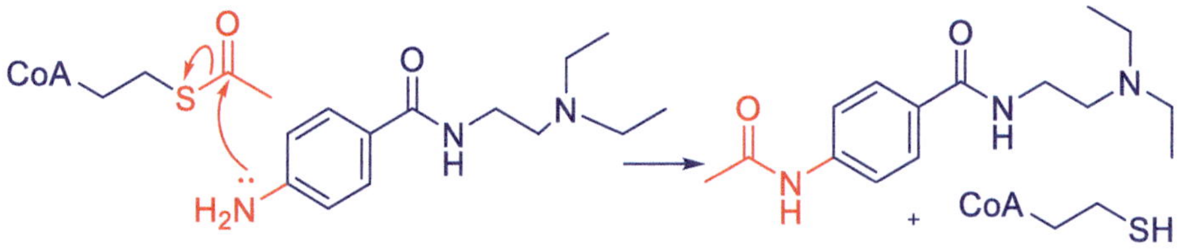

procainamide (**166**)
Clog *P*: 1.42

acetyl-procainamide (**167**)
Clog *P*: 1.64

3.4.5.6 Methylation

Methylation is a relatively minor conjugative pathway in Phase II metabolism. Functional groups that underdo methylation include primary, secondary, and tertiary amines; aromatic amines (anilines); phenols; and aromatic sulfhydryl groups (thiophenols). Amino acids methionine and cysteine act as methyl donors. Drugs including adrenaline, dopamine, histamine, melatonin, 6-mercaptopurine, morphine (**68**), nicotine, and serotonin are all prone to methylation during their Phase II metabolism. Methylation products are less polar than the parent drugs or their metabolites. The purpose of methylation here is to deactivate the biological activity.

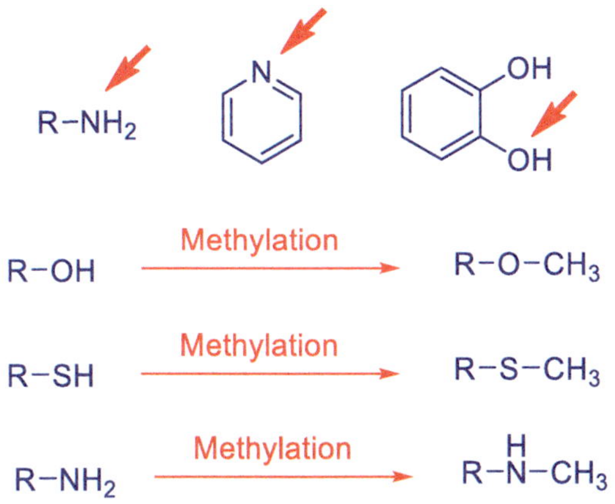

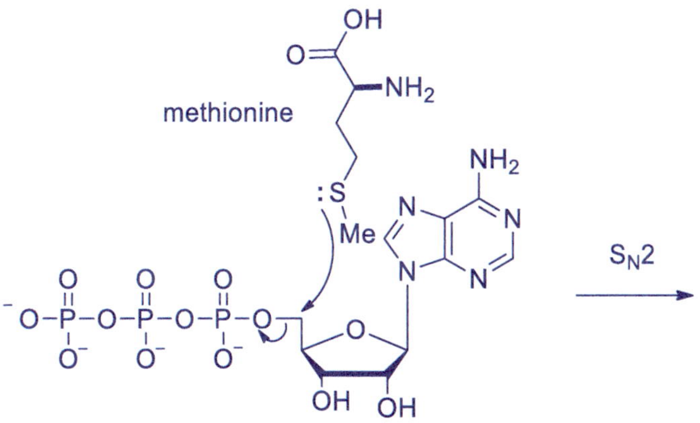

adenosine 5'-triphosphate (ATP, **168**)

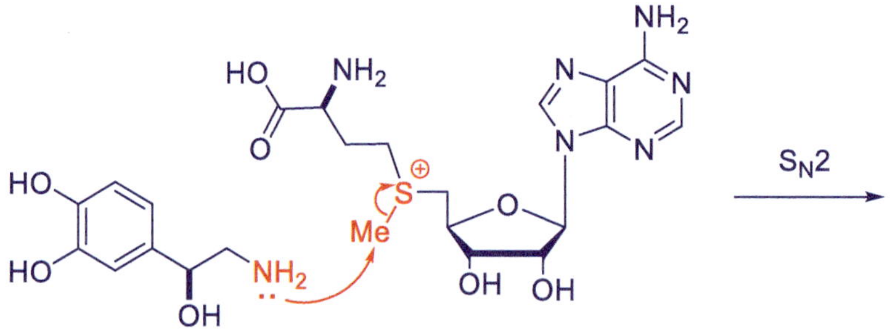

norepinephrine (**169**) S-adenosyl methionine (SAM, **170**)

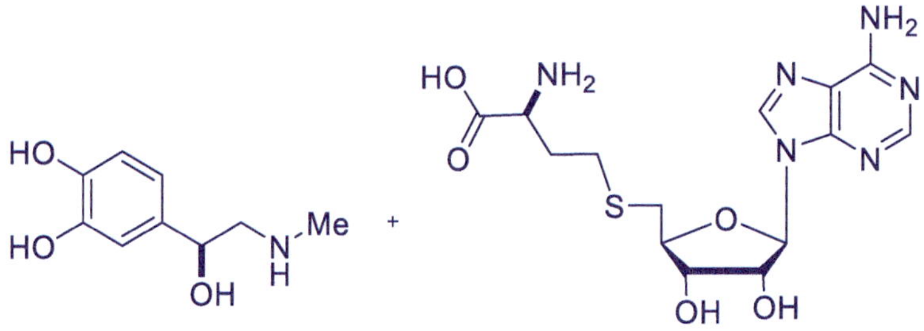

epinephrine (adrenaline, **171**) S-adenosyl homocysteine (SAH, **172**)

Mother Nature does S_N2 reactions with ease and precision. The reaction between methionine and ATP (**168**) gives rise to S-adenosyl methionine (SAM, **170**) via an S_N2 reaction. SAM is a good electrophile that is capable of providing a methyl group to nucleophiles. If the nucleophile happens to be norepinephrine (NE, **169**), the S_N2 reaction between norepinephrine (**169**) and SAM (**170**) would give rise to epinephrine (adrenaline, **171**). Concurrently, SAM (**170**) is converted to S-adenosyl homomethionine (SAH, **172**).[126]

Methylation of catechols is catalyzed by the enzyme catechol O-methyl transferase (COMT). For instance, the phenol group on α-methyldopa is methylated by SAM (**170**) with the aid of COMT.[127]

3.4.5.7 Ribonucleoside/Nucleotide Synthesis

Ribonucleoside/nucleotide synthesis is important during Phase II metabolism for the activation of many purine and pyrimidine antimetabolites used in cancer chemotherapy.

Discovered in 1957, fluorouracil (5-fluororuracil, 5-FU, **176**) is one of the oldest and most widely used cytotoxic agents. Its MOA is inhibition of *de novo* thymidine synthesis and disruption of DNA replication by "disguising" itself as a "legitimate" building block. It can only be given by IV because it is degraded rapidly and the therapeutic index is narrow and its tissue distribution is not selective for tumor. Initial attempts to abrogate the issues via the prodrug strategy arrived at 5′-deoxy-5-fluorouridine (5′-DFUR, doxifluridine, **175**), which has a higher selectivity toward tumor than 5-FU (**176**). 5′-DFUR (**175**) may be given orally albeit in large dose, but is associated with adverse effects such as diarrhea. Roche eventually came up with capecitabine (Xeloda, **173**) in the 1990s. Given orally, prodrug capecitabine (**173**) is metabolized to 5-FU (**176**) in three steps. First, capecitabine (**173**) is converted to 5′-deoxy-5-fluorocytidine (5′-DFCR, **174**) by carboxylesterase (CES). In human, CES is mainly expressed in the liver and intestines. Under the influence of cysteine deaminsae, 5′-DFCR (**174**) is oxidized to 5′-DFUR (**175**). Finally, 5′-DFUR (**175**) is hydrolyzed to 5-FU (**176**) by thymidine phosphorylase.[128] When the toxic 5-FU (**176**) is released, it is at the striking distance of the tumor. The prodrug strategy transformed an IV and toxic drug to an oral drug with fewer adverse effects.

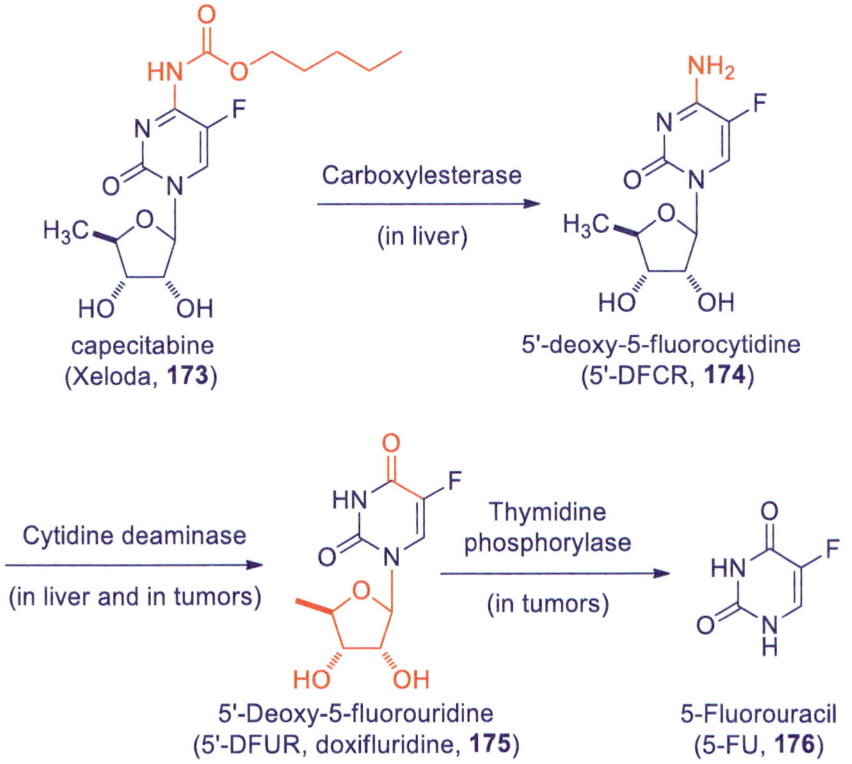

Four approaches to address Phase II metabolism issues include: (i) mask the site of conjugation; (ii) remove the site of conjugation; (iii) introduce electron-withdrawing groups near the site of conjugation; or (iv) create steric hindrance near the site of conjugation.

3.5 Excretion

Excretion or elimination is the process by which drugs or metabolites are irreversibly transferred from internal to external environment through renal or non-renal route. Most drugs are excreted in urine either as unchanged drugs or drug metabolites. In Figure 3.3 on PK parameters, the period when the drug concentration rises from 0 to C_{max} is the absorption phase. After C_{max} is reached, the drug's journey begins with the post-absorption phase where elimination rate is faster than absorption rate. And finally, it goes through the elimination phase where there is no significant absorption.

With regard to the kinetics of elimination, there are first-order kinetics and zero-order kinetics.

First-order kinetics is also known as non-linear kinetics. For first-order kinetics, constant fraction of drug is eliminated per unit time and the rate of elimination is proportional to plasma concentration. Its clearance and half-life remains constant. Most of the drugs follow first-order kinetics.

Zero-order kinetics is also known as linear kinetics. For zero-order kinetics, constant amount of the drug is eliminated per unit time and the rate of elimination is independent of plasma concentration. Its clearance is more at low concentrations and less at high concentrations and its half-life is less at low concentrations and more at high concentrations. Very few drugs follow pure zero-order kinetics. Any drug at high concentrations (when metabolic or elimination pathway is saturated, e.g., alcohol) may show zero-order kinetics.

Drugs and their metabolites are eliminated from either renal excretion (urine) or nonrenal excretion.

3.5.1. Renal Excretion

Renal excretion is the major route of excretion. It consists of three stages: glomerular filtration; tubular reabsorption; and tubular secretion.

3.5.1.1 Glomerular Filtration

Glomerular filtration is a non-selective and unidirectional process. Both ionized and unionized drugs are filtered except those bound to plasma proteins. It depends on PPB and renal blood flow. It does not depend on the lipid solubility because all substances (whether water-soluble or lipid-soluble) can cross fenestrated glomerular membrane.

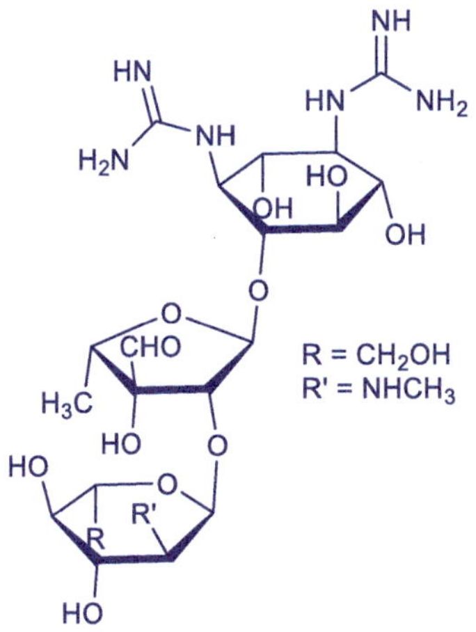

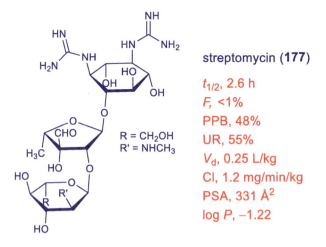

streptomycin (**177**)

$t_{1/2}$, 2.6 h
F, <1%
PPB, 48%
UR, 55%
V_d, 0.25 L/kg
Cl, 1.2 mg/min/kg
PSA, 331 Å^2
log *P*, −1.22

All unbound drug in plasma is filtered in the glomerulus, which is only significant for very polar compounds with log *D* <0. Some compounds are actively secreted into urine along the proximal tubule. Unionized drug can undergo passive re-absorption from urine into blood along the length of the nephron (net excretion may be zero). Drug that is bound to plasma proteins is not filtered. Drugs eliminated by renal excretion include aminoglycosides such as streptomycin (**177**), β-lactams, sulfonamides, quinolones, nitrofurans, and polymyxins.

3.5.1.2 Passive Tubular Re-absorption

Most substances (99%) are re-absorbed across renal tubular cells if they are un-ionized and lipid-soluble. Highly ionized and nonlipid-soluble drugs (1%) stay. It occurs after the glomerular filtration of drugs and takes place all along the renal tubules. Re-absorption results in increase in the half-life of the drug.

3.5.1.3 Tubular Secretion

Tubular secretion does not depend on lipid solubility or PPB. In the nephron, separate pumps are present for acidic and basic drugs: i.e., organic acid transport and organic base transport in the proximal tubules. Drugs utilizing the same transporter may show drug interactions, for instance, probenecid decreases the excretion of penicillin and increases the excretion of uric acid. Exogenous substances such as penicillin are removed, whereas endogenous substances like uric acid are retained by these pumps. If a drug is lipid-soluble, more of it will be reabsorbed and less will be excreted. The opposite is true for lipid-insoluble drugs.

3.5.2. Nonrenal Excretion

Nonrenal excretion include biliary excretion; hepatic elimination; gastrointestinal excretion; pulmonary excretion; salivary excretion; mammary excretion; skin/dermal excretion; and genital excretion.

3.5.2.1 Biliary Excretion

In the liver, drugs can be secreted into the bile. Transporters in the basolateral and canalicular membranes of hepatocytes mediate uptake into the hepatocyte and efflux into bile. Biliary clearance is commonly higher in rats/mice than in dog/man. Bile collects in gall-bladder, then released into intestine upon food intake. Drug may then be reabsorbed via a process known as enterohepatic re-circulation (EHC).

3.5.2.2 Hepatic Elimination

Not surprisingly, drug dosages must be reduced for patients with liver deficiency. Macrolides, lincosamides (**178**), rifampicin, tetracyclines (po) are prone to undergo hepatic elimination.

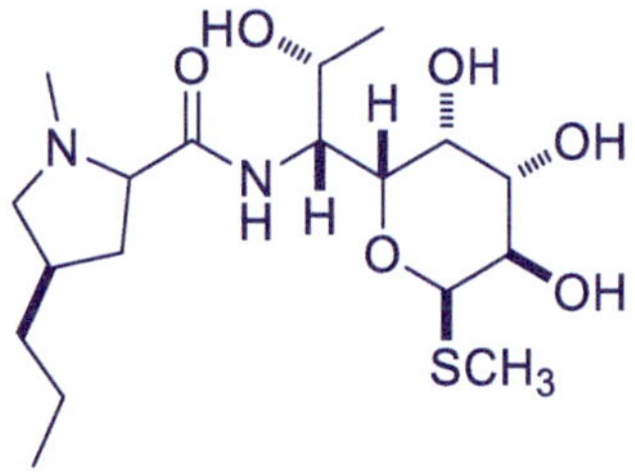

lincosamide (**178**)

3.5.2.3 Gastrointestinal Excretion

Excretion of drugs through gastrointestinal tract (GIT) route usually occurs after parenteral administration. Water-soluble and ionized forms of weakly acidic and basic drugs are excreted in GIT. They are reabsorbed into systemic circulation and undergo recycling. Nicotine and quinine are excreted in stomach.

3.5.2.4 Pulmonary Excretion

Gaseous and volatile substances such as general anesthetics (Halothane, for example) are absorbed through lungs by simple diffusion. Intact gaseous drugs are excreted but not metabolites. Alcohol which has high solubility in blood and tissues are excreted slowly by lungs.

3.5.2.5 Salivary Excretion

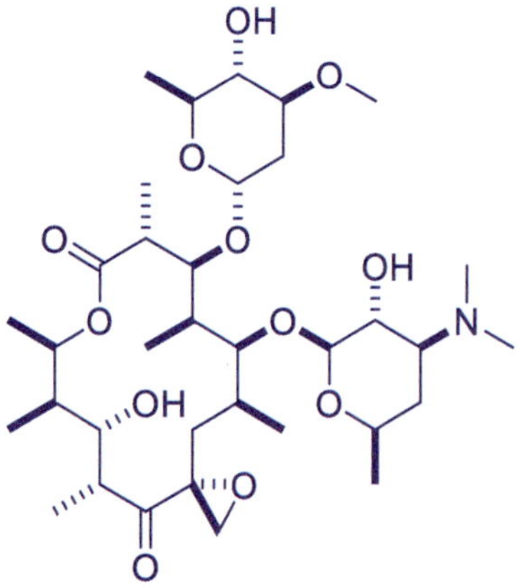

oleandomycin (**179**)

The pH of saliva varies from 5.8 to 8.4. Unionized lipid-soluble drugs are excreted passively. The bitter after-taste in the mouth of a patient is indication of salivary excretion of a drug. Some basic drugs inhibit saliva secretion and are responsible for mouth dryness. Compounds excreted in saliva are: spiramycin, phenytoin (**182**), zalcitabine, verapamil (**79**), caffeine, theophylline (**61**), and oleandomycin (**179**).

3.5.2.6 Skin Excretion

Drugs excreted through skin via sweat may lead to urticaria and dermatitis. Compounds like benzoic acid, salicylic acid, alcohol, and heavy metals like lead, mercury, and arsenic are excreted in sweat.

3.6 Prodrugs

Gerhard Domagk's Prontosil (**137**) is a prodrug, which is converted to the active sulfanilamide (**138**) by nitro-reductase from intestinal bacteria. Some drugs are inactive as such and need conversion in the body to one or more active metabolites. Such a drug is called a prodrug. Prodrugs currently constitute 5% of known drugs and a larger percentage of new drugs. The prodrug may offer advantages over the active form in being more stable, having better bioavailability or other desirable PK properties or less side effects and toxicity. Some prodrugs are activated selectively at the site of action.

Why prodrugs? There are at least five categories of some prodrugs that may be advantageous over the parent drugs according to Rautio.[129] (i) Overcoming formulation and administration problems; (ii) Overcoming absorption barriers; (iii) Overcoming distribution problems; (iv) Overcoming metabolism and excretion problems; (v) Overcoming toxicity problems.

However, prodrugs are not panacea. For example, even if a more lipophilic prodrug helped the drug to permeate biological membranes, the polar active drug is still more prone to form secondary metabolites and subsequently eliminated faster. Therefore, prodrugs are best to serve as the last resort. Rather pursuing an inherently bioavailable drug is more profitable.

3.6.1 Overcoming Formulation and Administration Problems

Anticonvulsant drug phenytoin (Dilantin, **182**) is erratically absorbed after both oral and parental administrations. It has an aqueous solubility of 25 μg/mL. The initial prodrug hydroxymethyl-phenytoin (**181**) had improved aqueous solubility. To further boost the aqueous solubility, phosphate prodrug fosphenyltoin (Cerebyx, **180**) was prepared. It has an aqueous solubility of 140 mg/mL, excellent for IV administration. After being given *in vivo*, fosphenyltoin (**180**) is converted to **181** by alkaline phosphatase. Subsequently, hydroxymethyl-phenytoin (**181**) undergoes spontaneous chemical hydrolysis to deliver phenytoin (**182**).[130]

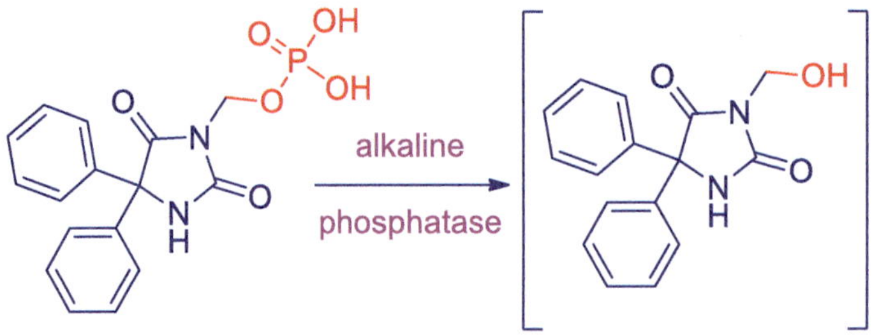

fosphenyltoin (Cerebyx, **180**)
phosphate prodrug: aq. sol: 140 mg/mL

181

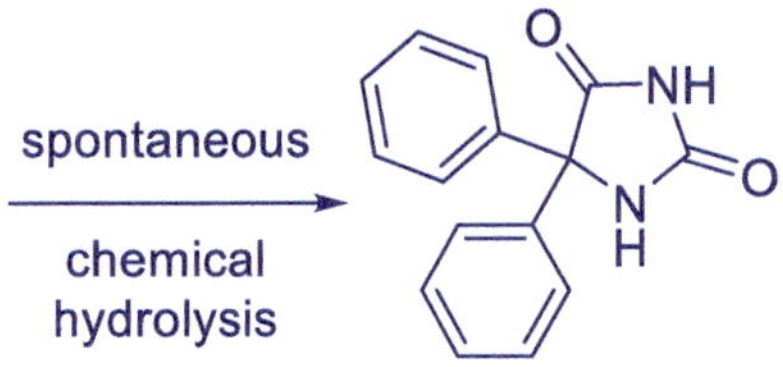

phenytoin
(Dilantin, **182**)
aq. sol: 25 μg/mL

The aforementioned example is not an isolated case. Many prodrugs enjoyed elevated aqueous solubility thus improved parenteral administration. A list of commercially available prodrugs and their solubility in comparison to those of the parent drugs is shown in Table 3.12.

Table 3.12 Commercially available prodrugs and their solubility.

Name of APIs	Aq. solubility (mg/mL)
Clindamycin Clindamycin-2-PO_4	0.2 150
Chloramphenicol Succinate sodium	2.5 500
Metronidazole Dimethylglycinate	10 200
Phenytoin Glyceride of phenytoin	0.03 2.26

Another example of improving aqueous solubility of a drug via the prodrug strategy may be found in camptothecin (**183**). Isolated from the Chinese joy tree, camptothecin (**183**) is a topoisomerase I inhibitor, but has almost no aqueous solubility. Topoisomerase I is a nuclear enzyme that plays a critical role in DNA replication and transcription. Upjohn prepared carbamate prodrug irinotecan (Camptosar, **184**), which has a reasonable aqueous solubility. It is relatively stable and results in an extended exposure and prolonged release of the parent drug. *In vivo*, SN36 (**185**), the active species, is released by metabolism of irinotecan (**184**) by carboxylesterase-2 (CES-2).[131]

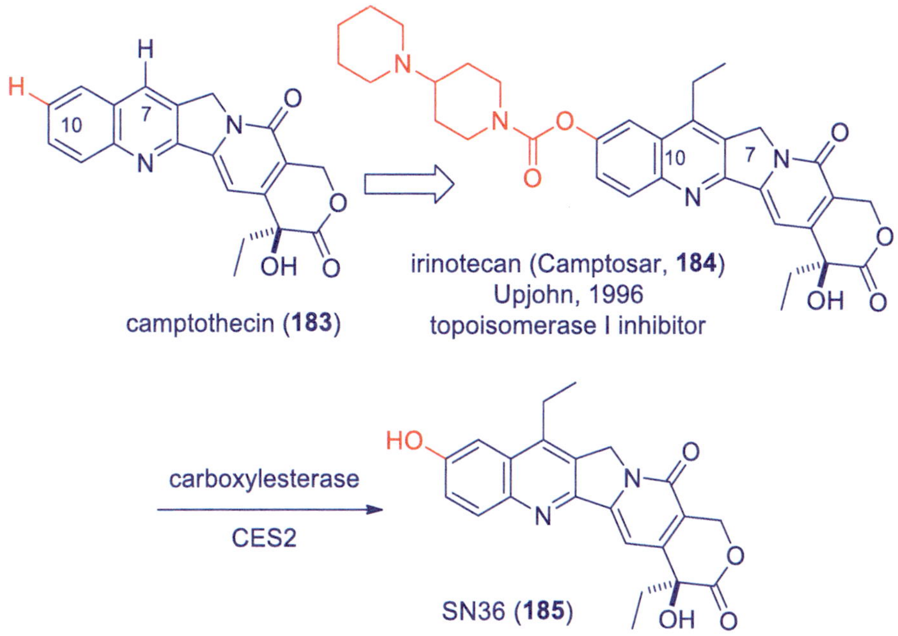

camptothecin (**183**)

irinotecan (Camptosar, **184**)
Upjohn, 1996
topoisomerase I inhibitor

SN36 (**185**)

3.6.2. Overcoming Absorption Barriers

For an oral drug, its lipophilicity needs to strike a balance so that it has a reasonable aqueous solubility yet lipophilic enough to penetrate the cell membrane. The success of oseltamivir (Tamiflu, **186**) is a case in point.

Influenza A viruses have two types on the basis of the antigenic properties of the envelop-associated surface glycoprotein molecules, namely, the hemagglutinin (HA) and the neuraminidase (NA). Neuraminidase (NA) is present in all influenza types and shares high sequence homology. Since it plays such an essential role in the viral life cycle and high conservation of its active site, it is an ideal target for drugs with broad-spectrum anti-influenza activity. Although this epiphany was arrived in the 1960s, little progress was made until high-resolution complex structures of neuraminidase with sialic acid became available in the late 1980s.

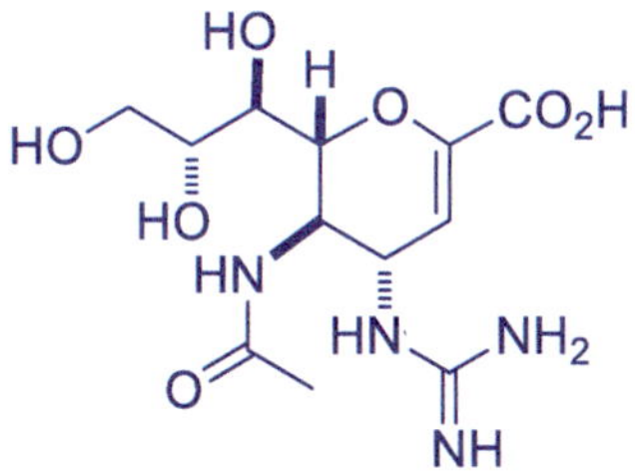

zanamivir (Relenza, **186**)

Based on an old low μM drug 2-deoxy-2,3-dehydro-*N*-acetyl-neutaminic acid (DANA, Neu5Ac2en), zanamivir (Relenza, **186**) was discovered and gained the FDA approval in 2000. However, like DANA, zanamivir (**186**) has a dihydropyran core structure and is highly polar. As a consequence, it has to be given by inhalation.

Gilead decided to use the metabolically more robust noncarbohydrate cyclohexene template as the bioisostere of the dihydropyran core structure. Surprisingly, the 3-pentyl group in the carbocyclic series contributed a considerable amount of binding energy via its hydrophobic interactions. They arrived at GS-4071 (**188**), which still had poor oral availability. Gratifyingly, the prodrug strategy saved the day. The corresponding ethyl ester, oseltamivir (Tamiflu, **187**), was found to be fivefold higher for its bioavailability than the parent drug GS-4071 (**188**). Apparently, more lipophilic prodrug penetrates the cell membrane more readily than the more polar parent drug. *In vivo*, oseltamivir (**187**) is hydrolyzed by endogenous CES-1.[132]

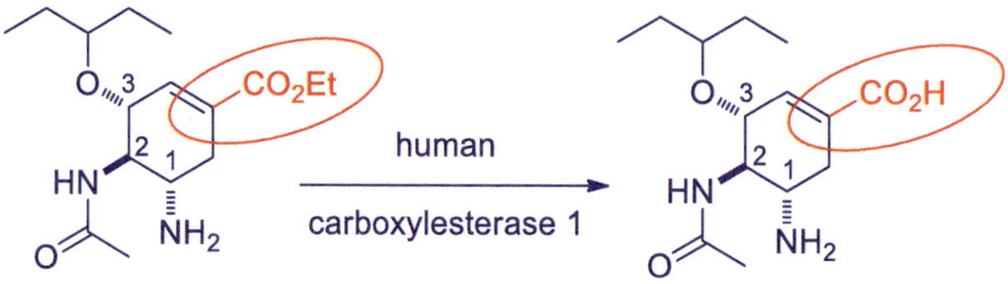

oseltamivir (Tamiflu, **187**) GS-4071 (Ro-64-0802, **188**)

Ester prodrug "masks" the polarity of the carboxylic acid and overcomes the absorption barriers. Ximelagatran (Exanta, **190**), a direct thrombin inhibitor, is a double prodrug of melagatran (**189**). Inspired by hirudin, a direct thrombin inhibitor, was isolated from the European medicinal leech's salivary glands, and a dipeptide as a weak direct thrombin inhibitor, AstraZeneca arrived at melagatran (**189**). Regrettably, it is highly ionic, with an oral bioavailability of less than 3–7% in man, although its bioavailability was greater than 50% in dogs. Although it can only be given parenterally, just like hirudin, AstraZeneca determined to make an oral thrombin inhibitor, which would be much better than injections. Two changes in the melagatran (**189**) molecule did the trick. One was transforming the original carboxylic acid to the corresponding ethyl ester, a very common tactic in medicinal chemistry. Another less common tactic was transforming the original amidine, a strong base, to hydroxyamidine, a nearly neutral fragment. In essence, ximelagatran (**190**) is a double prodrug of melagatran (**189**). As a consequence, the oral bioavailability was increased to 18–20% in man, which was good enough to be given orally.[133]

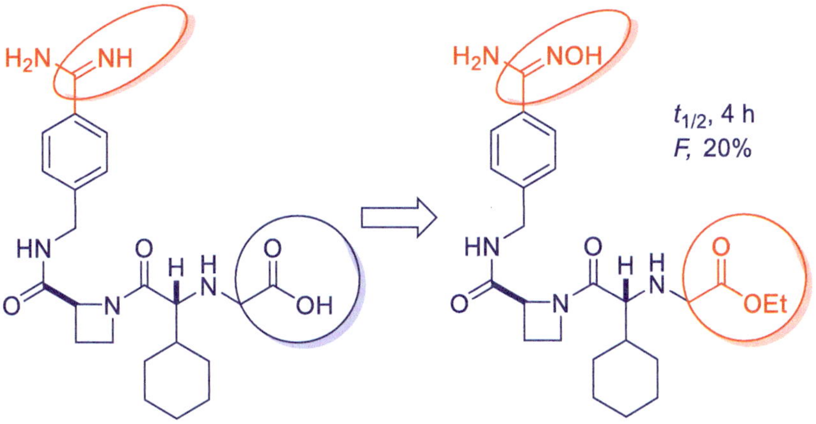

melagatran (**189**) ximelagatran (Exanta, **190**)

A drawback of ximelagatran is its severe liver toxicity in a small population of patients, which was one of the reasons why in 2004 the FDA rejected the drug for licensure in the United States. In 2006, AstraZeneca voluntarily withdrew ximelagatran from the market after reports of liver damage during additional trials.

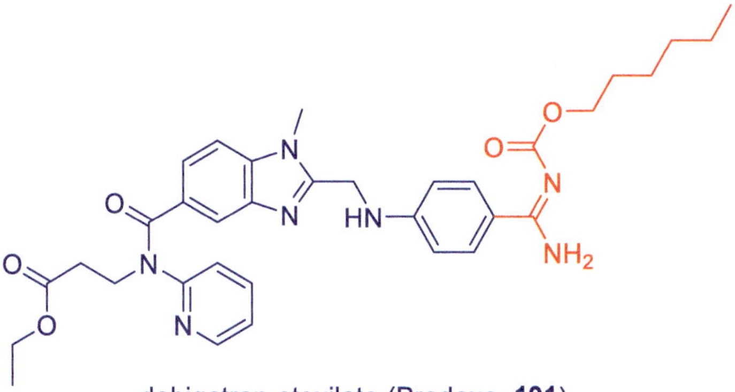

dabigatran etexilate (Pradaxa, **191**)
2010, Boehriger-Ingelheim
A direct thrombin inhibitor
F% = 7%

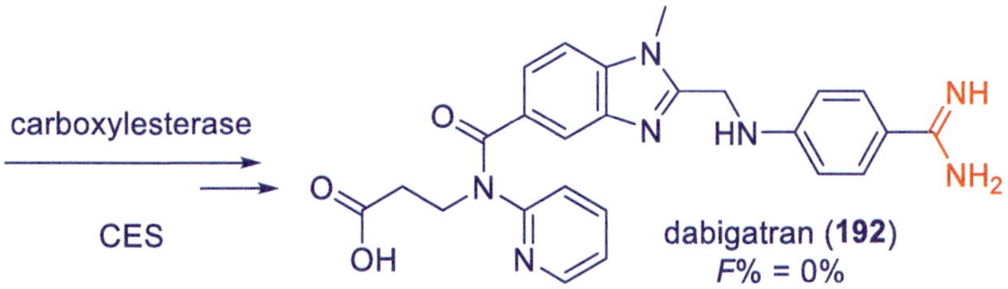

dabigatran (**192**)
F% = 0%

Dabigatran (**192**) also has a strongly basic amidine functional group. Rather than converting the amidine to the hydroxylamidine as in ximelagatran (Exanta, **190**), a team at Boehringer–Ingelheim sought to prepare the corresponding dabigatran etexilate (**191**). After transforming the acid to the ethyl ester, the resulting dabigatran etexilate (Pradaxa, **191**) became orally bioavailable, although *F*% is only 7.2%.[134]

3.6.3. Overcoming Distribution Problems

We already discussed the scenario concerning bioactivation of levodopa (**66**) to dopamine (**67**). However, armed with three polar groups, dopamine (**67**) is too polar to enter the brain via BBB. However, its precursor levodopa (**66**) does thanks to the amino acid transport located on BBB. Once in the brain, levodopa (**66**) is metabolized to dopamine (**67**). L-type amino acid transporter (LAT1) is responsible for mediating transporting amino acids such as levodopa (**66**).[135]

Carrier-mediated cellular uptake helps some drugs absorbed better. Internal H^+-coupled peptide transporter-1 (PEPT1) is one of those carriers. Midodrine (Amatine, **193**), an oral drug for orthostatic hypotension, is a PEPT1 substrate thus benefits from carrier-mediated cellular uptake. It is almost completely absorbed after oral administration and converted into its active form desglymidodrine (**194**), which is an α1-receptor agonist.[136]

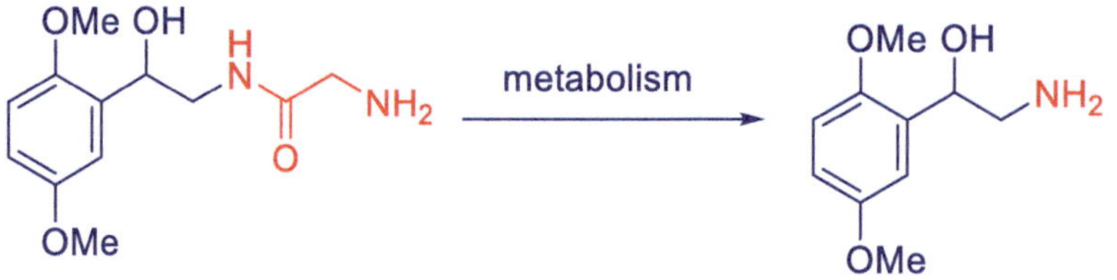

midodrine (Amatine, **193**)
1996, substrate of PEPT1

desglymidodrine (**194**)
α1-receptor agonist

3.6.4. Overcoming Metabolism and Excretion Problems

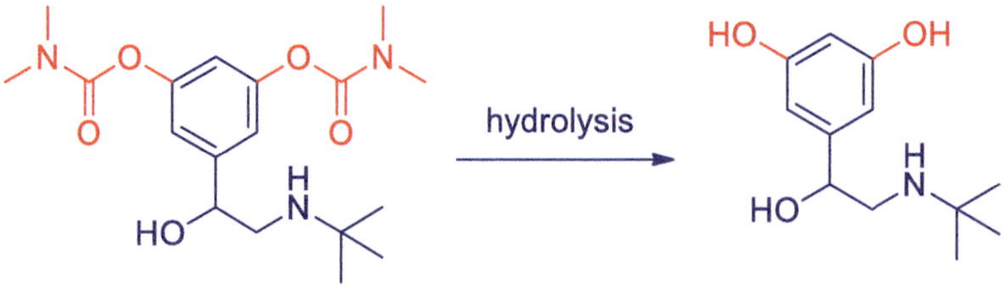

bambuterol (Bambec, **195**)

terbutaline (Bronclyn, **196**)

Terbutaline (Bronclyn, **196**), a β2 adrenergic agonist to treat asthma, is susceptible to rapid and extensive presystemic metabolism because it possesses two phenol groups that are readily oxidized. Its dimethylcarbamate derivative, bambuterol (Bambec, **195**), in contrast, is stable to pre-systemic elimination and is concentrated by lung tissue after absorption from the gastrointestinal tract (GIT). The prodrug bambuterol (**195**) is hydrolyzed to terbutaline (**196**) primarily by butyrylcholinesterase, and lung tissue is capable of this metabolic pathway.[137]

3.6.5. Overcoming Toxicity Problems

When acyclovir (Zovirax, **198**) is given orally, 19% of the dosed drug is excreted in the urine. Its valine derivative valacyclovir (Valtrex, **197**) saw 63% of acyclovir (**198**) in the urine. The prodrug valacyclovir (**197**) also has three to fivefold higher systemic oral bioavailability than the parent drug. Human peptide transporter 1 (hPEPT1)-mediated uptake of valacyclovir (**197**) into intestinal cells could be the underlying mechanism for this absorption. Once absorbed, valacyclovir (**197**) is first hydrolyzed to acyclovir (**198**) by human valacyclovirase, an α-amino acid ester hydrolase capable of activating valacyclovir and other prodrugs. Even acyclovir (**198**) is not active itself, but needs to be phosphorylated to be active *in vitro*. To that end, acyclovir (**198**) undergoes mono-phosphorylation by viral thymidine kinase, which is 3,000 times faster than in un-infected cells to offer a prodrug **199**. Further phosphorylation by cellular kinase then delivered the actual active triphosphate nucleotide **200**. With tongue in cheek, valacyclovir (**197**) may be called a pro-prodrug.[138]

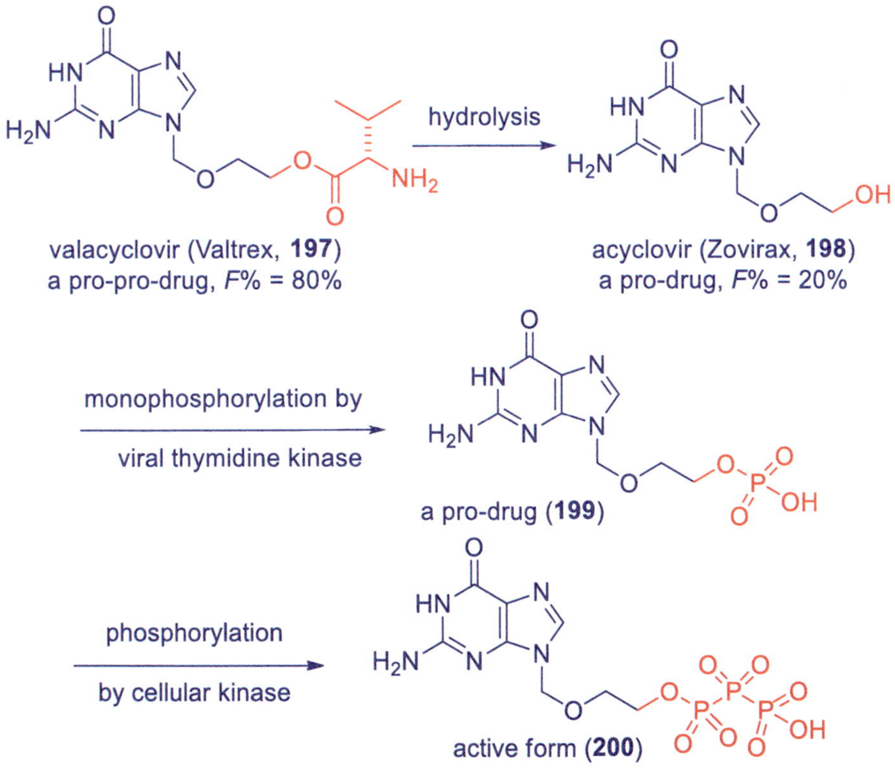

valacyclovir (Valtrex, **197**)
a pro-pro-drug, *F*% = 80%

acyclovir (Zovirax, **198**)
a pro-drug, *F*% = 20%

a pro-drug (**199**)

active form (**200**)

In the same vein, sofosbuvir (Sovaldi, **201**) is not active *in vitro*, but it becomes an active HCV NS5B polymerase inhibitor *in vivo* under the influence of a battery of enzymes in human body via a relatively complex activation pathway.[139] It is a prodrug of 2′-α-fluoro-2′-β-methyl nucleoside PSI-6130 (**203**), which is a potent and selective inhibitor of HCV NS5B polymerase that exhibits antiviral activity in cell culture systems and is clinically efficacious. The first of this class advanced into clinical trials, **203** exhibits an EC_{90} = 4.6 μM in an HCV replicon assay.[140] However, **203** demonstrated only modest oral bioavailability in preclinical studies. PSI-6206 (**202**) is a metabolite of **203** by deamination of the cytosine. While it was observed in both hepatocytes *in vitro* and rhesus monkeys, it was inactive in HCV replicon assays at concentrations of up to 100 μM. However, its triphosphate form, PSI-7409 (**209**), is a potent inhibitor of HCV polymerase in an enzymatic assay. Moreover, the mean half-life of PSI-7409 (**209**) in primary human hepatocytes is 38 h, considerably longer than that of the triphosphate derivative of PSI-6130 (**203**), which is 4.7 h, suggesting the potential of PSI-6206 (**202**) if the triphosphate form could be effectively delivered to cells.

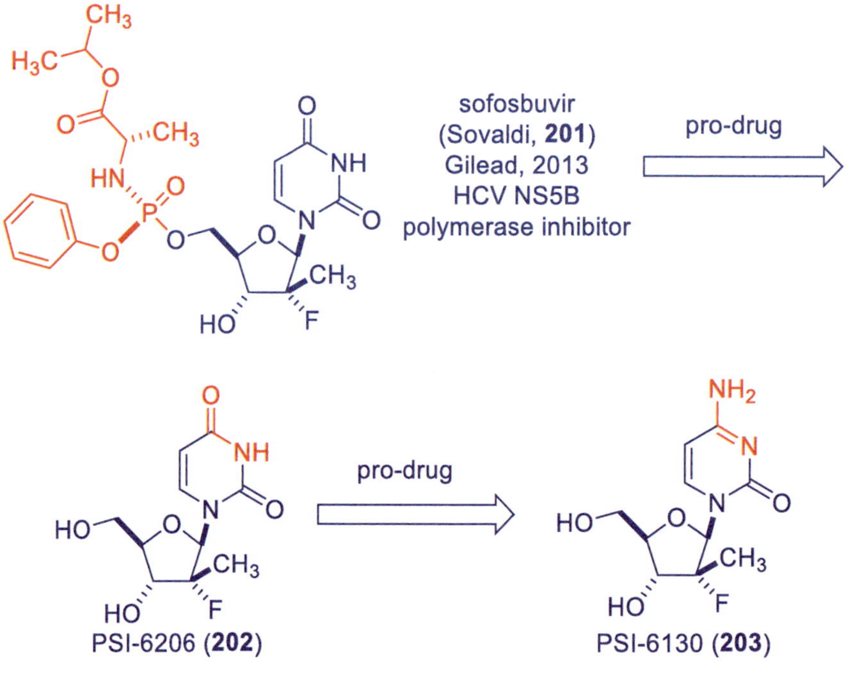

The key enzymes initially involved in the metabolism of sofosbuvir (**201**) are human cathepsin A (CatA) and carboxylesterase 1 (CES1), which are responsible for the hydrolysis of the carboxyl ester between the alaninyl moiety and the isopropyl alcohol.[141,142] This stereospecific reaction gives rise to the corresponding carboxylic acid **204**. A nonenzymatic intramolecular nucleophilic attack then results in the formation of an alaninyl phosphate intermediate **205**, which undergoes a rapid chemical reaction to hydrolyze the cyclic phosphate to a linear phosphate as carboxylic acid **206**. The next step is speculated to involve the histidine triad nucleotide-binding protein 1 (Hint 1)

enzyme in which the alaninyl phosphate intermediate is deaminated to form a monophosphate nucleotide **207**. The final two steps involve consecutive phosphorylation reactions mediated by cellular kinases, uridine monophosphate–cytidine monophosphate (UMP–CMP) kinase and nucleoside diphosphate kinase (NDPK), producing the diphosphate nucleotide **208** and subsequently the active triphosphate nucleotide **209**.

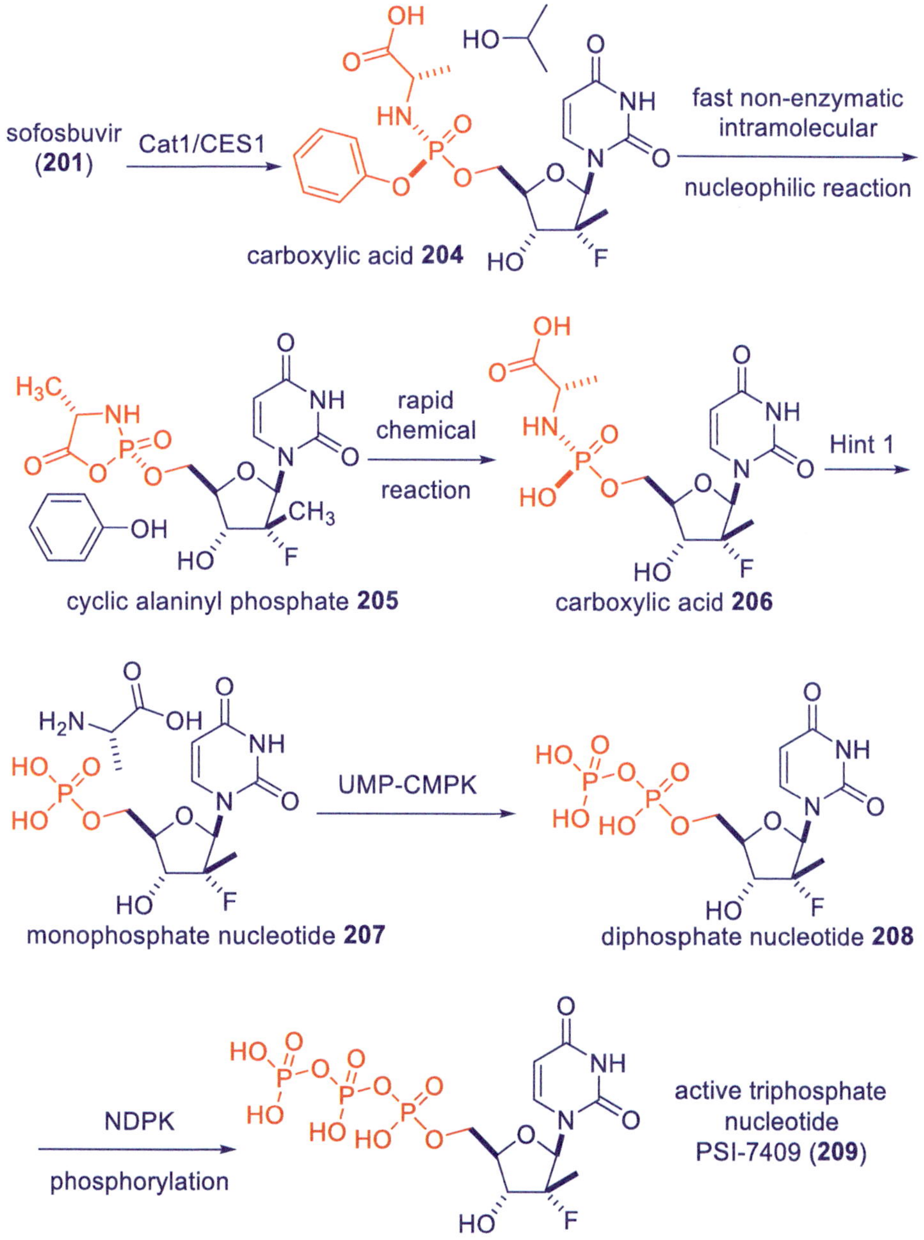

3.7 Further Reading

Birkett, D. J. *Pharmacokinetics Made Easy.* 2nd ed. McGraw-Hill Education: Australia, **2009**, 120 pp.

Coleman, M. D. *Human Drug Metabolism: An Introduction.* 2nd ed. Wiley: Chichester, England, **2010**, 360 pp.

Han, C.; Davis, C. B.; Wang, B., eds. *Evaluation of Drug Candidates for Preclinical Development: Pharmacokinetics, Metabolism, Pharmaceutics, and Toxicology.* Wiley: Hoboken, NJ, **2010**, 289 pp.

Meanwell, N. A., ed. *Tactics in Contemporary Drug Design (Topics in Medicinal Chemistry 9).* Springer: Heidelberg, Germany, **2015**, 394 pp.

Rautio, J., ed. *Prodrugs and Targeted Delivery: Towards Better ADME Properties.* Wiley-VCH: Weinheim, **2011**. 496 pp.

Smith, D. A.; Allerton, C.; Kalgutkar, A.; van de Waterbeemd, H.; Walker, D. K. *Pharmacokinetics and Metabolism in Drug Design,* 3rd Revised and Updated Ed. Wiley-VCH: Weinheim, **2006**, 187 pp.

Tsaioun, K.; Kates, S. S., eds. *ADMET for Medicinal Chemists: A Practical Guide.* Wiley: Hoboken, NJ, **2011**, 516 pp.

3.8 References

1. Kola, I.; Landis, J. *Nat. Rev. Drug Discov.* **2004**, *3*, 711–716.
2. Bayliss, M. K.; Butler, J.; Feldman, P. L.; Green, D. V. S.; Taylor, A. J; Leeson, P. D.; Palovich, M. R. *Drug Discov. Today* **2016**, *21*, 1719–1727.
3. Leeson, P. D. *Adv. Drug Delivery Rev.* **2016**, *101,* 22–33.
4. Young, R. J. Physical Properties in Drug Design. In Meanwell, N. A., ed. *Top. Med. Chem. 9(Tactics in Contemporary Drug Design).* Springer: Heidelberg, Germany, **2015**, pp. 1–68.
5. Wenlock, M. C.; Leeson, P. D. *J. Med. Chem.* **2003**, *46*, 1250–1256.
6. Manallack, D. T.; Prankerd, R. J.; Yuriev, E.; Oprea, T. I.; Chalmers, D. K. *Chem. Soc. Rev.* **2013**, *42*, 485–496.
7. Tetko, I. V.; Poda, G. I. *J. Med. Chem.* **2004**, *47*, 5601–5604.
8. Young, R. C.; Mitchell, R. C.; Brown, T. H.; Ganellin, C. R.; Jones, M.; Rana, K. K.; Saunders, D.; Smith, I. R.; Score, N. E.; Wilks, T. J. *J. Med. Chem.* **1988**, *31*, 656–671.
9. Rafi, S. B.; Hearn, B. R.; Vedantham, P.; Jacobson, M. P.; Renslo, A. R. *J. Med. Chem.* **2012**, *55*, 3163–3169.
10. (a) El Tayar, N.; Mark, A. E.; Vallat, P.; Brunne, R. M.; Testa, B.; van Gunsteren, W. F. *J. Med. Chem.* **1993**, *35*, 23757–23764. (b) Mackman, R. L.; Steadman, V.

A.; Dean, D. K.; Jansa, P.; Poullennec, K. G.; Appleby, T.; Austin, C.; Blakemore, C. A.; Cai, R.; Cannizzaro, C.; et al. *J. Med. Chem.* **2018**, *61*, 9473–9499.
11. Alex, A.; Millan, D. S.; Perez, M.; Wakenhut, F.; Whitlock, G. A. *MedChemComm* **2011**, *2*, 669–674.
12. Abraham, M. H.; Ibrahima, A.; Zissimosa, A. M.; Zhao, Y. H.; Comer, J.; Reynolds, D. P. *Drug Discov. Today* **2002**, *7*, 1056–1063.
13. Giordanetto, F.; Tyrchan, C.; Ulander, J. *ACS Med. Chem. Lett.* **2017**, *8*, 139–142.
14. Kelder, J.; Grootenhuis, P. D. J.; Bayada, D. M.; Delbressine, L. P. C.; Ploemen, J.-P. *Pharm. Res.* **1999**, *16*, 1514–1519.
15. Veber, D. F.; Johnson, S. R.; Cheng, H.-Y.; Smith, B. R.; Ward, K. W.; Kopple, K. D. *J. Med. Chem.* **2002**, *45*, 2615–2623.
16. Rigby, J. W.; Scott, A. K.; Hawksworth, G. M.; Petrie, J. C. *Br. J. Pharmacol.* **1985**, *20*, 327–331.
17. Lipinski, C. A.; Lombardo, F.; Dominy, B. W.; Feeney, P. J. *Adv. Drug Delivery Rev.* **2001**, *46*, 3–26.
18. Lipinski, C. A. *Drug Discov. Today Technol.* **2004**, *1*, 337–341.
19. (a) DeGoey, D. A.; Chen, H.-J.; Cox, P. B.; Wendt, M. D. *J. Med. Chem.* **2018**, *61*, 2636–2651. (b) Deeks, E. D. *Drugs* **2014**, *74*, 195–206.
20. (a) Congreve, M.; Carr, R.; Murray, C.; Jhoti, H. *Drug Discov. Today* **2003**, *8*, 876–877. (b) Jhoti, H.; Williams, G.; Rees, D. C.; Murray, C. W. *Nat. Rev. Drug Discov.* **2013**, *12*, 644–645.
21. Hopkins, A.; Groom, C. R.; Alex, A. *Drug Discov. Today* **2004**, *9*, 430–431.
22. Abad-Zapatero, C. *Exp. Opin. Drug Discov.* **2007**, *2*, 469–488.
23. Abad-Zapatero, C.; Perišić, O.; Wass, J.; Bento, P.; Overington, J.; Al-Lazikani, B.; Johnson, M. E. *Drug Discov. Today* **2010**, *15*, 804–811.
24. Kenny, P. W.; Montanari, C. A. *J. Comput. Aided Mol. Des.* **2012**, *24*, 1–13.
25. Kenny, P. W.; Leitão, A.; Montanari, C. A. *J. Comput. Aided Mol. Des.* **2014**, *28*, 699–710.
26. Leeson, P. D.; Springthorpe, B. *Nat. Rev. Drug Discov.* **2007**, *6*, 881–890.
27. Keserü, G. M.; Makara, G. M. *Nat. Rev. Drug Discov.* **2009**, *8*, 203–212.
28. Tarcsay, Á.; Nyíri, K.; Keserű, G. M. *J. Med. Chem.* **2012**, *55*, 1252–1260.
29. Shultz, M. D. *Bioorg. Med. Chem. Lett.* **2013**, *23*, 5980–5991.
30. Shultz, M. D. *Bioorg. Med. Chem. Lett.* **2013**, *23*, 5992–6000.
31. (a) Meanwell, N. A. *Chem. Res. Toxicol.* **2016**, *29*, 564–616. (b) Johnson, T. W.; Gallego, R. A.; Edwards, M. P. *J. Med. Chem.* **2018**, *62*, 6401–6420.
32. Rezvanfar, M. A.; Rahimi, H. R.; Abdollahi, M. *Exp. Opin. Drug Metab. Toxicol.* **2012**, *8*, 1231–1245.
33. Amidon, G. L.; Lennernäs, H.; Shah, V. P.; Crison, J. R. *Pharm. Res.* **1995**, *12*, 413–420.
34. (a) Walker, M. A. Improving Solubility via Structural Modification. In Meanwell, N. A., ed. *Top. Med. Chem. 9(Tactics in Contemporary Drug Design)*. Springer: Heidelberg, **2015**, pp. 69–106. (b) Walker, M. A. *Bioorg. Med. Chem. Lett.* **2017**, *27*, 5100–5108.
35. Zimmermann, J.; Buchdunger, E.; Mett, H.; Meyer, T.; Lydon, N. B. *Bioorg. Med. Chem. Lett.* **1997**, *7*, 187–192.

36. Hennequin, L. F.; Stokes, E. S.; Thomas, A. P.; Johnstone, C.; Plé, P. A.; Ogilvie, D. J.; Dukes, M.; Wedge, S. R.; Kendrew, J.; Curwen, J. O. *J. Med. Chem.* **2002**, *45*, 1300–1312.
37. Ding, C.; Tian, Q.; Li, J.; Jiao, M.; Song, S.; Wang, Y.; Miao, Z.; Zhang, A. *J. Med. Chem.* **2018**, *61*, 760–776.
38. Ishikawa, H.; Hashimoto, Y. *J. Med. Chem.* **2011**, *54*, 1539–1554.
39. Stepan, A. F.; Subramanyam, C.; Efremov, I. V.; Dutra, J. K.; O'Sullivan, T. J.; DiRico, K. J.; McDonald, W. S.; Won, A.; Dorff, P. H.; Nolan, C. E.; et al. *J. Med. Chem.* **2012**, *55*, 3414–3424.
40. Lovering, F.; Bikker, J.; Humblet, C. *J. Med. Chem.* **2009**, *52*, 6752–6756.
41. Lovering, F. *MedChemCom* **2013**, *4*, 515–519.
42. Wang, H. L.; Katon, J.; Balan, C.; Bannon, A. W.; Bernard, C.; Doherty, E. M.; Dominguez, C.; Gavva, N.; Gore, V.; Ma, V.; et al. *J. Med. Chem.* **2007**, *50*, 3528–3539.
43. Doherty, E. M.; Fotsch, C.; Bannon, A. W.; Bo, Y.; Chen, N.; Dominguez, C.; Falsey, J.; Gavva, N. R.; Katon, J.; Nixey, T.; et al. *J. Med. Chem.* **2007**, *50*, 3415–3527.
44. Wenglowsky, S.; Moreno, D.; Rudolph, J.; Ran, Y.; Ahrendt, K. A.; Arrigo, A.; Colson, B.; Gloor, S. L.; Hastings, G. *Bioorg. Med. Chem. Lett.* **2012**, *22*, 912–915.
45. Goosen, C.; Laing, T. J.; Du, P. J.; Goosen, T. C.; Flynn, G. L. *Pharm. Res.* **2002**, *19*, 13–19.
46. Yokoawa, F.; Wang, G.; Chan, W. L.; Ang, S. H.; Wong, J.; Ma, I.; Rao, S. P. S.; Manjunatha, U.; Lakshminarayana, S. B.; Herve, M.; et al. *ACS Med. Chem. Lett.* **2013**, *4*, 451–455.
47. Cernak, T.; Schönherr, H. *Angew. Chem. Int. Ed.* **2013**, *52*, 12256–12267.
48. Barsanti, P. A.; Aversa, R. J.; Jin, X.; Pan, Y.; Lu, Y.; Elling, R.; Jain, R.; Knapp, M.; Lan, J.; Lin, X.; Rudewicz, P.; Sim, J.; Taricani, L.; Thomas, G.; Yue, Q. *ACS Med. Chem. Lett.* **2015**, *6*, 37–41.
49. Johansson, A.; Löberg, M.; Antonsson, M.; von Unge, S.; Hayes, M.; Judkins, R.; Ploj, K.; Benthem, L.; Linden, D.; Brodin, P.; et al. *J. Med. Chem.* **2016**, *59*, 2497–2511.
50. Volpe, D. A. *Fut. Med. Chem.* **2011**, *3*, 2063–2077.
51. Reis, J. M.; Sinko, B.; Serra, C. H. *Mini-Rev. Med. Chem.* **2010**, *10*, 1071–1076.
52. Siepmann, J.; Siepmann, F.; Florence, A. T. *Drugs Pharm. Sci.* **2009**, *188*, 117–154.
53. (a) Dobson, P. D.; Kell, D. B. *Nat. Rev. Drug Discov.* **2008**, *7*, 205–220. (b) Nigam, S. *Nat. Rev. Drug Discov.* **2015**, *14*, 29–44.
54. Jain, K. S.; Kathiravan, M. K.; Somani, R. S.; Shishoo, C. J. *Bioorg. Med. Chem. Lett.* **2007**, *15*, 1181–1205.
55. Leishman, D. J.; Rankovic, Z. Drug Discovery vs. hERG. In Meanwell, N. A., ed. *Top. Med. Chem. 9(Tactics in Contemporary Drug Design).* Springer: Heidelberg, **2015**, pp. 225–259.
56. Rampe, D.; Brown, A. M. *J. Pharmacol. Toxicol. Methods* **2013**, *68*, 13–22.
57. Deng, Q.; Lim, Y.-H.; Anand, R.; Yu, Y.; Kim, J.-H.; Zhou, W.; Zheng, J.; Tempest, P.; Levorse, D.; Zhang, X.; Greene, S.; Mullins, D.; Culberson, C.;

Sherborne, B.; Parker, E. M.; Stamford, A.; Ali, A. *Bioorg. Med. Chem. Lett.* **2015**, *25,* 2958–2962.
58. Tang, H.; Zhu, Y.; Teumelsan, N.; Walsh, S. P.; Shahripour, A.; Priest, B. T.; Swensen, A. M.; Felix, J. P.; Brochu, R. M.; Bailey, T.; et al. *ACS Med. Chem. Lett.* **2016**, *7,* 697–701.
59. Huscroft, I. T.; Carlson, E. J.; Chicchi, G. G.; Kurtz, M. M.; London, C.; Raubo, P.; Wheeldon, A.; Kulagowski, J. J. *Bioorg. Med. Chem. Lett.* **2006,** *16,* 2958–2962.
60. Coen, M. *Drug Metab. Rev.* **2015**, *47,* 29–44.
61. Scheen, A. J. *Drugs* **2015**, *75*, 33–59.
62. Giraldo, V. *Biochem. Soc. Trans.* **2014**, *42,* 1460–1464.
63. Smith, D. A. *Med. Res. Rev.* **1996**, *16,* 243–266.
64. Lode, E. *Eur. J. Clin. Microbiol. Infect. Dis.* **1991**, *10*, 807–812.
65. Rankovic, Z. *J. Med. Chem.* **2015**, *58*, 2584–2608.
66. (a) Lipinski, C. A. Drew University Medical Chemistry Special Topics Course. July **1999**. (b) Fernandes, T. B.; Segretti, M. C. F.; Polli, M. C.; Parise-Filho, R. *Lett. Drug Des. Discov.* **2016**, *13,* 999–1006.
67. Johnson, T. W.; Richardson, P. F.; Bailey, S.; Brooun, A.; Burke, B. J.; Collins, M. R.; Cui, J. J.; Deal, J. G.; Deng, Y.-L.; Dinh, D.; et al. *J. Med. Chem.* **2014**, *57,* 4720–4744.
68. Juliano, R. L.; Ling, V. *Biochim. Biophys. Acta* **1976**, *455*, 152–162.
69. Didziapetris, R.; Japertas, P.; Avdeef, A.; Petrauskas, A. *J. Drug Target.* **2003**, *11,* 391–406.
70. Sikic, B.I.; Fisher, G. A.; Lum, B. L.; Halsey, J.; Beketic-Oreskovic, L.; Chen, G. *Cancer Chemother. Pharmacol.* **1997**, *40(Suppl.)*, S13–S19.
71. Lai, Y. *Transporters in Drug Discovery and Development: Detailed Concepts and Best Practice*. Woodhead Publishing: Cambridge, **2013**, 780 pp.
72. Shchekotikhin, A. E.; Shtil, A. A.; Luzikov, Y. N.; Bobrysheva, T. V.; Buyanov, V. N.; Preobrazhenskaya, M. N. *Bioorg. Med. Chem.* **2005**, *13*, 2285–2291.
73. Lukesh, J. C.; Carney, D. W.; Dong, H.; Cross, R. M.; Shukla, V.; Duncan, K. K.; Yang, S.; Brody, D. M.; Brutsch, M. M.; Radakovic, A.; Boger, D. L. *J. Med. Chem.* **2017**, *60*, 7591–7604.
74. Cragg, G. M. *Med. Res. Rev.* **1998**, *18,* 315–331.
75. Ojima, I.; Slater, J. C.; Michaud, E.; Kuduk, S. D.; Bounaud, P.-Y.; Vrignaud, P.; Bissery, M.-C.; Veith, J. M.; Pera, P.; Bernacki, R. J. *J. Med. Chem.* **1996**, *39,* 3889–3896.
76. Jansson, P. J.; Yamagishi, T.; Arvind, A.; Seebacher, N.; Gutierrez, E.; Stacy, A.; Maleki, S.; Sharp, D.; Sahni, S.; Richardson, D. R. *J. Biol. Chem.* **2015**, *15*, 9588–9603.
77. Trainor, G. L. *Exp. Opin. Drug Discov.* **2007,** *2*, 51–64.
78. Ascenzi, P.; Fanali, G.; Fasano, M.; Pallottini, V.; Trezza, V. *J. Mol. Struct.* **2014**, *1077*, 4–13.
79. Zhang, F.; Xue, J.; Shao, J.; Jia, L. *Drug Discov. Today* **2012**, *17,* 475–485.
80. Bohnert, T.; Gan, L.-S. *J. Pharm. Sci.* **2013**, *102,* 2953–2994.
81. Wendt, M. D.; Shen, W.; Kunzer, A.; McClellan, W. J.; Bruncko, M.; Oost, T. K.; Ding, H.; Joseph, M. K.; Zhang, H.; Nimmer, P. M.; et al. *J. Med. Chem.* **2006**, *49*, 1165–1181.

82. Edmondson, S. D.; Mastracchio, A.; Mathvink, R. J.; He, J.; Harper, B.; Park, Y. J.; Beconi, M.; Di Salvo, J.; Eiermann, G. J.; et al. *J. Med. Chem.* **2006**, *49*, 3614–3627.
83. Pevarello, P.; Fancelli, D.; Vulpetti, A.; Amici, R.; Villa, M.; Pittala, V.; Vianello, P.; Cameron, A.; Ciomei, M.; Mercurio, C.; Bischoff, J. R.; Roletto, F.; Varasi, M.; Brasca, M. G. *Bioorg. Med. Chem. Lett.* **2006**, *16*, 1084–1090.
84. Benet, L. Z.; Hoener, B.-A. *Clin. Pharmacol. Ther.* **2002**, *71,* 115–121.
85. Liu, X.; Wright, M.; Hop, C. E. C. A. *J. Med. Chem.* **2014**, *57*, 8238–8248.
86. Halford, F. J. *Anesthesiol.* **1943**, *4,* 67–69.
87. Lennernaes, H. *Clin. Pharmacokinet.* **2003**, *42,* 1141–1160.
88. Wyvratt, M. J.; Patchett, A. A. *Med. Res. Rev.* **1985**, *5,* 483–531.
89. Nichols, A. I.; Lobello, K.; Guico-Pabia, C. J.; Paul, J.; Preskorn, S. H. *J. Clin. Psychopharmacol.* **2009**, *29,* 383–386.
90. Kamath, J.; Handratta, V. *Exp. Rev. Neurother.* **2008**, *8,* 1787–1797.
91. Chen, C. *Drugs R&D* **2007**, *8,* 301–314.
92. Lynch, T.; Price, A. *Am. Fam. Physician* **2007**, *76,* 391–396.
93. Lewis, D. F. V. *Inflammopharmacol.* **2003**, *11,* 43–73.
94. Kaspera, R.; Naraharisetti, S. B.; Tamraz, B.; Sahele, T.; Cheesman, M. J.; Kwok, P.-Y.; Marciante, K.; Heckbert, S. R.; Psaty, B. M.; Totah, R. A. *Pharmacogenet. Genomics* **2010**, *20,* 619–629.
95. Schemiedlin-Ren, P.; Edwards, D. J.; Fitzsimmons, M. E.; He, K.; Lown, K. S.; Wpster, P. M.; Thummel, K. E.; Fisher, J. M.; Hollenberg, P. F.; Watkins, P. B. *Drug Metab. Dispos.* **1997**, *25*, 1228–1233.
96. Zhou, S.; Chan, E.; Lim, L. Y.; Boelsterli, U. A.; Li, S. C.; Wang, J.; Zhang, Q.; Huang, M.; Xu, A. *Curr. Drug Metab.* **2004**, *5*, 415–442.
97. Shrinivas, N. R. *J. App. Biopharm. Pharmacokinet.* **2013**, *1*, 44–55.
98. Gerntholtz, T.; Pascoe, M. D.; Botha, J. F.; Halkett, J.; Kahn, D. *Eur. J. Clin. Pharmacol.* **2004,** *60,* 143–148.
99. King, J. R.; Wynn, H.; Brundage, R.; Acosta, E. P. *Clin. Pharmacokinet.* **2004**, *43,* 291–310.
100. Paulson, S. K.; Hribar, J. D.; Liu, N. W. K.; Hajdu, E.; Bible, R. H.; Piergies, A.; Karim, A. *Drug Metab. Dispos.* **2000**, *28*, 308–314.
101. Rosenblum, S. B.; Huynh, T.; Afonso, A.; Davis, H. R.; Yumibe, N. *J. Med. Chem.* **1998**, *41*, 973–980.
102. Pirmohamed, M.; Kitteringham, N. R.; Guenthner, T. M.; Brekenridge, A. M.; Park, B. K. *Biochem. Pharmacol.* **1992**, *43*, 1675–1682.
103. Guengerich, F. P.; Johnson, W. W.; Shimada, T.; Ueng, Y. F.; Yamazaki, H.; Langouët, S. *Mutat. Res.* **1998**, *402,* 121–128.
104. Smith, D. A. *Curr. Top. Med. Chem.* **2011**, *11*, 467–481.
105. Remy, C. N. *J. Biol. Chem.* **1961**, *236*, 2999–3005.
106. (a) Gingell, R.; Bridges, J. W.; Williams, R. T. *Xenobiotica* **1971**, *1*, 143–156. (b) Gingell, R.; Bridges, J. W. *Xenobiotica* **1973**, *9*, 599–604.
107. Boelsterli, U. A.; Ho, H. K.; Zhou, S.; Leow, K. Y. *Curr. Drug Metab.* **2006**, *7,* 715–727.
108. Rosenkranz, H. S.; Mermelstein, R. *Mutat. Res.* **1983**, *114*, 217–267.

109. Amano, T.; Fukami, T.; Ogiso, T.; Hirose, D.; Jones, J. P.; Taniguchi, T.; Nakajima, M. *Biochem. Pharmacol.* **2018**, *151*, 69–78.
110. Lewis, J.; Shimmon, R.; Fu, S. *J. Anal. Toxicol.* **2013**, *37*, 179–181.
111. Somogyi, A. A.; Reinhard, H. A.; Bochner, F. *Br. J. Clin. Pharmacol.* **1996**, *41*, 175–179.
112. El-Masri, H. A.; Portier, C. J. *Drug Metab. Dispos.* **1998**, *26*, 585–594.
113. Savi, P.; Pereillo, J. M.; Uzabiaga, M. F.; Combalbert, J.; Picard, C.; Maffrand, J. P.; Pascal, M.; Herbert, J. M. *Thromb. Haemost.* **2000**, *84,* 891–896.
114. Pereillo, J. M.; Maftouh, M.; Andrieu, A.; Uzabiaga, M.-F.; Fedeli, O.; Savi, O.; Pascal, M.; Herbert, J.-M.; Maffrand, J. P.; Picard, C. *Drug Metab. Dispos.* **2002**, *30*, 1288–1295.
115. Savi, P.; Labouret, C.; Delesque, N.; Guette, F.; Lupker, H.; Herbert, J. M. *Biochem. Biophys. Res. Commun.* **2001**, *283*, 379–383.
116. Herrmann, M. L.; Schleyerbach, R.; Kirschbaum, B. J. *Immunopharmacol.* **2000**, *47*, 273–289.
117. Leach, A. G. Tactics to Avoid Inhibition of Cytochrome P450s. In Meanwell, N. A., ed. *Top. Med. Chem. 9(Tactics in Contemporary Drug Design).* Springer: Heidelberg, **2015**, pp. 107–158.
118. Grady, M. A.; Gasperoni, T. L.; Kirkpatrick, P. *Nat. Rev. Drug Discov.* **2003**, *2*, 427–428.
119. Penning, T. D.; Talley, J. J.; Bertenshaw, S. R.; Carter, J. S.; Collins, P. W.; Docter, S.; Graneto, M. J.; Lee, L. F.; Malecha, J. W.; Miyashiro, J. M.; et al. *J. Med. Chem.* **1997**, *40,* 1347–1365.
120. Gleeson, M. P. *J. Med. Chem.* **2008,** *51*, 817–834.
121. Nagashima, S.; Hondo, T.; Nagata, H.; Ogiyama, T.; Maeda, J.; Hoshii, H.; Kontani, T.; Kuromitsu, S.; Ohga, K.; Orita, M.; et al. *Bioorg. Med. Chem.* **2009**, *17*, 6926–6936.
122. Chen, M.; LeDuc, B.; Kerr, S.; Howe, D.; Williams, D. A. *Drug Metab. Dispos.* **2010**, *38*, 368–375.
123. Moise, P. A.; Birmingham, M. C.; Schentag, J. J. *Drugs Today* **2000**, *36*, 229–244.
124. Potter, D. W.; Hinson, J. A. *Mol. Pharmacol.* **1986**, *30,* 33–41.
125. Reidenberg, M. M.; Drayer, D. E.; Levy, M.; Warner, H. *Clin. Pharmacol. Ther.* **1975**, *17,* 722–730.
126. Struck, A.-W.; Thompson, M. L.; Wong, L. S.; Micklefield, J. *ChemBioChem.* **2012,** *13*, 2642–2655.
127. Lautala, P.; Ulmanen, I.; Taskinen, J. *Mol. Pharmacol.* **2001**, *59*, 393–402.
128. (a) Tabata, T.; Katoh, M.; Tokudome, S.; Hosakawa, M.; Chiba, K.; Nakajima, M.; Yokoi, T. *Drug Metab. Dispos.* **2004**, *32*, 762–767. (b) Reigner, B.; Blesch, K.; Weidekamm, E. *Clin. Pharmacokinet.* **2001**, *40*, 85–104. (Xoleda).
129. Rautio, J., ed. *Prodrugs and Targeted Delivery: Towards Better ADME Properties.* Wiley-VCH: Weinheim, **2011**.
130. Varia, S. A.; Schuller, S.; Sloan, K. B.; Stella, V. J. *J. Pharm. Sci.* **1984**, *73,* 1068–1073.
131. Slatter, J. G.; Schaaf, J.; Sams, J. P.; Feenstra, K. L.; Johnson, M. G.; Bombardt, P. A.; Cathcart, K. S.; Verburg, M. T.; Pearson, L. K.; Compton, L. D.; et al. *Drug Metab. Dispos.* **2000**, *28*, 423–433.

132. (a) McClellan, K.; Perry, C. M. *Drugs* **2001**, *61,* 263–283. (b) Lew, W.; Wang, M. Z.; Chen, X.; Rooney, J. F.; Kim, C. Neuraminidase Inhibitors as Anti-Influenza Agents. In De Clercq, E., ed. *Antiviral Drug Strategies*. Wiley-VCH: Weinheim, **2001**.
133. (a) Gustafsson, D.; Bylund, R.; Antonsson, T.; Nilsson, I.; Nystroem, J.-E.; Eriksson, U.; Bredberg, U.; Teger-Nilsson, A.-C. *Nat. Rev. Drug Discov.* **2004**, *3*, 649–659. (b) Gustafsson, D. *Semin. Vasc. Med.* **2005**, *5*, 227–234.
134. Blech, S.; Ebner, T.; Ludwig-Schwellinger, E.; Stangier, J.; Roth, W. *Drug Metab. Dispos*. **2008**, *36*, 386–399.
135. Uchino, H.; Kanai, Y.; Kim, D. K.; Wempe, M. F.; Chairoungdua, A.; Morimoto, E.; Anders, M. W.; Endou, H. *Mol. Pharmacol.* **2002**, *61*, 729–737.
136. Tsuda, M.; Terada, T.; Irie, M.; Katsura, T.; Niida, A.; Tomita, K.; Fujii, N.; Inui, K.-i. *J. Pharmacol. Exp. Therapeut.* **2006**, *318,* 455–460.
137. Sitar, D. S. *Clin. Pharmacokinet.* **1996**, *31,* 246–256.
138. Beutner, K. R.; Friedman, D. J.; Forszpaniak, C.; Andersen, P. L.; Wood, M. J. *Antimicrob. Agents Chemother.* **1995**, *39,* 1546–1553.
139. Sofia, M. J. *Antiviral Chem. Chemother.* **2011**, *22*, 23–49.
140. Rodriguez-Torres, M. *Exp. Rev. Anti-Infect. Ther*. **2013**, *1*, 1269–1279.
141. Murakami, E.; Tolstykh, T.; Bao, H.; Niu, C.; Steuer, H. M.; Bao, D.; Chang, W.; Espiritu, C.; Bansal, S.; Lam, A. M.; Otto, M. J.; Sofia, M. J.; Furman, P. A. *J. Biol. Chem.* **2010**, *285*, 34337–34347.
142. Ma, H.; Jiang, W.-R.; Robledo, N.; Leveque, V.; Ali, S.; Lara-Jaime, T.; Masjedizadeh, M.; Smith, D. B.; Cammack, N.; Klumpp, K.; Symons, J. *J. Biol. Chem.* **2007,** *282*, 29812–29820.

4

Bioisosteres

In this chapter, we discuss what medicinal chemists do every day: We take the original hits and improve them to lead compounds, drug candidates, and ultimately drugs.

How do we do that?

Bioisosteres!

The concept of bioisosteres is a good guide for investigating the structure–activity relationship (SAR) and many other relevant drug-likeness attributes. Educating ourselves with what bioisosteres have been successful in the literature provides us better guidance and avoids detours.

4.1 Introduction to Bioisosteres

4.1.1 Definition

According to the International Union of Pure and Applied Chemistry (IUPAC), a bioisostere is "a compound resulting from the exchange of an atom or of a group of atoms with another, broadly similar, atom or group of atoms. The objective of a bioisosteric replacement is to create a new compound with similar biological properties to the parent compound. The bioisosteric replacement may be physiochemically or topologically based."[1]

Here, "bioisostere" and "isostere" are used interchangeably.

4.1.2 Utility of Bioisosteres

Rather than going through the lengthy history of how the concept of bioisosteres has evolved over the last century like every other review on this topic, we will discuss the applications immediately.

In a traditional sense, bioisosteres are applied to improve potency by SAR. Moreover, bioisosteres have also been used to enhance selectivity; to alter physical properties; to improve permeability; to reduce or redirect metabolism; to eliminate or modify a toxicophore; and to provide intellectual properties (novelty). These applications are exemplified one by one beneath:

Medicinal Chemistry for Practitioners, First Edition. Jie Jack Li.

4.1.2.1 To Improve Potency

Minute structural changes could result in considerably different biological responses. Bioisosteres may give rise to vastly more potent/efficacious drugs.

An early utility of bioisosteres was exemplified by thiazide diuretics. In 1957, Novello at Merck discovered chlorothiazide (Diuril, **1**) as a new diuretic. This was revolutionary because all diuretics before it were mercury-containing drugs known as "mercurials," which were associated with substantial toxicities. Chlorothiazide (**1**), although less toxic than mercurials, had 250 mg twice a day prescription. As its dosage increased, its carbonic anhydrase inhibition began to manifest and elevation of bicarbonate excretion ensued.

One year later in 1958, de Stevens at Ciba simply reduced one double bond on chlorothiazide (**1**) and obtained hydrochlorothiazide (HydroDiuril, **2**). Not only was it 10–20 times more potent than the parent drug **1** in its ability to promote sodium excretion from the body, it also had a lower order of toxicity. The increased renal excretion of water and sodium was accompanied by a very pronounced increase in chloride excretion because it is a considerably weaker carbonic anhydrase inhibitor than the prototype drug **1**.[2] Furthermore, de Stevens and colleagues carried out extensive SAR investigations employing different bioisosteres (although the term "bioisostere" *per se* was not popularly known then). It was found that a lipophilic substituent at position C-3 gave very potent compounds with trichlormethiazide (**3**) found to be 100-fold more active than the parent drug **1** and cyclopenthiazide (**4**) 1,000-fold more potent than the prototype drug **1**.[3]

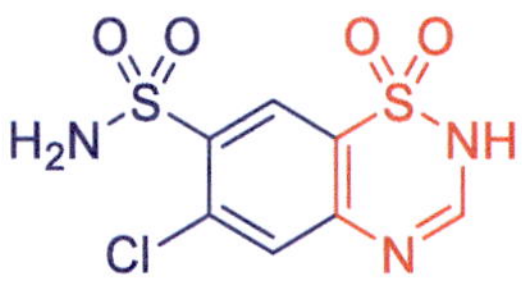

chlorothiazide (Diuril, **1**) x 1

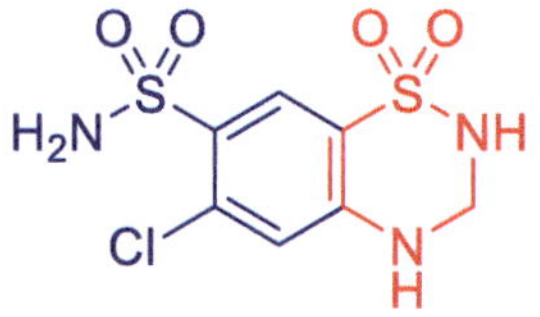

hydrochlorothiazide (HydroDiuril, **2**) x 10

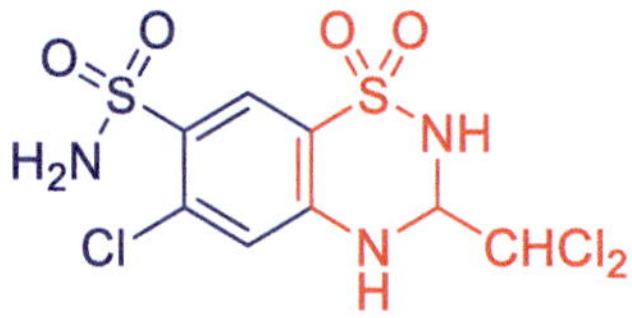

trichlormethiazide (**3**) x 100

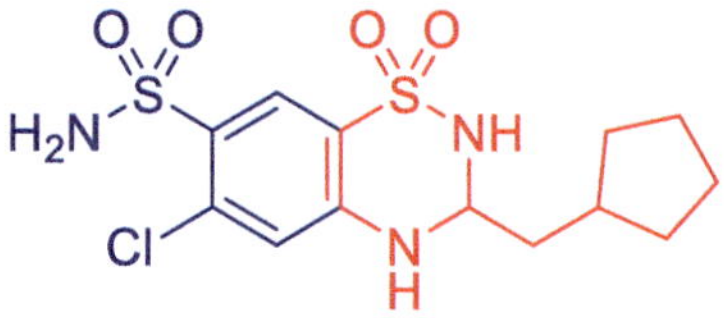

cyclopenthiazide (**4**) x 1000

Phospholipase D2 (PLD2) selective inhibitor **5** was not a very potent compound with an IC_{50} of 11.8 μM for the cellular assay. Remarkably, when one hydrogen atom was substituted with an (*S*)-methyl isostere, the resulting PLD2 inhibitor **6** gained a stunning 590-fold increase in potency in cellular assay.[4]

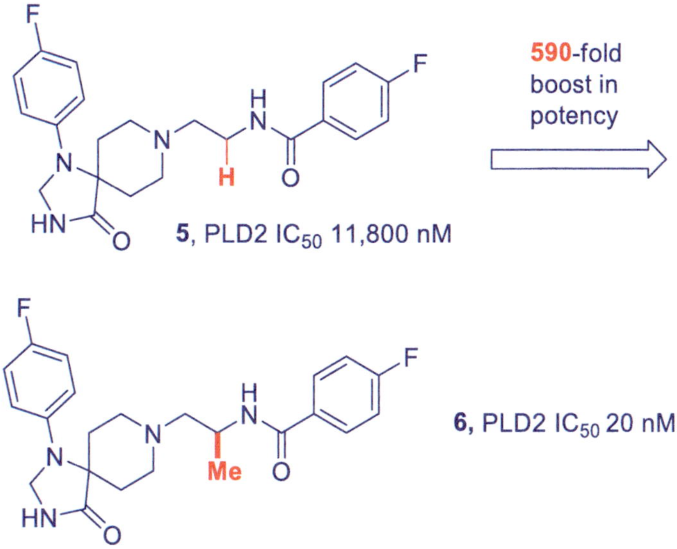

5, PLD2 IC_{50} 11,800 nM

6, PLD2 IC_{50} 20 nM

Throughout this chapter, we will see time and again that utility of bioisosteres offers a useful tool for SAR exploration to improve the potency of compounds.

4.1.2.2 To Enhance Selectivity

In addition to improving potency, bioisosteres have been used to enhance selectivity of a less selective compound.

With diphenhydramine (Benadryl) as a starting point, Lilly embarked on a journey to search for selective serotonin reuptake inhibitors (SSRIs) as efficacious and safe antidepressants. Before the emergence of SSRIs, tricyclic antidepressants were reasonably efficacious in combating depression, but their significant toxicities prevented the majority of sufferers from finishing the course of treatment.

Lilly chemists arrived at a juncture where propanolamine analog **7** showed promise. It had a K_i of 1,371 nM for the inhibition of serotonin (SET) reuptake and K_i of 2.4 nM for the inhibition of norepinephrine (NE) reuptake. In other words, analog **7**'s selectivity was exactly the opposite as what was desired for SSRIs. Employing 4-trifluoromethyl group as a bioisostere of the 2-methoxyl group on the parent compound **7** produced a new compound **8** with a favorable selectivity. The resulting analog was fluoxetine (Prozac, **8**). Its K_i for SET reuptake inhibition was 17 nM, an 81-fold boost. Meanwhile, its K_i for NE reuptake inhibition was 2,703 nM, a 1,126-fold drop from the prototype compound **7**.[5] Fluoxetine (**8**) is a very selective SSRI endowed with both efficacy and safety in treating depression. On the market since the early 1990s, Prozac (**8**) revolutionized our perception of mental disorder in general, and depression treatment in particular.

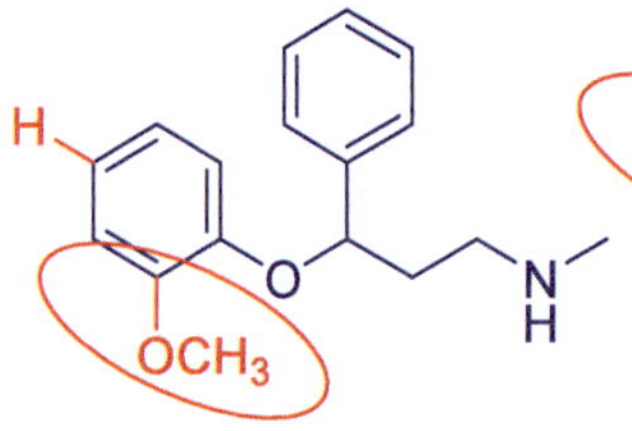

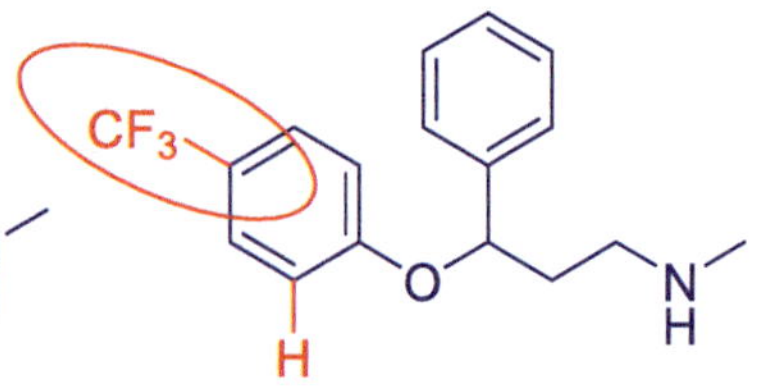

	nisoxerine (**7**)	fluoxetine (Prozac, **8**)	
inhibition of SET uptake K_i (nM)	1,371	17	81x↑
inhibition of NE uptake K_i (nM)	2.4	2,703	1,126x↓

4.1.2.3 To Alter Physical Properties

Many isosteres exist in the literature to alter physical properties.

N,N-Dimethyl-4-(*p-tert*-butyl-phenylbutanamine (**9**) is not a drug-like molecule. For one thing, its aqueous solubility is on par with that of brick dust. Additionally, the basic dimethylamine group makes the compound highly amphiphilic (think "detergent"). Last, but not the least, the left-hand portion of amine **9** is so hydrophobic that the *t*-butylphenyl group is subjected to extensive metabolism by CYP450 enzymes.

Metabolically more stable and "lipophilicity neutral" oxetane has been applied as the isostere to improve aqueous solubility and metabolism. Even having the oxetane ring in the middle of molecule **10** offered a greatly enhanced aqueous solubility. For derivatives **11** and **14**, replacing the greasy *t*-butyl group with an oxetane resulted in more "balanced" polarity and more *hydro*philic molecules with higher water solubility (>4 mg/mL). On the other hand, having the oxetane ring adjacent to the tertiary amine considerably lowered the pK_a values. Compounds **12** and **13** have pK_a values of 8.0 and 7.2, respectively, and solubility of 25 μM/mL and 57 μM/mL, respectively.[6]

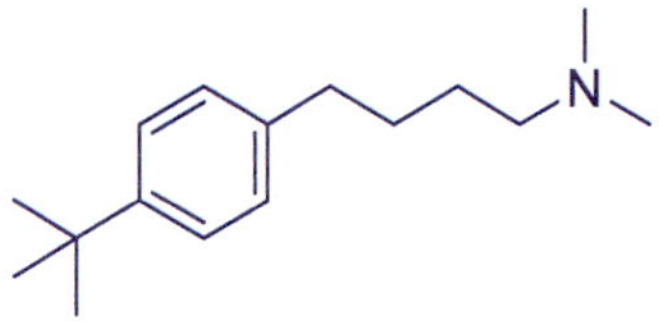

9, pK_a of N: 9.9
Sol: <1 μg/mL

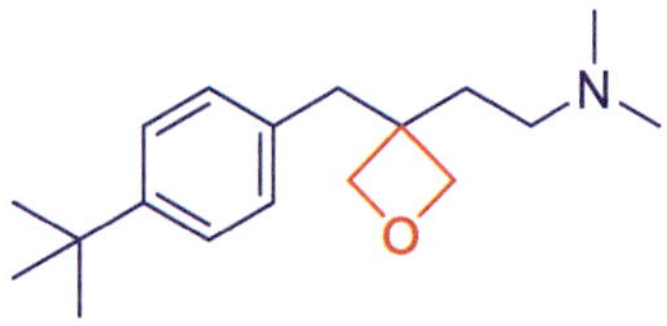

10, pK_a of N: 9.2
Sol: 4,100 μg/mL

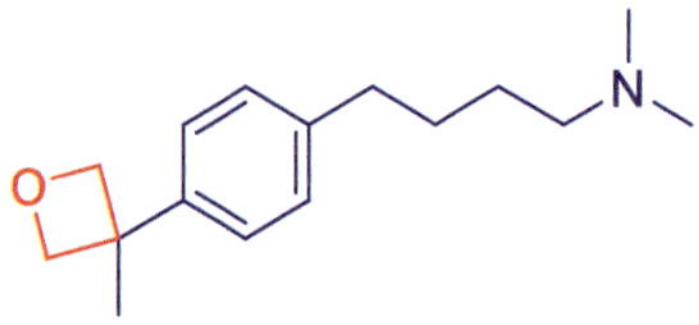

11, pK_a of N: 9.9
Sol: 4,400 μg/mL

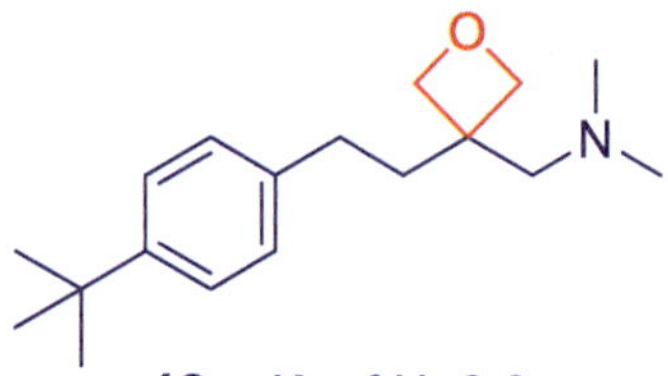

12, pK_a of N: 8.0
Sol: 25 μg/mL

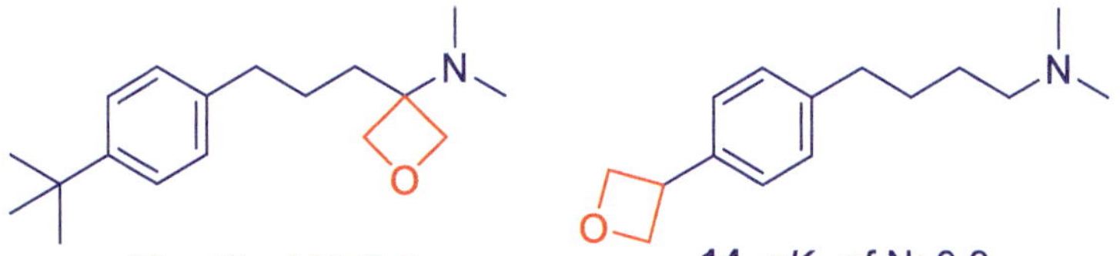

13, pK_a of N: 7.2
Sol: 57 μg/mL

14, pK_a of N: 9.9
Sol: 4,000 μg/mL

Since its introduction by Carreira and coworkers in 2006, oxetane has gained widespread acceptance as a useful isostere in medicinal chemistry. When grafted to the molecular architecture, oxetane often offers unique (often superior) physiochemical properties and biochemical profiles.[7]

4.1.2.4 To Improve Permeability

One of the drug design quandary is that a drug-like molecule should have a balanced lipophilicity. Too polar, the molecule will not permeate the cell membrane. Too greasy, it will not dissolve well in water. Intramolecular hydrogen bonding frequently boosts cell membrane penetration because it makes the molecule more lipophilic and easier to permeate cell membrane. For instance, amido-carbamates **15** and **16** have identical polar surface areas (PSAs), yet compound **16** is four times more cell-permeable in Caco-2 assay than **15** by virtue of the intramolecular hydrogen bond as shown beneath.[8] Caco-2 assay is one of the most popular cellular methods to measure permeability.

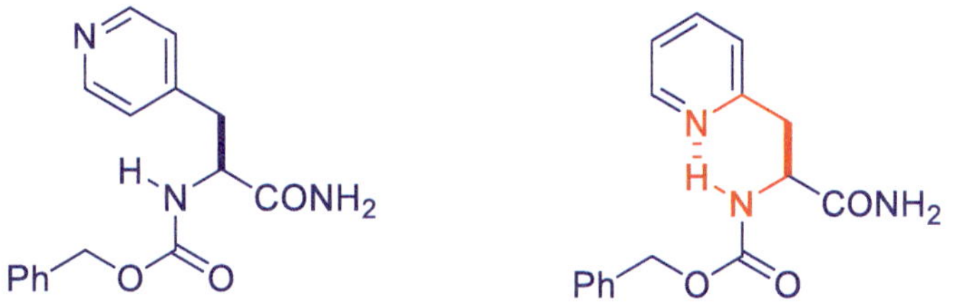

15, $P_{app(\text{A to B})}$ = 43 nm/s

16, $P_{app(\text{A to B})}$ = 177 nm/s
4x more cell-permeable (Caco-2)

En route to the discovery of razaxaban as a highly potent, selective, and orally bioavailable factor Xa inhibitor, a series of anilides **17**–**19** were scrutinized for their cell permeability employing Caco-2 permeability assay. While the very polar nitrile **17** has negligible permeability, the *ortho*-fluorobenzamide **19** has an exponentially superior cell membrane penetration in comparison to both **17** and **18**. The fact that compound **19** can form an intramolecular hydrogen bonding undoubtedly helps to elevate the permeability.[9]

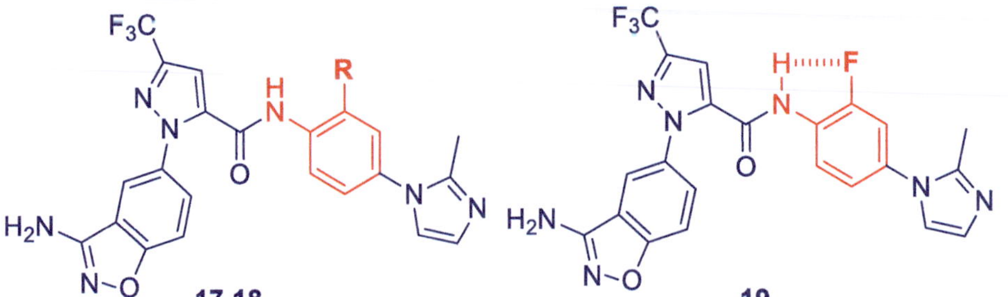

17,18

19

Compound	R	Caco-2 permeability (×10⁻⁶ cm/s)
17	CN	<0.1
18	H	0.82
19	F	7.41

4.1.2.5 To Reduce or Redirect Metabolism

The methyl group is susceptible to oxidative metabolism by CYP450 enzymes. Cl, F, and Br atoms have all been used as bioisosteres to replace the methyl group and provide more robust resistance to CYP oxidation.

The two methyl groups on OPC-4392 (**20**) are readily oxidized to the corresponding alcohols and acids. The more polar metabolites are significantly less efficacious as D_2 antagonists. Replacing the two methyl groups with two metabolically more resistant chlorine atoms led to OPC-14597 (aripiprazole, Abilify, **21**) that is 67-fold more efficacious than the parent compound.[10]

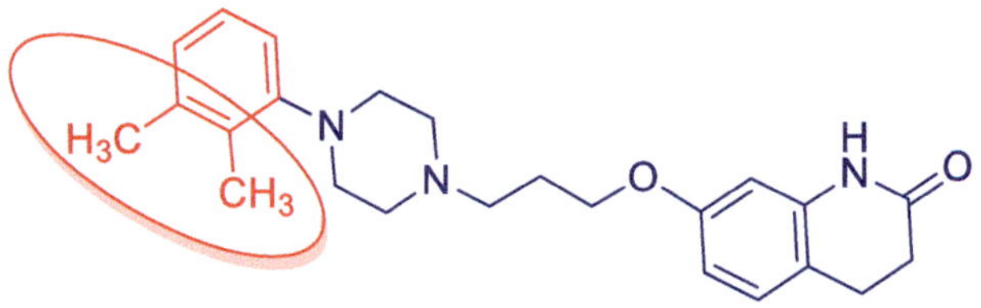

OPC-4392 (**20**), ED_{50} = 41.3 μmol/kg, p.o.

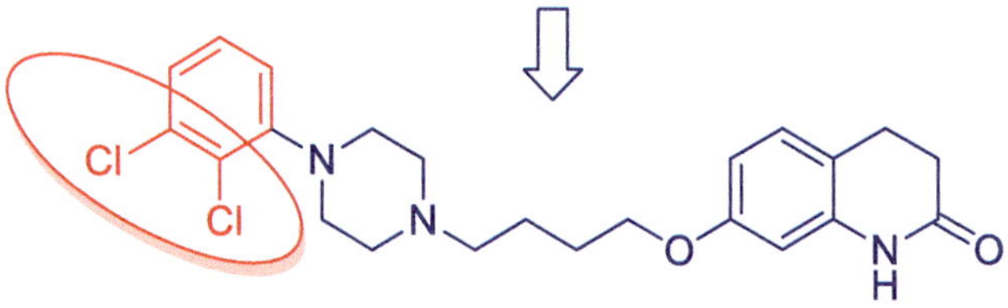

aripiprazole (Abilify, **21**), ED_{50} = 0.6 μmol/kg, p.o.
Otsuka/BMS, 2002

4.1.2.6 To Reduce Toxicity

Isostere may improve a drug's safety profile.

Ticlopidine hydrochloride (Ticlid, **22**) by Castaigne S. A. was an anticoagulant riddled with toxicities, most significantly, thrombotic thrombocytopenic purpura (TTP), a blood disorder where tiny blood clots form in small arteries throughout the body, destroying red blood cells and causing anemia. The disorder can also cause fever, kidney failure, slurred speech, confusion, disorientation, and coma. Extensive efforts were made to fix the serious issue using isosteres of H_a. When H_a was either methyl or ethyl, the isosteres were efficacious but still had toxicities. Eventually, replacing H_a with a methyl ester as the isostere led to clopidogrel sulfate (Plavix, **23**), which is both efficacious and safe. It went on to become the second best-selling drugs in many years. Interestingly, it was later discovered that the mechanism of action for clopidogrel (**23**) is acting as an antagonist of $P2Y_{12}$ receptor.[11]

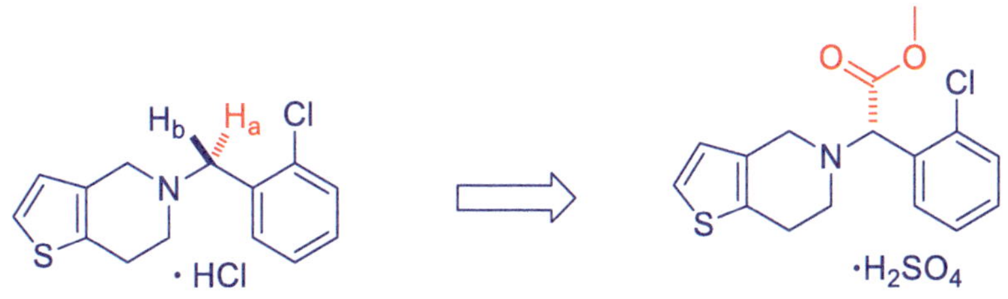

ticlopidine hydrochloride (Ticlid, **22**)
Castaigne S. A./Syntex, 1979
anticoagulant

clopidogrel sulfate (Plavix, **23**)
Sanofi/Bristol–Myers Sqibb, 1993
$P2Y_{12}$ antagonist

4.1.2.7 To Provide Intellectual Property

Pharmaceutical industry has become increasingly competitive and "druggable" targets get rapidly crowded with patents. Isosteres offer opportunities to create novelty. Shionogi's HMG-CoA inhibitor rosuvastatin (**25**) serves to exemplify the power of isosteres in providing intellectual property.

While all other synthetic statins invariably contain an isopropyl substituent, Bayer's cerivastatin sodium (**24**) even has two sandwiching the nitrogen atom on the pyridine. To produce their own statin, Watanabe et al. at Shionogi chose pyrimidine as an isostere of pyridine for the core structure. Meanwhile, they used methanesulfonamide as a unique isostere for one of the two isopropyl substituents on **24**. The outcome was rosuvastatin (Crestor, **25**), which was bestowed with novel intellectual property for Shionogi. Moreover, it is not a CYP450 3A4 substrate, thus devoid of drug–drug interactions (DDIs) with drugs also metabolized by CYP3A4.[12]

Nearly every endeavor in medicinal chemistry involves isosteres. Since a tome on isosteres alone would be too voluminous for this book, we focus only on some most important classes of isosteres.

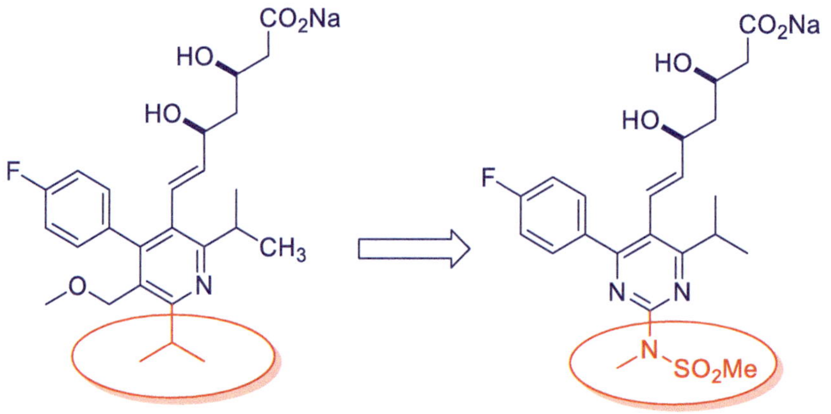

cerivastatin sodium (Baycol, **24**)
CYP450 3A4 substrate

rosuvastatin (Crestor, **25**)
NOT CYP450 3A4 substrate

4.2 Deuterium, Fluorine, and Chlorine Atoms as Isosteres of Hydrogen

Deuterium, fluorine, and chlorine atoms, as well as the methyl group, are popular isosteres of hydrogen.

4.2.1 Deuterium

Kinetic isotope effects (KIEs) are an important chapter in physical organic chemistry. As far as deuterium is concerned, the C–D bond could be 10-fold stronger than the corresponding C–H bond. Therefore, it is logical to use deuterium to replace hydrogen if the C–H bond is vulnerable to metabolism. The key for this strategy to succeed is that the C–H bond cleavage must be the rate-determining step (RDS).[13]

Cystic fibrosis (CF) is a genetic disease caused by mutations to the gene coding the cystic fibrosis transmembrane conductance regulator (CFTR) protein in many epithelial cells and blood cells. Vertex's ivacaftor (Kalydeco, **26**) is the first medicine marketed in 2011 to treat the underlying cause of CF in people with the G551D mutation in the CFTR gene. *In vitro* and clinical studies indicate that the drug is extensively metabolized, primarily by CYP3A4. Following oral administration in humans, the majority (87.8%) of ivacaftor (**26**) is eliminated in the feces after metabolic conversion. The major metabolites M1 (**27**) and M6 (**28**) accounted for approximately 65% of the total dose eliminated with 22% as M1 (**27**) and 43% as M6 (**28**). M1 (**27**) has approximately one-sixth the potency of the parent drug and is considered pharmacologically active. M6 (**28**) has less than 1/50th the potency of the parent drug and is not considered pharmacologically active.[14]

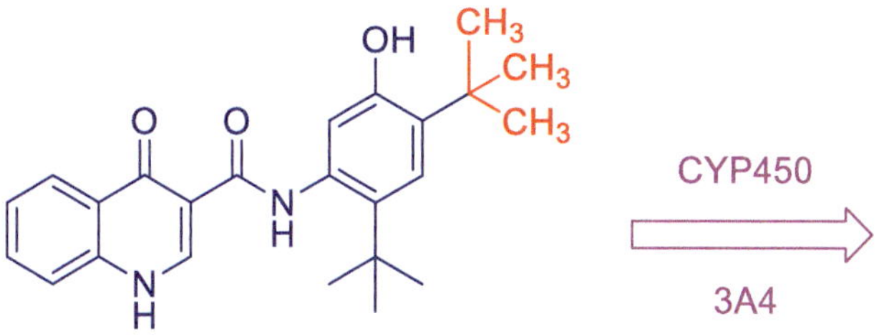

ivacaftor (Kalydeco, **26**)
CFTR Potentiator, 2012

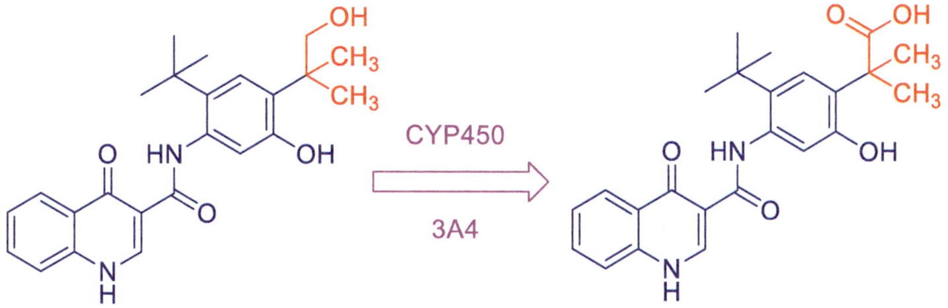

M1: hydroxymethyl-ivacaftor (**27**) M6: ivacaftor carboxylate (**28**)

Since the *tert*-butyl groups on ivacaftor (**26**) is subject to CYP3A4 metabolism to produce either less active or inactive metabolites, it was sensible that Concert Pharmaceuticals prepared the corresponding d_9-ivacaftor (CTP-656, **29**) to retard the metabolic oxidation. Indeed, CTP-656 (**29**) has a significantly longer half-life to enable a qd (once daily) regimen, while the prototype drug **26** has to be taken bid (twice daily). The Phase II clinical trials data was good enough to entice Vertex to buy the right to CTP-656 (**29**) from Concert in 2017.[15]

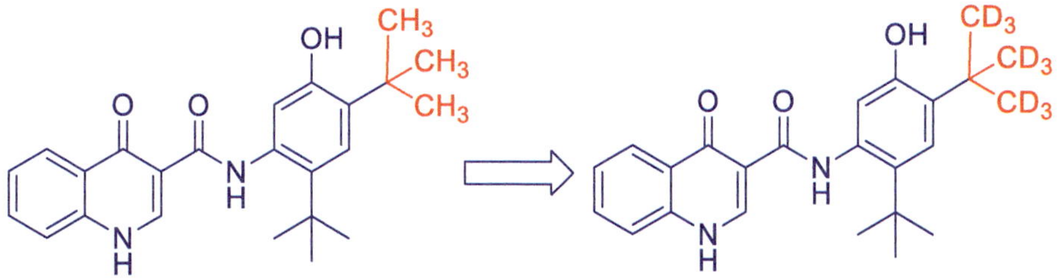

ivacaftor (Kalydeco, **26**) CTP-656 (**29**)

Also in 2017, the FDA was sufficiently convinced to approve the first deuterated drug molecule, deutetrabenazine (Austedo, **31**), which is useful in treating chorea associated with Huntington's disease and tardive dyskinesia.

Tetrabenazine (**30**), a vesicular monoamine transporter 2 (VMAT2) inhibitor, is an old drug to treat involuntary movement. In 2008, the drug as a racemic mixture was approved by the FDA to treat chorea as a central monoamine-depleting agent. Deutetrabenazine (**31**) is a deuterated analog of tetrabenazine (**30**) with the two methoxyl groups on the parent drug replaced by a pair of trideuteromethoxyl groups, thereby altering the rate of metabolism to afford greater tolerability and an improved dosing regimen. By prolonging the residence time of the active drug species in plasma, greater efficacy is achieved. Deutetrabenazine (**31**) is taken twice a day rather than thrice a day for tetrabenazine (**30**). One group of authors claimed that deutetrabenazine (**31**) is better tolerated than tetrabenazine (**30**) based on an "indirect tolerability comparison,"[16] although another group offered some cautionary observations and suggestions vis-à-vis the limitations of indirect comparison.[17]

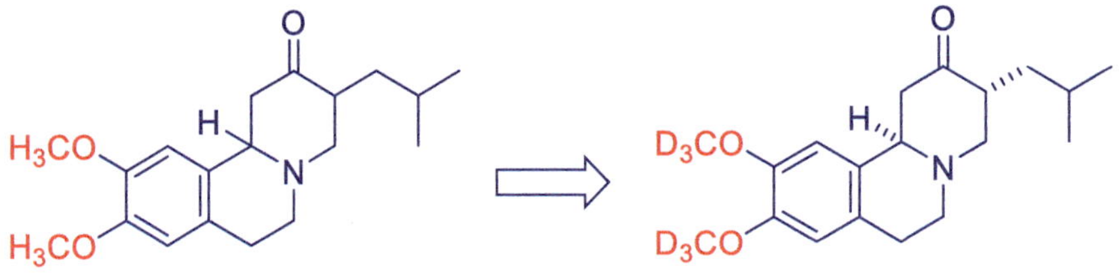

tetrabenazine (**30**)
human $t_{1/2}$, 4.8 h

deutetrabenazine (Austedo, **31**)
human $t_{1/2}$, 8.6 h

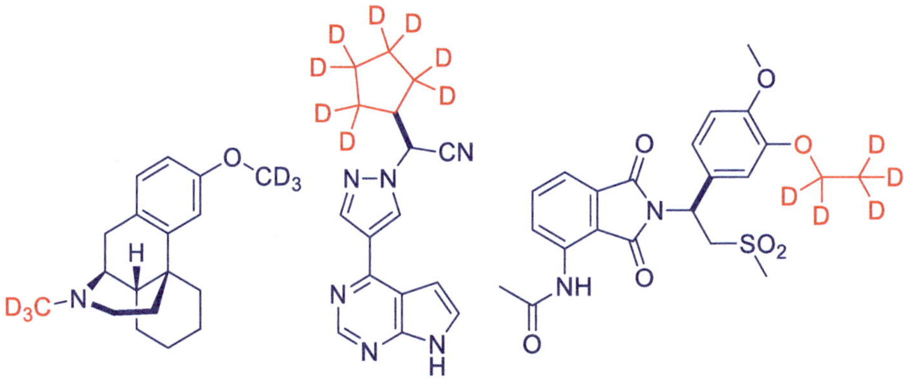

dextromethorphan-d_6 (**32**) deuterated ruxolitinib (**33**) deuterated apremilast (**34**)

The first FDA approval of a deuterium drug was a momentous event that fueled much enthusiasm to the field. To date, several deuterated drugs are currently going through different phases of clinical trials. They include dextromethorphan-d_6 (**32**) as an antagonist of *N*-methyl-D-aspartate (NMDA) glutamate receptor in Phase III for treating Alzheimer's disease (AD) agitation symptoms; CTP-543 (deuterated ruxolitinib, **33**) as a Janus kinase/signal transducer and activator of transcription proteins (JAK/STAT) inhibitor in Phase II to treat alopecia aerate; BMS-986165 as an allosteric tyrosine kinase 2 (Tyk2) inhibitor in Phase II to treat psoriasis; and CTP-730 [deuterated apremilast

(Otazla), **34**] as a phosphodiesterase-4 (PDE4) inhibitor in Phase I to treat inflammatory disease; etc.[18]

4.2.2. Fluorine

The first fluorine-containing drug, 9α-fluorocortisone (Florinef, **35**), was approved by the FDA in 1955.[19] The failure of fluorine-containing drugs to emerge earlier before then was possibly due to the ill-conceived notion that they might be toxic from the limited experience with fluoroacetic acid. 9-Fluorocortisone (**35**) was indeed an improvement over the parent cortisol, which was readily oxidized to cortisone (**36**). Apparently, the presence of the α-fluorine atom retarded the oxidation of the 11-hydroxyl group. Other fluorinated steroid drugs, fluticasone propionate, flunisolide, triamcinolone, and many more subsequently became available on the market in the late 1950s and onward.

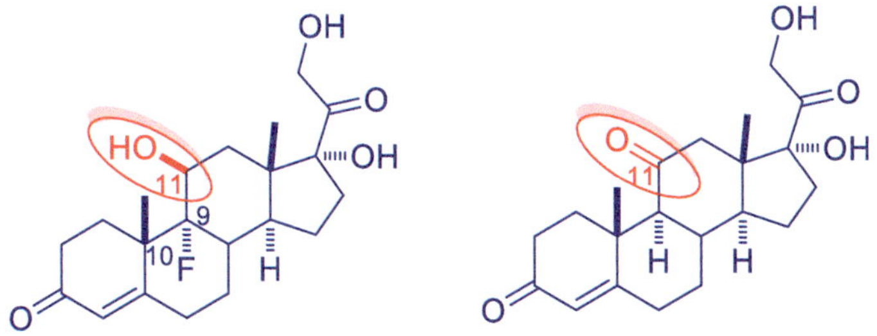

9α-fluorocortisone (Florinef, **35**) cortisone (**36**)

5-Fluorouracil (5-FU, **38**) is an important anti-proliferative cancer chemotherapy. The fluorine atom at the C-5 position may be viewed as an isostere of hydrogen at the C-5 position on uracil (**37**). "Disguising" itself as uracil (**37**), 5-fluorouracil (**38**) interrupts the DNA synthesis by blocking thymidylate synthase, although alternative MOAs have been forwarded as well.[20]

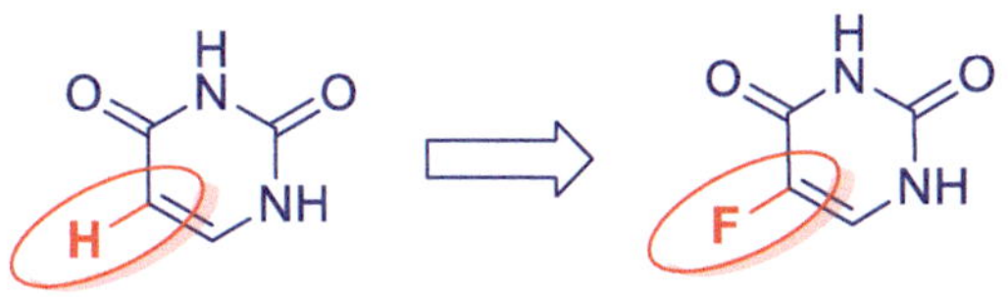

uracil (**37**) 5-fluorouracil (**38**)

Fluorine as an isostere for hydrogen also revolutionized another class of drugs, namely, quinolone antibiotics. George Lesher at Sterling–Winthrop discovered nalidixic acid (**39**) in 1946. It inhibits DNA synthesis and replication by binding to topoisomerase and DNA gyrase. Nalidixic acid (**39**) and pipemidic acid (**40**) are known as the first-generation quinolone antibiotics. While efficacious, they have poor pharmacokinetics (PK) and are extensively metabolized, thus have minimal serum levels.

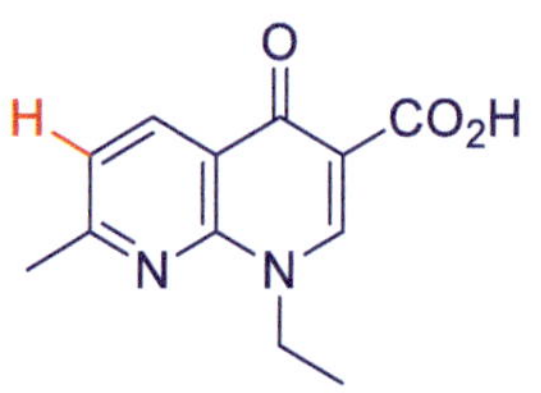

nalidixic acid (**39**)

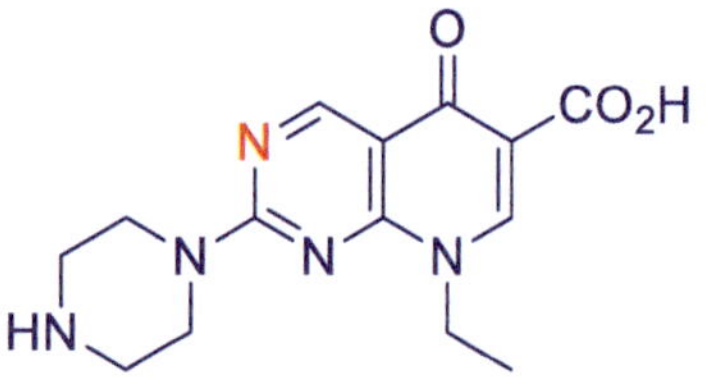

pipemidic acid (**40**)

Installation of the fluorine isostere at the ortho-position to the piperizine ring brought second-generation fluoroquinolone antibiotics. Since the fluorine atom blocked the metabolism-labile spot, norfloxacin (Noflo, **41**) and ciprofloxacin (Cipro, **42**) have longer half-lives and better PK profiles thus increased Gram-negative and systemic activity.

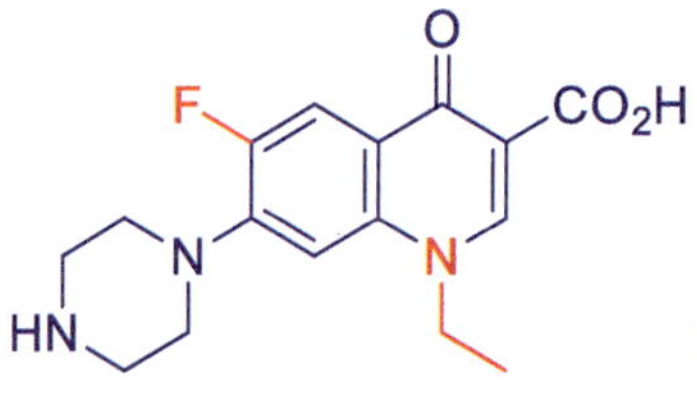

norfloxacin (Noflo, **41**)

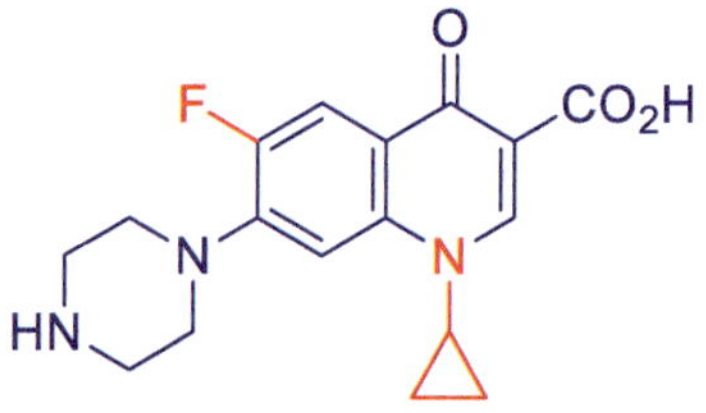

ciprofloxacin (Cipro, **42**)

If one fluorine is good, more fluorine atoms must be better. Indeed, the third-generation fluoroquinolones fleroxacin (Quinodis, **43**) and tosufloxacin (Ozex, **44**), each with three fluorine atoms, have expanded activity against Gram-positive bacteria and atypical pathogens.[21]

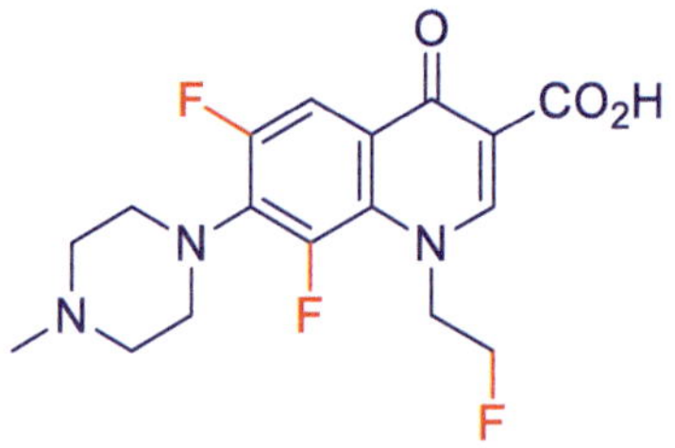

fleroxacin (Quinodis, **43**)

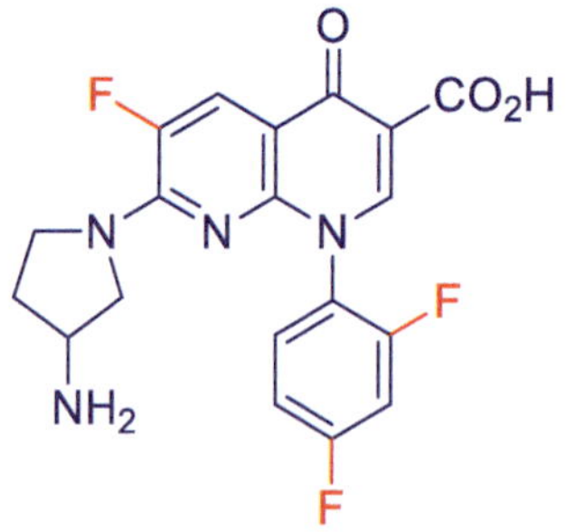

tosufloxacin (Ozex, **44**)

Applying fluorine as an isostere for hydrogen to block metabolic "weak" spots is nowadays a routine practice. A textbook example is the case in the discovery of eztimibe (Zetia, **46**). In an effort to discover cholesterol absorption inhibitors, Rosenblum et al. came across β-lactam SCH-48461 (**45**). The drug suffered from four points of metabolism: phenyl oxidation, benzyl oxidation, and two points of demethylation.

Using a fluorine atom as the isostere for hydrogen to block the phenyl oxidation and another fluorine atom to replace one methoxyl group, along with remedying two other points of metabolism, they arrived at SCH-58235 (eztimibe, Zetia, **46**). While SCH-48461 (**45**)′s ED_{50} is 2.2 mg/kg/day, eztimibe (**46**)′s ED_{50} is 0.04 mg/kg/day, a 55-fold boost of efficacy.[22]

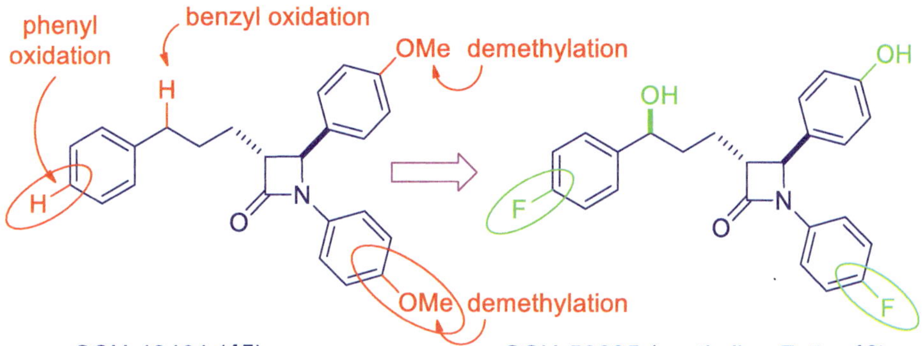

SCH-48461 (**45**)
ED_{50} = 2.2 mg/kg/day

SCH-58235 (ezetimibe, Zetia, **46**)
ED_{50} = 0.04 mg/kg/day

Pentacyclic mono-fluoro-compound **47** was a potent mitogen-activated protein (MAP) kinase 2 (MK2) inhibitor with an IC_{50} of 3 nM. It suffered from poor bioavailability with an AUC value of 121 nmol/h/L. Modulating absorption, distribution, metabolism and excretion (ADME) properties by fluorination resulted in pentacyclic di-fluoro-compound **48** with improved oral exposure, which was as potent as the prototype drug but with significantly increased AUC value of 3,486 nmol/h/L, a nearly 30-fold improvement.[23] It is possible that an intramolecular hydrogen bond on **48** was largely responsible to its cell permeation and thus its bioavailability.

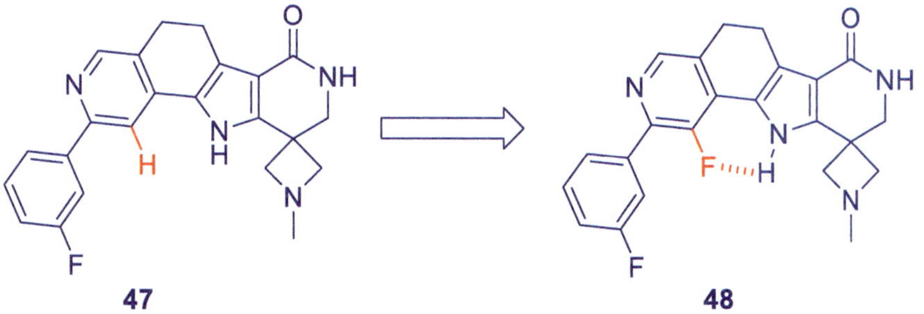

47
rat AUC: 121 nmol/h/L

48
rat AUC: 3,486 nmol/h/L

The fluorine isostere may impact “clearance via glucuronidation.” Indole acid **49**, a direct activator of 5′-adenosine monophosphate-activated protein kinase (AMPK), is a clinical candidate as a potential treatment of diabetic nephropathy. While it is reasonably efficacious with an EC_{50} value of 5.6 nM, it is cleared via a combination of Phase II metabolism (glucuronidation) and renal excretion (CL_{int} = 14 mL/min/10^6 cells).

Indeed, the major metabolite is the acyl glucuronide conjugate by uridine glucuronosyl transferase (UGT) isoforms. Successful optimization of metabolic and renal clearance led to difluoro-indole acid **50** (log *D* = 1.3), which is more acidic than the prototype drug **49** (log *D* = 2.0) since fluorine is more electronegative than both chlorine and hydrogen. Although difluoro-indole acid **50** (EC_{50} = 35 nM) is less efficacious than the parent drug **49**, its clearance has been significantly reduced (CL_{int} = 2.8 mL/min/10^6 cells).[24]

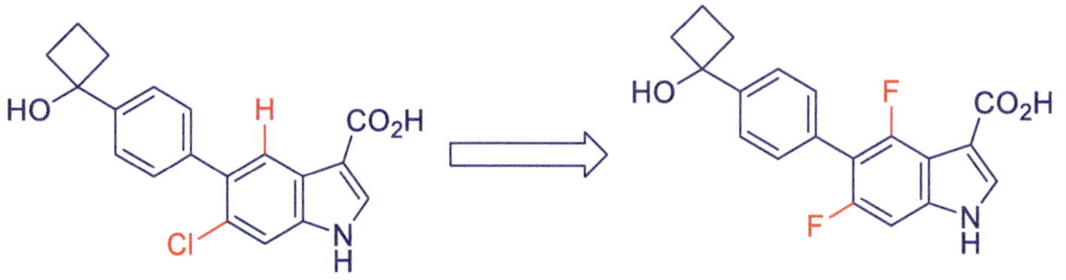

49

EC_{50}, 5.6 nM, log *D*, 2.0

CL_{int} = 14 μL/min/10^6 cells

50

EC_{50}, 35 nM, log *D*, 1.3

CL_{int} = 2.8 μL/min/10^6 cells

Not only has fluorine been employed as an isostere to improve the pharmacokinetics of drugs, it has been shown to be able to boost efficacy as well. Azepan-2-one **51** was a γ-secretase inhibitor that reduced amyloid-β production in a cell-based assay albeit with a modest efficacy (EC_{50} = 170 nM). Replacing two hydrogen atom on the lactam moiety led to difluorolactam **52**, which saw 43-fold increase of efficacy with an EC_{50} values of 4 nM. Meanwhile, the dimethyl (where the two fluorine atoms are) analog has an improved efficacy as well with an EC_{50} values of 15 nM, a 12-fold enhancement from the parent compound **51**.[25]

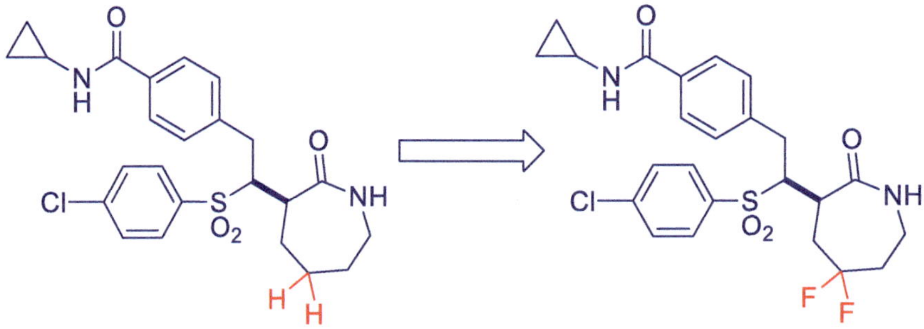

51

EC_{50} = 170 nM

52

EC_{50} = 4 nM

En route to the discovery of linezolid (Zyvox, **54**), it was discovered that the fluorine isostere was superior in terms of both potency and PK in comparison to the prototype **53**.[26]

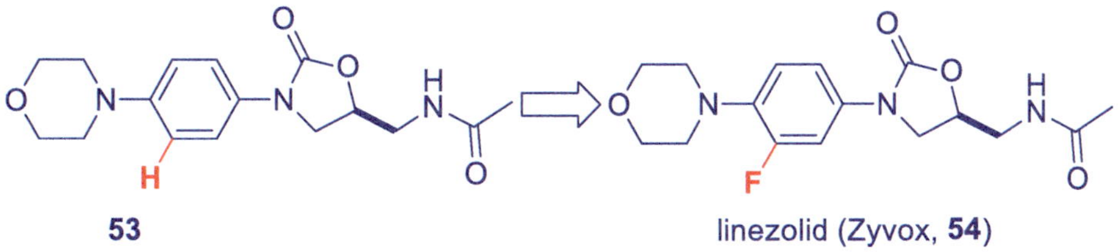

Amide **55** was an orally bioavailable agonist of the human thrombopoietin (TPO) receptor. Although this compound had promising potency and pharmacokinetic properties, the presence of the 5-unsubstituted-2-aminothiazole was of significant concern, owing to the propensity for such thiazoles to undergo metabolic activation to generate potentially hepatotoxic species. Employing fluorine as the isostere for the hydrogen at the C-5 position afforded amide **56** of 5-fluoro-2-aminothiazole. It abolished the reactive metabolism liability associated with the prototype drug **55**. Addition of a fluorine significantly minimized the hydrolysis of the amide bond and the resulting amide **56** had an improved hepatic safety profile in rodent toxicology studies.[27]

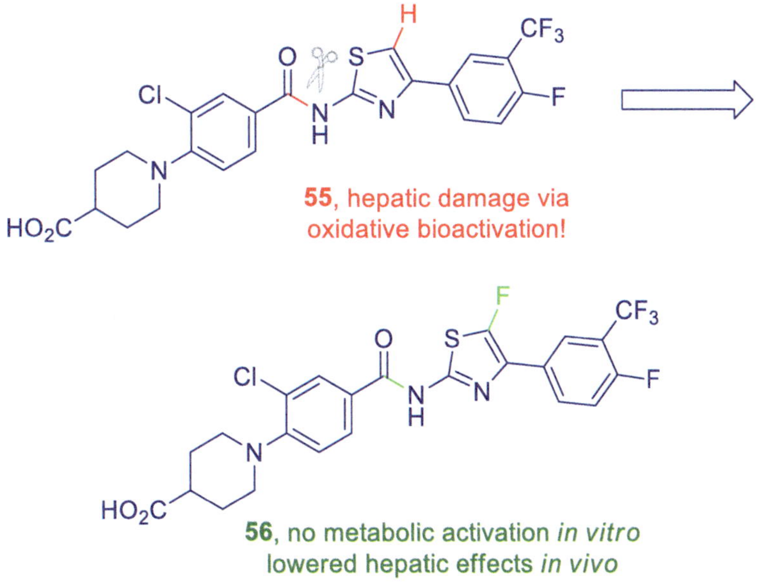

More on fluorine-containing isosteres in Section 4.3.4.

4.2.3. Chlorine

Like fluorine, chlorine atom has been used as a metabolically more stable isostere to replace hydrogen as well. In addition, chlorine as an isostere for hydrogen could also improve potency.

Indole-3-glyoxamide **57** (EC_{50} = 152.9 nM) was discovered as an inhibitor of human immunodeficiency virus-1 (HIV-1) infectivity deploying a phenotypical screen. Derivatives of compound **57** were found to interfere with the HIV-1 entry process by stabilizing a conformation of the virus gp120 protein not recognized by the host cell CD4 receptors. The 4-fluoro-analog **58** was over 50-fold more potent (EC_{50} = 2.59 nM) than prototype **57** in the pseudo-type assay as inhibitors of HIV-1 attachment. Meanwhile, the 4-chloro-analog **59** was over 35-fold more potent (EC_{50} = 4.3 nM) than prototype **57**.[28]

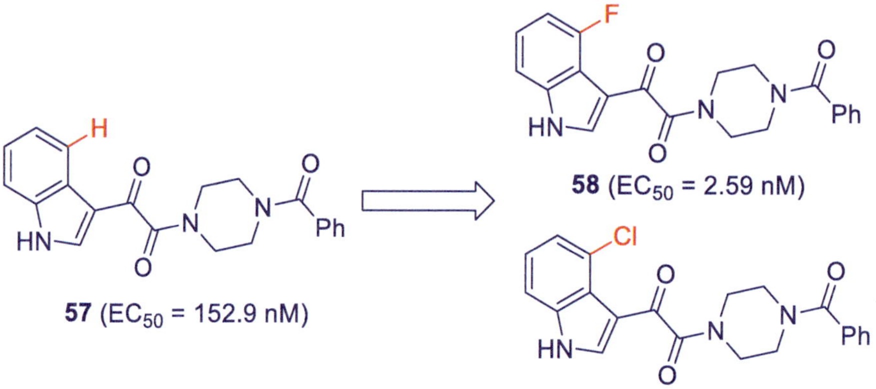

57 (EC_{50} = 152.9 nM)

58 (EC_{50} = 2.59 nM)

59 (EC_{50} = 4.3 nM)

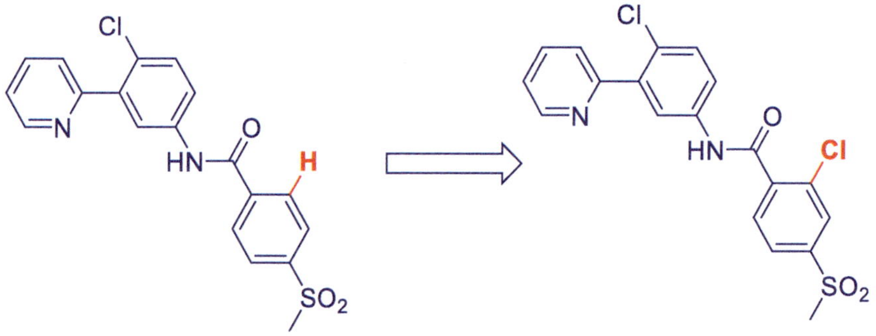

60, Clog *P* = 3.2
sol. at pH 6.5 = 0.5
sol. at pH 1 = 420
Gli LUC S12 IC_{50}, 40 nM

vismodegib (Erivedge, **61**), Clog *P* = 4.0
sol. at pH 6.5 = 9.5
sol. at pH 1 > 3,000
Gli LUC S12 IC_{50}, 13 nM

Chlorine isostere for hydrogen could improve solubility despite its high Clog *P* on occasions. In the process of discovering vismodegib (Erivedge, **61**), Genentech chemists came upon amide **60** as a potent inhibitor of the hedgehog pathway. Replacing the ortho-hydrogen on **60** (Clog *P* = 3.2) with the chlorine atom produced vismodegib (**61**, Clog *P* = 4.0). Although the lipophilicity increased, vismodegib (**61**) is markedly more soluble than the parent compound **60**.[29] One could speculate that the ortho-chlorine hindered the free rotation of the phenyl and the amide bond. As a consequence, vismodegib (Erivedge, **61**) is less flat and harder to pack into crystalline lattices. See Section 3.2.2 for more insight in modulating drugs′ solubility.

4.2.4. Methyl

Methyl as an isostere for hydrogen has been employed to improve selectivity. Before Zimmermann et al. at Novartis discovered the first receptor tyrosine kinase inhibitor imatinib (Gleevec), they arrived at phenylaminopyrimidine **62**. Although it was a potent v-Abl kinase inhibitor in enzymatic assays, it was not selective against c-Src, PKCα, and PKCδ. Installation of the "flag methyl group" led to phenylaminopyrimidine **63**. The methyl isostere abolished activities against c-Src, PKCα, and PKCδ and bestowed **63** with a remarkable selectivity.[30]

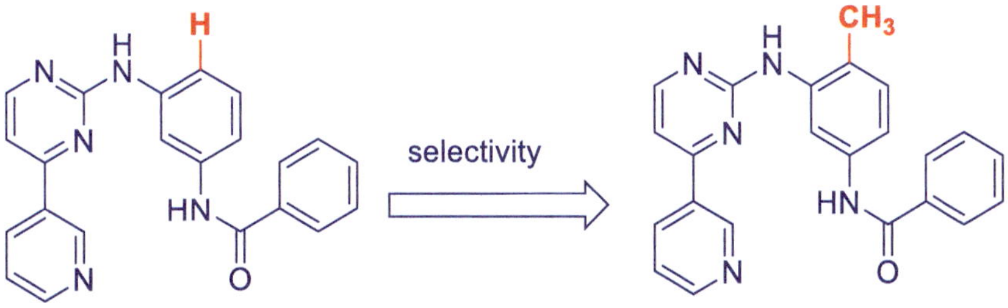

phenylaminopyrimidine **62**
IC_{50}, v-Abl-K, 0.4 μM
IC_{50}, c-Src, 15.7 μM
IC_{50}, PKCα, 1.2 μM
IC_{50}, PKCδ, 23 μδM

phenylaminopyrimidine **63**
IC_{50}, v-Abl-K, 0.4 μM
IC_{50}, c-Src, >100 μM
IC_{50}, PKCα, 72 μM
IC_{50}, PKCδ, >500 μM

By now, the profound methyl effects on drugs′ potency have been amply documented.[31] While "the magic methyl" does not deliver magic every single time, it has had enough successes to warrant a consideration to use the methyl as an isostere of hydrogen when applicable.

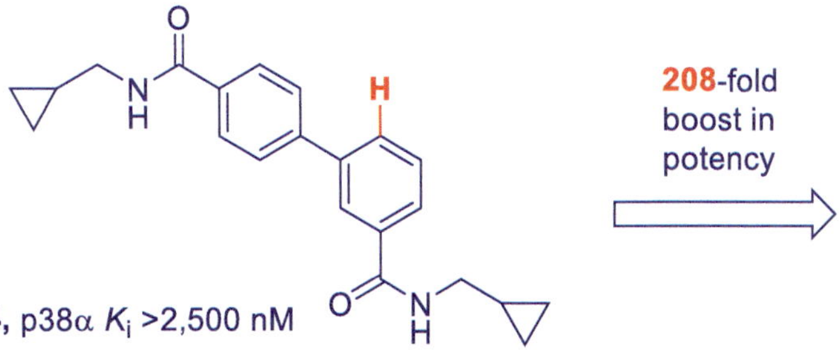

64, p38α K_i >2,500 nM

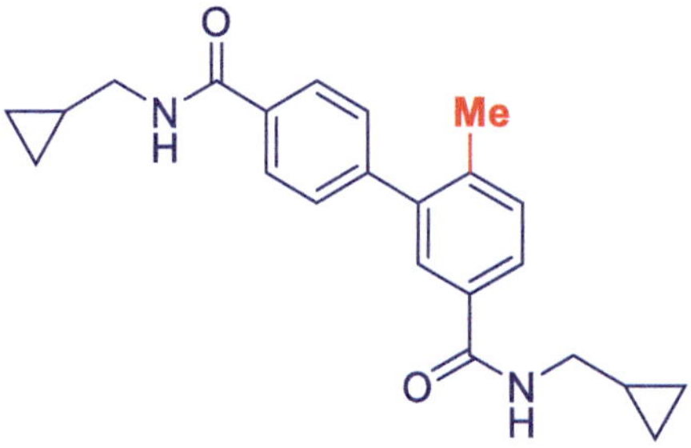

65, p38α K_i = 12 nM

Example 1, biphenyl-amide **64** is a p38α MAP kinase inhibitor with a moderate potency. Exchanging the *ortho*-hydrogen on **64** with a methyl group delivered biphenyl-amide **65** with a 208-fold elevation of potency.[32] Computational results indicated that the torsional twist induced by the *ortho*-methyl group leads to a low-energy conformation that more closely resembles the conformer observed in the X-ray crystal structure of the protein–inhibitor complex. The dihedral angle of the biaryl bond in **64** was calculated to be 50°, whereas installation of an *ortho*-methyl on **65** twists this dihedral angle out to 65°.

Example 2, in addition to methylation on an sp^2 carbon, installation of a methyl group on an sp^3 carbon has a profound impact on potency as well from time to time. A compounding effect is observed with doubly methylated analog **67**, where each newly installed methyl group leads to a surprising 1067-fold increase in potency in comparison to the parent drug **66** as sphingosine-1-phosphate receptor 1 ($S1P_1$) inhibitors.[33]

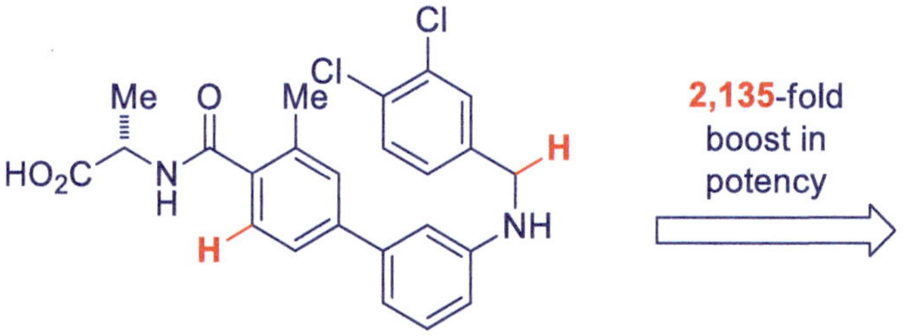

66, $S1P_1$ IC_{50} 4,270 nM

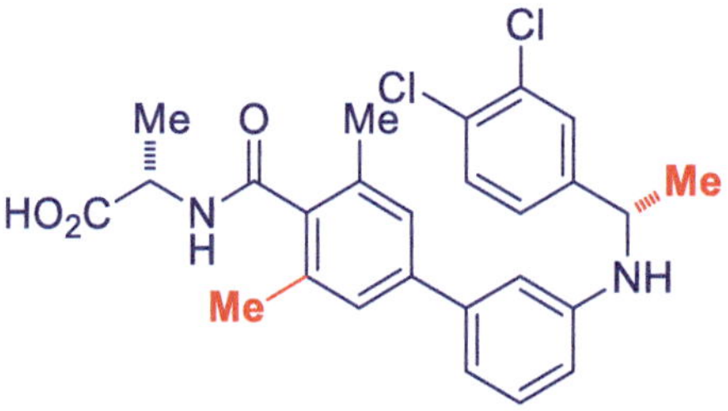

67, $S1P_1$ IC_{50} 2 nM

The aforementioned two examples are exceptions, rather than the rules for the methyl isostere. In fact, Jorgensen's group analyzed 2,100 examples of the outcome of using the methyl group as a hydrogen isostere. It is just as likely to decrease affinity as it is to increase affinity (~50:50).[34]

4.2 Alkyl Isosteres

After a meal, human body has to digest certain amount of fat—even for vegetarians since plants have fat as well. Avocado, in fact, is rich in fat. Digestion includes a process where fat is catabolized and energy is generated. Many fat molecules are linear or branched aliphatic acids and they are catabolized via a mechanism known as β-oxidation. As shown in the following scheme, the whole process takes four steps. The initial of each cycle of fatty acid β-oxidation is the dehydrogenation of an acyl-CoA to the corresponding *trans* 2-enoyl-CoA; The second step is the hydration of *trans* 2-enoyl-CoA to form L-3-hydroxyacyl-CoA by enoyl CoA hydrase (ECH, also referred to as crotonase); The third step of the β-oxidation cycle, the oxidation of the hydroxyacyl-CoA to a 3-keto-acyl-CoA is catalyzed by L-3-hydroxyacyl-CoA dehydrogenase (HAD), with nicotinamide adenine dinucleotide (NAD^+) as the cofactor; and the fourth and the last step is reversible cleavage reaction of 3-keto-acyl-CoA to yield acetyl-CoA and an acyl-CoA molecule that has been shortened by two carbon atoms.[35]

This β-oxidation cycle is repetitive and keeps going until the long-chain fatty acid is consumed to give the final products.

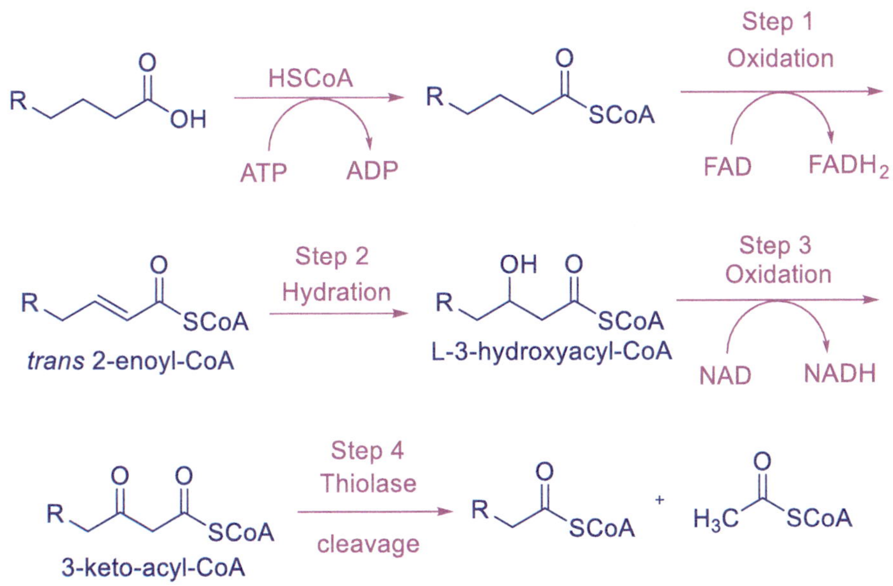

If a drug, such as prostaglandins, contains a linear or branched aliphatic acid, it is understandably subjected to the same metabolism of β-oxidation cycle and its bioavailability may suffer.

4.3.1 O for CH_2

One of the tactics to impede β-oxidation of aliphatic acid in drugs is to use oxygen as the isostere for the CH_2 group at the β-position of the carboxylic acid.

Naturally occurring prostacyclin (PGI_2) is inherently unstable and is not orally bioavailable. Its carbacyclin analog iloprost (**68**) as a prostaglandin mimic is bioavailable in man with a biological half-life of 20–30 min and showed inhibition of *ex vivo* ADP-induced platelet aggregation and vasodilating effects. Its relatively short duration of action after oral application was due to a rapid metabolism, most likely involving β-oxidation of the upper aliphatic acid side chain. Replacing the methylene group in the 3-position with an oxygen atom as its isostere would, in theory, prevent the β-oxidation. Indeed, cicaprost (**69**) proved to be at least five times more effective than iloprost (**68**) and its hypotensive action lasted two to three times longer after oral application.[36]

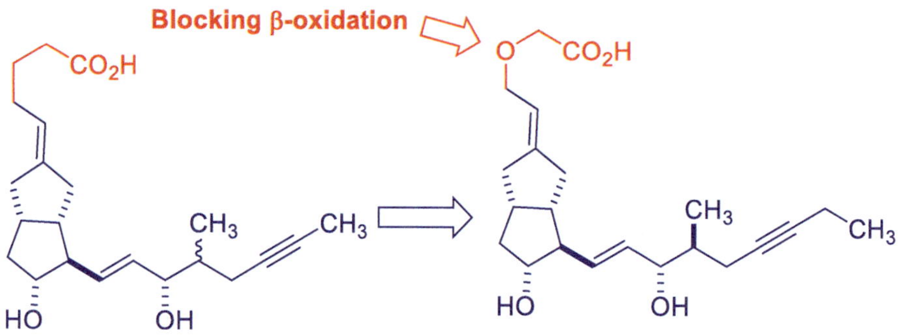

Iloprost (**68**)
$t_{1/2}$ = 25 min

cicaprost (**69**)
$t_{1/2}$ = 60 min

4.3.2 Cyclopropane as an Alkyl Isostere

Cyclopropane is a unique structure. On the one hand, it is aliphatic because all three carbons are sp^3 hybridized. On the other hand, since the bond angles are 60°, it behaves more like an aromatic group. Nonetheless, it is widely known that cyclopropane is more resistant to CYP450 metabolic oxidation than its linear brethren, *n*-propyl, *n*-ethyl, or methyl group.

β-blockers (β_1 andrenoceptor antagonists) have been an important class of cardiovascular drugs since the 1960s. Metoprolol (**72**) had a relatively short duration of action and an elevated "first pass" hepatic deactivation that was responsible for its low bioavailability. CYP450 2D6 was known to be the major isoform to carry out the metabolism. Employing cyclopropylmethyl as an isostere for the methyl group on metoprolol (**72**) gave rise to betaxolol (Kerlon, **73**), which was found to exhibit an appropriate preclinical pharmacological and human pharmacokinetics with elevated oral bioavailability and prolonged half-life for the treatment of chronic cardiovascular disease such as hypertension and angina.[37]

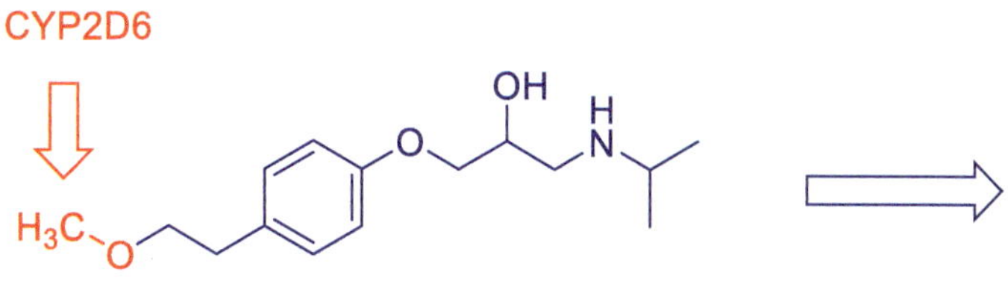

metoprolol (**72**, log *P* 2.0)
$t_{1/2}$ 3 h, *F* 38%

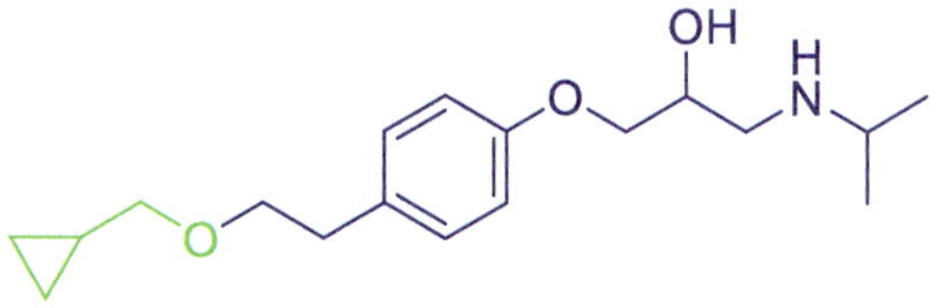

betaxolol (Kerlon, **73**, log *P* 2.8)
$t_{1/2}$ 14–22 h, *F* 89%

In the 1980s, nucleoside reverse transcriptase inhibitors (NRTIs) such as AZT were the only hope to save lives of AIDS patients infected by the HIV virus. However, initial NRTIs were invariably associated with high toxicity and low bioavailability. Non-nucleoside reverse transcriptase inhibitors (NNRTIs) represented a giant step forward thanks to their improved efficacy and PK profiles. Boehringer Ingelheim found a class of tricyclic diazepinones as their initial hits. Hit-to-lead (H2L) and lead optimization exercises led to the ethyl derivative **74** and the cyclopropyl analog **75**. Whereas ethyl derivative **74** was more potent in both enzymatic and cellular assays than **75**, and it was more soluble as well. Nonetheless, the cyclopropyl analog **75** was selected as the drug candidate because it was more bioavailable due to the fact that cyclopropane was more resistant to metabolism, while the ethyl group was prone to undergo N-dealkylation.[38] Nevirapine (Viramune, **75**) was the first NNRTI approved by the FDA in 1996.

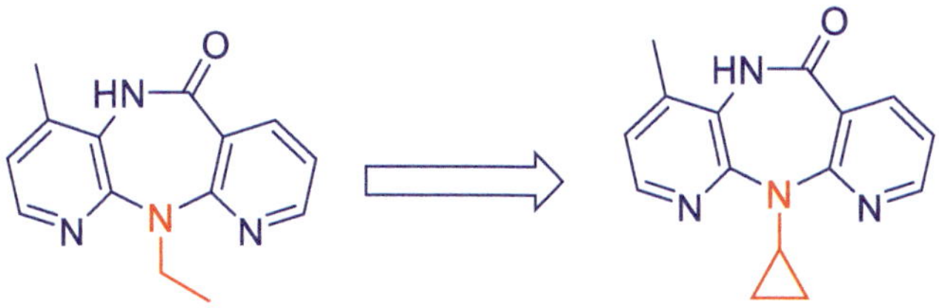

	74	nevirapine (Viramune, **75**)
HIV-1 RT enzymatic assay,	IC_{50} = 35 nM	IC_{50} = 84 nM
Cellular culture assay,	IC_{50} = 30 nM	IC_{50} = 40 nM
Solubility,	0.17 mg/mL	0.10 mg/mL
Bioavailability	+	++

4.3.3 Silicon as an Isostere of Carbon

Silicon as an isostere for carbon is controversial because there is no FDA-approved silicon-containing drugs. Is it due to silicon's inherent deficiency or there has not been one that is good enough to put on the market? I would like to believe the latter is the case.

One of the arguments against utilizing silicon to replace carbon is that it does not offer much advantage, yet it is more lipophilic. This is not always true because silicon can be advantageous when applied appropriately. For instance, haloperidol (Haldol, **76**) was the gold standard for schizophrenia treatment before the emergence of atypical antipsychotics. It was remarkably efficacious but was associated with extrapyramidal symptoms (EPS). One of its metabolic pathways involves dehydration of its piperidinol group to afford the tetrahydropiperidine and then the corresponding pyridinium metabolite that is neurotoxic. Deploying silicon as an isostere of the carbon gives rise to sila-haloperidol (**77**) that blocks the CYP oxidation to pyridinium on haloperidol. As a consequence, sila-haloperidol (**77**) is metabolized in a more conventional fashion involving the ring-opening, dealkylation, and ring hydroxylation.[39]

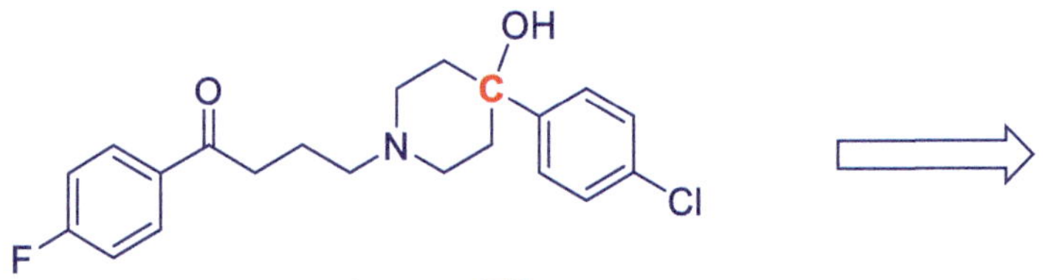

haloperidol (Haldol, **76**)
$t_{1/2}$, 18 h, *F%*, 60%

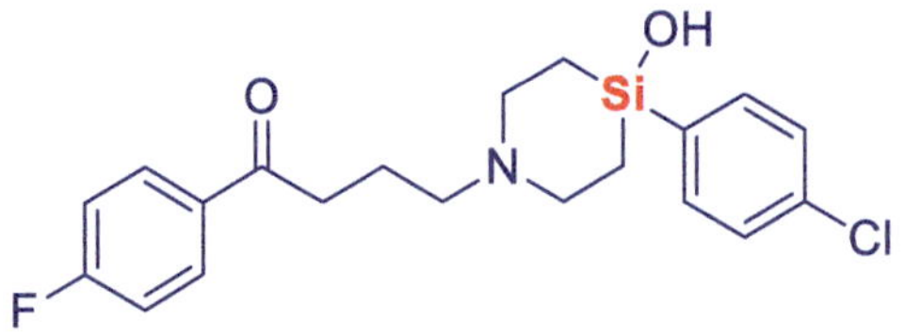

sila-haloperidol (**77**)
blocks the CYP oxidation
to pyridinium on haloperidol

Approximately nine silicon-containing drugs have been tried on human in clinics, although several of them have already been terminated.[40] Topoisomerase I inhibitor camptothecin (**78**) was isolated from Chinese joy tree. It has great potentials as a cancer therapy, except it has a minimal aqueous solubility. Taking advantage of the prodrug strategy, Upjohn arrived at irinotecan (Camptosar) and GSK succeeded in topotecan (Hycamtin). They were approved by the FDA in 1996 and 2007, respectively. In order to improve upon camptothecin (**78**)'s bioavailability, a silicon-containing drug karenitecin (cositecan, **79**) was prepared and developed to treat ovarian epithelial cancer. Regrettably, its Phase III clinical trials were terminated because the efficacy was not statistically significant.

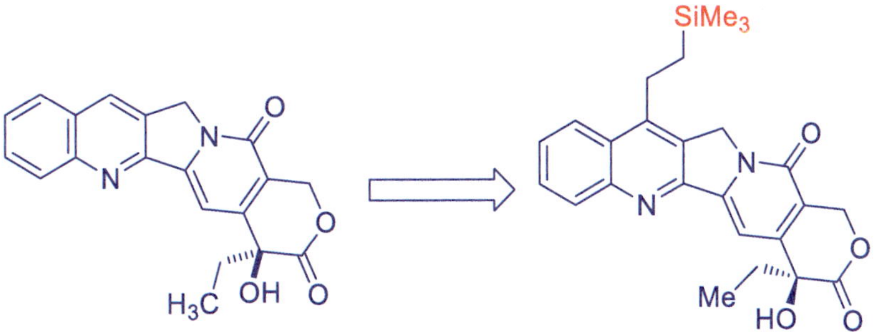

camptothecin (**78**)

karenitecin (cositecan, **79**)
Phase III, 2014
efficacy not stastically significant

Until one day, one silicon-containing drug is approved by the FDA for treating one human disease, naysayers will always have doubts about Si as an isostere for C.

4.3.4 t-Butyl Isosteres

t-Butyl group occupies an unique position in medicinal chemistry. Whereas it is bulky to provide steric hindrance and protect neighboring functional groups from metabolism, it is frequently the victim of metabolism (e.g., ω-oxidation) itself since it is so lipophilic. Some *t*-butyl isosteres have proven to be more resistant to oxidative metabolism.

Phosphatidylinositol-3-kinase (PI3K) inhibitor **80** is selective for PI3Kα, but its clearance in female Sprague–Dawley rats is high, Clp = 39 mL/min/kg. Replacing one of the three methyl groups on the *t*-butyl group with trifluoromethyl group resulted in NVP-BYL719 (alpelisib, **81**). It has a fourfold improvement for its clearance (Clp = 10 mL/min/kg) and its ADME is good enough to qualify it as a drug candidate to move to clinical trials. So far, it has performed well in Phase III trials and is shown to be active in MCF-7 (PIK3CA)-altered xenograft solid tumors.[41]

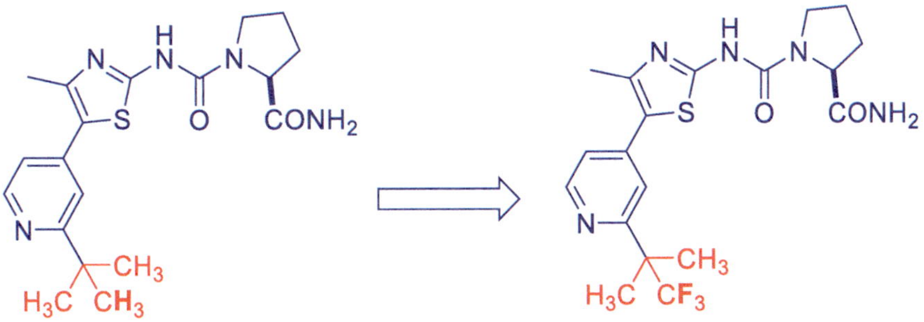

80
$t_{1/2}$, 3.4 h
Cl_p = 39 mL/min/kg

alpelisib (**81**)
$t_{1/2}$, 2.9 h
Cl_p = 10 mL/min/kg

The lipophilic *tert*-butyl group tends to be oxidized to the corresponding alcohol by CYP450, known as ω-oxidation, likely via a proton-abstraction of the sp^3 hybridized methyl group. While switching one of the three methyl groups with polar hydroxyl, cyano, and acid groups normally increases microsomal stability, sometimes they are not tolerated for maintaining activities. To keep a similar polarity, (trifluoromethyl)-cyclopropyl (Cp–CF_3) has proven to be better for the purpose of maintaining similar lipophilicity. Finasteride (Propecia, **82**) is a case in point. In addition to the 6α-oxidation for **82**, its *t*-butyl group is metabolically oxidized as well, rendering it with a relatively short half-life of 63 min in human liver microsome (HLM). Replacing the *tert*-butyl group on **82** with the Cp–CF_3 group as an isostere, the derivative **83** has an elongated half-life of 114 min.[42] This moderate increase is consistent with a caveat of this approach: replacing *tert*-butyl with Cp–CF_3 is not expected to substantially reduce metabolism at distal soft-spots such as 6α-oxidation.

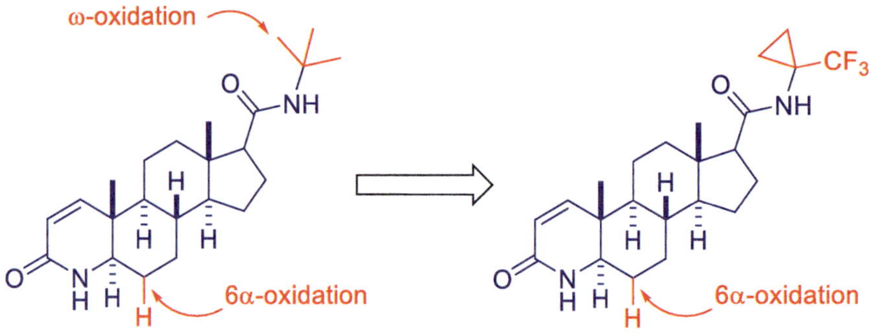

finasteride (Propecia, **82**)
HLM $t_{1/2}$, 63 min

83
HLM $t_{1/2}$, 114 min

Recently, trifluoromethyl oxetanes have been evaluated as a *tert*-butyl isostere.[43]

Since we are seeing so many fluorine atoms for *t*-butyl isosteres, it is an opportune time to address the age-old question: How many fluorine atom can a drug have?

We know Teflon does not make a good drug. How about drugs with many fluorine atoms? Most cholesterol ester transfer protein (CETP) inhibitors happen to provide some clues. The first CETP inhibitor, Pfizer's torcetrapib (**84**) with 9 fluorine atoms failed the colossal Phase III clinical trials in 2006 due to hypertension side effect. Lilly halted their Phase III clinical trials on CETP inhibitor evacetrapib (**85**) with *only* 6 fluorine atoms in 2016 due to its lack of efficacy. Merck abandoned CETP Inhibitor anacetrapib (**86**) with 10 fluorine atoms in 2017, although there was some evidence of efficacy in its Phase III clinical trials. Interestingly, 40% of anacetrapib (**86**) was still in the body three months after the patient stopped taking it. The drug was detectable four years after stopping taking it!

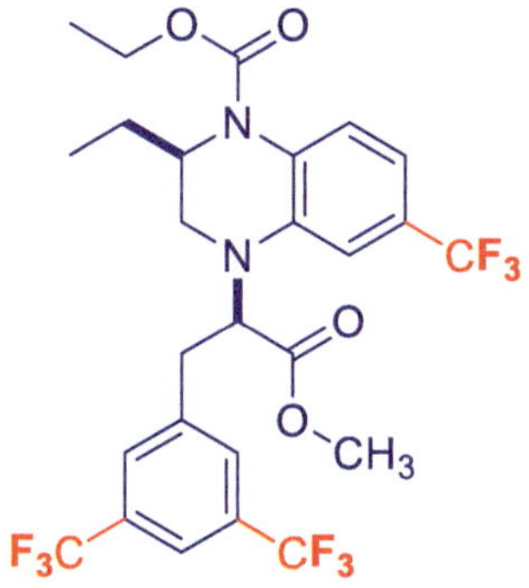

torcetrapib (**84**)
Pfizer, 2006

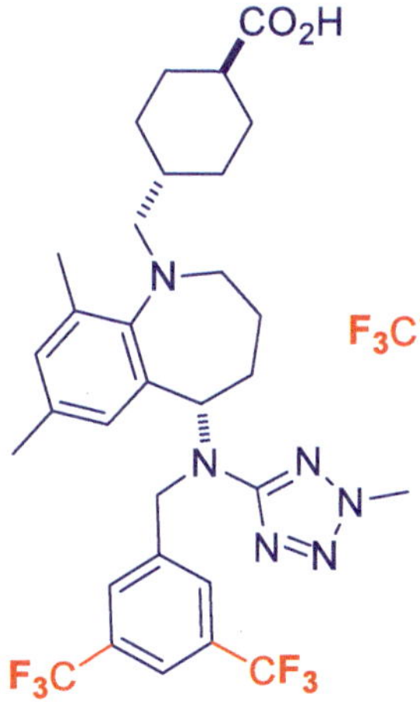

evacetrapib (**85**)
Lilly's, 2016

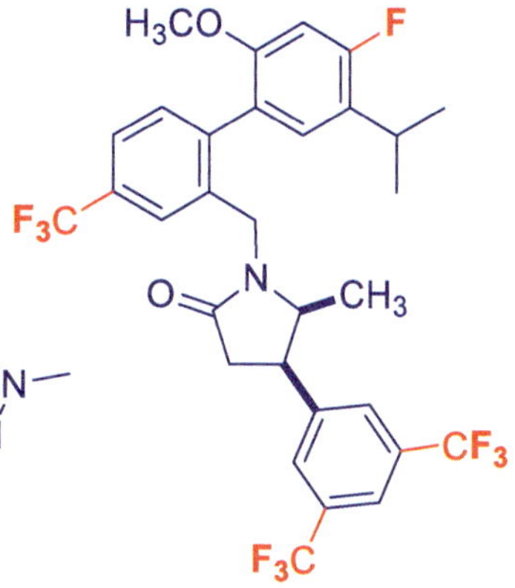

anacetrapib (**86**)
Merck, 2017

For the question how many fluorine atoms can a drug have, today's answer is seven. As tabulated below, lomitapide (Juxtapid, **87**), rolapitant (Varubi, **88**), and sitagliptin (Januvia, **89**) all possess six fluorine atoms. The trophy of possessing the most fluorine atoms (seven) goes to aprepitant (Emend, **90**), a drug marketed by Merck since 2003 to prevent nausea and vomiting brought upon by cancer chemotherapy. It is a substance P antagonist and a neurokinin 1 (NK_1) inhibitor.

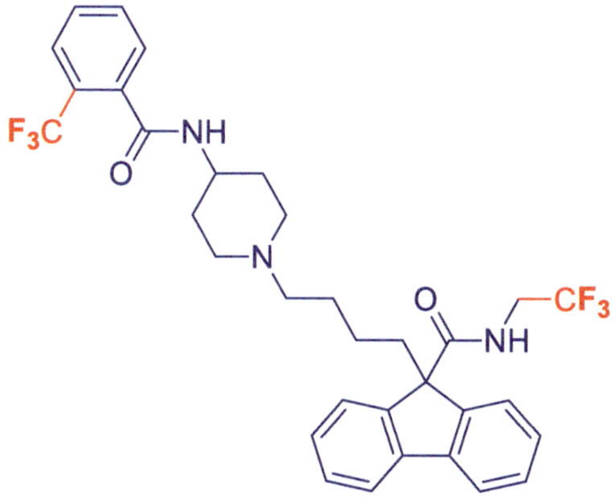

lomitapide (Juxtapid, **87**)
Aegerion, 2012
microsomal triglyceride transfer protein inhibitor

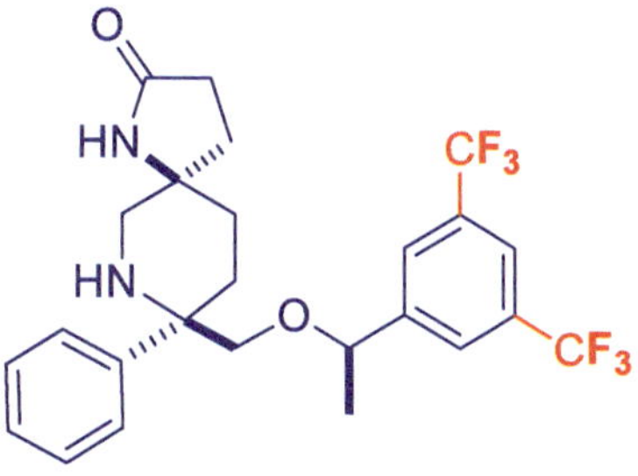

rolapitant (Varubi, **88**)
Tesaro, 2017
NK_1 inhibitor

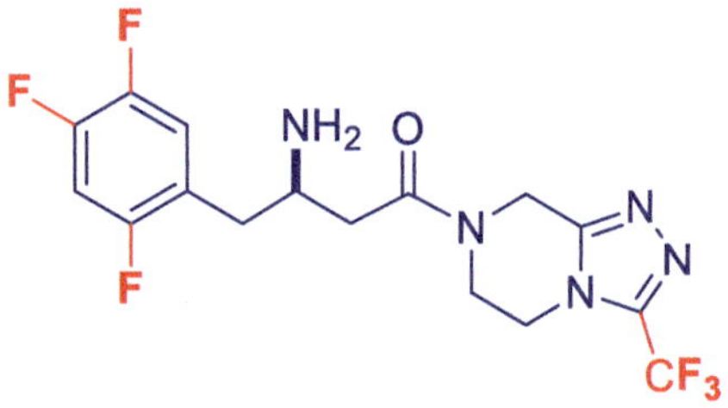

sitagliptin (Januvia, **89**)
Merck, 2006
DPP-4 inhiibtor

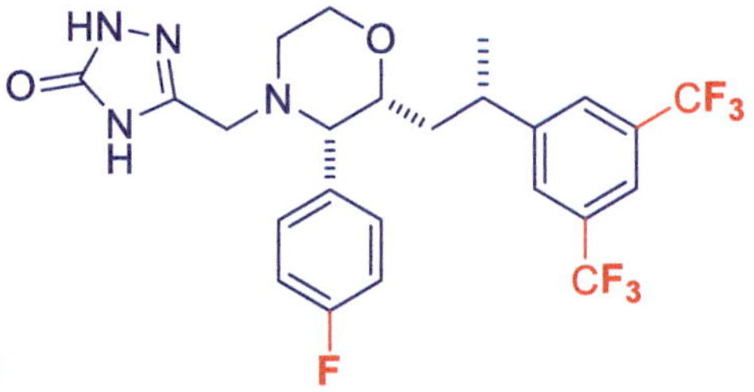

aprepitant (Emend, **90**)
Merck, 2003
Substance P antagonist
NK_1 receptor antagonist

4.4 Alcohol, Phenol, and Thiol Isosteres

4.4.1 Alcohol

4.4.1.1 RCF_2H as an Isostere of ROH

Isosteres for aliphatic alcohol are not numerous. Chief among them is difluoromethyl. Back in 1995, Erickson proposed that $-CF_2H$ may serve as a viable isostere for OH and NH since difluoromethyl group is able to act as a weak hydrogen bond donor.[44] The $CF_2H\cdots O{=}C$ interaction has an estimated binding energy of ~1.0 kcal/mol, while more traditional hydrogen bond interactions have a binding energy of 2–15 kcal/mol. The hydrogen bond for $CF_2H\cdots O$ distance is ~2.4 Å, which is on par to that of a normal D–H···A of 2–3 Å. Here, D stands for the hydrogen bond donor and A is the acronym for the hydrogen bond acceptor. Furthermore, since RCF_2H is a more lipophilic H-bond donor than OH or NH, the isosteric replacement has potential to improve membrane permeability

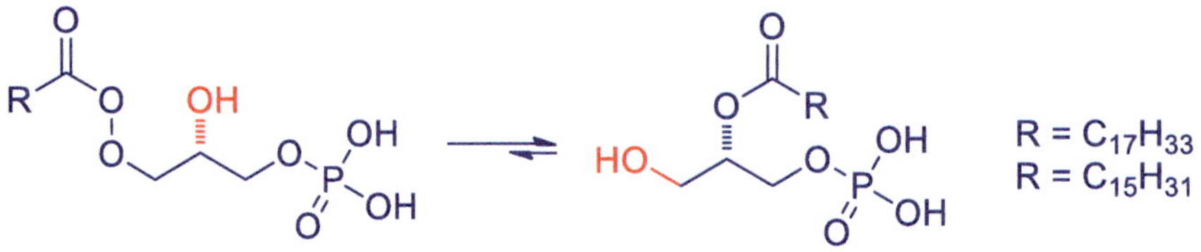

sn-1 LPA (**91**) *sn*-2 LPA (**92**)

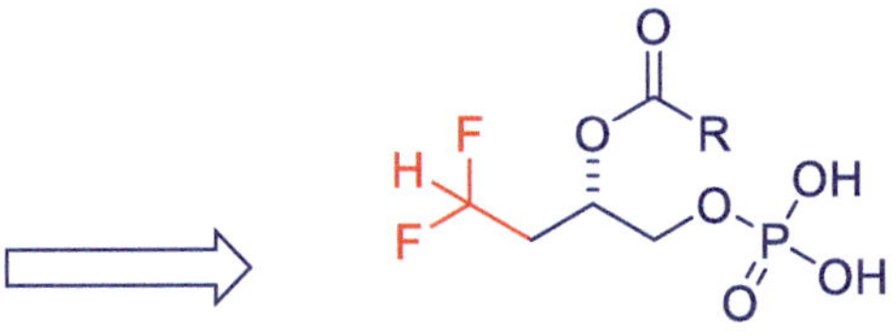

di-F-LPA analog **93**

In addition to being an agonist for the nuclear hormone receptor peroxisome proliferator-activated receptor-γ (PPAR-γ), lysophosphatidic acid (*sn*-1 LPA, **91**) also binds to and activates LPA receptors 1–4 (GPCRs). However, the tumor promoter ovarian cancer-activating factor is the isomer *sn*-2 LPA (**92**) rather than the more common **91** and the chemical equilibrium favors the latter by six-fold. As an isostere of the hydroxyl group, CF_2H analog **93** was designed to block migration of acyl moiety. Indeed, diF-LPA (**93**) was found to stimulate luciferase in CV-1 cells transfected with luciferase under control of a PPAR-responsive element. In addition, both **92** and **93** failed to interact with LPA receptors 1–3, offering an example of an isostere enhancing specificity.[45]

4.4.1.2 Sulfoximines

The sulfoximine group may be considered as the aza-analog of a sulfone. Its nitrogen atom is mildly basic with a pK_a of 2.7 when protonated. It is chemically stable and tetrahedrally hybridized thus amenable to incorporate into enzyme inhibitors. It is a good isostere for alcohol because it may serve as a hydrogen bond donor and acceptor.

Merck's L-700417 (**94**) is an HIV-1 protease inhibitor with an IC_{50} of 0.6 nM. A transition state mimic (TSM) for HIV-1 protease, **94**'s alcohol group interacts with catalytic aspartic acids in the active site to mimic the hydrated amide of the substrate, just like all other HIV-1 protease inhibitors do. While *silanediol* and *phosphinate* have been successfully employed as effective isosteres of TSMs, Vince et al. opted to use sulfoximine as the isosteric substitution of the alcohol group. The fruit of their labor was sulfoximine analog **95**, which has an IC_{50} of 2.5 nM, a fourfold drop of potency in the enzymatic assay, but still quite active. Interestingly, replacing the alcohol group with a sulfone offered inhibitor **96**, which still retains some potency with an IC_{50} of 21.1 nM.[46] Despite a 10-fold lower than that of the parent drug **94**, it is remarkable since sulfone is only a hydrogen bond acceptor with no hydrogen bond donating ability as the alcohol does.

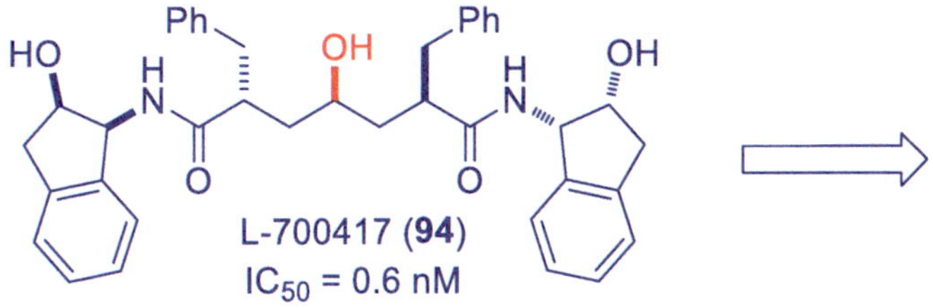

L-700417 (**94**)
IC_{50} = 0.6 nM

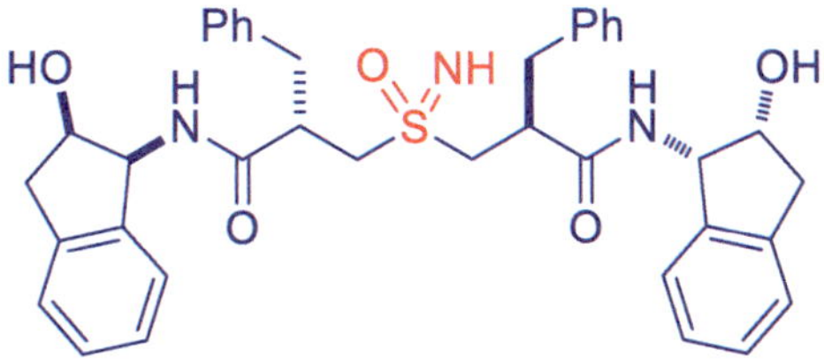

95, IC_{50} = 2.5 nM

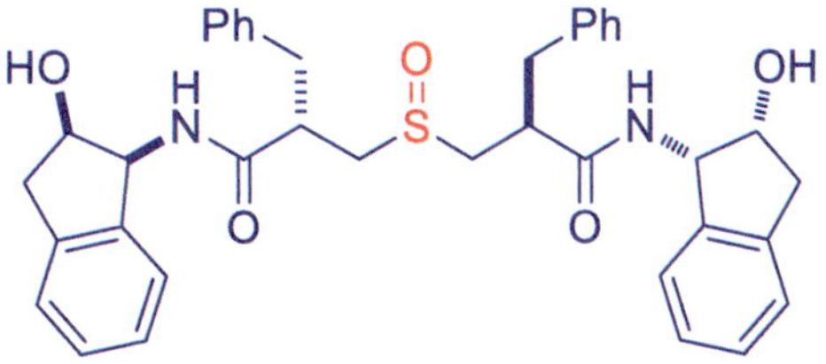

96, IC_{50} = 21.1 nM

Emboldened by their initial success with the sulfoximine as an isosteric substitution of the alcohol group, Vince et al. moved to attempt an encore on an HIV-1 protease inhibitor already on the market: Merck's indinavir (Crixivan, **97**, IC_{50} = 0.4 nM). They prepared and evaluated sulfoximine-based inhibitors for HIV-1 protease including **98**. Surprisingly, sulfoximine **98** is virtually inactive with no *in vitro* potency to speak of against the HIV-1 protease enzyme. The authors attributed to limited conformational flexibility of the peptidic template interfering with optimal binding based on docking studies. Indeed, the piperazine ring confers rigidity that prevents **98** from adopting appropriate conformation to mimic the transition state.[47]

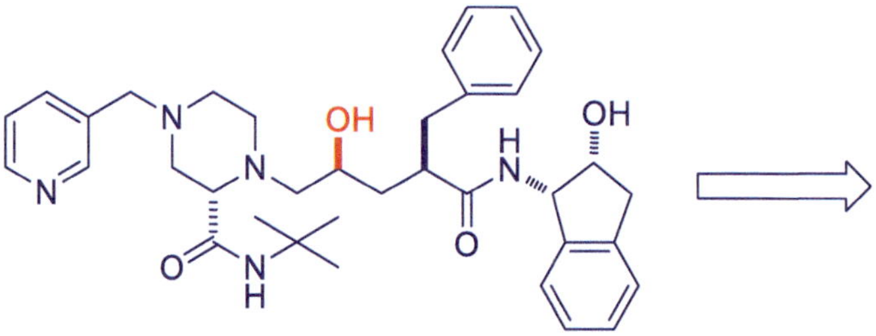

indinavir (Crixivan, **97**)
IC_{50} = 0.4 nM

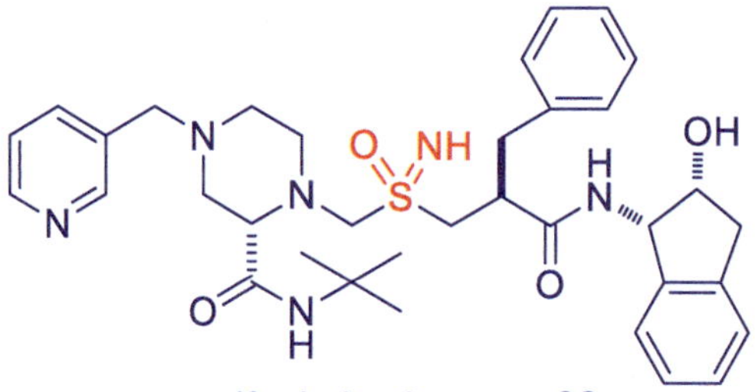

sulfoximine isostere **98**
IC_{50} = 100 μM, inactive

4.4.1.3 Amides and Sulfonamides

It is reasonable to surmise that amides and sulfonamides may serve as isosteres of alcohols since both of them are hydrogen bond donors.

Flavagline **99** is a remarkable anticancer agent with unique cardioprotective activities not seen in other chemotherapies. It directly modulates the activity of the scaffold proteins prohibitin-1 and prohibitin-2 and the translation initiation factor eIF4a. Bioisosteric modification of flavaglines provided the corresponding flavagline formamide **100** and flavagline mesylamide **101**. Regrettably, neither **100** nor **101** displayed any significant cytotoxicity in Hep3B and HuH7 cancer cell lines.[48]

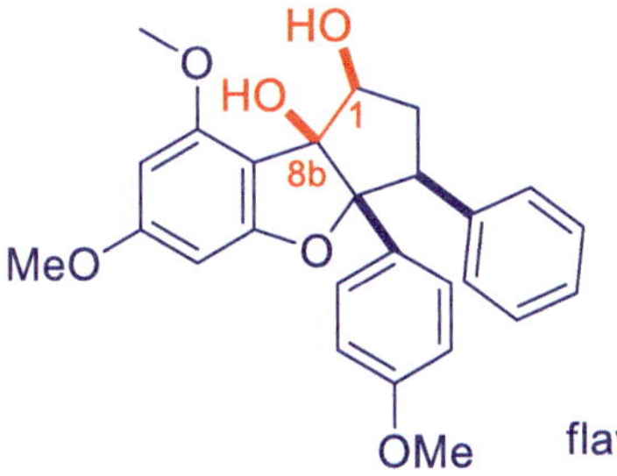

flavagline **99**

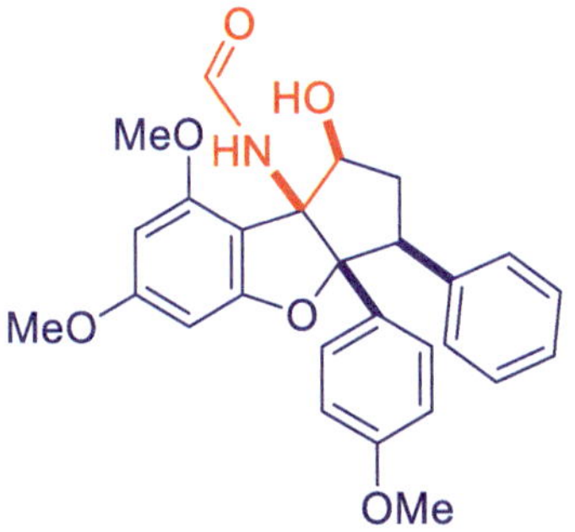

flavagline formamide **100**

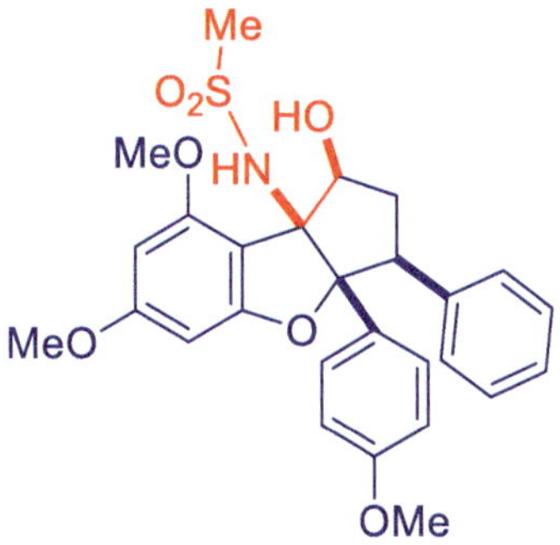

flavagline mesylamide **101**

This example is a somber reminder that a classical strategy in pharmacomodulation, i.e., the isosteric replacement of an alcohol by an acylamino or a mesylamino moiety may not always lead to active compounds. In the same vein, a mastery of bioisosteres helps you to succeed, but it is never a guarantee.

4.4.2 Phenol

As a polar group, phenols are readily glucuronidated (secondary metabolism) and subsequently excreted, thus tend to have lower bioavailability. On the other hand, catechols are substrates of catechol *O*-methyl transferase (COMT) and often become methylated and lose activity. Both phenols and catechols can be oxidized by CYP450 enzymes to the corresponding *ortho*- and *para*-quinones, which are reactive metabolites. Therefore, metabolically more stable isosteres are frequently sought to replace them. Pyrrole and pyrazole are popular isosteric substitutions for phenol in the forms of indole and indazole to address the metabolic soft spots exposed by phenol.

The bioisosterism between the pyrrole ring and the phenolic hydroxyl group is well known. Labetalol (Normodyne, **102**, ED_{30} = 25 mg/kg, po), an antagonist of adrenergic receptors, is used clinically as an antihypertensive agent. Indole–phenol bioisosterism led to a pyrrolo analog AY-28,925 (**103**, ED_{30} = 5 mg/kg, po) that is fivefold more active than the parent drug labetalol (**102**). It is speculated that intramolecular hydrogen bonding is the common feature for both **102** and **103**. True enough, the methylated indole derivative **104** is virtually inactive due to its lack of hydrogen bond donating ability.[49]

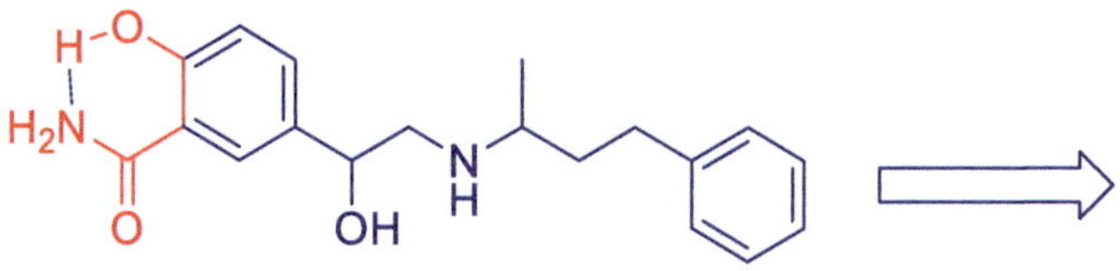

labetalol (Normodyne, **102**)
ED_{30} = 25 mg/kg, po

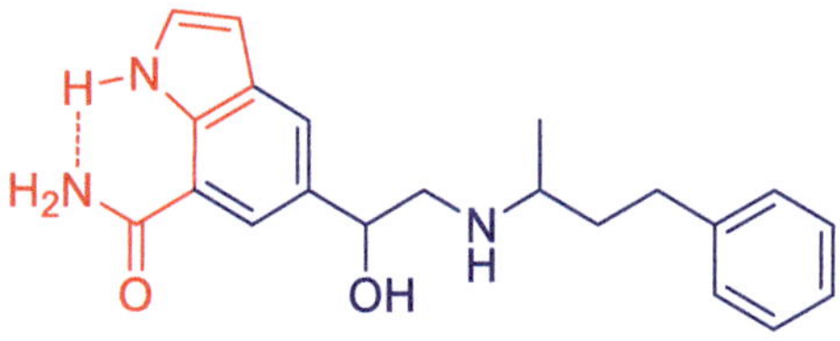

AY-28,925 (**103**)
ED_{30} = 5 mg/kg, po

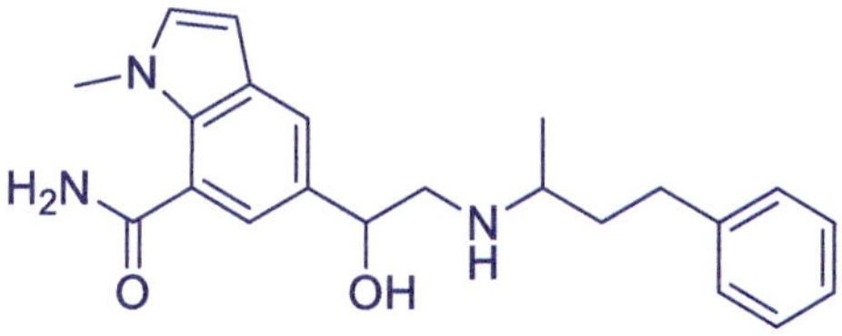

104
inactive

The Raf kinases, which are components of this cascade, are serine/threonine kinases that activate MEK1/2. Mutant B-Raf containing a V600E substitution (where B-Raf protein's 600th amino acid valine is replaced by glutamic acid) causes aberrant constitutive activation of this pathway and has high occurrence in several human cancers. Two B-Raf kinase inhibitors have been approved by the FDA to treat cancer: vemurafenib (Zelboraf, 2011) and dabrafenib (Tafinlar, 2013), with the former being the first marketed drug that was discovered employing fragment-based drug discovery (FBDD) strategy.

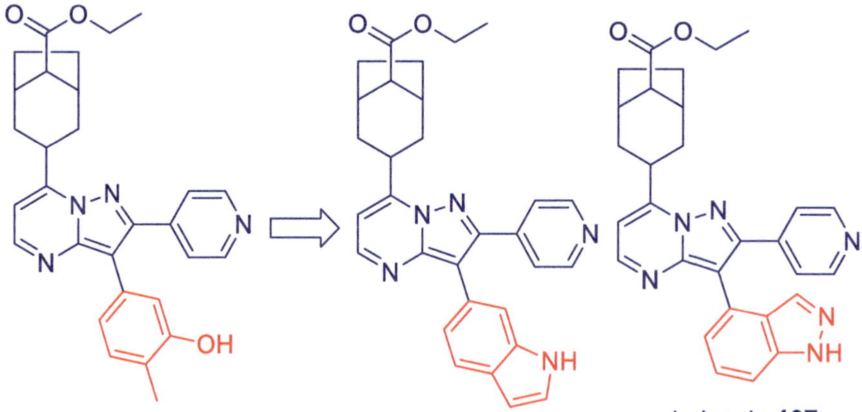

phenol **105**
B-raf, IC_{50}, 0.3 nM

indole **106**
IC_{50}, 36 nM

indazole **107**
IC_{50}, 2 nM

Phenol **105** is a B-raf inhibitor (IC_{50}, 0.3 nM). Bioisosterism was deployed to address the potential metabolic soft spot posed by the phenol. Although replacing the phenol functional group with hydrogen bond acceptors, not surprisingly, did not work, isosteric substitution with hydrogen bond donors offered indole **106** (IC_{50}, 36 nM) and indazole **107** (IC_{50}, 2 nM), respectively. More important, **107** potently inhibited cell proliferation at submicromolar concentrations in the A375 and WM266 cell lines and exhibited good therapeutic indices in cells.[50]

One of the key functions of a phenol group is serving as a hydrogen bond donor. Therefore, it is not surprising that the most popular surrogate for phenolic functions are NH group rendered acidic through the presence of an electron-withdrawing group. Phenol **108** is an *N*-methyl-D-aspartate (NMDA) receptor (NR) antagonist. While reasonably potent with an IC_{50} of 0.17 M, **108** has a low oral bioavailability. Preliminary pharmacokinetic and metabolic studies indicated that it had good permeability across membranes but had a short half-life *in vivo*. Further investigations revealed that the phenol moiety was subjected to rapid secondary hydroxylation and conjugation. My former colleagues at Parke–Davis sought to replace the phenol by heterocyclic NH-containing rings that were expected to slow metabolism and hence improve oral bioavailability.[51]

To that end, the corresponding indole, indazole, benzotriazole (**109**), indolone (**110**), oxindole, and isatin derivatives gave weaker NR1A/2B activity than the parent phenol (**108**). However, benzimidazolone derivative **111** and carbamate **112** were more potent. In particular, benzimidazolone derivative **111** was a very potent and selective NR1A/2B receptor antagonist. This compound demonstrated oral activity in a rodent model of Parkinson's disease at 10 and 30 mg/kg.[51]

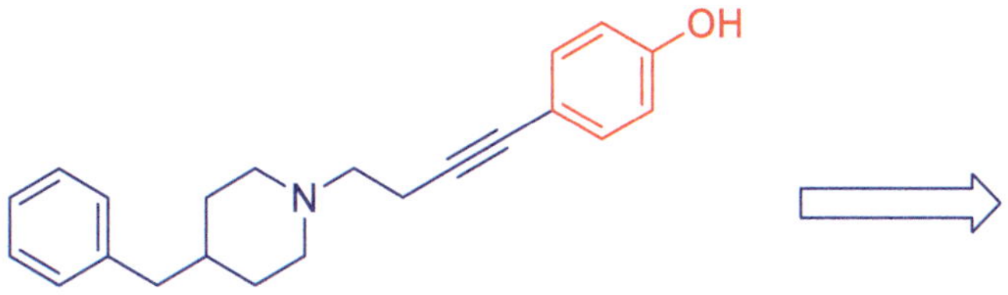

108, NMDA NR1A/2B receptors, IC_{50}, 0.17 M

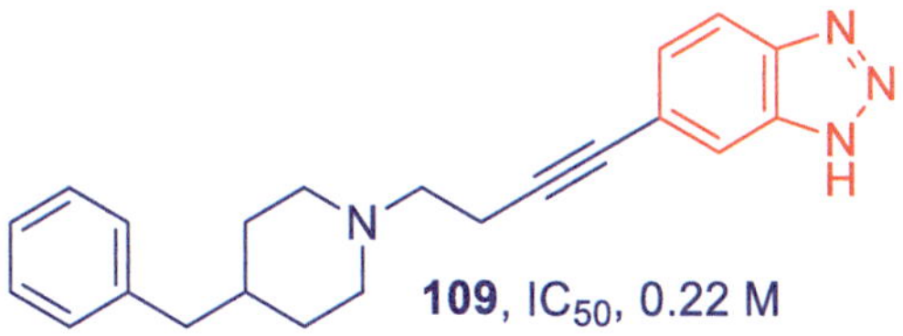

109, IC_{50}, 0.22 M

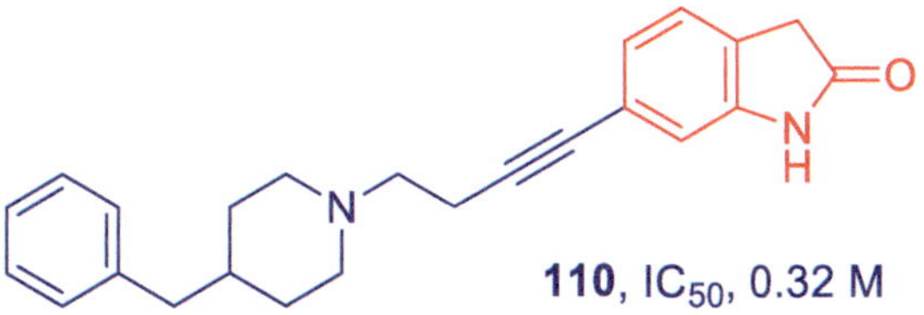

110, IC_{50}, 0.32 M

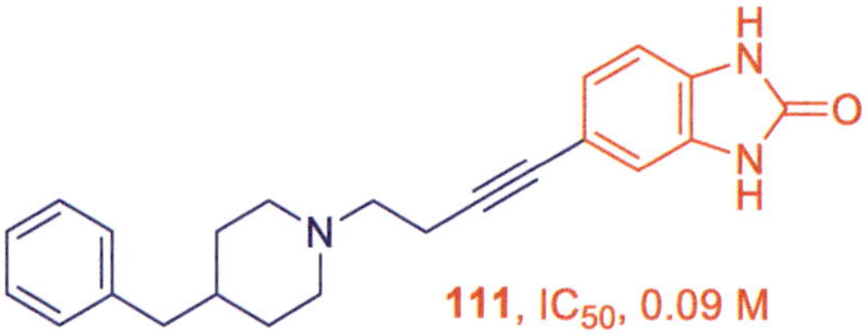

111, IC_{50}, 0.09 M

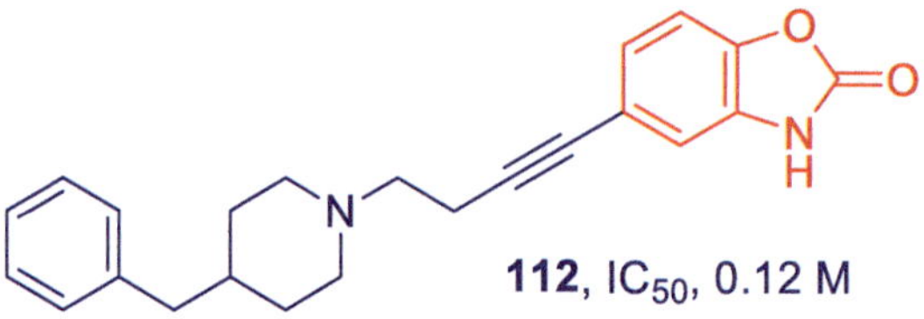

112, IC_{50}, 0.12 M

As far as catechol is concerned, all phenol isosteres apply as well. In addition, benzimidazole mimics catechol with an intramolecular bond.

Phenol **113** [3-(1-propylpiperidin-3-yl)phenol, 3-PPP], as a highly selective agonist for presynaptic brain dopamine receptors, may be viewed as a catechol isostere.[52] Due to the presence of the phenol functionality, 3-PPP (**113**) is plagued by low oral bioavailability and short duration of action. As a result, its clinical potential as antipsychotic and antiparkinsonian agent is severely limited. Attempts to address these issues led to a successful drug pramipexole (Mirapex, **114**). Its aminothiazole motif may be regarded as an isostere of the catechol group on dopamine. More interestingly, conjugated alkyne **115**, a selective D_3 dopamine receptor agonist, serves as an atypical, nonaromatic isostere of catechol.[53]

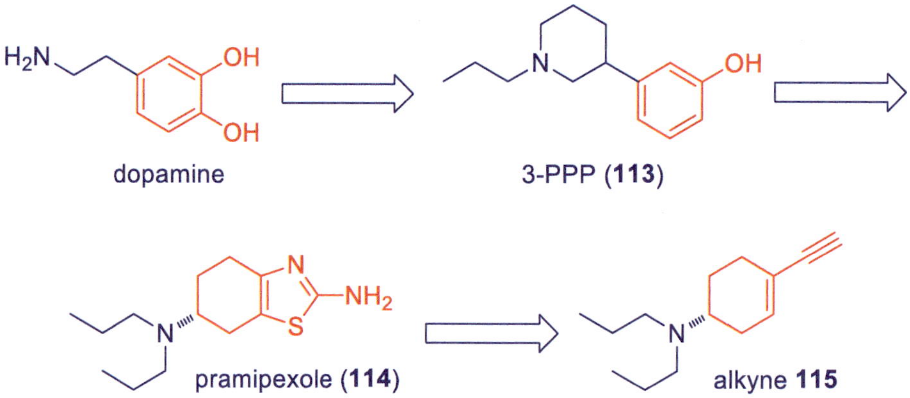

Benzoxazolone ring, resistant to metabolic oxidation, has been employed as a catechol mimic in the context of insulin-like growth factor 1 receptor kinase inhibitors.[54]

4.4.3 Thiol

Thiol is a structural alert. Although captopril (Capoten), the first commercial angiotensin converting enzyme (ACE) inhibitor to treat hypertension, has a thiol group as a zinc chelator, it is associated with a trio of shortcomings: short half-life; rashes, and loss of taste. Both alcohol and amine may serve as isosteres for the thiol functionality. In addition, difluoromethyl has also been successfully employed as an isosteric substituent for thiol.

In the context of hepatitis C virus (HCV) NS3 protease inhibitors, difluoromethyl group was investigated as an isostere for the lipophilic cysteine thiol group. Computationally, the van der Waals surface of 1,1-difluoroethane (HCF_2CH_3, 46.7 Å) is similar to that of methanethiol ($HSCH_3$, 47.1 Å). In addition, electrostatic potential maps indicated their surface similarities as well with negative potential around S lone pairs and the F atoms and positive potential around the H atoms.[54]

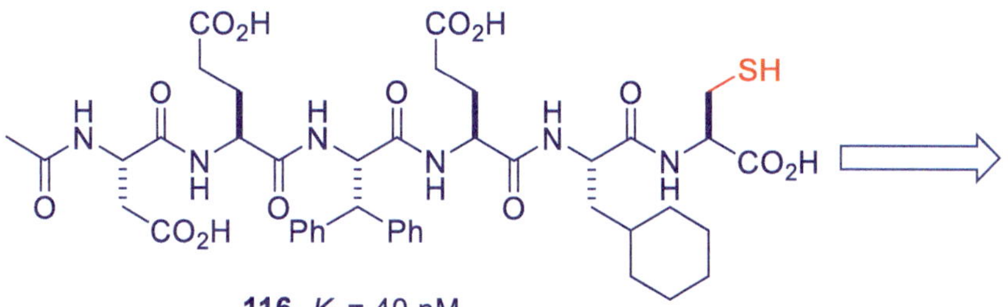

116, K_i = 40 nM

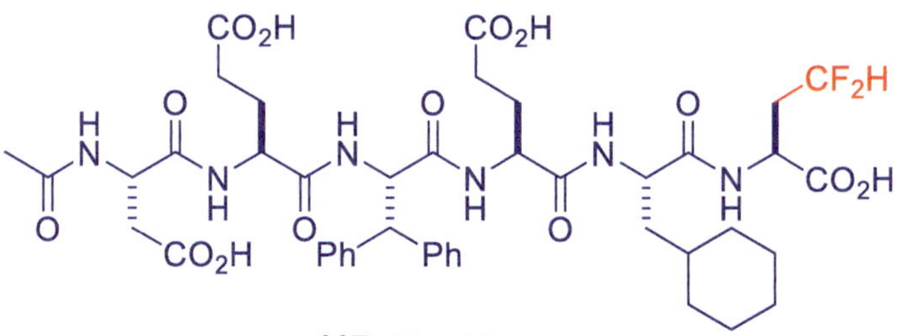

117, K_i = 30 nM

Hexapeptide **116** is an HCV NS3 protease inhibitor and the terminal Cys is P_1 of the natural substrate. Its thiol on cysteine has non-covalent contact with the phenyl ring of the S_1 Phe154. In order to replace the often problematic thiol functionality, its difluoromethyl derivative **117** was prepared and tested to be nearly equipotent as the parent drug **116**. The X-ray co-crystal structure of **117** with NS3/4A revealed that $-CF_2H$

almost completely donates H bond to C=O of Lys136, just as thiol does on **116**. One of the fluorine atoms is close to the 4-H of Phe154. Here, $-CF_2H$ works under such highly specific conditions as a Cys mimetic and underlines its general applicability as a non-reactive Cys surrogate.[55]

4.5 Carboxylic Acid and Derivative Isosteres

4.5.1 *Carboxylic Acid*

Carboxylic acids exhibit pK_a values of approximately 4.0 and are partially dissociated under physiological conditions. The carboxylic acid functionality is present in many endogenous ligands. It is also an important pharmacophore and more than 450 drugs contain the acid group. In addition to its metabolic vulnerability and toxicity on occasions, its polarity brings about several shortcomings including limited passive diffusion across cell membrane (ester prodrugs have frequently resorted to address this issue) and extensive secondary metabolism (glucuronidation and sulfation). Therefore, countless isosteres have been explored to mitigate its drawbacks while retaining its useful attributes.

Ballatore et al. published a superb review on carboxylic acid isosteres in 2013.[56]

4.5.1.1 Linear Carboxylic Acid Isosteres

Hydroxamic acids, phosphonic acids, phosphinic acids, sulfonic acids, phosphates, and even sulfonamides may be considered as linear carboxylic acid isosteres. Taurine (2-aminoethanesulfonic acid) may be viewed as an isostere of β-alanine (3-aminopropanoic acid). Moreover, acyl-sulfonamides, acyl-cyanamides, and sulfonyl ureas have all been explored as acid isosteres.

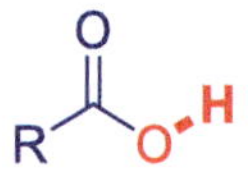

R = CH_3, pK_a = 4.76, log D = −1.65

Carboxylic acids

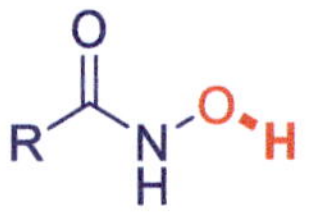

Hydroxamic acids

(R = CH_3, log D = −0.82)

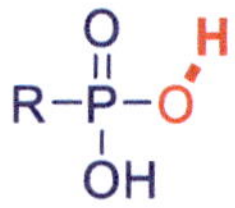

Phosphonic acids

(R = CH_3, log D = −2.13)

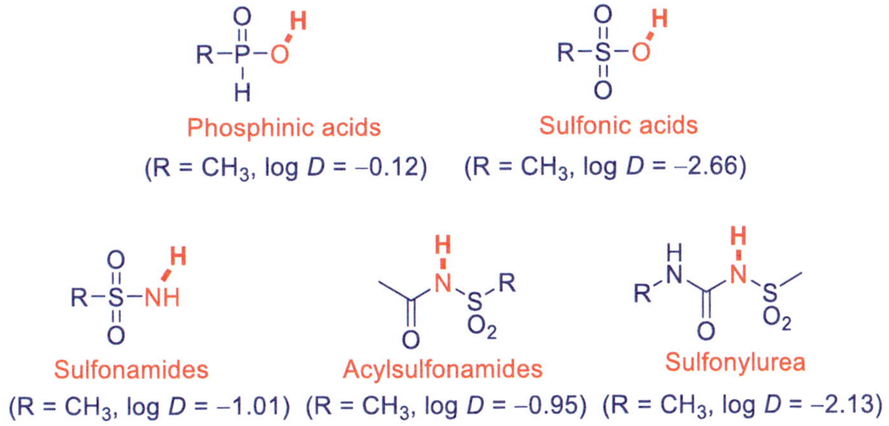

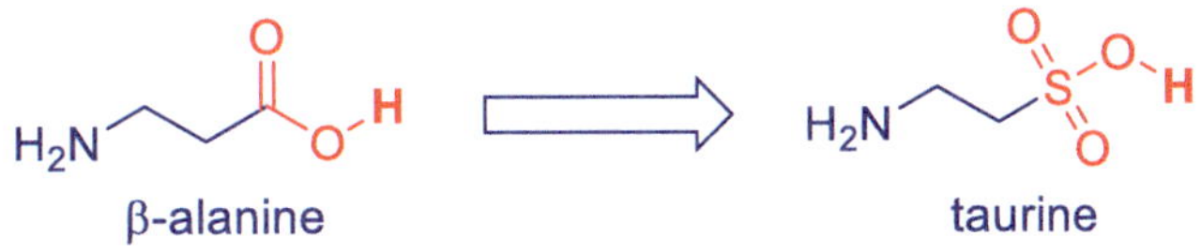

Abbvie's Bcl-2 inhibitor venetoclax (Venclexta) was discovered employing the FBDD strategy. From their FBDD using the "SAR by NMR" method, Abbvie screened 10,000 compound library with MW <215 at 1 mM concentration. One fragment from the first-site ligands was *p*-fluorophenyl-benzoic acid **118**.[57]

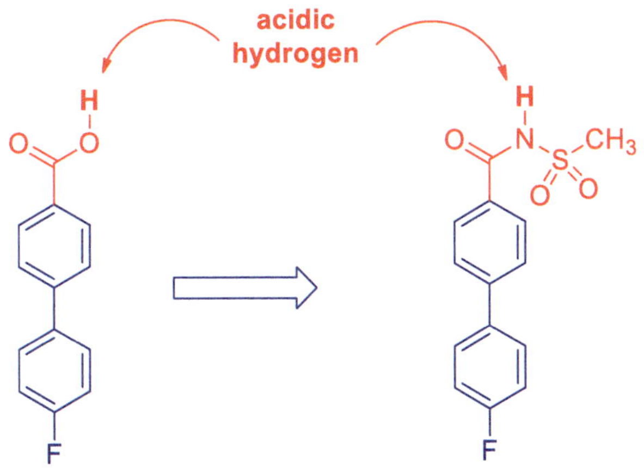

acid **118**, K_d = 300 μM acylsulfonamide **119**, K_d = 320 nM

Many attempts were made to tether this fragment with the second-site binders without success when the carboxylic acid was kept intact. A breakthrough was achieved when acylsulfonamide **119** was chosen as the isosteric substitution for the parent acid. Although **119** is a bit less active than the parent acid **118**, its acylsulfonamide group created a point of extension on sulfur to link it with the second-site binders. Acylsulfonamide provided a more effective vector to the Ile85 pocket and preserved acidity and interaction with Arg139 of Bcl-2. The practice led to a 36 nM Bcl-X_L inhibitor. Addressing the protein-binding shift issues associated with its binding to site 3 of human serum albumin (HSA-III) and other pharmacokinetic issues, venetoclax (ABT-199, Venclexta, **120**) was ultimately discovered as a potent Bcl-2 inhibitor (0.01 nM) after addition of the azaindole substituent that filled the P_4 pocket and captured a hydrogen bond. In 2016, it was approved by the FDA for treating chronic lymphocytic leukemia (CLL) with the 17p deletion.[57]

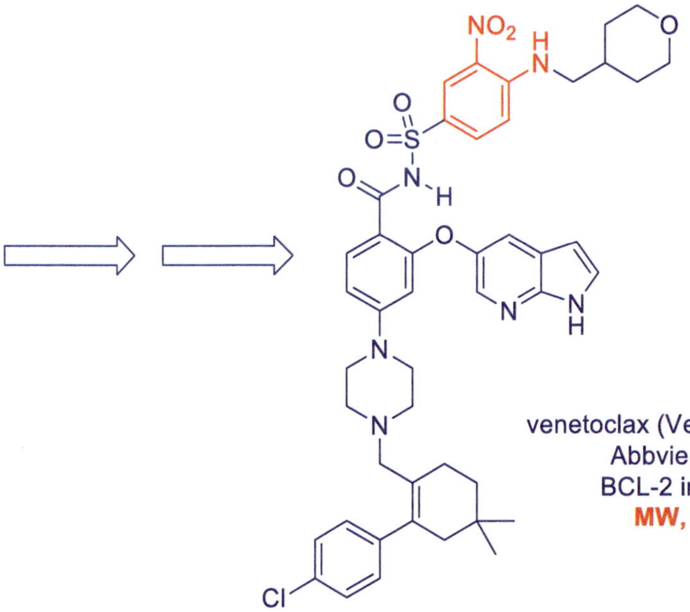

venetoclax (Venclexta, **120**)
Abbvie, 2016
BCL-2 inhibitor
MW, 868

Phenethanolamine aniline **121** is a lead compound for GSK's β_3 adrenergic receptor agonists program. Employing isosteric substitution of the carboxylic acid group, a series of functionally potent and selective β_3 adrenergic receptor agonists were identified as a potential treatment of obesity and type II diabetes. Although **121** was a reasonably potent agonist of human β_3 receptor with an IC_{50} <20 nM and a pEC_{50} value of 7.8, it still had appreciable activity against human β_1 and β_2 adrenergic receptors (β_1/β_3 ratio = 5). Several isosteres proved more potent and more selective than the parent carboxylic acid **121**. The pEC_{50} for acylsulfonamide **122** is 8.3 (β_1/β_3 ratio = 20); acylsulfonamide **123**, 9.1 (β_1/β_3 ratio = 1,000); sulfonylsulfonamide **124**, 8.8 (β_1/β_3 ratio = 79); sulfonylsulfonamide **125**, 8.8 (β_1/β_3 ratio >500); and sulfonylurea **126**, 7.8 (β_1/β_3 ratio = 25). One exception was sulfonylurea **126**, which is equipotent to the parent compound **121**.[58]

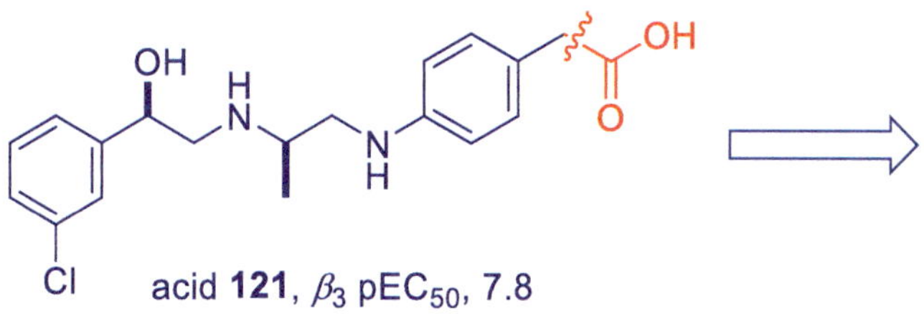

acid **121**, β_3 pEC$_{50}$, 7.8

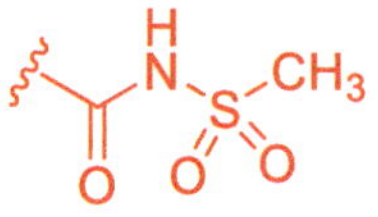

acylsulfonamide **122**
β_3 pEC$_{50}$, 8.3

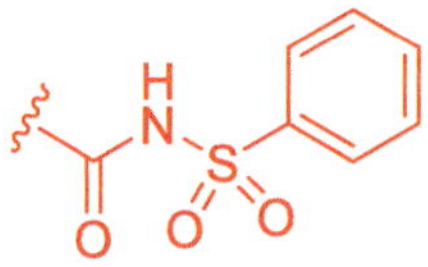

acylsulfonamide **123**
β_3 pEC$_{50}$, 9.1

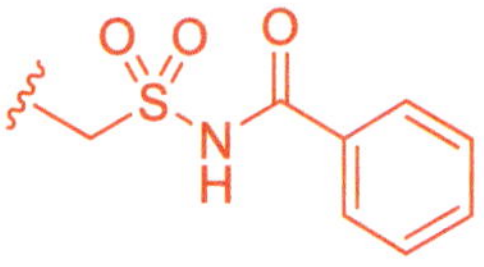

acylsulfonamide **124**
β_3 pEC$_{50}$, 8.0

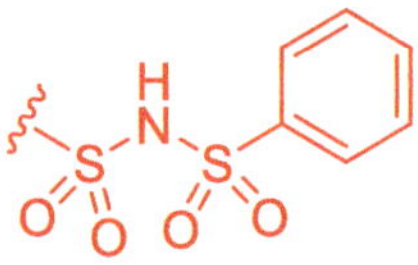

sulfonylsulfonamide **125**
β_3 pEC$_{50}$, 8.8

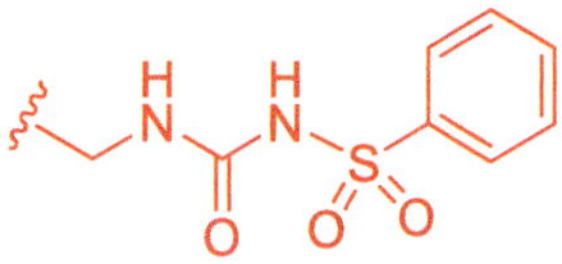

sulfonylurea **126**
β_3 pEC$_{50}$, 7.8

The *N*-acylsulfonamide group is a popular and well-known bioisostere for carboxylic acid because of their similar acidity. In the realm of prostaglandins, sulprostone (**128**) is an analog of prostaglandin E$_2$ (PGE$_2$, **127**), where *N*-acylsulfonamide replaced the carboxylic acid. Sulprostone (**128**) is a prostanoid receptor agonist being tested for gynecology applications.[59]

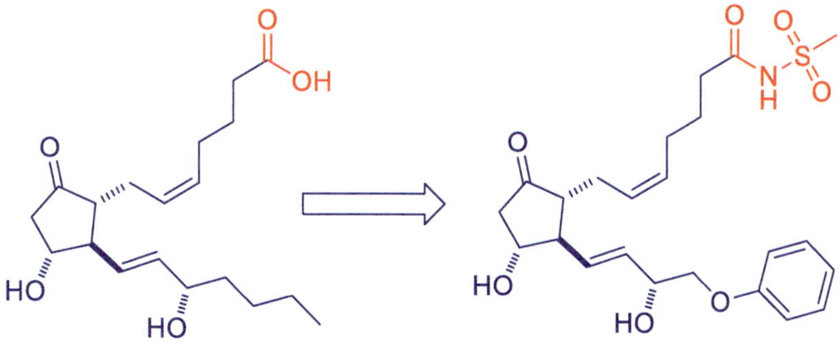

prostaglandin E2 (**127**)

sulprostone (**128**)

Also in the PGE$_2$ arena, acid **129** is a potent (K_i, 21 nM) and selective EP3 (among EP1–4) ligand with modest functional activity (IC$_{50}$ 580 nM). Replacing the acid with *N*-acylsulfonamide group afforded compounds **130** and **131**, respectively. The *N*-acylmethanesulfonamide **130** showed equipotent binding affinity with reduced antagonist activity. On the other hand, *N*-acylbenzenesulfonamide **131** was 40-fold more potent in the binding assay but had poor functional activity, which was attributed to protein binding since assay medium included 1% bovine serum albumin (BSA). Gratifyingly, decorating the phenyl ring with substitutions improved both potency and functional efficacy. Thus, the *N*-3,4-difluorobenzene-sulfonamide **132** and *N*-(3-cyanobenzene sulfonyl analog **133** proved to be 256-fold and 480-fold more potent in binding/function, respectively. The most optimized EP3 selective antagonist, *N*-3,4-difluorobenzene sulfonyl analog **132**, also exhibited potent *in vivo* efficacy, which was indicated as the inhibitory effect on the PGE_2-induced uterine contraction in pregnant rats.[60]

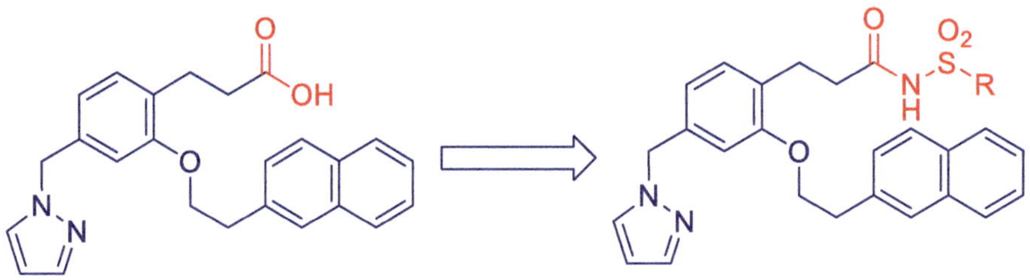

acid **129**, K_i (binding) = 21 nM
functional assay: IC_{50} = 580 nM

acylsulfonamide **130–133**

Compound	R	Binding K_i (nM)	Function IC_{50} (nM)
129	–	21	580
130	$-CH_3$	22	>10,000
131	–Ph	0.50	140
132	–3,4-diF-Ph	0.086	1.2
133	–3-CN-Ph	0.065	18

Cycling back to Bcl-2/Bcl-xL inhibitors, **134** effectively inhibited tumor growth but failed to achieve complete regression *in vivo*. Replacing the carboxylic acid on **134** with acylsulfonamide delivered bisacylsulfonamide **135** with vast improvement of potency and efficacy. It binds to Bcl-2 and Bcl-xL proteins with K_i values of <1 nM and inhibits cancer cell growth with IC_{50} values of 1–2 nM in four small-cell lung cancer cell lines sensitive to potent and specific Bcl-2/Bcl-xL inhibitors. Compound **135** is capable of achieving rapid, complete, and durable tumor regression *in vivo* at a well-tolerated dose schedule.[61]

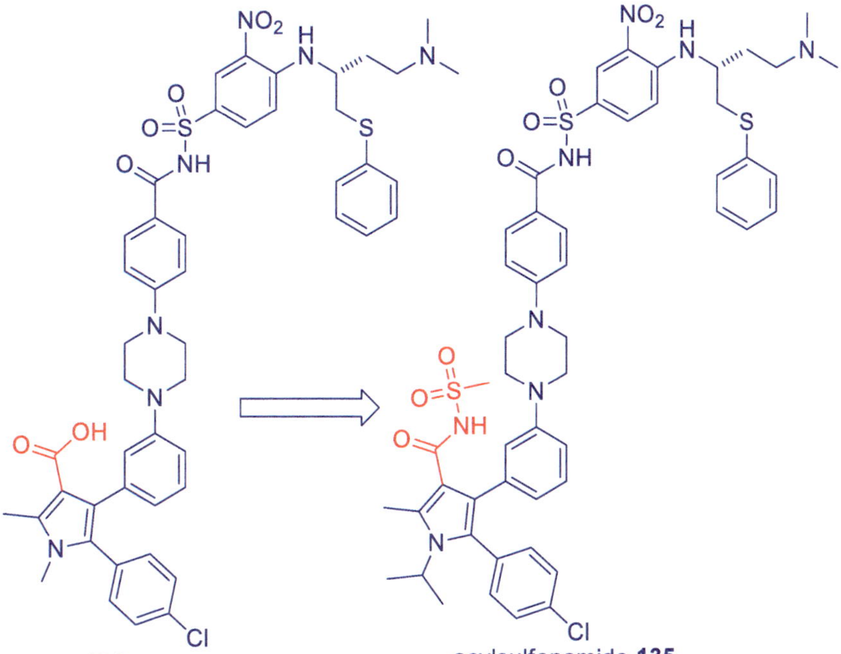

acid **134**
K_i = 1.3 nM to Bcl-2
K_i = 6 nM to Bcl-xL
IC_{50} = 61 nM in H146 cancer cell line
modest tumor growth inhibition

acylsulfonamide **135**
K_i < 1 nM to Bcl-2
K_i < 1 nM to Bcl-xL
IC_{50} = 1 nM in H146 cancer cell line
complete tumor growth inhibition

4.5.1.2 Cyclic Carboxylic Acid Isosteres

A plethora of cyclic carboxylic acid isosteres exist in the literature. Some of the more important examples are reviewed here.

γ-Aminobutyric acid (GABA) is one of the major inhibitory neurotransmitters in the mammalian central nervous system (CNS). When the GABA levels in the brain fall below a threshold level, convulsions begin. But GABA itself can only work when directly injected to the brain because it is a small, polar and hydrophilic molecule and it does not cross the BBB.

On the other hand, GABA aminotransferase (AT) has been a target to treat a wide variety of neurological disorders. Silverman sought to make more lipophilic GABA analogs as GABA-AT inhibitors and found that 2,6-difluorophenol fit the criteria nicely. While a "naked" phenol has a pK_a of 9.8, it drops to 7.1 when flanked by two fluorine atoms. Furthermore, fluorine atom is nearly as small as the hydrogen atom, therefore, may mimic the carboxylate carbonyl oxygen. At pH 8.5, the pH optimum for GABA aminotransferase, 2,6-difluorophenol would be completely ionized and may mimic a carboxylate ion.[62] To that end, Silverman prepared the para-analog **136** and the meta-analog **137**. Both compounds were found to be competitive inhibitors of GABA aminotransferase with K_i values of 6.3 and 11 μM, respectively. More important, the increased lipophilicity should facilitate crossing of the BBB.[63]

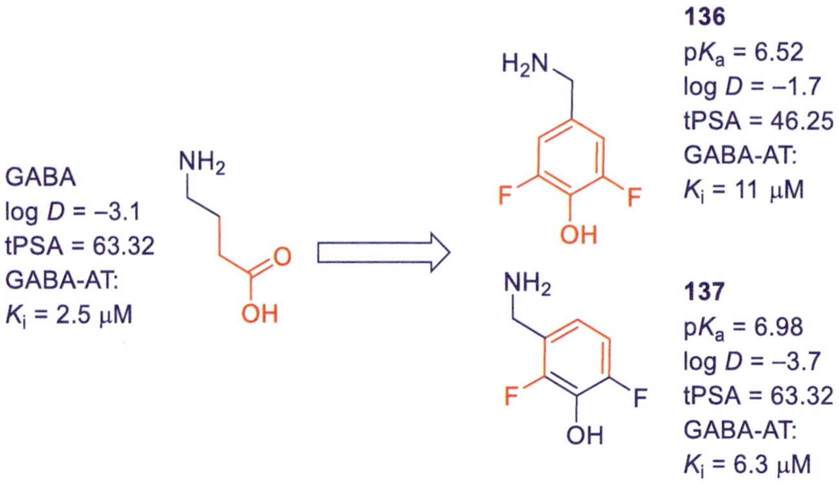

Isosteric substitution of acid with 2,6-difluorophenol also found success in the area of aldose reductase inhibitors as potential treatment of diabetes mellitus. Aldose reductase is the first and rate-limiting enzyme of the polyol pathway, reducing glucose to sorbitol using NADPH as cofactor. Pyrroloacetic acid **138** has an IC_{50} of 2.4 μM, but it is somewhat hydrophilic, with a log *P* value of 1.23. The 2,6-difluorophenol isostere **139** has a boosted lipophilicity with a log *P* value of 3.56. It was tested approximately five-time more potent than the parent acid **138**. Interestingly, inspection of low-energy conformation revealed that the distance of the geometric centers of the aromatic area of **139** and **138** (centroid) from the phenolic oxygen or the carbonyl were quite similar: 7.1 Å and 7.2 Å, respectively.[64]

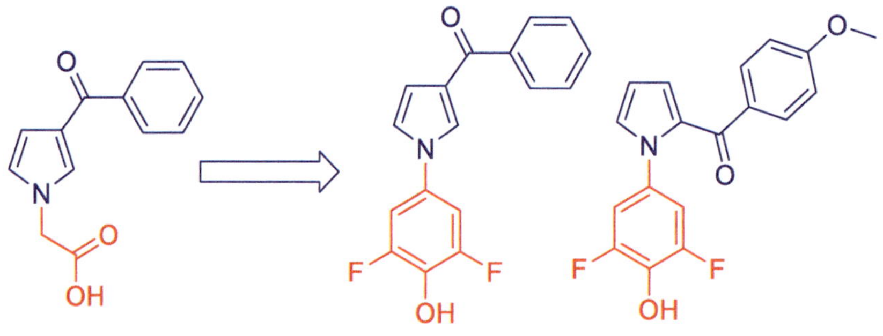

	138	**139**	**140**
Aldose reductase inhibition:			
IC_{50}	= 1.97 μM	396 nM	390 nM
log *P*	= 1.23	3.56	3.49

Additional efforts using 2,6-difluorophenol as an isosteric replacement of acid led to **140**. In addition to having similar potency and lipophilicity as those of **139**, the methoxyl substitution conferred a remarkable selectivity as well. Compound **140** was tested to be 72-fold more selective inhibiting rat lens aldose reductase receptor-2 in comparison to rat kidney aldose reductase receptor-1.[65]

For better or worse, 2,6-difluorophenol has become an obligatory isosteric replacement of carboxylic acid nowadays.

Oxetanes have become a mainstay in medicinal chemistry. For the household analgesic ibuprofen, oxetan-3-ol **141**, along with thietan-3-ol, and the corresponding sulfoxide and sulfone, were prepared and evaluated as the isosteric replacements of the carboxylic acid functionality. The acid on ibuprofen is negatively ionized in physiological conditions that is responsible for an insufficient passive diffusion across biological membranes. In stark contrast, oxetan-3-ol **141** and the sulfur derivatives are mostly neutral at physiological pH, and they are found to be comparatively more lipophilic and more permeable in parallel artificial membrane permeability assay (PAMPA) in comparison to ibuprofen. Given the relatively low acidity and high permeability, oxetan-3-ol **141** and the sulfur derivatives may be useful in the context of CNS drug design when isosteric replacement of the carboxylic acid is often mandated to improve the brain penetration of a drug candidate.[66]

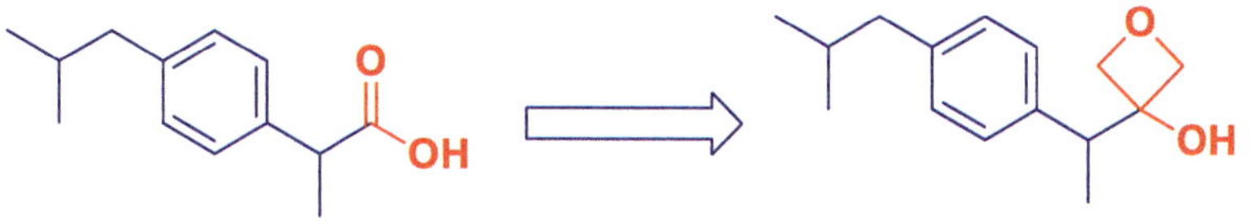

ibuprofen oxetan-3-ol **141**

A venerable isostere for acid is tetrazole (pK_a = 4.5–4.9). Herr reviewed medicinal chemistry and synthetic methods of 5-substituted-1*H*-tetrazoles as carboxylic acid isosteres in 2002.[67] Anionic tetrazoles are 10-fold more lipophilic than the corresponding carboxylate, yet they are more resistant to many biological metabolic degradation pathways, whereas carboxylate is prone to metabolism, especially secondary metabolism—conjugations.

Before the emergence of losartan (Cozaar, **147**), tetrazole was largely shied away as an acid isostere. Initially, carboxylic acid **142** as a nonpeptide angiotensin II receptor antagonist exhibited potent biological activity *in vitro*, but its effects were minimized upon oral administration. Isosteric replacements such as hydroxamic acid **143**, methylated hydroxamic acid **144**, acylsufonamide **145**, and sulfonamide **146** all proved inactive after oral dosage. Conspicuously, acylsufonamide **145** was very potent *in vitro* with an IC_{50} value of 83 nM, but showed little *in vivo* activity after oral administration. Gratifyingly, tetrazole **147** was not only drastically active *in vitro*, but also significantly active *in vivo* after oral administration.[68] Notably, five out of seven of the AT_1 receptor antagonists used clinically for the treatment of high blood pressure contain a tetrazole motif. The five drugs are valsartan (Diovan), irbesartan (Avapro), candesartan (Atacand), telmisartan (Micardis), and olmesartan (Benicar).

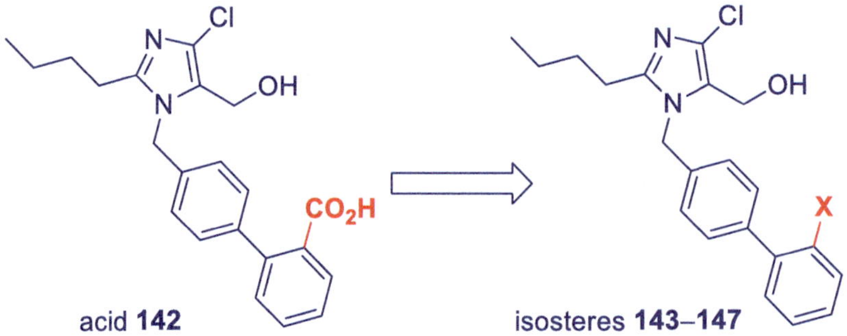

acid **142** isosteres **143–147**

Compound	X	pK_a	IC_{50} (μM)	iv dose	mg/kg, po
142	–CO_2H	5	0.23	3	11
143	–CONHOH	10.5	4.1	3	>30
144	–$CONHOCH_3$	10.9	2.9	10	Inactive
145	–$CONHSO_2Ph$	8.4	0.14	>3	30
146	–$NHSO_2CF_3$	4.5	0.083	10	100
147	Tetrazole	5.5	0.019	0.80	0.59

Clofibrate, as a peroxisome proliferator-activated receptor-α (PPARα) agonist, was one of the early drugs used to lower plasma cholesterol and triglyceride levels. But it is an ethyl ester prodrug of clofibric acid (**148**), clofibrate's active metabolite. Clofibric acid (**148**)'s isosteric substitution tetrazole **149** was prepared and was found to be ~2.5-times more effective inhibiting the enzyme 11β-hydroxysteroid dehydrogenase type 1 (11β-HSD1) than clofibric acid (**148**). The inhibition of 11β-HSD1 has been shown to reduce the glucose serum levels in diabetic patients due to the low production of cortisol.[69]

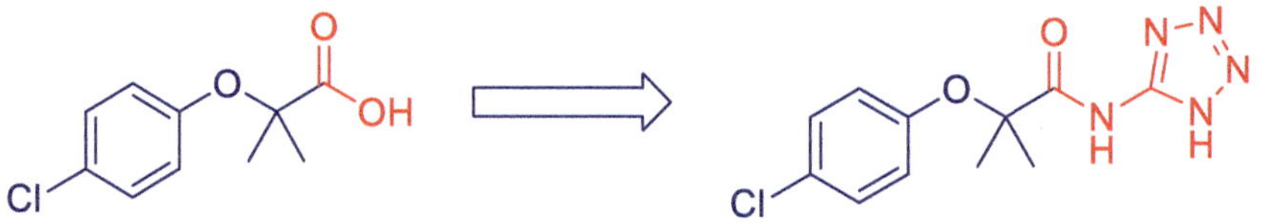

clofibric acid (**148**) tetrazole **149**

A team of Vertex chemists explored several isosteric replacements of the carboxylic acid of drug candidate VX787 (**150**). Compounds **150–158** were tested in two cellular assays: a phenotypic cell production assay (CPE) and a branched DNA assay (bDNA).[70]

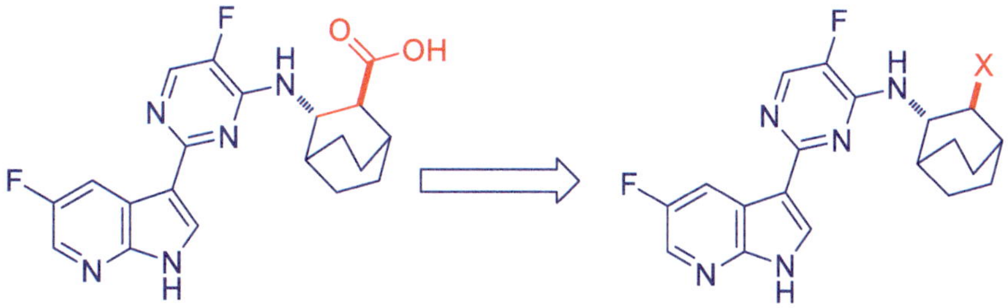

VX-787 (**150**) isosteres (**151–158**)

#	X	pK_a	CPE IC_{50} (nM)	bDNA EC_{99} (nM)	LLE*
150	$-CO_2H$	4.7	2	11	5.6
151	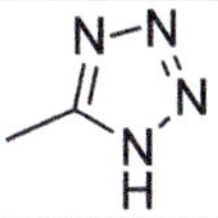	6.0	25	180	4.5
152	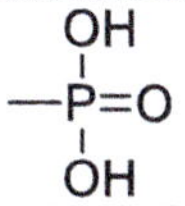	2.4	91	290	6.7
153	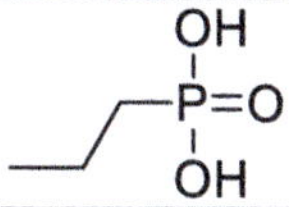	2.5	200	2,100	5.4
154	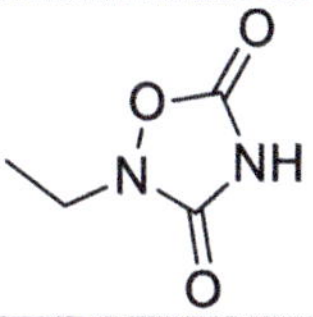	5.2	48	210	4.0
155	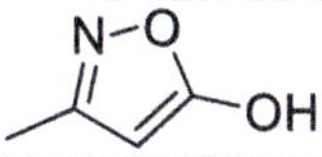	4.7	72	62	5.3
156	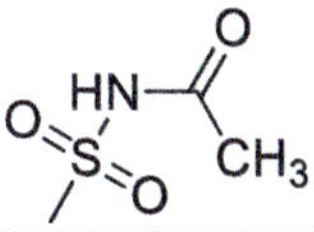	5.5	19	160	5.9
157	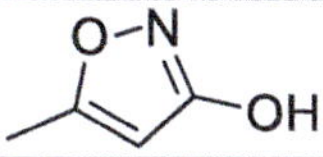	4.8	1	33	4.7
158	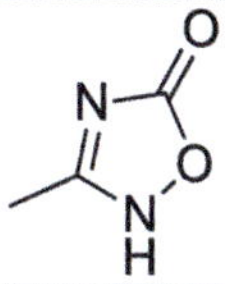	7.3	ND	140	3.9

*LLE = pEC_{50}(bDNA) – Clog *P*

As showcased by examples **154**, **155**, **157**, and **158**, many five-membered heterocycles are useful isosteric replacements of carboxylic acid. In addition to tetrazoles, they may be divided into the following four classes. Class **I**, isoxazoles (X = O) and isothiazoles (X = S); Class **II**, oxazolidinediones (X = O); thiazolidinediones (X = S); Class **III**, 5-oxo-1,2,4-oxadiazole (X = O, Y = O), 5-oxo-1,2,4-thiadiazole (X = S, Y = O), 5-thioxo-1,2,4-oxadiazole (X = O, Y = S); and Class **IV**, tetramic acids (X = NH), tetronic acids (X = O), and cyclopetane-1,3-dione (X = CH_2). A word of caution with regard to Classes **I**–**III**, the sulfur-containing isosteres are known structural alerts, may be associated with safety issues.[56]

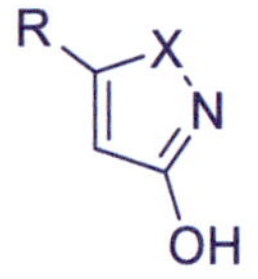

I
isoxazoles (X = O)
isothiazoles (X = S)

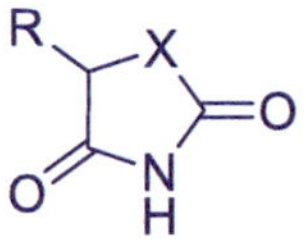

II
oxazolidinediones (X = O)
thiazolidinediones (X = S)

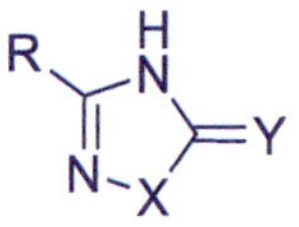

III
5-oxo-1,2,4-oxadiazole (X = O, Y = O)
5-oxo-1,2,4-thiadiazole (X = S, Y = O)
5-thioxo-1,2,4-oxadiazole (X = O, Y = S)

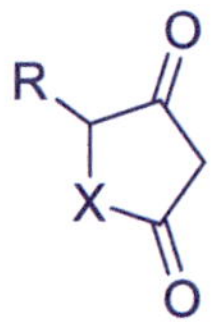

IV
tetramic acids (X = NH)
tetronic acids (X = O)
cyclopentane-1,3-dione (X = CH_2)

3-Hydroxy-isoxazole ring (Class **I**) has a comparable acidity to that of a carboxylic acid function and was one of the earlier isosteres employed to replace carboxylic acid. For instance, isosteric substitution of the acid group on γ-aminobutyric acid (GABA) led to conformationally more restrained agonist 4,5,6,7-tetrahydroisoxazolo[5,4-c]pyridin-3-ol (THIP, gaboxadol, **159**). As a GABA agonist, THIP (**159**) has a pK_a of 4.4, similar to that of GABA (4.23). It is also capable of penetrating the blood-brain barrier (BBB) and it is also stable *in vivo*.[71] Lundbeck took gaboxadol (**159**) to clinics as a potential treatment for insomnia. But it was hit with a double whammy: concerns over safety and efficacy, thus the clinical trials were discontinued.

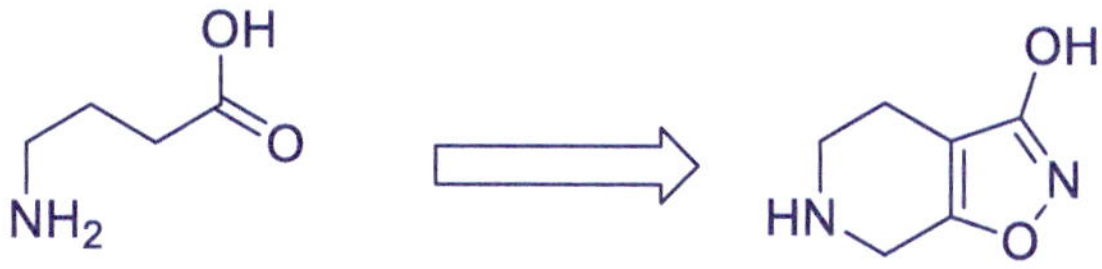

GABA THIP (**159**)

With regard classes **II**, **III**, and **IV**, a sample of their structure–property relationship is tabulated below from the data procured by Huryn, Ballatore and co-workers.[72]

	Aq. Sol. (μM)	log $D_{7.4}$	PAMPA (cm/s)	pK_a	PPB (% fu)
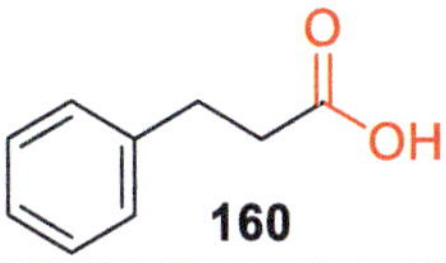 **160**	110.69	–0.49	1.66 * 10^{-6}	4.64	9.5
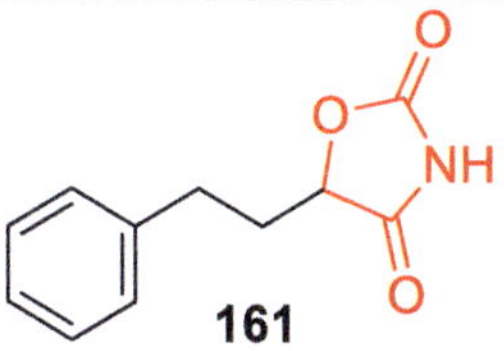 **161**	≥200	–0.16	2.46 * 10^{-6}	6.63	14
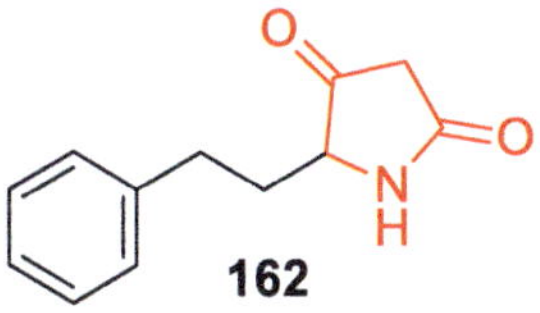 **162**	≥200	–0.35	2.50 * 10^{-6}	6.08	ND
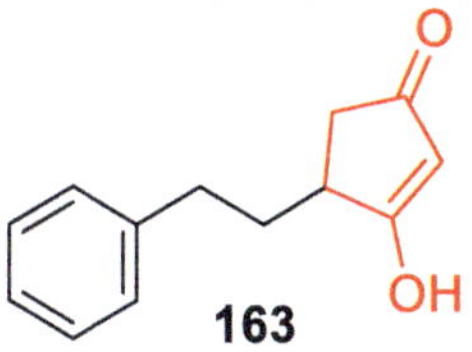 **163**	194.93	–0.70	2.12 * 10^{-6}	4.01	7.96

Along with tetrazole and trifluoromethylsulfonamide, squaric acid **165**, with a pK_a value of ~0.37, was examined as an isostere for carboxylic acid **165** in the context of angiotensin II antagonists. Biochemically, it was found to be approximately 10-fold more potent than the acid congener, although not as potent as the tetrazole derivative. Squaric acid **165** was tested active in Goldblatt hypertensive rats following oral administration. It showed a long-lasting effect, although its efficacy was lower than that of tetrazole analog.[73]

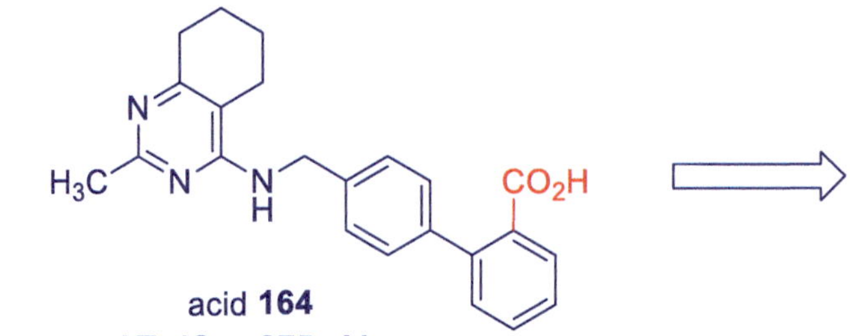

acid **164**
AT_1 IC_{50}, 275 nM

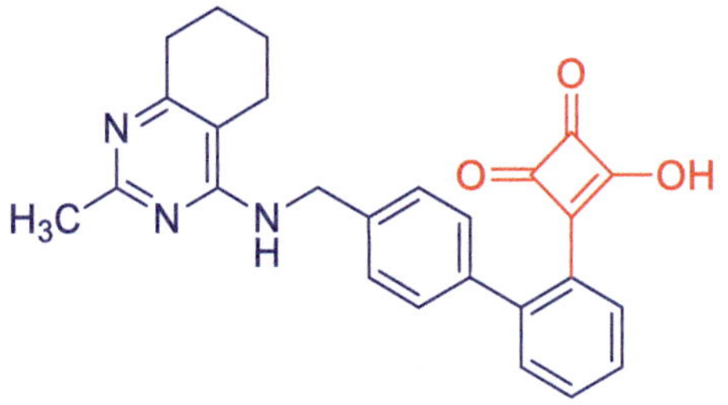

squaric acid **165**
AT_1 IC_{50}, 275 nM

4.5.2 Hydroxamic Acid

Three hydroxamic acid-containing drugs **166–168** have been approved by the FDA for marketing. All three of them are histone deacetylase (HDAC) inhibitors for the treatment of cutaneous T cell lymphoma (CTCL) and peripheral T-cell lymphoma (PTCL).[74]

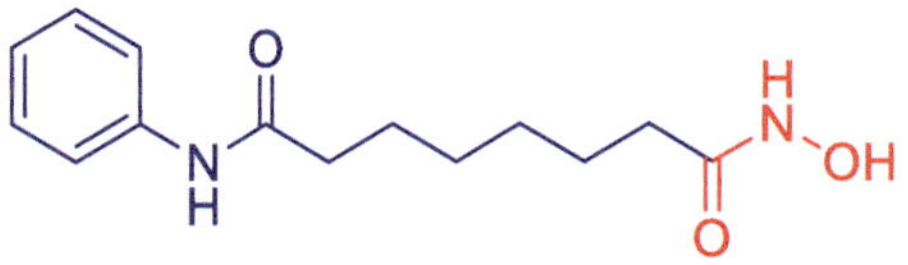

vorinostat (SAHA, Zolinza, **166**)
Merck, 2006
HDAC inhibitor

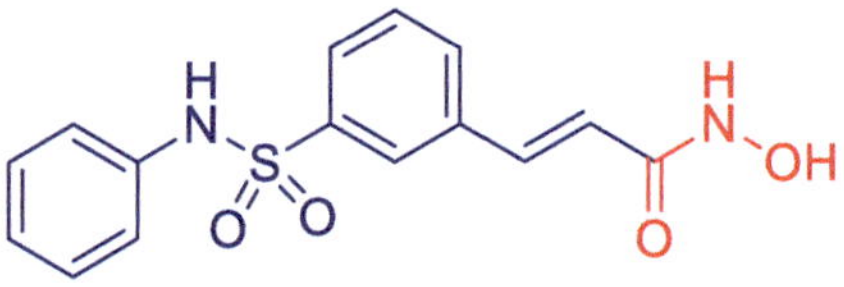

belinostat (Beleodaq, **167**)
TopoTarget, 2014
HDAC inhibitor

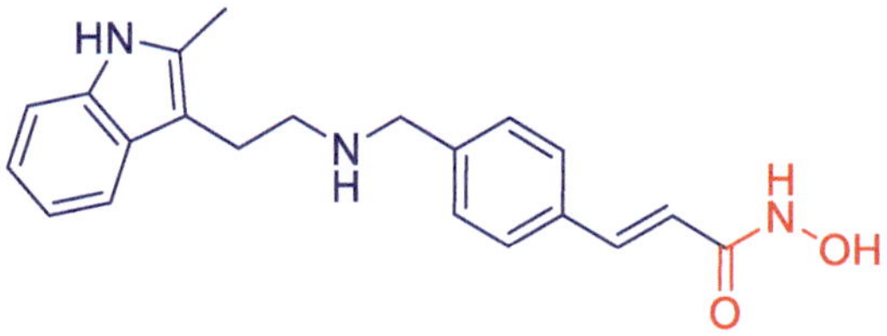

panobinostat (Farydak, **168**)
Novartis, 2015
HDAC inhibitor

Regrettably, all three of them have narrow therapeutic indices (TIs), i.e., these drugs′ pharmacological effects are closely associated with their toxicity with regard to their dosages. This is inherently associated with the hydroxamic acid functionality. As shown beneath, in addition to being hydrolyzed to the corresponding carboxylic acid, another major metabolic pathway of hydroxamic acid under physiological conditions is being sulfated or acetylated and then undergoing the Lossen rearrangement to afford the isocyanate intermediate. Our sophomore organic chemistry taught us that isocyanate is very reactive to nucleophiles to form covalent bonds, which might offer an explanation of the toxicities associated with narrow TIs of hydroxamic acid-containing drugs.

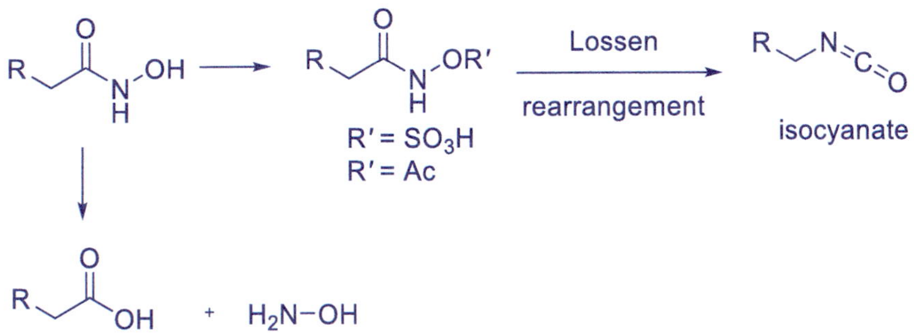

Tumor necrosis factor-α (TNF-α)-converting enzyme (TACE) is a membrane-bound zinc metalloprotease. Like many matrix metalloprotease (MMP) inhibitors, initial TACE inhibitors were hydroxamate ligands such as hydroxamates, reverse hydoxamates, and *N*-hydroxyureas.

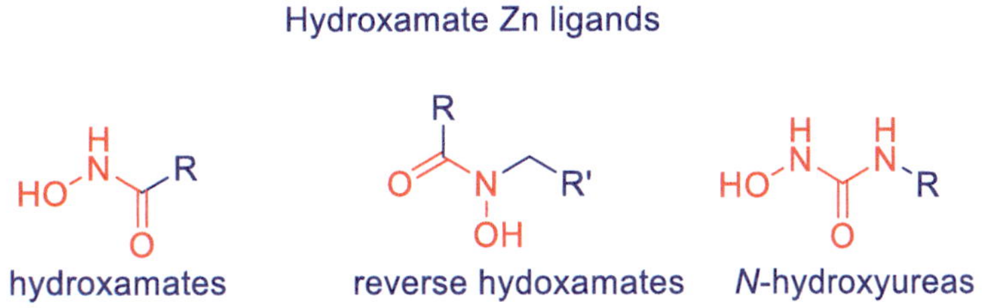

To steer away hydroxamate-associated poor pharmacokinetic and metabolic liabilities, many nonhydroxamate zinc-binding groups have been explored as zinc metalloprotease inhibitors. They include thiadiazoles, phosphonates, thiols, nitropyrimidines, 6*H*-1,3,4-thiadiazines, carboxylates, barbituates, and rhodanines. (For rhodanines, see Section 5.5 on the concept of PAINS, pan-assay interference compounds.)

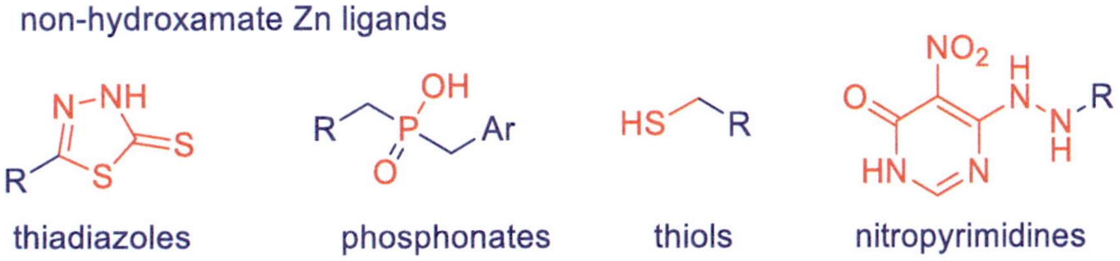

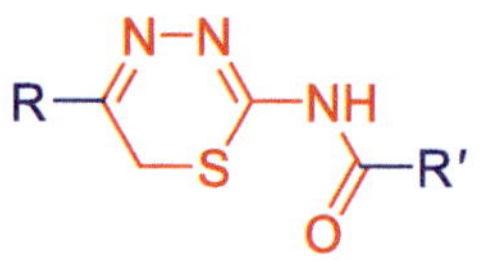

6*H*-1,3,4-thiadiazines

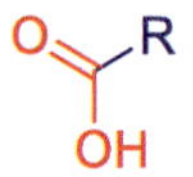

carboxylates

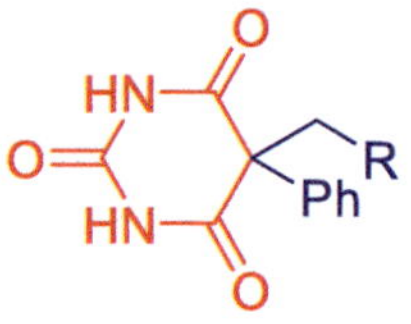

barbituates

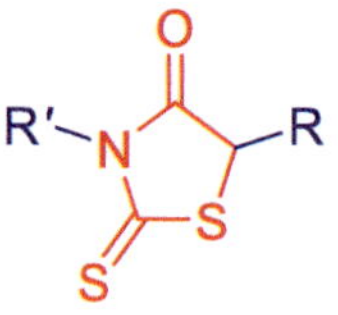

rhodanines

Sheppeck et al. at BMS sought to investigate more drug-like non-hydroxamate zinc-binding groups. Using hydroxamate IK682 (**169**) as a reference compound, they discovered that hydantoin **170**, triazolone **171**, and imidazolone **172**, with the P_1' fragment constant, showed good TACE inhibitory activities. It was expected that the relatively lower potency by the heterocyclic isosteres **170**–**172** will compensate for the intrinsic toxicities associated with hydroxamates such as **169**.[75]

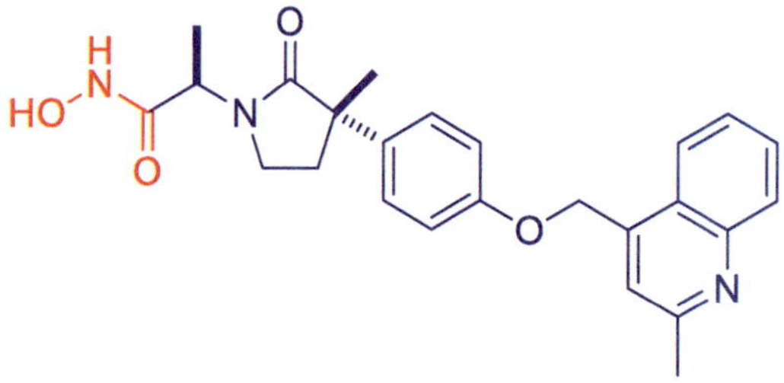

IK682 (**169**)
pTACE IC_{50}, 1 nM

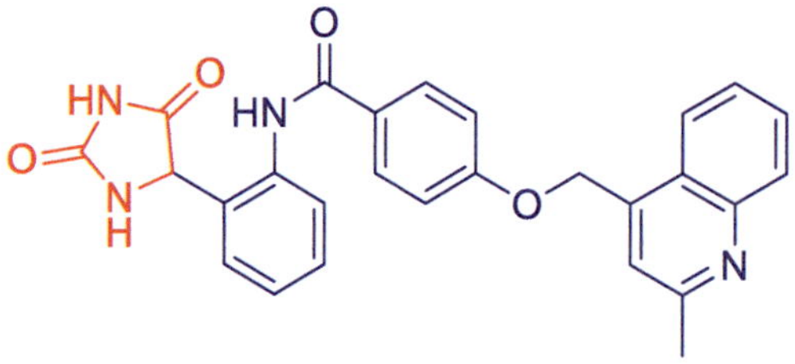

hydantoin **170**
pTACE IC_{50}, 14 nM

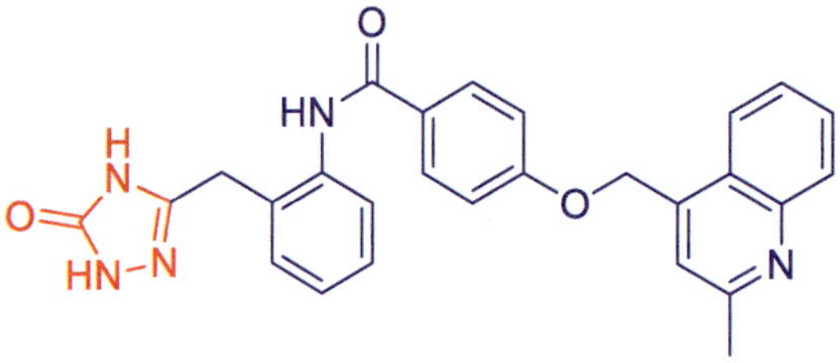

triazolone **171**
pTACE IC_{50}, 34 nM

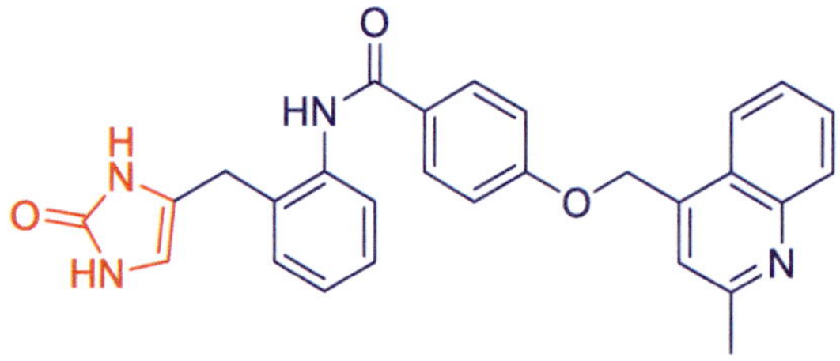

imidazolone **172**
pTACE IC_{50}, 9 nM

Celecoxib (Celebrex, **173**) is a well-known cyclooxygenase-2 (COX-2) inhibitor. In fact, it is the only COX-2 inhibitor that remains on the market today. It was theorized that dual inhibitors for COX-2 and 5-lipoxygenase would have improved efficacy since both of the oxygenases are associated with the arachidonic acid pathway. Incorporation of a strongly chelating bidentate ligand, hydroxamic acid, led to hydroxamate **174**, which was tested as a dual COX-2/5-lipoxygenase (5-LOX) inhibitor. To avoid known liabilities associated hydroxamates, the $CONCHF_2$ fragment was prepared as a cyclic hydroxamic acid mimetic on **175**.[76] Although there is a substantial buildup of negative potential around the two fluorine atoms, the aliphatic fluorine seldom acts as a hydrogen-bond acceptor, presumably because of its high electronegativity and low polarizability. It is plausible that the CHF_2 group may interact with positive charged region on the enzyme that may contribute to enhanced affinity and competitive reversible inhibition of COX and/or 5-LOX.

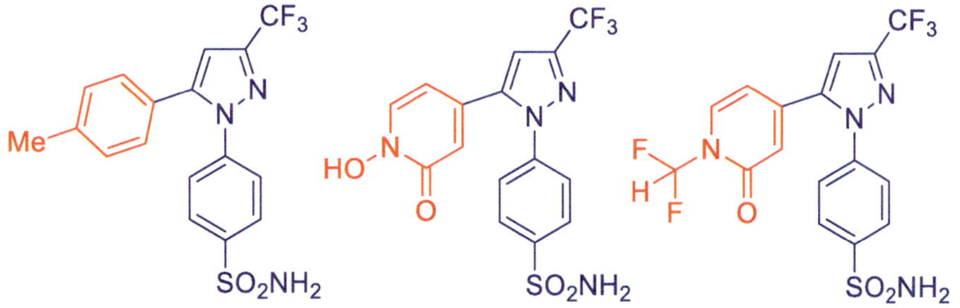

	celecoxib (Celebrex, **173**)	hydroxamate **174**	difluoromethyl derivative **175**
COX-1, IC_{50} (μM)	7.7	10.2	13.1
COX-2, IC_{50} (μM)	0.12	7.5	0.69
5-LOX, IC_{50} (μM)	NA	4.9	5.0
ED_{50} (mg/kg)	10.8	99.8	22.7

The resulting difluoromethyl derivative **175** was tested active *in vivo* anti-inflammatory carrageenan-induced rat foot paw edema model with an ED_{50} of 22.7 mg/kg. It compares favorably to the ED_{50} of 10.8 mg/kg for celecoxib (**173**). It is possible that difluoromethyl derivative **175** had better pharmacokinetic properties in comparison to hydroxamate **174**.[76]

4.5.3 Ester and Amide

Numerous isosteric replacements for amides and esters exist in medicinal chemistry. The table beneath lists some key representative amide and ester bioisosters.[77,78]

It may be a surprise to some, but a retroamide could be considered a viable bioisostere of an amide. The most successful utility of this maneuver might be the case of atenolol (**177**). ICI's β-adrenergic receptor antagonist (β-blocker) practolol (**176**) was plagued by idiosyncratic toxicities including skin lesion and lacrymal gland fibrosis.[79,80]

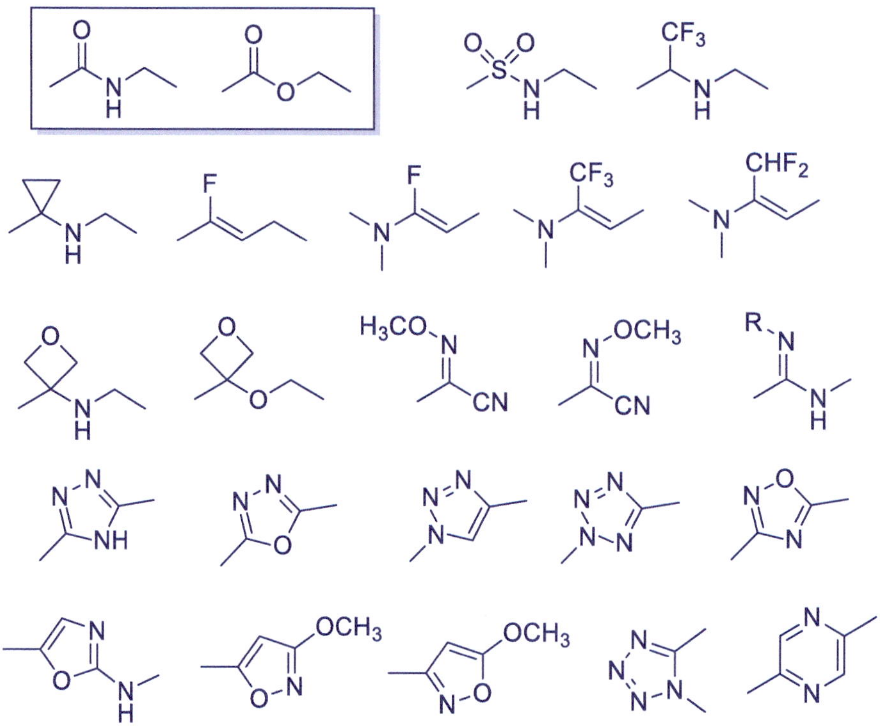

The ^{13}C-labeled practolol (**176**) binds to tissue proteins irreversibly presumably arisen from the metabolism.[81] Its anilide phenol ether moiety is prone to be oxidized to the corresponding quionone–imine reactive metabolite, which is readily captured by nucleophiles on tissue proteins. A simple switch of the anilide functional group on practolol (**176**) to the "retroamide" motif on atenolol (**177**) significantly lowered the toxicity level. Thirty years after its launch, atenolol (**177**), despite being widely prescribed, is an extraordinarily safe cardiovascular drug for treating hypertension.

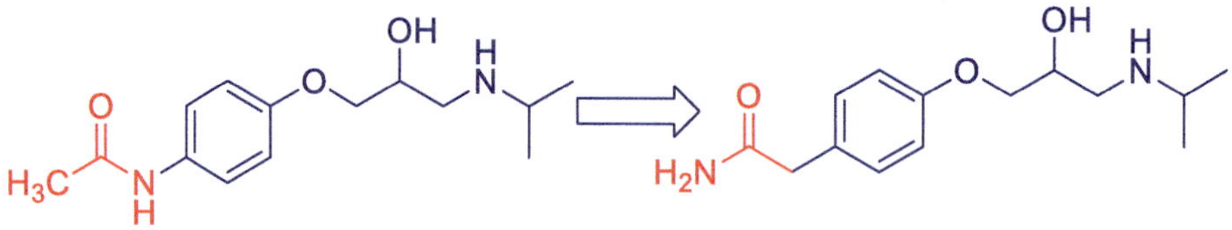

practolol (**176**) atenolol (**177**)

Since oxetane is "lipophilicity neutral," aminooxetane **179** may serve as a more lipophilic bioisostere of amide **178**.[82] More importantly, unlike other peptidic bond that is readily hydrolyzed by peptidases, aminooxetane is resistant to such onslaught by peptidases. Yet, the new class of pseudo-dipeptides still maintain the same H-bond donor/acceptor pattern.

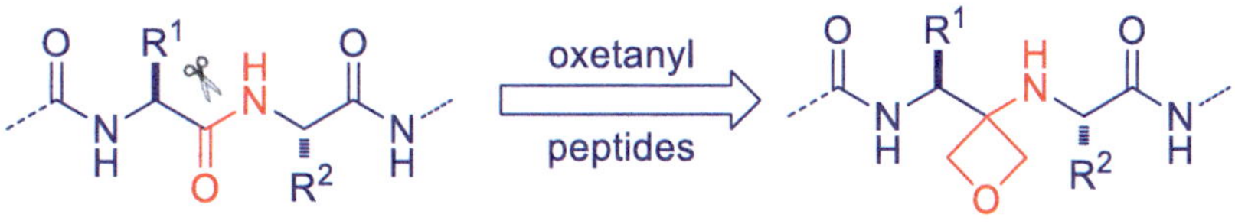

peptide (**178**) amino-oxetane (**179**)

On a separate note, trifluoroethylamine was a successful bioisostere of amide in the field of cathepsin K inhibitors for the treatment of osteoporosis.[83–85] Compound L-006235 (**180**) was bioavailable but had poor selectivity for cathepsin K versus cathepsins B, L, and S due to its lysosomotropic nature. The trifluoroethylamine substitution of the amide group led to L-873724 (**181**), which had good selectivity but low bioavailability because of its fast clearance as the consequence of CYP450 oxidation of the *t*-butyl group and the α-methylene of the nitrile group. Installation of a fluorine atom and a cyclopropyl group resulted in odanacatib (**182**). Regrettably, Merck discontinued its development in 2016 when increase of stroke was discovered during the Phase III trials.

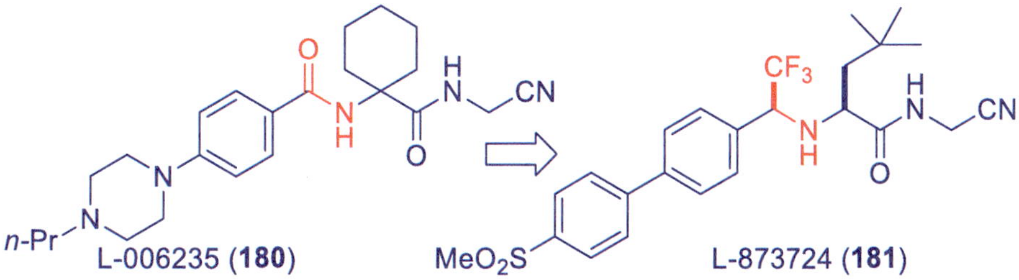

L-006235 (**180**) L-873724 (**181**)

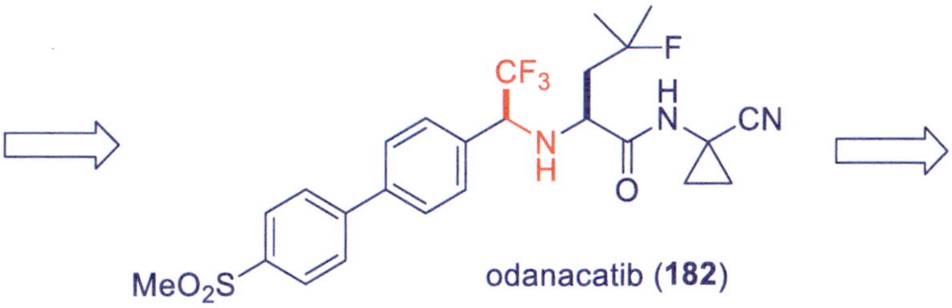

odanacatib (**182**)

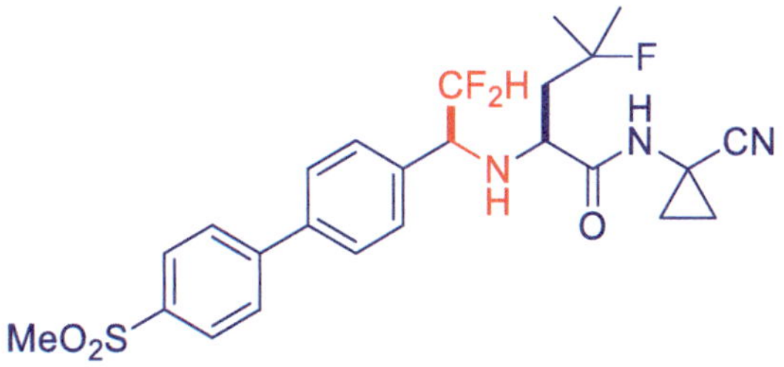

difluormethyl analogue (**183**)

Along a similar line, difluoromethyleneketone has been reported to serve as an isostere of amide.[86] For example, peptide **184** is a substrate of proteolytic enzymes of the scissile amide bond. Replacement of the amide bond with a difluoromethyleneketone has generated a number of potent transition-state type inhibitors (transition-state mimetics). The ability of difluoromethyleneketones to occupy additional binding sites on the leaving group side represents an advantage in terms of potentially increased affinity and selectivity. Difluoromethyleneketone retroamides **185** are the type E inhibitors for the inactivation of proteolytic enzymes.

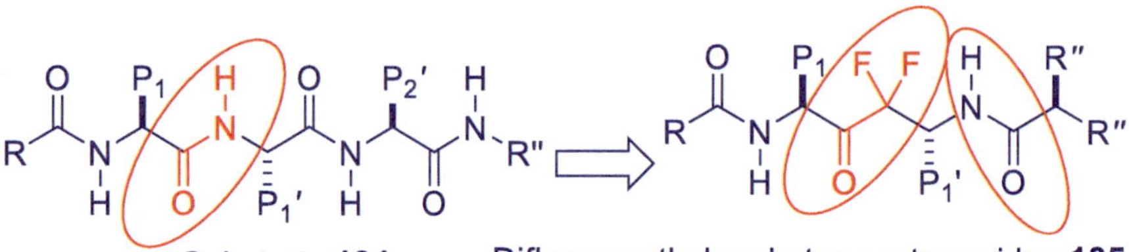

Substrate **184** Difluoromethyleneketone retroamides **185**
Type E Inhibitors

The most prevalent and fruitful isosteres of amides are probably heterocycles. For instance, a triazole to replace amide was successful for inhibitors of *Chlamydia trachomatis* infectivity.[87] Initially thiazolino 2-pyridone amide **186** attenuated *C. trachomatis* infectivity (EC_{50}, 60 nM) without affecting host cell or commensal bacteria viability. To replace the hydrolyzable amide bond on **186**, extensive SAR was carried out to evaluate the feasibility of bioisosteres including sulfonamide and numerous five-membered heterocycle rings. The heterocycles investigated included imidazole, oxazole, thiazole, oxadiazole, and all four possible triazoles. Triazole **187** saw five-fold boost of efficacy, leading to a highly potent 1,2,3-triazole-based infectivity inhibitor (EC_{50} 13 nM). The triazoles are more resistant to hydrolysis in comparison to the original amide bond. Furthermore, the same maneuver employing 1,2,3-triazole to replace an amide bond (IC_{50}, 6.0 μM) was successful in achieving low nanomolar HIV-1 viral infectivity factor (Vif) antagonists.[88] In contrast, the corresponding oxadiazole analog only had an IC_{50} of 6.8 μM.

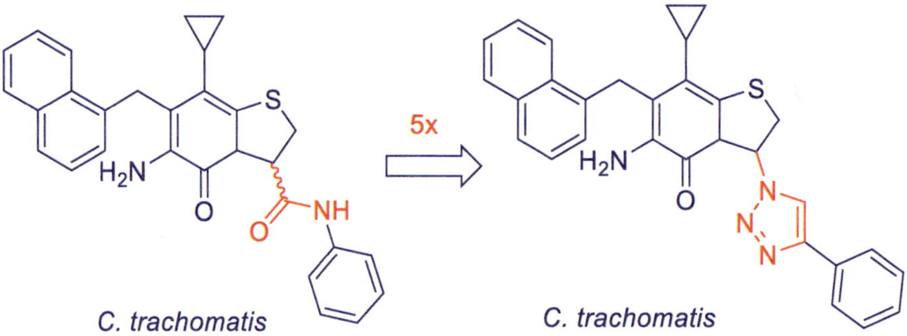

C. trachomatis
186, EC_{50} ~ 60 nM

C. trachomatis
187, EC_{50} ~ 13 nM

A potent and selective PPARδ modulator **188** displayed significantly improved pharmacokinetic properties relative to previously reported compounds. Scrutiny of the X-ray structure of **188** bound to PPARδ ligand-binding domain (LBD) revealed that its carbonyl and the *N*-methyl group adopt a thermodynamically less favorable *cis*-relationship. Therefore, it was reasoned that a "locked *cis*-amide" conformation would confer improved PPARδ binding. To that end, seven 5-membered heterocycle isosteres including thiazole, pyrazole, isoxazole, triazole, etc., were prepared to replace the amide. One of them, imidazole **189** was tested >10,000-fold selective for activation of PPARδ over both PPARα and PPARγ. In addition, imidazole **189** was selective against other nuclear receptors and CYP450 enzymes. Moreover, it had good pharmacokinetic properties to enable further investigations as potential treatment for Duchenne muscular dystrophy.[89]

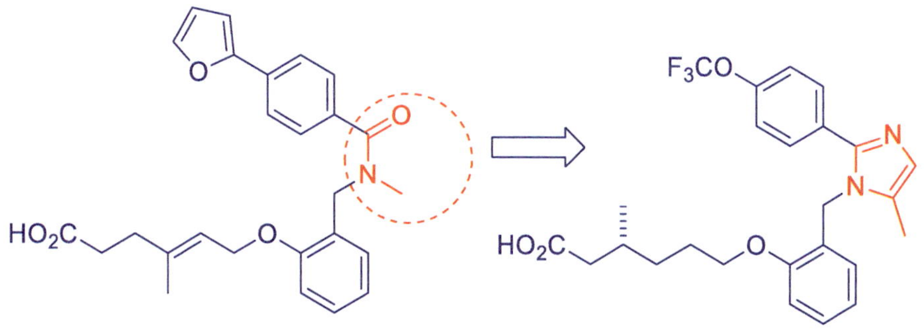

amide **188**
cis-amide conformation in protein-bound X-ray
PPARδ EC_{50} = 37 nM
PPARα EC_{50} = 6,100 nM

imidazole **189**
PPARδ EC_{50} = 0.4 nM
PPARα EC_{50} = 6,900 nM
good oral bioavailability

With regard to anilides, it was found, on occasions, to be advantageous to convert the conformationally flexible amide bond to a rigid "tied up" heterocycles. $\alpha_4\beta_1$ Integrin very late antigen-4 (VLA-4) antagonist **190** was potent but was poorly absorbed (*F*%, 0.7%). It was speculated that the hydrophilic nature of the anilide bond might be responsible for the poor bioavailability.[90] Excellent potency was retained for the benzoxazole and the benzimidazole derivatives where a hydrogen bond acceptor was appropriately positioned to mimic the amide bond oxygen. The most fruitful permutation was arrived when the hydrogen bond donor (N–H) was deleted in the form of oxazole **191**, which was identified as a potent, specific, and bioavailable VLA-4 antagonist (*F*%, 17%).

As far as esters are concerned, they are readily hydrolyzed by esterases. Therefore, esters are often replaced by the corresponding amides.

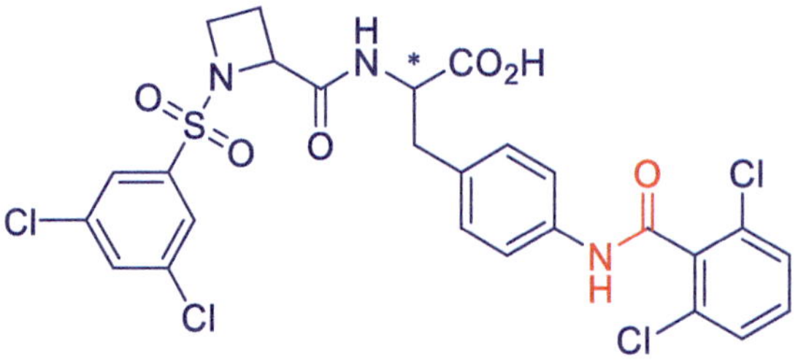

anilide (**190**), poor oral bioavailability (*F*%, 0.7%)

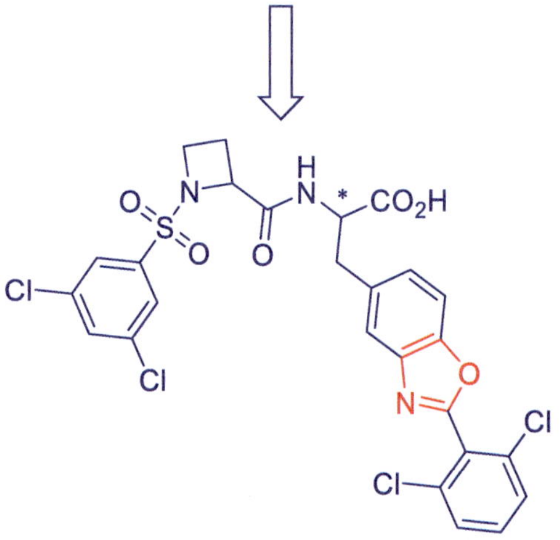

benzoxazole (**191**, *F*%, 17%)

Not surprisingly, heterocycles[91] have been shown to be productive bioisosteres for the ester functional group as well.

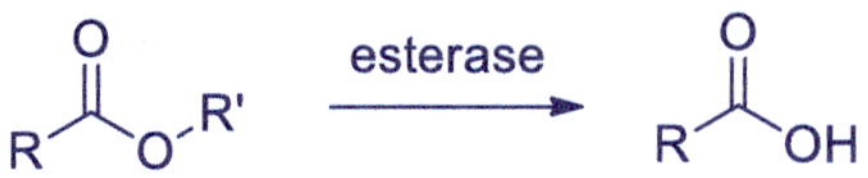

Good replacements:

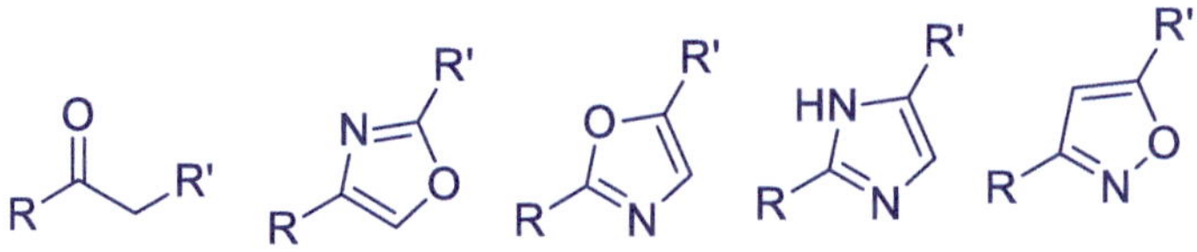

1,4-Benzodiazepine ester **192** is a novel γ-aminobutyric acid type A ($GABA_A$) receptor ligand. While it showed anxiolytic-like effects with reduced sedative/ataxic liabilities, its efficacy was low even at the dose of 30 mg/kg. Gratifyingly, 1,3-oxazole analog **193** was discovered from a series of six bioisosteres with significantly improved pharmacokinetics and pharmacodynamics properties even at the 10 mg/kg dose as compared to **192**.[92]

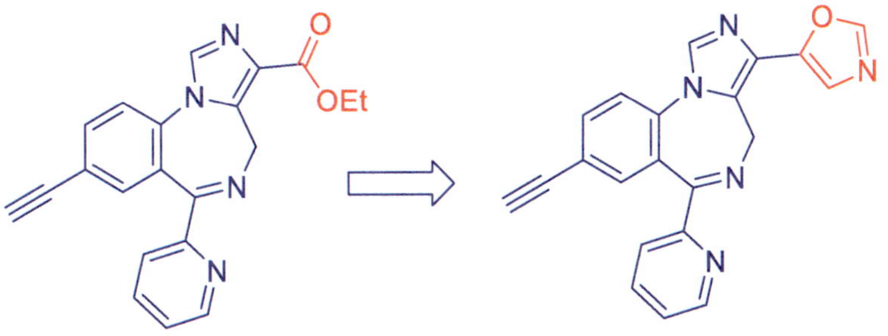

ester **192** oxazole ester bioisostere **193**

4.5.4 Urea, Guanidine, and Amidine

There is nothing inherently wrong with urea-containing drugs because quite a few FDA-approved drugs do contain the urea moiety. Three kinase inhibitors, sorafenib (Nexavar, **194**), regorafenib (Stivarga, **195**), and lenvatinib (Lenvima, **196**), have been in the arsenal of targeted anticancer drugs.

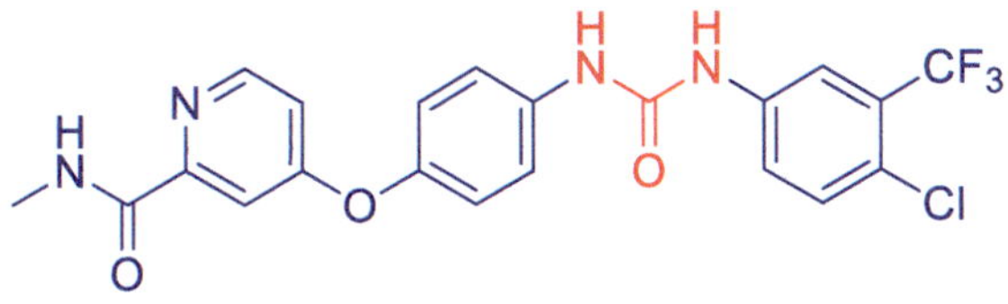

sorafenib (Nexavar, **194**)
Bayer/Onyx, 2007
PDGF and VEGF inhibtor

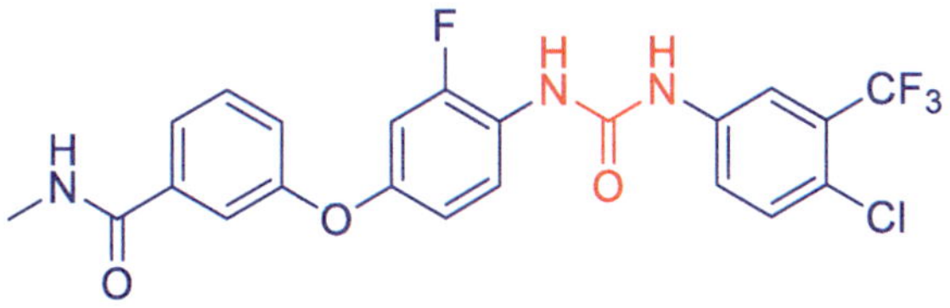

regorafenib (Stivarga, **195**)
Bayer, 2012
VEGFR/TIE2 inhibior

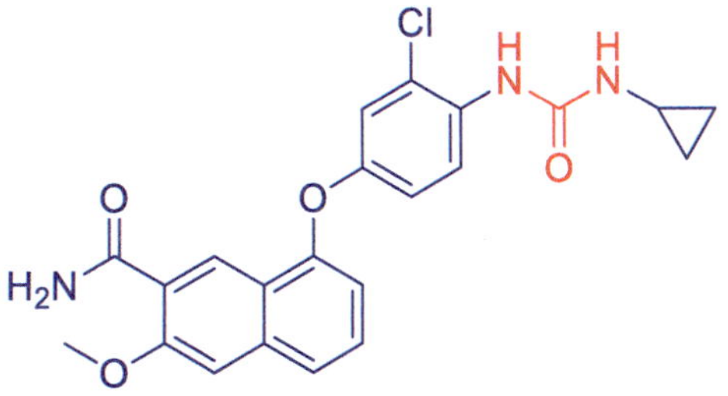

lenvatinib (Lenvima, **196**)
Eisai, 2015
VEGFR and FGFR inhibitor

However, ureas make several hydrogen bonds by behaving as both a hydrogen bond donor and an acceptor. Too many hydrogen bonding is a prescription for poor aqueous solubility, among other shortcomings. Therefore, urea isosteres are often sought to alter the drug's pharmacological and physio-chemical properties.

Guanidine, with a pK_a value of 13, is too basic to cross cell membrane thus often has very poor oral bioavailability. The first logic isostere is thiourea. In pursuit of their histamine-2 selective antagonists to treat ulcer, Smith-Kline French chemists initially arrived at guanylhistamine (**197**), which may be considered as a urea isostere. To ameliorate the basicity issue, its thiourea isostere, burimamide (**198**) was prepared. Unfortunately, thiourea **198** was not bioavailable.

An additional methyl on the imidazole moiety afforded metiamide (**199**) with bioavailability, but it was proven to cause agranulocytosis despite the fact that it was ten-time more potent than burimamide (**198**). It turns out that agranulocytosis, a dangerous depression of the production of infection-fighting white cells in bone marrow, is frequently associated with thioureas. This is why thioureas are considered structural alerts, which have potential to cause toxicities. Luckily, Smith-Kline French eventually came up with cimetidine (Tagamet, **200**). Substituting the thiourea with cyanoguanidine as an isostere obliterated the agranulocytosis issue. Cimetidine (Tagamet, **200**) went on to become the first blockbuster drug, ever.[93]

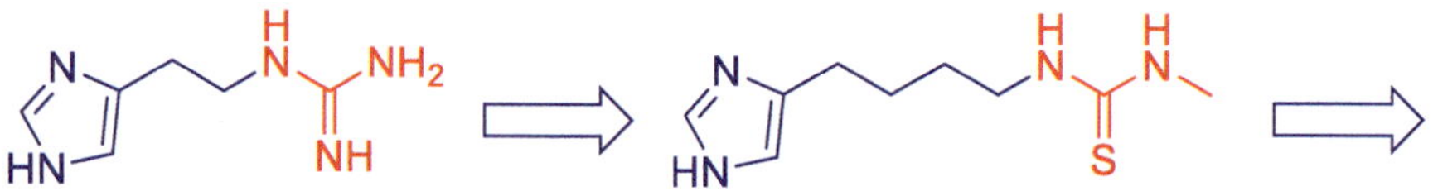

guanylhistamine (**197**)
pK_a = 13, too basic

burimamide (**198**)
NOT Bioavailable

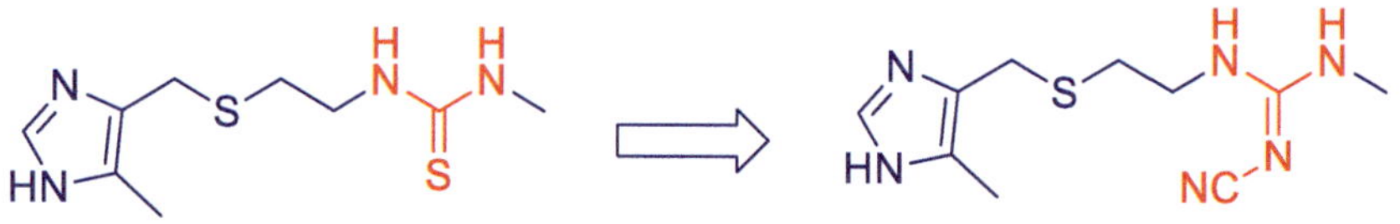

metiamide (**199**)
caused agranulucytosis

cimetidine (Tagamet, **200**)
the first blockbuster drug

Using Smith-Kline French's burimamide (**198**) as their jumping point, Allen & Hanbury, under Sir David Jack's leadership, sought to find a "me-too" histamine-2 selective antagonist to treat ulcer. They succeeded in ranitidine (Zantac, **201**) where 1,1-diamino-2-nitroethene served as an isosteric substitution of the thiourea. Ranitidine (**201**) is more selective and has a longer half-life, an overall better drug.[94] Finally, Yamanouchi/Merck's famotidine (Pepcid, **202**) took advantage of sulfamoyl-amidine as a thiourea isostere. Famotidine (**202**) is 30-fold more potent than cimetidine (**200**).[95] Ironically, on the left of the molecule is a guanidine attached to the thiazole ring.

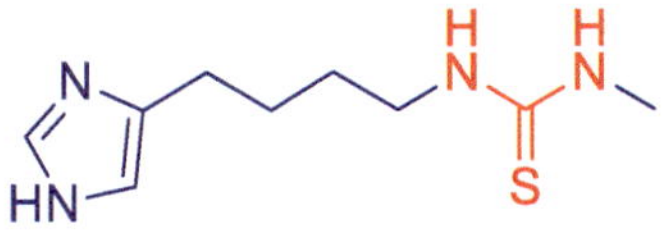

burimamide (**198**)
NOT Bioavailable

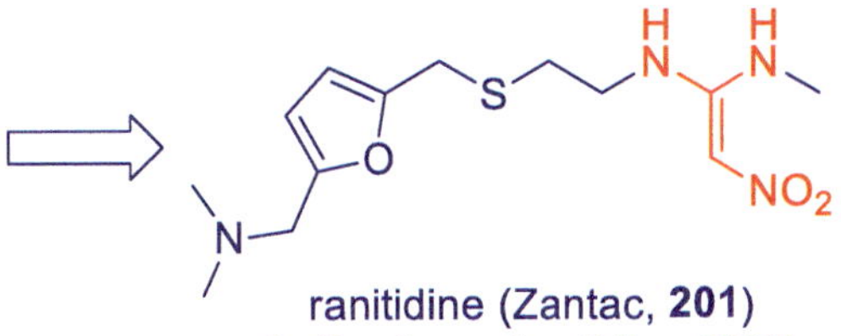

ranitidine (Zantac, **201**)
better than cimetidine (**200**)

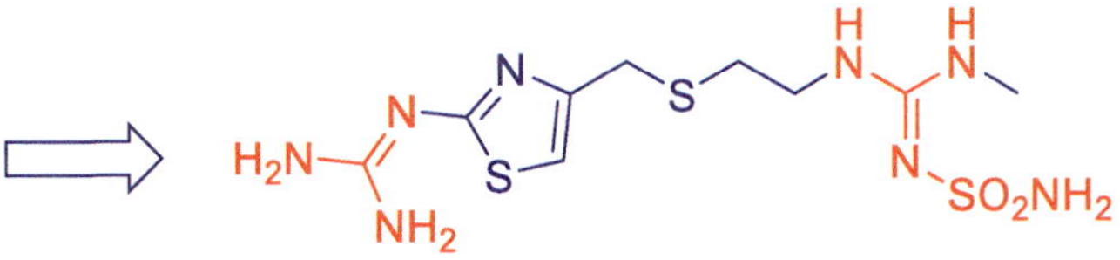

famotidine (Pepcid, **202**)
$t_{1/2}$, 3.5 h, *F*%, <45%

Amidine, with a pK_a value of 12, is nearly as basic as guanidine. Therefore, it is too basic to cross cell membrane thus often has low oral bioavailability. For example, bis-guanidine pentamidine (**203**), as an anti-infective, showed little oral bioavailability and had to be given either by injection or inhalation.

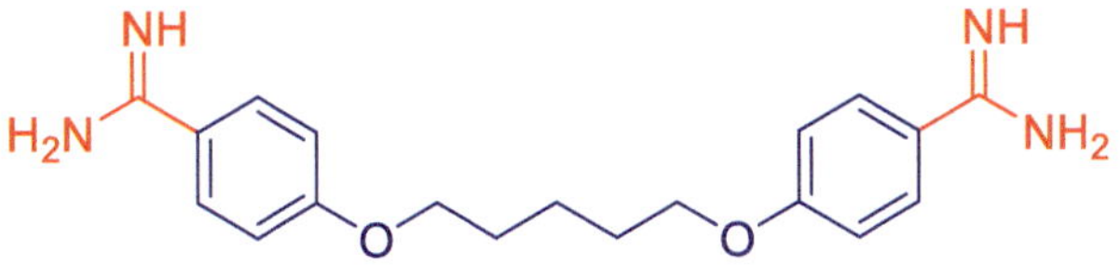

pentamidine (**203**)
$t_{1/2}$ = 3 d
F% negligible

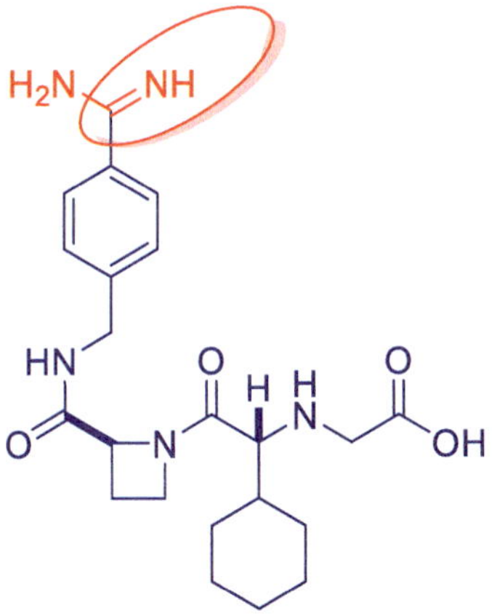

melagatran (**204**)
F% = 3–7%

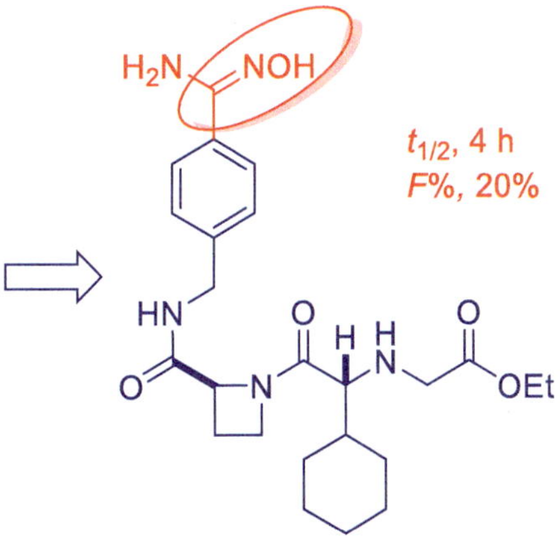

$t_{1/2}$, 4 h
F%, 20%

ximelagatran (Exanta, **205**)
AstraZeneca
direct thrombin inhibitor

Dipeptide melagatran (**204**) is a direct thrombin inhibitor discovered by AstraZeneca in 1999. With an amidine group as an appendage, the drug formed a zwitterion with the carboxylic acid. The highly ionic compound only had 3–7% bioavailability in humans. Converting amidine on melagatran (**204**) to the corresponding hydroxyamidine and making the ethyl ester of the acid led to ximelagatran (Exanta, **205**), which had an oral bioavailability of 18–20%. Ximelagatran (**205**) was marketed in a dozen countries before it was pulled off the market due to hepatoxicity.[96] Meanwhile, Boehringer Ingelheim's dabigatran etexilate (Pradaxa, **206**) is a P2Y12 receptor antagonist. It employed a carbamate prodrug to mask its amidine group to reduce basicity.[97]

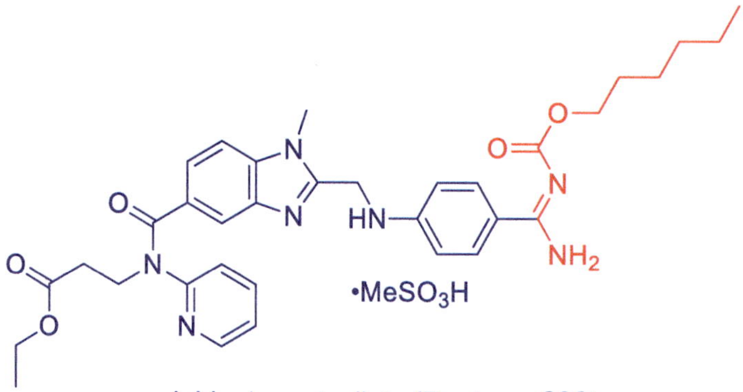

dabigatran etexilate (Pradaxa, **206**)
Boehringer Ingelheim, 2008
$P2Y_{12}$ receptor antagonist

4.6 Scaffold Hopping

4.6.1. Phenyl

Phenyl ring is important in biology. Two amino acids have the benzene ring: phenylalanine (Phe, F) and tyrosine (Tyr, Y). The third amino acid, tryptophan (Trp, W), has an indole ring, which contains a benzene ring. As a consequence, sp^2-rich ligands tend to have tighter binding to proteins since they have amino acids as their building blocks. However, too many aromatic rings increase π-stacking and decrease solubility. Therefore, many isosteres have been experimented to add more sp^3 characteristics.

Sulfonamides **207** and avagacestat (**208**) are γ-secretase inhibitors with similar potencies.[98] Employing bicyclo[1.1.1]pentane (BCP) as a bioisostere for the fluorophenyl moiety, the F_{sp3} count more than doubled from 0.25 to 0.52. F_{sp3} is the fraction of sp^3 hybridized heavy atoms, an alternative to number of aromatic rings (#Ar). This maneuver translated into the practical advantage of improved kinetic and thermodynamic aqueous solubility, and increased membrane permeability and metabolic stability. The Phase II clinical trials of avagacestat (**208**) for treating Alzheimer's disease (AD) was discontinued due to lack of efficacy.

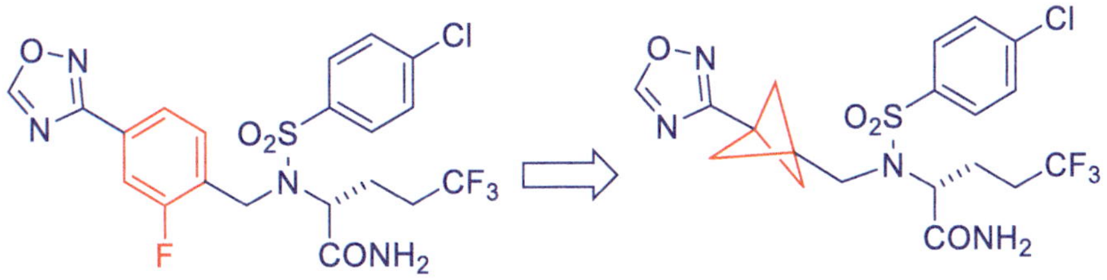

207, pIC_{50} = 9.65
log D = 4.7
F_{sp3} = 0.25
LE = 0.40
LLE = 4.95
LELP = 11.75
sol pH 7.4 = 0.9 μM

avagacestat (**208**), pIC_{50} = 9.65
log D = 3.8
F_{sp3} = 0.52
LE = 0.43
LLE = 5.95
LELP = 8.83
29.4 μM

Resveratrol (**209**) has garnered much attention in medical research for the last two decades. But the progress has been hampered since its bioavailability is too low: with three phenol group, resveratrol (**209**) goes through rapid first-pass metabolism to its glucuronide and sulfate conjugates. Replacing one of the phenyl ring with bicycle[1.1.1]pentane (BCP) resulted in BCP-resveratrol (**210**). The alcohol on **210** is nearly neutral in comparison to acidic phenol on **209**. In addition, the BCP portion has all sp^3 carbons. BCP-resveratrol (**210**) has a 32-fold boost of aqueous solubility than resveratrol (**209**) and is more bioavailable.[99]

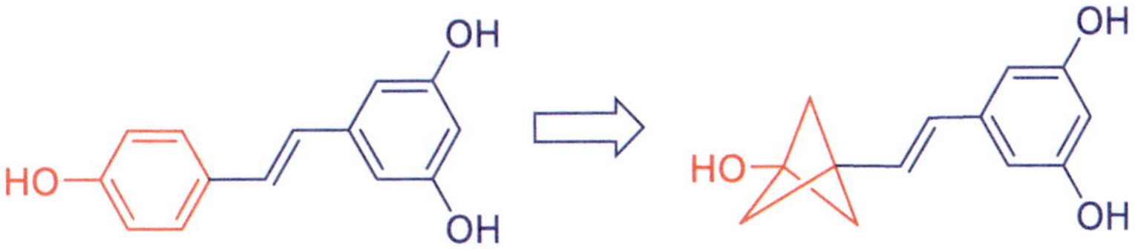

Resveratrol (**209**)
log $D_{7.4}$ = 1.9
solubility at pH 7.4 = 619 μg/mL
$t_{1/2}$ = 0.19 h
C_{max} = 273 ng/mL
AUC = 47.5 ng h/kg

BCP-resveratrol (**210**)
log $D_{7.4}$ = 2.9
19 μg/mL
$t_{1/2}$ = 2.6 h
C_{max} = 942 ng/mL
AUC = 587 ng h/kg

Darapladib (**211**), an inhibitor of lipoprotein-associated phospholipase A_2 ($LpPLA_2$), was in Phase III clinical trials as a treatment of atherosclerosis. But it has suboptimal physicochemical properties including high molecular weight and high property forecast index (PFI). Replacing one phenyl ring with bicyclo[1.1.1]pentane (BCP) gave rise to analog **212**. Although analog **212** is slightly less potent than the parent compound **211**, it is bestowed to superior physicochemical properties. It has an improved permeability of 705 nm/s from 203 nm/s for **211**. It has a nine-fold increase in kinetic solubility. As a consequence, analog **212** has a lower PFI value.[100] PFI stands for Property Forecast Index

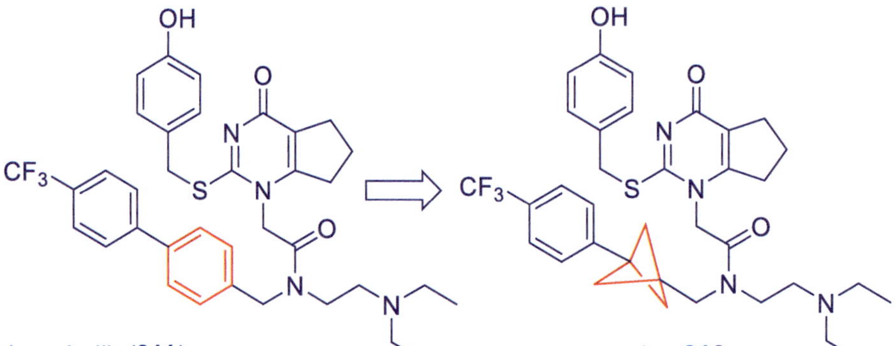

darapladib (**211**)
pIC$_{50}$ = 10.2
AMP = 230 nm/s
kinetic solubility = 8 μM
Chrom log $D_{7.4}$ = 6.3
PFI = 10.3

analog **212**
pIC$_{50}$ = 9.4
AMP = 750 nm/s
kinetic solubility = 74 μM
Chrom log $D_{7.4}$ = 7.0
PFI = 10.0

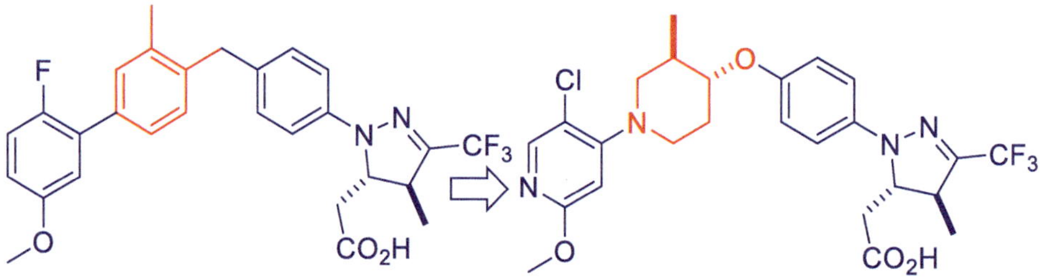

213
GPR40 hEC$_{50}$ = 33 nM
GPR40 mEC$_{50}$ = 31 nM
Clog P = 7.7
PPARγ = 2.6 μM

214
GPR40 hEC$_{50}$ = 70 nM
GPR40 mEC$_{50}$ = 63 nM
Clog P = 5.9
PPARγ >47 μM

BCP is not the only isostere for the phenyl ring. Many sp^3-rich rings have been explored to replace it. For instance, piperidine was successful to replace a phenyl ring in a GPR40 agonist project. G protein-coupled receptor 40 (GPR40), also known as free fatty acid receptor 1 (FFAR1), is a G$_q$-protein-coupled receptor (GPCR) that is predominately expressed in pancreatic β-cells and is also found in the GI tract and brain. GPR40 agonists have shown promise as treatments for type 2 diabetes mellitus (T2DM). One of BMS's lead compounds, **213**, was potent both *in vitro* and *in vivo* with a good pharmacokinetic profile. But some *in vitro* off-target activity (PPARγ) was observed and its pharmacology was observed *in vivo*. Extensive optimization led to piperidine **214** with good potency and selectivity against PPARγ. Moreover, it had improved PK, glucose-lowering efficacy, and safety profile.[101]

5.6.2 Biphenyl

Biphenyl moieties are common for non-peptide angiotensin II receptor antagonists as represented by losartan (Cozaar, **147**). Separately, biphenyl **215** was one of BMS′s lead compounds for their coagulant factor Xa project. Its P_1 fragment as the methoxyphenyl group occupied the S_1 pocket while its biphenyl moiety projected into S_4 pocket. X-Ray co-crystal of **215** with FXa indicated that ortho-substituted biphenyl moiety adopted a perpendicular conformation. Phenylcyclopropanes were explored as biphenyl mimics as represented by **216** whose potency was markedly improved compared to the biphenyl—a general phenomenon across several paired analogs—and cyclopropylmethyl exhibited lower lipophilicity. X-Ray co-crystal of **216** with FXa confirmed perpendicular conformation increased potency, which appeared to be a function of optimized hydrophobic interactions with S_4 and slightly reduced strain in the bound geometry.[102]

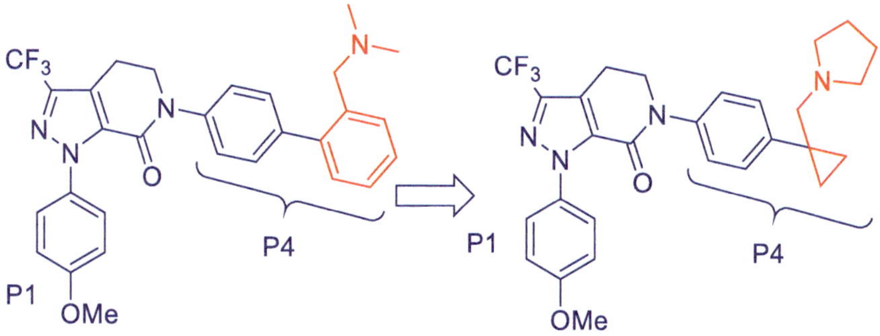

215
FXa: K_i = 0.3 nM
log *P*: 5.94
Clog *P*: 5.09

216
FXa: K_i = 0.021 nM
log *P*: 5.3
Clog *P*: 4.88

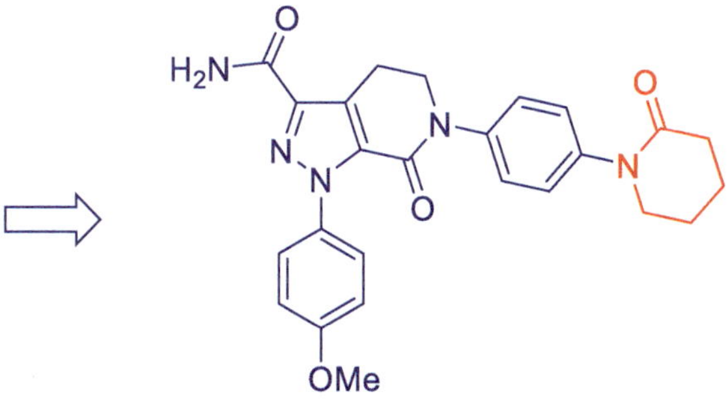

apixaban (Eliquis, **217**)
FXa: K_i = 0.08 nM
log *P*: 2.02
Clog *P*: 1.89
$t_{1/2}$: 5.8 h
F%: 58%

Much effort, including aforementioned investigation culminated to the discovery of apixaban (Eliquis, **217**), has been made to explore biphenyl isosteres. With 1-phenylpiperidin-2-one as a bioisostere for biphenyl, apixaban (**217**) has a much lower lipophilicity (log $P = 2.02$) and is more drug-like.

We have learned a tremendous amount of biphenyl isosteres from the experience in the field of factor Xa inhibitors. Dozens of biphenyl isosteres appeared in the literature as the P_4 fragments to occupy the S_4 pocket. Some examples are shown below:[103]

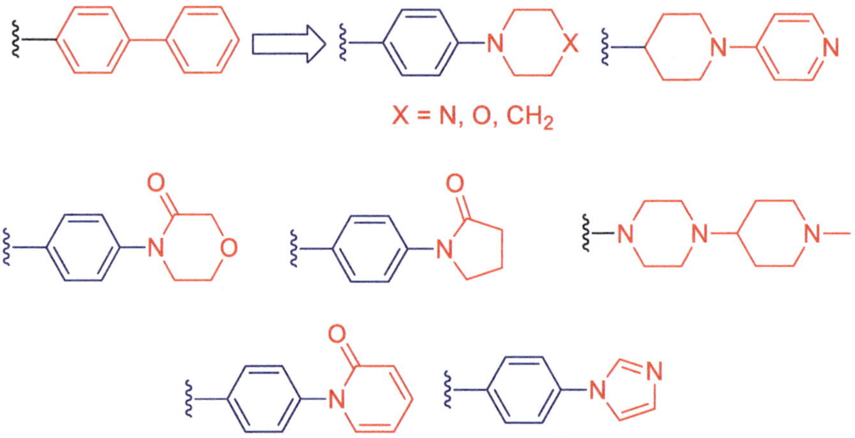

4.6.3. N for CH in Aromatic Rings

When applied appropriately, using N to replace CH in aromatic rings has done wonders to drug discovery. The bioisosterism has improved drugs' *in vitro* binding affinity, *in vitro* functional activity, *in vitro* PK/ADME profile, *in vitro* safety profile, and *in vivo* pharmacological profile. Pennington and Moustaka published an excellent review on this subject in 2017 with a title: *The Necessary Nitrogen Atom: A Versatile High-Impact Design Element for Multi-parameter Optimization.*[104]

218, Cdc7 IC_{50} = 2,700 nM **219**, Cdc7 IC_{50} = 9.0 nM

In a drastic example, replacing CH on an indole ring of compound **218** with a nitrogen atom to give **219** resulted in a 300-fold improvement of *biochemical potency*. As a cell division cycle-7 (Cdc7) kinase inhibitor, indole **218** was not potent in enzymatic assay with an IC_{50} of 2.7 μM. Miraculously, azaindole **219** was very potent with an IC_{50}

of 9.0 nM. Indole **218** preferred a biaryl dihedral angle greater than 150°, while azaindole **219** favored a biaryl dihedral angle of 0°.[105]

Indazole **220** was an inhibitor of fibroblast growth factor receptor (FGFR) identified from a scaffold-hopping approach based on a known FGFR inhibitor (AZD4547). After an N for CH maneuver, the resultant *aza*-indazole **221** gained 11-fold boost of enzymatic potency (from 3.3 nM to 0.3 nM). More remarkably, *aza*-indazole **221** had a 190-fold improvement in *cellular assays* using the H1581 cell line. Furthermore, it also showed significant antitumor activity in an FGFR-driven H1581 xenograft model.[106]

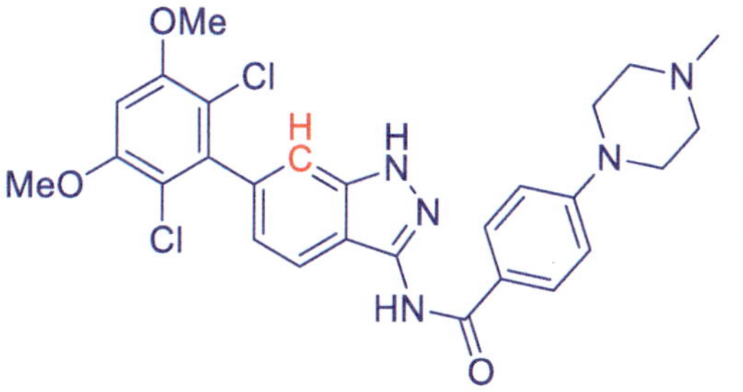

indazole **220**
enzymatic potency
FGFR1 IC_{50} = 3.3 nM
cellular potency
H1581 cell IC_{50} = 320 nM

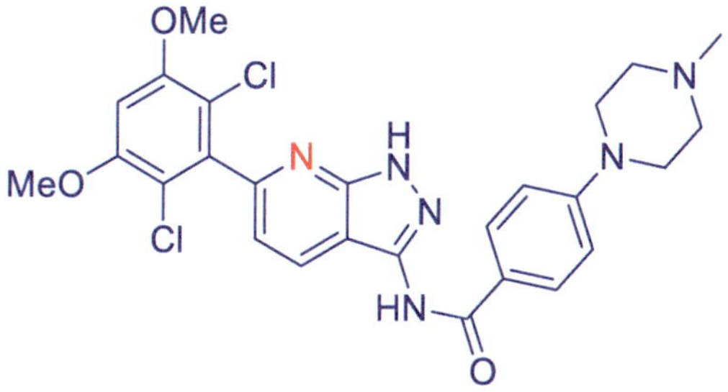

aza-indazole **221**
FGFR1 IC_{50} = 0.30 nM
H1581 cell IC_{50} = 1.7 nM

Reference 104 compiled several examples where the N for CH switches improved target selectivity. Sometimes, the N for CH switch may result in functional switch. It may interconvert GPCR agonists and antagonists.[107] There are also cases of transformations of inverse agonist to antagonist; partial agonist to antagonist; or antagonist to partial agonist.[104] In addition to impact *in vitro* functional activity, the N for CH switch may modulate a drug's *in vitro* PK/ADME profile as well.

Isoquinoline **222**, with no potential for an internal H-bond, has a log *D* of 2.0 at pH 7.4, very poor apical to basolateral permeability, and high efflux (ratio of BA/AB = 79-fold. Not unexpectedly, quinoline **223**, with an intramolecular H-bond, has a measured log *D* value of 3.2 that is 1.2 unit higher than that of **222**. Quinoline **223** has a *solubility* of 55 μM/mL in aqueous media that is nine-fold more soluble than that of the parent compound **222**.[108]

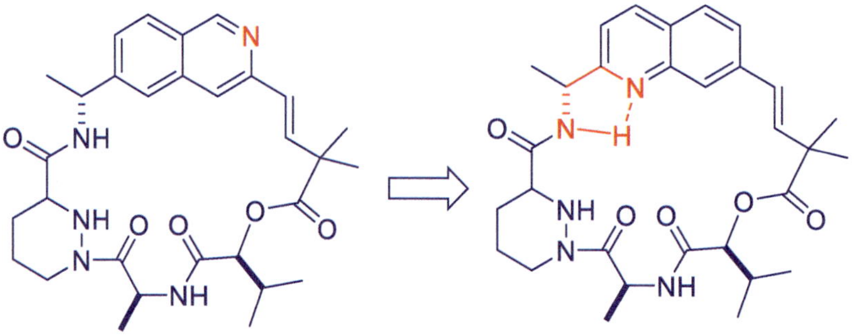

isoquinoline **222**
without intramolecular H-bond
log *D* = 2.0 (pH 7.4)
Caco-2, *AB*/*BA* = 0.1/7.9 x 10^{-6} cm^{-1}
Solubility, 6 μg/mL

quinoline **223**
with intramolecular H-bond
log *D* = 3.2 (pH 7.4)
Caco-2, *AB*/*BA* = 17/47 x 10^{-6} cm^{-1}
Solubility, 55 μg/mL

CP-533,536 (**224**) is a selective and nonprostanoid EP2 receptor agonist (EP2 stands for prostaglandin E2). Switching one CH to N led to omidenepag (OMD, **225**), which is 15-fold more potent than its progenitor CP-533,536 (**224**). However, OMD (**225**)'s cell membrane permeability was insufficient and the permeability rate measured with the parallel artificial membrane permeability assay (PAMPA) was merely 0.9 × 10^{-6} cm/s. Thankfully, the isopropyl ester prodrug, omidenepag isopropyl (OMDI, **226**), had adequate cell membrane permeability with a permeability rate of 2.8 × 10^{-5} cm/s. After showing efficacy in lowering intraocular pressure (IOP) following ocular administration in ocular normotensive monkeys, omidenepag isopropyl (**226**) was selected as a clinical candidate for the treatment of glaucoma.[109]

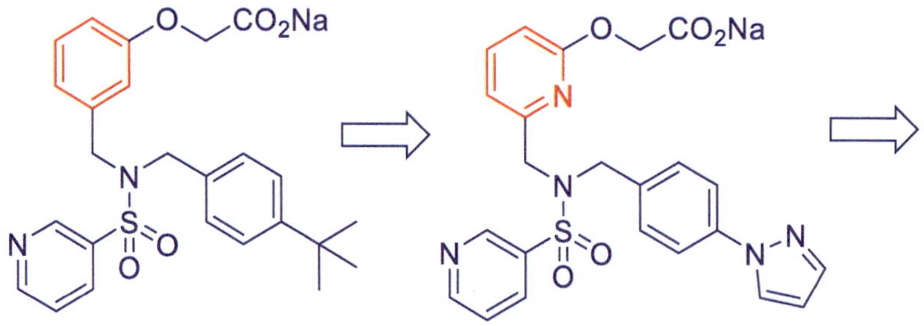

CP-533,536 (**224**)

h-EP2 EC_{50}: 17 nM

OMD (**225**)

h-EP2 EC_{50}: 1.1 nM
PAMPA: 0.9 x 10^{-6} cm/s

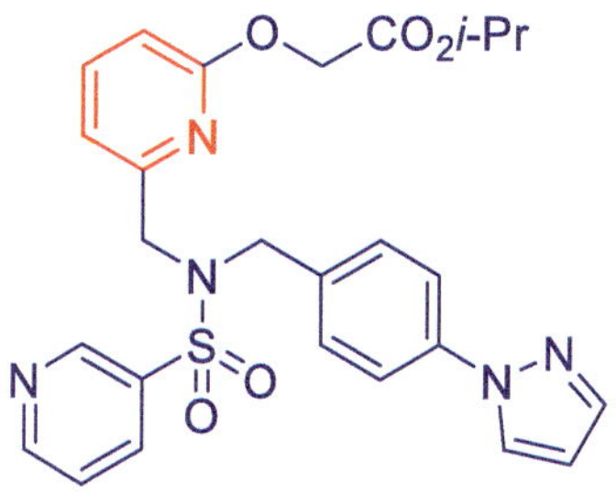

OMDI (**226**)
PAMPA: 2.8×10^{-5} cm/s

The N for CH switch has been employed to fix protein binding (plasma protein shift) issue. For example, olefin **227** as a selective estrogen receptor degrader (SERD) had an excellent potency toward lowering steady-state ERα levels but was highly protein-bound in diluted mouse plasma (f_u = 0.30%). The N for CH switch offered many pyridyl analogs. One of them, 2-pyridyl analog **228** exhibited an 11-fold lower protein binding (f_u = 3.2%). Apparently, reduction of the molecule's lipophilicity was beneficial in reducing protein binding.[110]

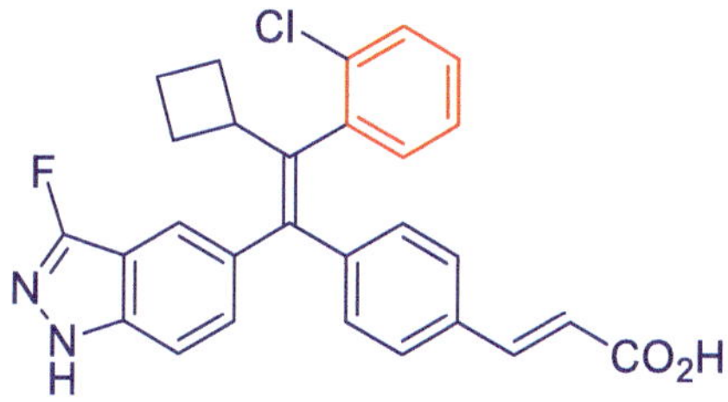

227
MCF-7 ERα
EC_{50} = 0.40 nM
E_{max} = 90%
f_u = 0.30%

11-fold decrease of protein binding

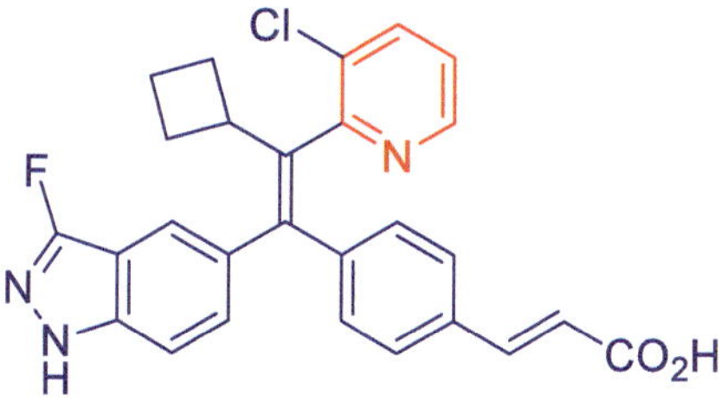

228
MCF-7 ERα EC_{50} = 0.30 nM
E_{max} = 90%
f_u = 3.2%

The N for CH switch has been employed to fix metabolic stability issue.[104] Furthermore, the N for CH switch has also been applied to improve *in vitro* safety profile including CYP inhibition (potential DDI) and hERG activity (potential cardiotoxicity). The following example decreased mutagenicity via the N for CH switch.

Compound **229** with a naphthalene scaffold is a Kelch-like ECH-associated protein 1 (KEAP1)/nuclear factor (erythroid-derived 2)-like (NFR2) inhibitor. But its core structure, 1,4-diaminonaphthalene scaffold, is a structural alert with a potential of causing mutagenicity. Indeed, a mini-Ames assay in *Salmonella typhimurium* and *Escherichia coli* confirmed that **229** is positive, inducing reverse mutations at histidine locus of strains of the two bacteria. Many isosteres have been prepared to replace the core structure. One of the quinoline analogs, compound **230** with a 1,4-isoquinoline scaffold was tested to have lower potential to cause mutagenicity without sacrificing potency, metabolic stability, or solubility.[111]

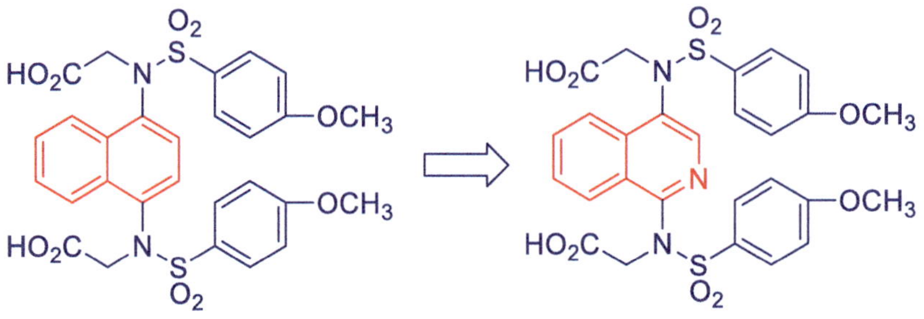

229
IC_{50} = 25 nM
K_d = 20 nM
aqueous solubility: 440 μM
mutagenicity: positive

230
IC_{50} = 60 nM
K_d = 102 nM
380 μM
negative

5.6.4. *Isosterism Between Heterocycles*

Scaffold hopping by switching different heterocycles may result in significant differences in potency, ADME, and safety profile, sometimes with stunning consequences. A matched pair of melanin concentrating hormone receptor 1 (MCHR1) agonists **231** and **232** have comparable potency. Yet **232** is more than 200,000-fold more soluble than **231** in DMSO at pH 7.4.[112] The profound difference is apparently the consequence of different dipole moments of the two regioisomeric oxadiazoles.

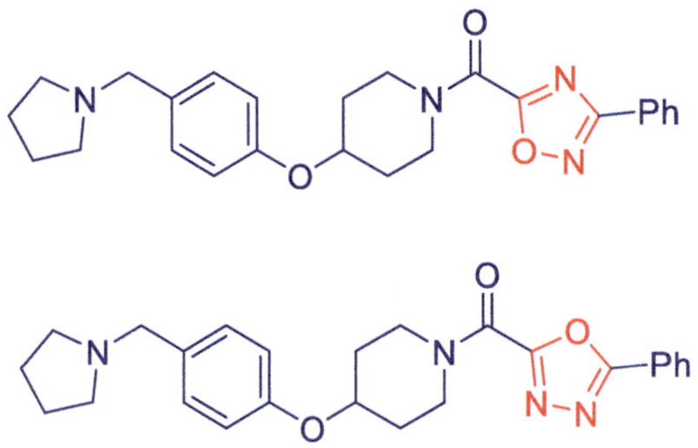

231
IC_{50}, 43 nM
log *D*, 3.2
LLE, 3.9
sol <0.005 μM

232
IC_{50}, 75 nM
log *D*, 2.2
LLE, 5.2
sol >100 μM

Plexxikon's vemurafenib (Zelboraf, **233**), a B-Raf kinase inhibitor for treating treat metastatic melanoma carrying the BRAF V600E mutation,[113] was the first marketed drug discovered from the FBDD approach. An interactive visual application of the novel scaffold fingerprint (SFP)-based tool to search for potential bioisosteres of vemurafenib

(**233**) having a set of required pharmacophore features and substitution pattern gave rise to 15 different scaffolds shown below:[114]

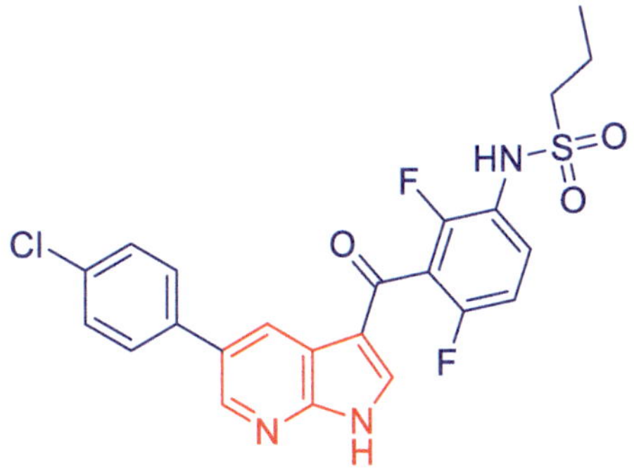

vemurafenib (Zelboraf, **233**)
Roche/Plexxikon, 2011
B-Raf; A/B/C-Raf inhibitor
1st marketed drug from FBDD

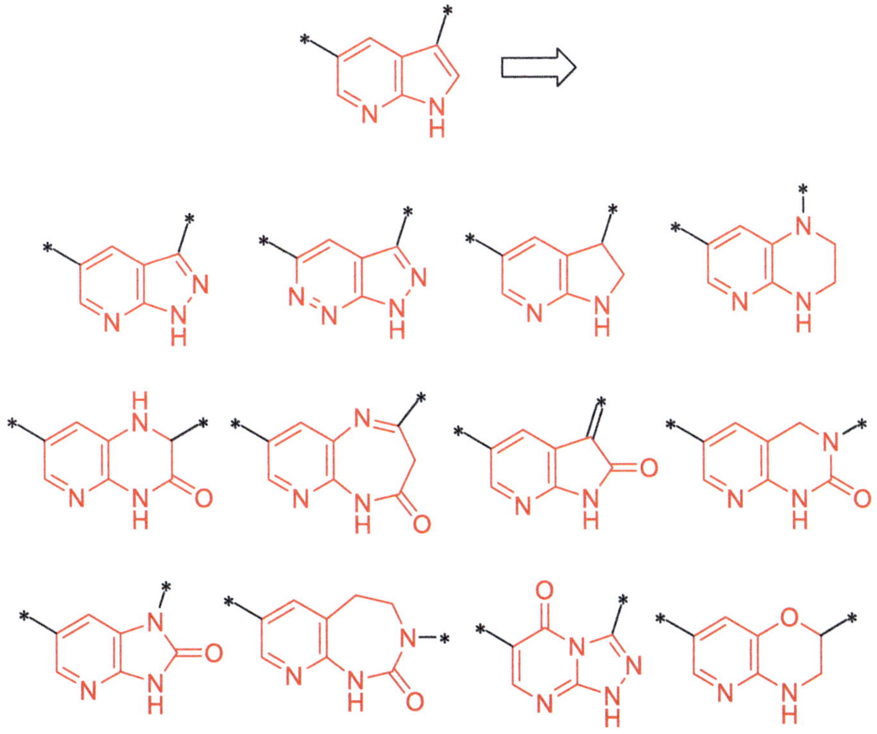

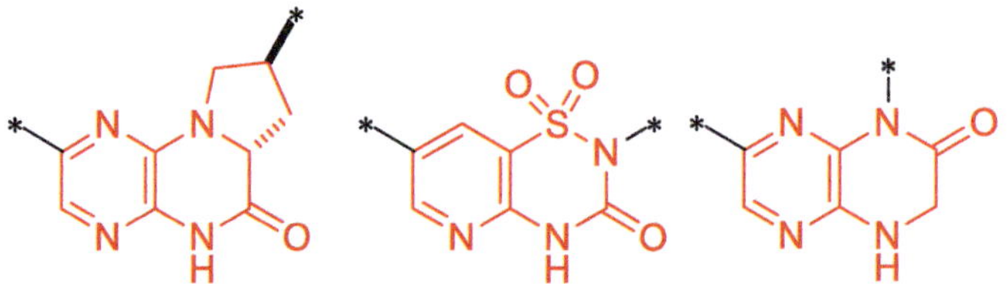

As mentioned earlier, bioisosterism is a good way to create novel intellectual properties. For instance, in the area of PDE5 inhibitors for treating erectile dysfunction (ED), Bayer successfully carried out a scaffold hopping exercise to arrive at a novel core structure. Bayer's vardenafil (Levitra, **235**) is a direct result of scaffold hopping from Pfizer's sildenafil (Viagra, **233**). Moving one nitrogen atom to the bridgehead converted the original pyrazole ring to an imidazole ring. As a consequence, vardenafil (**235**) is seven-fold more potent than the parent sildenafil (**234**), although its bioavailability is lower than its progenitor.[115]

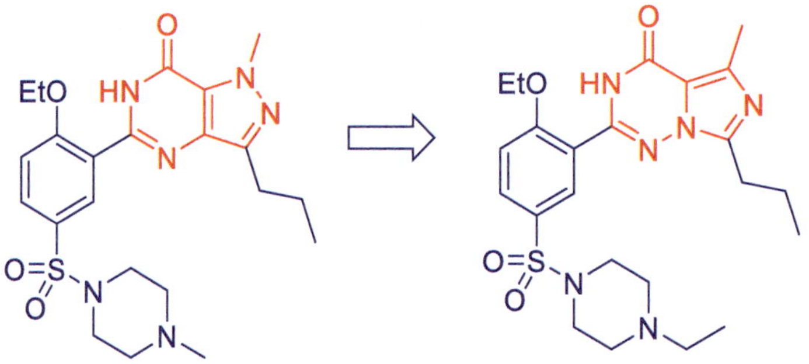

Sildenafil (Viagra, **234**)
Pfizer, 1998
PDE5 IC_{50}, 5.1 nM
$t_{1/2}$, 3.8 h
F%, 41%

Vardenafil (Levitra, **235**)
Bayer/GSK, 2003
PDE5 IC_{50}, 0.7 nM
$t_{1/2}$, 4.7 h
F%, 15%

In one case, one simple heterocyclic scaffold hopping transformed a hit from DNA-encoded library (DEL) to a drug candidate (DC). DEL screening against receptor interacting protein (RIP1) kinase provided GSK′481 (**236**) as a decent hit with potent biochemical and cellular activities. Extensive SAR effort led to identification of GSK2982772 (**237**). It is remarkable how little difference between the DC and the DEL hit. GSK2982772 (**237**) is now in Phase IIa clinical studies for psoriasis, rheumatoid arthritis, and ulcerative colitis.[116]

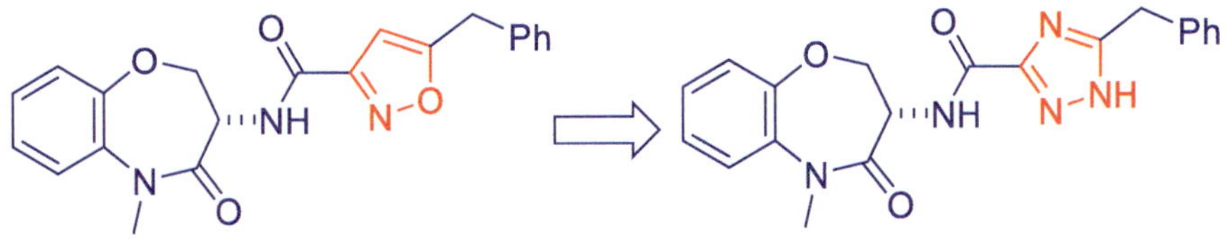

GSK'481 (**236**)
RIP1 IC_{50} = 1.6 nM
U937 cell, IC_{50} = 10 nM
log *D* = 5.9
AUC = 0.38 mg h/mL

GSK2982772 (**237**)
RIP1 IC_{50} = 1.0 nM
U937 cell, IC_{50} = 6.3 nM
log *D* = 3.8
AUC = 2.3 mg h/mL

Thiazole **238** showed strong antiproliferative activity against a panel of five cancer cell lines. It is a synthetic inhibitor of tubulin polymerization. It is well-known that the trimethoxyphenyl skeleton is the characteristic structural requirement to maximize activity in a large series of inhibitors of tubulin polymerization, such as colchicine and podophyllotoxin. The bioisosteric equivalence between thiazole and 1,2,4-triazole prompted the scaffold hopping to give rise to triazole **239**, which are more potent in many cell lines including CCRF-CEM and HeLa cells. The compound could be a new antimitotic agent with clinical potential.[117]

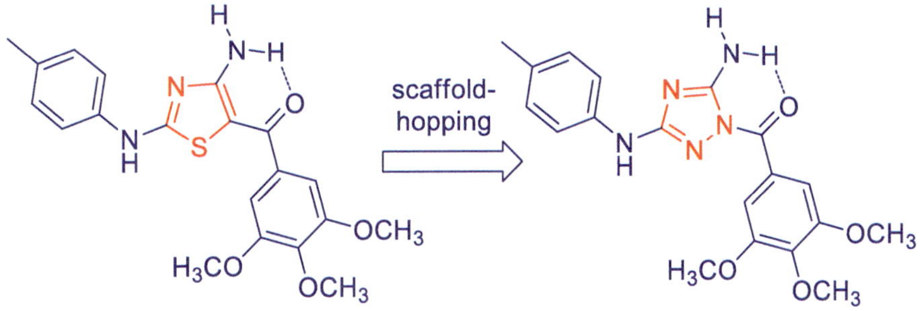

238
CEM IC_{50} = 41 nM
HeLa IC_{50} = 86 nM

239
CEM IC_{50} = 0.21 nM
HeLa IC_{50} = 3.20 nM

4.7 Peptide Isosteres

Peptides would have been ideal drugs, after all, they are building blocks of the targets, if it were not for a small detail, namely, bioavailability. Peptides are generally too polar to permeate across biological barriers such as intestinal lumen and mucosa. In addition, since linear peptides adopt an open saw-tooth conformation, they are prone to be recognized and subsequently cleaved by proteases. As a consequence, they normally have low oral bioavailability (<1–2%) and short *in vivo* half-lives (<30 min). Thankfully, last several decades accumulated a substantial amount of knowledge in the field of peptidomimetics to elevate their oral bioavailability.

4.7.1 Cyclization

Cyclization of linear peptides may be achieved by several means: side-chain-to-side-chain and side-chain-to-backbone are among the most popular methods. But the most dramatic effect is obtained from backbone cyclization to generate peptidomimetic derivatives by covalently interconnecting atoms in the backbone (N and/or C) to form a ring. A case in point is found in the arena of MC4R agonists.

The tetrapeptide sequence His-Phe-Arg-Trp (**240**), derived from melanocyte-stimulating hormone (RMSH) and its analogs, causes a decrease in food intake and elevates energy utilization upon binding to the melanocortin-4 receptor (MC4R). It may be viewed as an endogenous ligand, but it has poor membrane permeability and is extensively metabolized. In 2008, Hoffman et al. prepared a library of 16 backbone cyclic peptidomimetic derivatives. One of them, cyclic pentapeptide **241** (BL3020-1) from Phe-D-Phe-Arg-Gly-NH_2, was selective in activating the MC4R. More importantly, it has favorable transcellular penetration through enterocytes and enhanced intestinal metabolic stability. This peptide was detected in the brain following oral administration to rats with an 8% bioavailability, a 105 min half-life, and V_D of 2.1 L/kg after oral administration. Backbone cyclization was shown here to produce a potential drug lead for treating obesity.[118]

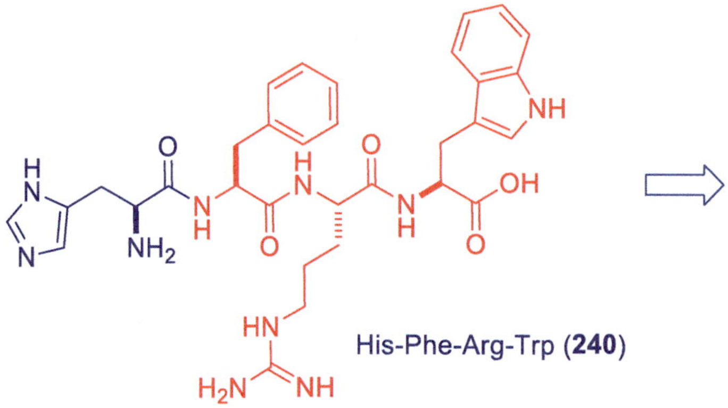

His-Phe-Arg-Trp (**240**)

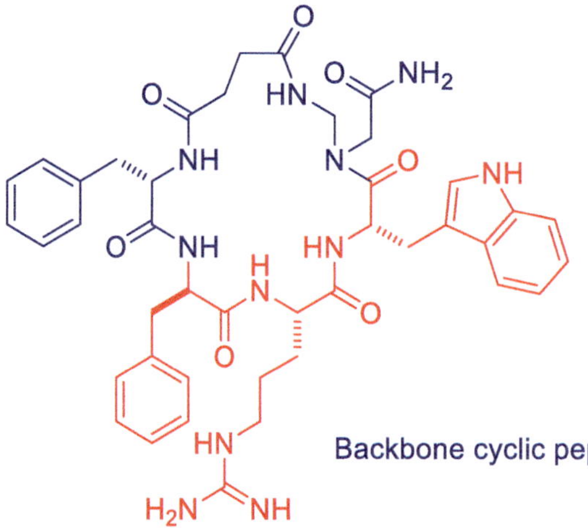

Backbone cyclic peptide **241**

It was not appreciated in 2008 that intramolecular hydrogen bonding has a profound impact on a cyclic peptide's cell membrane permeability. Cyclic peptide **241** certainly has several potential sites for such an intramolecular hydrogen bonding.

In addition to backbone cyclization, side-chain cyclization has been widely employed in peptidomimetic design. As shown beneath, cyclization of the aromatic sidechain on the native peptide would provide constraint peptidomimetics that may potentially have improved potency, efficacy, selectivity, stability, and absorption.[119]

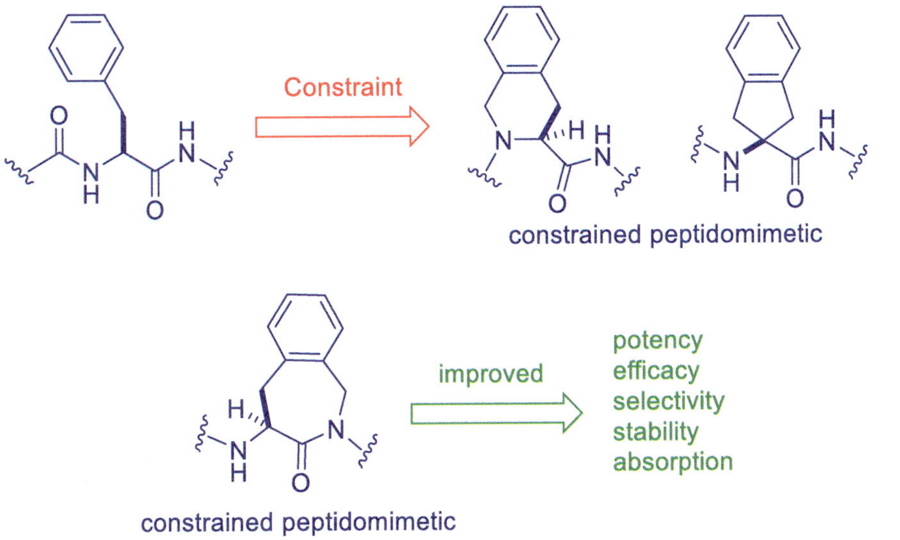

4.7.2. Intramolecular Hydrogen Bonding

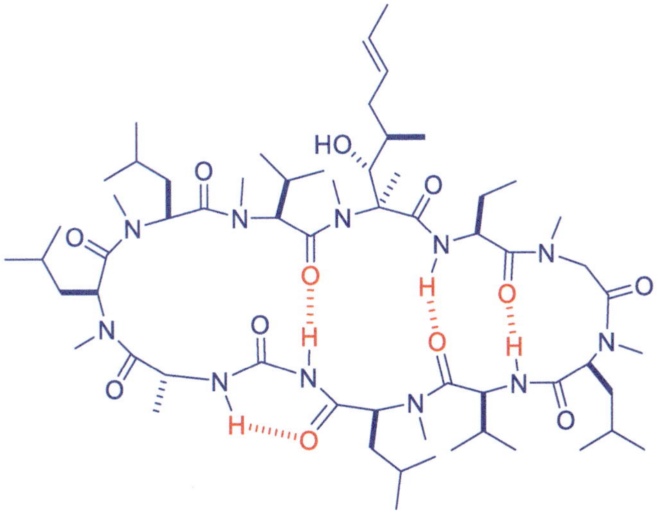

cyclosporine A (**242**)
(F% = 29%)

Why cyclosporine A (**242**) is orally bioavailable with *F*% of 29%? In Chapter 3, we attributed it to the chameleon effect from the four intramolecular hydrogen bonds. In fact, all four NH groups are engaged as intramolecular hydrogen bond donors.

The Lokey peptide (1NMe$_3$, **243**) with an *F*% = 28% and a terminal half-life of 2.8 h (168 min) is an example of the powerful impact of intramolecular hydrogen bonding. The cyclic hexapeptide was designed with specific *N*-methylation pattern (see Section 4.7.3) to simultaneously reduce hydrogen bond donor count of the compound and promote intramolecular hydrogen bonding network.[120]

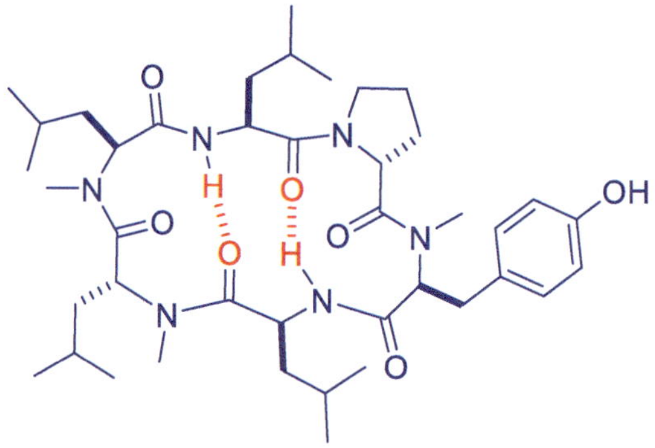

Lokey peptide
(1NMe$_3$, **243**)
(*F*% = 28%)

4.7.3. N-Methylation

The simplest approach to obliterate hydrogen bonding is to remove the hydrogen bond donors by N-alkylation, especially N-methylation of the amide bonds of peptides. Both cyclosporine A (**242**) and Lokey peptide (1NMe3, **243**) are beneficiaries of this strategy.[121]

For cyclosporine A (**242**), its lack of externally oriented NH groups, lipophilic side chains (especially the four leucines) and structural motifs all may help to increase its oral bioavailability. However, Mother Nature has bestowed N-methylation for seven out of the 11 amide groups and N-methylation significantly contributed to cyclosporine A (**242**)'s oral bioavailability as the consequence of evolution.

Kessler and coworkers studied the impact of N-methylation on oral bioavailability related to somatostatin-related Veber–Hirschmann peptide. They systemically generated a library of 30 compounds with varying methylation of the secondary amides contained in the starting macrocycle. Although eight out of the 30 derivatives were active for some members of the somatostatin receptors, only tri-N-methylated somatostatin analog **245** exhibited permeability across the Caco-2 cell membrane (68% increase compared to non-N-methylated) and was orally bioavailable in rats (*F*% = 10%). The non-N-methylated peptide and the other N-methylated peptides were not bioavailable at all. Comparison of the bioavailable derivative **245** and the Veber–Hirschmann peptide revealed that the conformation was not altered by N-methylation, and all the *externally oriented* NH groups were N-methylated as shown below.[122]

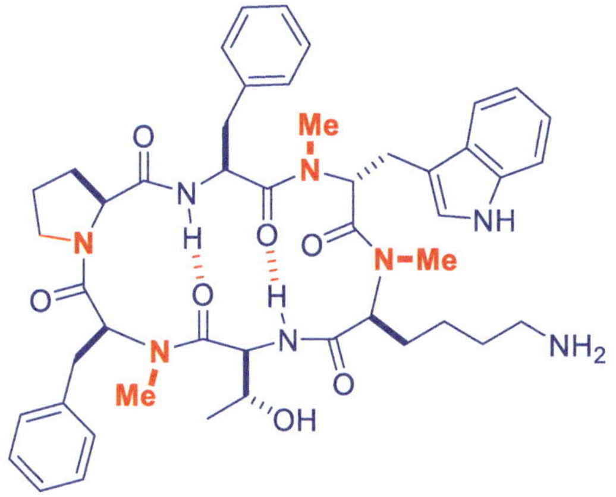

tri-N-methylated
somatostatin analog **245**
(*F*% = 10%)

4.7.4. Shielding

Cyclosporine A (**242**) is the poster child of bioavailable cyclic peptide natural products. But it is certainly not the only one. Sanguinamide A (**246**), a hexapeptide marine natural product, has a 7% oral bioavailability in rats. With a molecular weight of 721 and Clog *P* of 5.5, both out of the Lipinski's rule of five space, there are three possible contributing factors to sanguinamide A (**246**)'s bioavailability. (i). The thiazole heterocycle ring, as an amide isostere, rigidified the structure, imposing a single conformation; (ii). The presence of two transannular hydrogen bonds; (iii). The *shielding* of the polar amides from the solvent by the lipophilic side chains.[123]

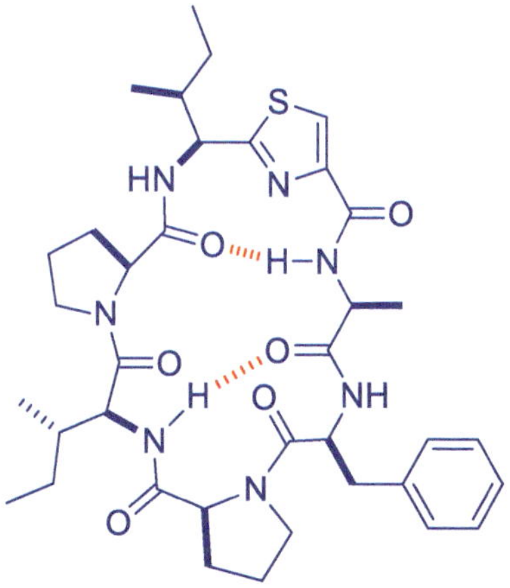

sanguinamide A (**246**)
(7%)

To increase the shielding effect of side chains, danamide D (**247**) and danamide F (**248**) have been designed to improve oral bioavailability. The methyl group of alanine on sanguinamide A (**246**) has been replaced with a bulkier *t*-butyl group. The elevated shielding effect gives rise to improved bioavailability for 21% and 51% for danamide D (**247**) and danamide F (**248**), respectively.

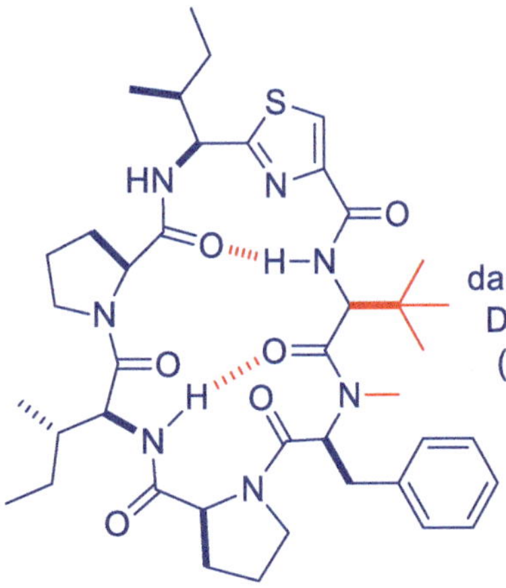

danamide D (**247**) (**21%**)

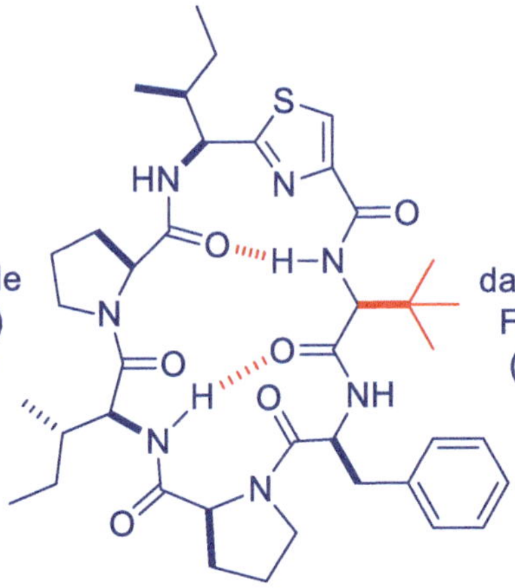

danamide F (**248**) (**51%**)

Many other tactics exist including conformational interconversion, peptoids, unusual amino acids, and more.[124]

To conclude in closing this chapter, many additional isoteres exist in addition to the ones reviewed in this chapter. Every day, novel isosteres are being created in medicinal chemistry labs around the world as reflected in new journal articles. I hope to update and expand them in the second edition of this book in due course.

4.8 Further Reading

- Meanwell, N. A. *J. Med. Chem.* **2018,** *61*, 5822–5880.
- Meanwell, N. A. The Influence of Bioisosteres in Drug Design: Tactical Applications to Address Developability Problems. In Meanwell, N. A., Ed. *Top. Med. Chem. 9(Tactics in Contemporary Drug Design)*. **2015,** pp. 283–381.
- Meanwell, N. A. *J. Med. Chem.* **2011,** *54*, 2529–2591.

4.9 References

1. *IUPAC Compendium of Chemical Terminology.* **1998**, *70,* 1129.
2. de Stevens, G.; Werner, L. H.; Halamandaris, A.; Ricca, S., Jr. *Experientia* **1958**, *14*, 463.
3. de Stevens, G. *J. Med. Chem.* **1991**, *34*, 2665–2670.
4. O'Reilly, M. C.; Scott, S. A.; Brown, K. A.; Oguin, T. H.; Thomas, P. G.; Daniels, J. S.; Morrison, R.; Brown, H. A.; Lindsley, C. W. *J. Med. Chem.* **2013**, *56*, 2695–2699.
5. Wong, D. T.; Bymaster, F. P.; Engleman, E. A. *Life Sci.* **1995**, *57*, 411–441.
6. Wuitschik, G.; Rogers-Evans, M.; Mueller, K.; Fischer, H.; Wagner, B.; Schuler, F.; Polonchuk, L.; Carreira, E. M. *Angew. Chem. Int. Ed.* **2006**, *45*, 7736–7739.
7. (a) Burkhard, J. A.; Wuitschik, G.; Rogers-Evans, M.; Mueller, K.; Carreira, E. M. *Angew. Chem. Int. Ed.* **2010**, *49*, 9052–9067. (b) Wuitschik, G.; Carreira, E. M.;

Wagner, B.; Fischer, H.; Parrilla, I.; Schuler, F.; Rogers-Evans, M.; Mueller, K. *J. Med. Chem.* **2010**, *53,* 3227–3246.

8. Rafi, S. B.; Hearn, B. R.; Vedantham, P.; Jacobson, M. P.; Renslo, A. R. *J. Med. Chem.* **2012**, *55*, 3163–3169.
9. Quan, M. L.; Lam, P. Y. S.; Han, Q.; Pinto, D. J. P.; He, M. Y.; Li, R.; Ellis, C. D.; Clark, C. G.; Teleha, C. A.; Sun, J.-H.; Wexler, R. R.; et al. *J. Med. Chem.* **2005**, *48,* 1729–1744.
10. Oshiro, Y.; Sato, S.; Kurahashi, N.; Tanaka, T.; Kikuchi, T.; Tottori, K.; Uwahodo, Y.; Nishi, T. *J. Med. Chem.* **1998**, *41*, 658–667.
11. Maffrand, J. P.; Eloy, F. *J. Heterocycl. Chem.* **1976**, *13*, 1347–1349.
12. Watanabe, M.; Koike, H.; Ishiba, T.; Okada, T.; Seo, S.; Hirai, K. *Bioorg. Med. Chem.* **1997**, *5*, 437–444.
13. DeWitt, S. H.; Maryanoff, B. E. *Biochem.* **2018**, *57*, 472–473.
14. Zha, J.; Zhang, J.; Ordonez, C. *J. Clin. Pharmacol.* **2011**, *51*, 1358–1359.
15. Thayer, A. M. *Chem. Eng. News* March 6, **2017**.
16. Claassen, D. O.; Carroll, B.; De Boer, L. M.; Wu, E.; Ayyagari, R.; Gandhi, S.; Stamler, D. *J. Clin. Mov. Disord.* **2017**, *4*, 3.
17. Rodrigues, F. B.; Duarte, G. S.; Costa, J.; Ferreira, J. J.; Wild, E. J. *J. Clin. Mov. Disord.* **2017**, *4*, 9.
18. (a) Schmidt, C. *Nat. Biotechnol.* **2017**, 35, 493–494. (b) For the latest review on applications of deuterium in medicinal chemistry, see: Pirali, T.; Serafini, M.; Cargnin, S.; Genazzani, A. A. *J. Med. Chem.* **2019**, *62*, 5276–5297.
19. Fried, J.; Sabo, E. F. *J. Am. Chem. Soc.* **1954**, *76,* 1455–1456.
20. Ghoshal, K.; Jacob, S. T. *Biochem. Pharmacol.* **1997**, *53,* 1569–1575.
21. Siporin, C. *Ann. Rev. Microb.* **1989**, *43*, 601–627.
22. Rosenblum, S. B.; Huynh, T.; Afonso, A.; Davis, H. R.; Yumibe, N. *J. Med. Chem.* **1998**, *41*, 973–980.
23. Velcicky, J.; Schlapbach, A.; Heng, R.; Revesz, L.; Pflieger, D.; Blum, E.; Hawtin, S.; Huppertz, C.; Feifel, R.; Hersperger, R. *ACS Med. Chem. Lett.* **2018**, *9*, 392–396.
24. Edmonds, D. J.; Kung, D. W.; Kalgutkar, A. S.; Filipski, K. J.; Ebner, D. C.; Cabral, S.; Smith, A. C.; Aspnes, G. E.; Bhattacharya, S. K.; Borzilleri, K. A.; et al. *J. Med. Chem.* **2018**, *61*, 2372–2383.
25. Kitas, E. A.; Galley, G.; Jakob-Roetne, R.; Flohr, A.; Wostl, W.; Mauser, H.; Alker, A. M.; Czech, C.; Ozmen, L.; David-Pierson, P.; Reinhardt, D.; Jacobsen, H. *Bioorg. Med. Chem. Lett.* **2008**, *18*, 304–308.
26. Brickner, S. J.; Hutchinson, D. K.; Barbachyn, M. R.; Manninen, P. R.; Ulanowicz, D. A.; Garmon, S. A.; Grega, K. C.; Hendges, S. K.; Toops, D. S.; Ford, C. W.; Zurenko, G. E. *J. Med. Chem.* **1996**, *39*, 673–679.
27. Antipas, A. S.; Blumberg, L. C.; Brissette, W. H.; Brown, M. F.; Casavant, J. M.; Doty, J. L.; Driscoll, J.; Harris, T. M.; Jones, C. S.; McCurdy, S. P.; Mitton-Fry, M.; et al. *Bioorg. Med. Chem. Lett.* **2010**, *20,* 4069–4072.
28. Meanwell, N. A.; Wallace, O. B.; Fang, H.; Wang, H.; Deshpande, M.; Wang, T.; Yin, Z.; Zhang, Z.; Pearce, B. C.; James, J.; et al. *Bioorg. Med. Chem. Lett.* **2009**, *19,* 1977–1981.

29. Robarge, K. D.; Brunton, S. A.; Castanedo, G. M.; Cui, Y.; Dina, M. S.; Goldsmith, R.; Gould, S. E.; Guichert, O.; Gunzner, J. L.; Halladay, J.; et al. *Bioorg. Med. Chem. Lett.* **2009**, *19,* 5576–5581.
30. Zimmermann, J.; Buchdunger, E.; Mett, H.; Meyer, T.; Lydon, N. B. *Bioorg. Med. Chem. Lett.* **1997**, *7*, 187–192.
31. Schönherr, H.; Cernak, T. *Angew. Chem. Int. Ed.* **2013,** *52,* 12256–12267.
32. Angell, R.; Aston, N. M.; Bamborough, P.; Buckton, J. B.; Cockerill, S.; deBoeck, S. J.; Edwards, C. D.; Holmes, D. S.; Jones, K. L.; Laine, D. I.; Patel, S.; Smee, P. A.; Smith, K. J.; Somers, D. O.; Walker, A. L. *Bioorg. Med. Chem. Lett.* **2008**, *18,* 4428–4432.
33. Quancard, J.; Bollbuck, B.; Janser, P.; Angst, D.; Berst, F.; Buehlmayer, P.; Streiff, M.; Beerli, C.; Brinkmann, V.; Guerini, D.; et al. *Chem. Biol.* **2012**, *19,* 1142–1151.
34. Leung, C. S.; Leung, S. S. F.; Tirado-Rives, J.; Jorgensen, W. L. *J. Med. Chem.* **2012**, *55,* 4489–4500.
35. Kim, J.-J. P.; Battaille, K. P. *Curr. Opin. Struct. Bol.* **2002,** *12,* 721–728.
36. Skuballa, W.; Schillinger, E.; Stuerzebecher, C.-S.; Vorbureggen, H. *J. Med. Chem.* **1986,** *29,* 313–315.
37. Manoury, P. M.; Binet, J. L.; Rousseau, J.; Lefevre-Borg, F. M.; Cavero, I. G. *J. Med. Chem.* **1987**, *30,* 1003–1011.
38. Grozinger, K.; Proudfoot, J.; Hargrave, K. Discovery and Development of Nevirapine. In Chorghade, M. S. ed. *Drug Discovery and Development, Vol. 1: Drug Discovery.* Wiley: Weinheim, Germany, **2006**, pp. 353–363.
39. Johansson, T.; Weidolf, L.; Popp, F.; Tacke, R.; Jurva, U. *Drug Metab. Dispos.* **2010**, *38,* 73–83.
40. (a) Ramesh, R.; Reddy, D. S. *J. Med. Chem.* **2018**, *61*, 3779–3798. (b) Franz, A. K.; Wilson, S. O. *J. Med. Chem.* **2013**, *56*, 388–405.
41. Furet, P.; Guagnano, V.; Fairhurst, R. A.; Imbach-Weese, P.; Bruce, I.; Knapp, M.; Fritsch, C.; Blasco, F.; Blanz, J.; Aichholz, R.; Hamon, J.; Fabbro, D.; Caravatti, G. *Bioorg. Med. Chem. Lett.* **2013**, *23*, 3741–3748.
42. Barnes-Seeman, D.; Jain, M.; Bell, L.; Ferreira, S.; Cohen, S.; Chen, X.-H.; Amin, J.; Snodgrass, B.; Hatsis *ACS Med. Chem. Lett.* **2013**, *4,* 514–516.
43. Mukherjee, P.; Pettersson, M.; Dutra, J. K.; Xie, L.; am Ende, C. W. *ChemMedChem* **2017**, *12*, 1574–1577.
44. Erickson, J.; McLoughlin, J. I. *J. Org. Chem.* **1995**, *60*, 1626–1631.
45. Xu, Y.; Qian, L.; Pontsler, A. V.; McIntyre, T. M.; Prestwich, G. D. *Tetrahedron* **2004**, *60*, 43–49.
46. Lu, D.; Vince, R. *Bioorg. Med. Chem. Lett.* **2007**, *17*, 5614–5619.
47. Raza, A.; Sham, Y. Y.; Vince, R. *Bioorg. Med. Chem. Lett.* **2008**, *18*, 5406–5410.
48. Zhao, Q.; Tijeras-Raballand, A.; de Gramont, A.; Raymond, E.; Desaubry, L. *Tetrahedron Lett.* **2016**, *57*, 2943–2944.
49. Asselin, A. A.; Humber, L. G.; Crosilla, D.; Oshiro, G.; Wojdan, A.; Grimes, D.; Heaslip, R. J.; Rimele, T. J.; Shaw, C. C. *J. Med. Chem.* **1986**, *26*, 1009–1015.
50. Di Grandi, M. J.; Berger, D. M.; Hopper, D. W.; Zhang, C.; Dutia, M.; Dunnick, A. L.; Torres, N.; Levin, J. I.; Diamantidis, G.; Zapf, C. W.; et al. *Bioorg. Med. Chem. Lett.* **2009**, *19*, 6957–6961.

51. Wright, J. L.; Gregory, T. F.; Kesten, S. R.; Boxer, P. A.; Serpa, K.; Meltzer, L. T.; Wise, L. D.; Espitia, C. S.; Konkoy, C. S.; Whittemore, E. R.; Woodward, R. M. *J. Med. Chem.* **2000**, *43*, 3408–3419.
52. Hacksell, U.; Arvidsson, L. E.; Svensson, U.; Nilsson, J. L. G.; Sanchez, D.; Wikstroem, H.; Lindberg, P.; Hjorth, S.; Carlsson, A. *J. Med. Chem.* **1981**, *24*, 1475–1482.
53. Hübner, H.; Haubmann, C.; Utz, W.; Gmeiner, P. *J. Med. Chem.* **2000**, *43*, 756–762.
54. Blum, G.; Gazit, A.; Levitzki, A. *J. Biol. Chem.* **2003**, *278*, 40442–40454.
55. Narjes, F.; Koehler, K. F.; Koch, U.; Gerlach, B.; Colarusso, S.; Steinkühler, C.; Brunetti, M.; Altamura, S.; Francesco, R. C.; Matassa, V. G. *Bioorg. Med. Chem. Lett.* **2002**, *12*, 701–704.
56. Ballatore, C.; Huryn, D. M.; Smith, A. B, III. *ChemMedChem* **2013**, *8*, 385–395.
57. Wendt, M. D.; Shen, W.; Kunzer, A.; McClellan, W. J.; Bruncko, M.; Oost, T. K.; Ding, H.; Joseph, M. K.; Zhang, H.; Nimmer, P. M.; et al. *J. Med. Chem.* **2006**, *49*, 1165–1181.
58. Uehling, D. E.; Donaldson, K. H.; Deaton, D. N.; Hyman, C. E.; Sugg, E. E.; Barrett, D. G.; Hughes, R. G.; Reitter, B.; Adkison, K. K.; Lancaster, M. E.; et al. *J. Med. Chem.* **2002**, *45*, 567–583.
59. Schaaf, T. K.; Bindra, J. S.; Eggler, J. F.; Plattner, J. J.; Nelson, A. J.; Johnson, M. R.; Constantine, J. W.; Hess, H.-J.; Elger, W. *J. Med. Chem.* **1981**, *24*, 1353–1359.
60. Asada, M.; Obitsu, T.; Kinoshita, A.; Nakai, Y.; Nagase, T.; Sugimoto, I.; Tanaka, M.; Takizawa, H.; Yoshikawa, K.; Sato, K.; Narita, M.; Ohuchida, S.; Nakai, H.; Toda, M. *Bioorg. Med. Chem. Lett.* **2010**, *20*, 2639–2643.
61. Aguilar, A.; Zhou, H.; Chen, J.; Liu, L.; Bai, L.; McEachern, D.; Yang, C.-Y.; Meagher, J.; Stuckey, J.; Wang, S. *J. Med. Chem.* **2013**, *56*, 3048–3067.
62. Chebib, M.; Johnston, G. A. R.; Mattsson, J. P.; Rydstrom, K.; Nilsson, K.; Qiu, J.; Stevenson, S. H.; Silverman, R. B. *Bioorg. Med. Chem. Lett.* **1999**, *9*, 3093–3098.
63. Qiu, J.; Stevenson, S. H.; O'Beirne, M. J.; Silverman, R. B. *J. Med. Chem.* **1999**, *42*, 329–332.
64. Nicolaou, I.; Zika, C.; Demopoulos, V. J. *J. Med. Chem.* **2004**, *47*, 2706–2709.
65. Chatzopoulou, M.; Mamadou, E.; Juskova, M.; Koukoulitsa, C.; Nicolaou, I.; Stefek, M.; Demopoulos, V. J. *Bioorg. Med. Chem. Lett.* **2011**, *19*, 1426–1433.
66. Lassalas, P.; Oukoloff, K.; Makani, V.; James, M.; Tran, V.; Yao, Y.; Huang, L.; Vijayendran, K.; Monti, L.; Trojanowski, J. Q.; Lee, V. M.-Y.; Kozlowski, M. C.; Smith, A. B.; Brunden, K. R.; Ballatore, C. *ACS Med. Chem. Lett.* **2017**, *8*, 864–868.
67. Herr, J. R. *Bioorg. Med. Chem.* **2002**, *10*, 3379–3393
68. Carini, D. J.; Duncia, J. V.; Aldrich, P. E.; Chiu, A. T.; Johnson, A. L.; Pierce, M. E.; Price, W. A.; Santella, J. B., III; Wells, G. J.; et al. *J. Med. Chem.* **1991**, *34*, 2525–2547.
69. Navarrete-Vazquez, G.; Alaniz-Palacios, A.; Hidalgo-Figueroa, S.; Gonzalez-Acevedo, C.; Avila-Villarreal, G.; Estrada-Soto, S.; Webster, S. P.; Medina-Franco, J. L.; Lopez-Vallejo, F.; Guerrero-Alvarez, J.; et al. *Bioorg. Med. Chem.* **2013**, *23*, 3244–3247.

70. Boyd, M. J.; Bandarage, U. K.; Bennett, H.; Byrn, R. R.; Davies, I.; Gu, W.; Jacobs, M.; Ledeboer, M. W.; Ledford, B.; Leeman, J. R.; et al. *Bioorg. Med. Chem.* **2015**, *25*, 437–444.
71. Krogsgaard-Larsen, P. *J. Med. Chem.* **1981**, *24*, 1377–1383.
72. Lassalas, P.; Gay, B.; Lasfargeas, C.; James, M. J.; Tran, V.; Vijayendran, K. G.; Brunden, K. R.; Kozlowski, M. C.; Thomas, C. J.; Smith, A. B., III; Huryn, D. M.; Ballatore, C. *J. Med. Chem.* **2016**, *59*, 3183–3203.
73. Soll, R. M.; Kenney, W. A.; Primeau, J.; Garrick, L.; McCaully, R. J.; Colatsky, T.; Oshiro, G.; Park, C. H.; Hartupee, D.; White, V.; McCallum, J.; Russo, A.; Dinish, J.; Wojdan, A. *Bioorg. Med. Chem. Lett.* **1993**, *3*, 757–760.
74. Zhang, L.; Lei, J.; Shan, Y.; Yang, H.; Song, M.; Ma, Y. *Mini-Rev. Med. Chem.* **2013**, *13*, 1999–2013.
75. Sheppeck, J. E., II; Gilmore, J. L.; Tebben, A.; Xue, C.-B.; Liu, R.-Q.; Decicco, C. P.; Duan, J. J.-W. *Bioorg. Med. Chem. Lett.* **2007**, *17*, 72769–2774.
76. Chowdhury, M. A.; Abdellatif, K. R. A.; Dong, Y.; Das, D.; Suresh, M. R.; Knaus, E. E. *J. Med. Chem.* **2009**, *52*, 1525–1529.
77. Patani, G. A.; LaVoie, E. J. *Chem. Rev.* **1996**, *96*, 3147–3176.
78. Choudhary, A.; Raine, R. T. *ChemBioChem* **2011**, *12*, 1801–1807.
79. Petersen, K.-U. *Arzneim.-Forsch.* **2002**, *52*, 423–429.
80. Sim, E.; Stanley, L.; Gill, E. W.; Jones, A. *Biochem. J.* **1988**, *251*, 323–323.
81. Le Count, D. In Bindra, J. S.; Lednicer, D. eds. *Chronicles of Drug Discovery.* Wiley: New York, NY, **1982**, pp. 113–132.
82. McLaughlin, M.; Yazaki, R.; Carreira, E. M. *Org. Lett.* **2014**, *16,* 4070–4073.
83. Black, W. C.; Bayly, C. I.; Davis, D. E.; Desmarais, S.; Falgueyret, J.-P.; Leger, S.; Li, C. S.; Masse, F.; McKay, D. J.; Palmer, J. T.; Percival, M. D.; Robichaud, J.; Tsou, N.; Zamboni, R. *Bioorg. Med. Chem. Lett.* **2005**, *15*, 4741–4744.
84. Gauthier, J. Y.; Chauret, N.; Cromlish, W.; Desmarais, S.; Duong, L. T.; Falgueyret, J.-P.; Kimmel, D. B.; Lamontagne, S.; Leger, S.; LeRiche, T.; et al. *Bioorg. Med. Chem. Lett.* **2008**, *18*, 923–928.
85. Isabel, E.; Mellon, C.; Boyd, M. J.; Chauret, N.; Deschenes, D.; Desmarais, S.; Falgueyret, J.-P.; Gauthier, J. Y.; Khougaz, K.; Lau, C. K.; et al. *Bioorg. Med. Chem. Lett.* **2011**, *21*, 920–923.
86. Schirlin, D.; Baltzer, S.; Altenburger, J. M.; Tarnus, C.; Remy, J. M. *Tetrahedron* **1996**, *52,* 305–318.
87. Good, J. A. D.; Kulen, M.; Silver, J.; Krishnan, K. S.; Bahnan, W.; Nunez-Otero, C.; Nilsson, I.; Wede, E.; de Groot, E.; Gylfe, A.; Bergstroem, S.; Almqvist, F. *J. Med. Chem.* **2017**, *60,* 9393–9399.
88. Mohammed, I.; Kummetha, I. R.; Singh, G.; Sharova, N.; Lichinchi, G.; Dang, J.; Stevenson, M.; Rana, T. M. *J. Med. Chem.* **2016**, *59,* 7677–7682.
89. Lagu, B.; Kluge, A. F.; Tozzo, E.; Fredenburg, R.; Bell, E. L.; Goddeeris, M. M.; Dwyer, P.; Basinski, A.; Senaiar, R. S.; Jaleel, M.; et al. *ACS Med. Chem. Lett.* **2018**, *9*, 935–940.
90. Lin, L. S.; Lanza, T. J.; Castonguay, L. A.; Kamenecka, T.; McCauley, E.; Van Riper, G.; Egger, L. A.; Mumford, R. A.; Tong, X.; MacCoss, M.; Schmidt, J. A.; Hagmann, W. K. *Bioorg. Med. Chem. Lett.* **2004**, *14*, 2331–2334.

91. For example, Lewis, R.T.; Macleod, A. M.; Merchant, K. J.; Kelleher, F.; Sanderson, I.; Herbert, R. H.; Cascieri, M. A.; Sadowski, S.; Ball, R. G.; Hoogsteen, K. *J. Med. Chem.* **1995**, *38*, 923–933.
92. Poe, M. M.; Methuku, K. R.; Li, G.; Verma, A. R.; Teske, K. A.; Stafford, D. C.; Arnold, L. A.; Cramer, J. W.; Jones, T. M.; Cerne, R.; et al. *J. Med. Chem.* **2016**, *59*, 10800–10806.
93. Ganellin, C. R. Cimetidine. In Bindra, J. S.; Lednicer, D. eds. *Chronicles of Drug Discovery: Volume 1.* Wiley: New York, NY, **1983**, pp. 1–38.
94. Bradshaw, J. Ranitidine. In Lednicer, D., Ed. *Chronicles of Drug Discovery: Volume 3.* Wiley: New York, NY, **1993**, pp. 45–81.
95. Yanagisawa, I.; Hirata, Y.; Ishii, Y. *J. Med. Chem.* **1987**, *30*, 1787–1793.
96. Sorbera, L. A.; Bayes, M.; Castaner, J.; Silvestre, J. *Drugs Fut.* **2001**, *26*, 1155–1170.
97. Siddique, A.; Shantsila, E.; Lip, G. Y. H. *Therapy* **2008**, *5*, 793–796.
98. Stepan, A. F.; Subramanyam, C.; Efremov, I. V.; Dutra, J. K.; O'Sullivan, T. J.; DiRico, K. J.; McDonald, W. S.; Won, A.; Dorff, P. H.; Nolan, C. E.; et al. *J. Med. Chem.* **2012**, *55*, 3414–3424.
99. Goh, Y. L.; Cui, Y. T.; Pendharkar, V.; Adsool, V. A. *ACS Med. Chem. Lett.* **2017**, *8*, 516–520.
100. Measom, N. D.; Down, K. D.; Hirst, D. J.; Jamieson, C.; Manas, E. S.; Patel, V. K.; Somers, D. O. *ACS Med. Chem. Lett.* **2017**, *8*, 43–48.
101. Shi, J.; Gu, Z.; Jurica, E. A.; Wu, X.; Haque, L. E.; Williams, K. N.; Hernandez, A. S.; Hong, Z.; Gao, Q.; Dabros, M.; et al. *J. Med. Chem.* **2018**, *61*, 681–694.
102. Qiao, J. X.; Cheney, D. L.; Alexander, R. S.; Smallwood, A. M.; King, S. R.; He, K.; Rendina, A. R.; Luettgen, J. M.; Knabb, R. M.; Wexler, R. R.; Lam, P. Y. *Bioorg. Med. Chem. Lett.* **2008**, *18*, 4118–4123.
103. Patel, N. R.; Patel, D. V.; Murumkar, P. R.; Yadav, M. R. *Eur. J. Med. Chem.* **2016**, *121*, 671–698.
104. Pennington, L. D.; Moustakas, D. T. *J. Med. Chem.* **2017**, *60*, 3552–3576.
105. Bryan, M. C.; Falsey, J. R.; Frohn, M.; Reichelt, A.; Yao, G.; Bartberger, M. D.; Bailis, J. M.; Zalameda, L.; San Miguel, T.; Doherty, E. M.; Allen, J. G. *Bioorg. Med. Chem. Lett.* **2013**, *23*, 2056–2060.
106. Zhao, B.; Li, Y.; Xu, P.; Dai, Y.; Luo, C.; Sun, Y.; Ai, J.; Geng, M.; Duan, W. *ACS Med. Chem. Lett.* **2016**, *7*, 629–634.
107. Dosa, P. I.; Amin, E. A. *J. Med. Chem.* **2016**, *59*, 810–840.
108. Mackman, R. L.; Steadman, V. A.; Dean, D. K.; Jansa, P.; Poullennec, K. G.; Appleby, T.; Austin, C.; Blakemore, C. A.; Cai, R.; Cannizzaro, C.; et al. *J. Med. Chem.* **2018**, *61*, 9473–9499.
109. Iwamura, R.; Tanaka, M.; Okanari, E.; Kirihara, T.; Odani-Kawabata, N.; Shams, N.; Yoneda, K. *J. Med. Chem.* **2018**, *61*, 6869–6891.
110. Govek, S. P.; Nagasawa, J. Y.; Douglas, K. L.; Lai, A. G.; Kahraman, M.; Bonnefous, C.; Aparicio, A. M.; Darimont, B. D.; Grillot, K. L.; Joseph, J. D.; et al. *Bioorg. Med. Chem. Lett.* **2015**, *25*, 5163–5167.
111. Richardson, B. G.; Jain, A. D.; Potteti, H. R.; Lazzara, P. R.; David, B. P.; Tamatam, C. R.; Choma, E.; Skowron, K.; Dye, K.; Siddiqui, Z.; et al. *J. Med. Chem.* **2018**, *61*, 8029–8047.

112. Johansson, A.; Löberg, M.; Antonsson, M.; von Unge, S.; Hayes, M.; Judkins, R.; Ploj, K.; Benthem, L.; Linden, D.; Brodin, P.; et al. *J. Med. Chem.* **2016**, *59*, 2497–2511.
113. Bollag, G.; Hirth, P.; Tsai, J.; Zhang, J.; Ibrahim, P. N.; Cho, H.; Spevak, W.; Zhang, C.; Zhang, Y.; Habets, G.; et al. *Nature* **2010**, *467*, 596–599.
114. Rabal, O.; Amr, F. I.; Oyarzabal, J. *J. Chem. Inform. Model.* **2015**, *55*, 1–18.
115. Rezvanfar, M. A.; Rahimi, H. R.; Abdollahi, M. *Exp. Opin. Drug Metab. Toxicol.* **2012**, *8*, 1231–1245.
116. Harris, P. A.; Berger, S. B.; Jeong, J. U.; Nagilla, R.; Bandyopadhyay, D.; Campobasso, N.; Capriotti, C. A.; Cox, J. A.; Dare, L.; Dong, X.; et al. *J. Med. Chem.* **2017**, *60*, 1247–1261.
117. Romagnoli, R.; Baraldi, P. G.; Salvador, M. K.; Prencipe, F.; Bertolasi, V.; Cancellieri, M.; Brancale, A.; Hamel, E.; Castagliuolo, I.; Consolaro, F.; et al. *J. Med. Chem.* **2014**, *57*, 6795–6808.
118. Hess, S.; Linde, Y.; Ovadia, O.; Safrai, E.; Shalev, D. E.; Swed, A.; Halbfinger, E.; Lapidot, T.; Winkler, I.; Gabinet, Y.; et al. *J. Med. Chem.* **2008**, *51*, 1026–1034.
119. Van der Poorten, O.; Knuhtsen, A.; Sejer Pedersen, D.; Ballet, S.; Tourwe, D. *J. Med. Chem.* **2016**, *59*, 10865–10890.
120. Räder, A. F. B.; Reichart, F.; Weinmueller, M.; Kessler, H. *Bioorg. Med. Chem.* **2018**, *26*, 2766–2773.
121. Rand, A. C.; Leung, S. S. F.; Eng, H.; Rotter, C. J.; Sharma, R.; Kalgutkar, A. S.; Zhang, Y.; Varma, M. V.; Farley, K. A.; Khunte, B.; Limberakis, C.; Price, D. A.; Liras, S.; Mathiowetz, A. M.; Jacobson, M. P.; Lokey, R. S. *MedChemComm* **2012**, *3*, 1282–1289.
122. Biron, E.; Chatterjee, J.; Ovadia, O.; Langenegger, D.; Brueggen, J.; Hoyer, D.; Schmid, H. A.; Jelnick, R.; Gilon, C.; Hoffman, A.; et al. *Angew. Chem. Int. Ed.* **2008**, *47*, 2595–2599.
123. Nielsen, D. S.; Hoang, H. N.; Lohman, R.-J.; Hill, T. A.; Lucke, A. J.; Craik, D. J.; Edmonds, D. J.; Griffith, D. A.; Rotter, C. J.; Ruggeri, R. B.; et al. *Angew. Chem. Int. Ed.* **2014**, *53*, 12059–12063.
124. Price, D. A.; Mathiowetz, A. M.; Liras, S. Designing Orally Bioavailable Peptide and Peptoid Macrocycles. In Marsault, E.; Peterson, M. L. eds. *Practical Medicinal Chemistry with Macrocycles*. Wiley: Hoboken, NJ, **2017**, pp. 59–76.

5

Structural Alerts for Toxicity

Structural alerts are functional groups or fragments of drugs that have potential to cause toxicity. As a medicinal chemist, being aware of structural alerts for toxicophores helps to prioritize compounds to make. Many reviews on structural alerts have been published.[1–7]

The concept of structural alerts is controversial. While medicinal chemists are more and more conscientious of the potential peril of "ugly" structures, some believe it has gone too far, especially the proliferation and overly reliance of quantitative structure–activity relationship (QSAR) computational models.[8] But knowledge is power: drug discovery has become such an expensive enterprise that having the knowledge of structural alerts with potential toxicity is helpful in ranking and prioritizing compounds. Without impeding creativity and innovation, the chances of success are better if we could replace structural alerts with "beautiful" structures when everything else is equal. Rather than viewing structural alerts as "black-and-white" and "forbidden," when in doubt, we need to let data speak.

If one perceives structural alerts as a dogma, then the knowledge does more harm than good for successful drug design. Therefore, for each structural alert discussed here, exceptions are also presented as a counter-argument so that the message is clear: structural alerts are merely alerts.

5.1 Reactive Electrophiles

Amino acids are building blocks of peptides and proteins such as enzymes and receptors. Many amino acids possess nucleophiles: thiol on cysteine (C), hydroxyl on serine (S), and carboxylate on aspartic acid (D) are just a few examples. If a drug contains reactive electrophiles, they may react with the endogenous nucleophiles to form covalent bonds, which may result in toxicity. Here is a list of common nucleophiles encountered in human body:

1. Amines, $R–NH_2$, R_2NH, R_3N
2. Alcohols, R–OH
3. Sulfides, $R–S^{\ominus}$
4. Thiols, R–SH
5. Carboxylates, $R\text{-}CO_2^{\ominus}$

Medicinal Chemistry for Practitioners, First Edition. Jie Jack Li.

5.1.1. Alkylating Agents

Primary halides, with the exception of the fluorides, are alkylating agents—good leaving groups when attacked by the nucleophiles via the S_N2 mechanism. However, the fluorine atoms on 2-fluoroacetic acid and α-fluoroketones are activated thus good enough leaving groups ripe for nucleophilic attack. In contrast, trifluoromethyl and difluoromethyl groups are not leaving groups at all, hence not alkylating agents.

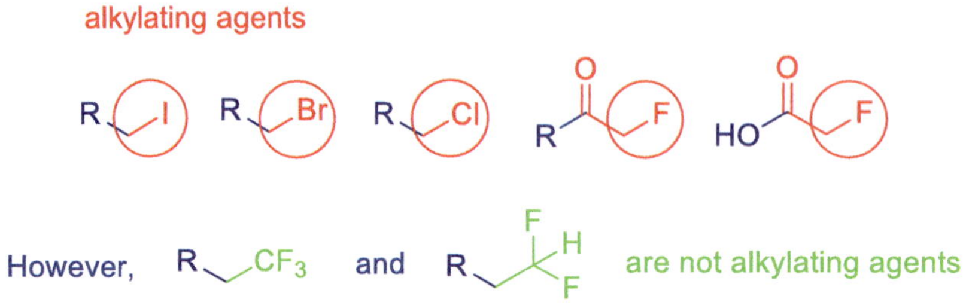

Dimethyl sulfate (Me_2SO_4) and methyl iodide (MeI) are routinely used in organic synthesis as methylating agents and must be handled with care. Inhalation of a copious amount of either has caused death via drowning since all polar groups in the lung would be methylated.

Alkylating agents are a double-edged sword. Ironically, the first chemotherapy to treat cancer owed its genesis to chemical weapon mustard gas (**1**). It was discovered in 1943 that mustard gas killed more white cells than normal cells. American soldiers were subjected to mustard gas intentionally as an experimentation of cancer treatment. Since gases are not convenient to administer, mechlorethamine (Nitrogen mustard, **2**) was invented as an intravenous injection. It was followed by chlorambucil (Leukeran, **3**), cyclophosphamide (Cytoxan, **4**), and busulfan (Myleran, **5**). Compounds **2**–**4** are collectively known as nitrogen mustards.

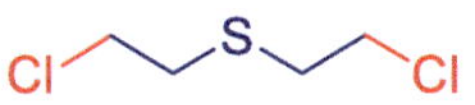

mustard gas (**1**)

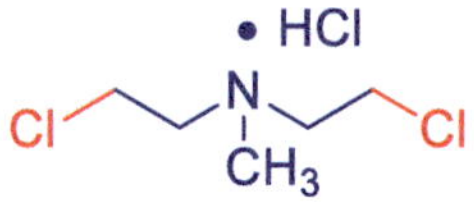

mechlorethamine (Nitrogen mustard, **2**)

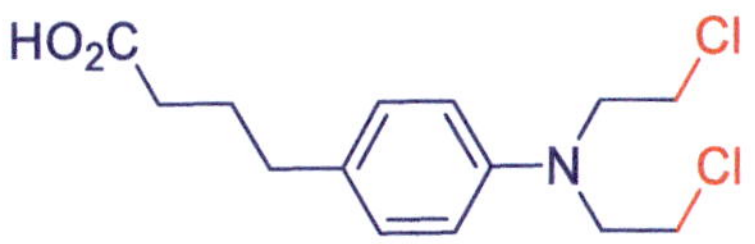

chlorambucil (Leukeran, **3**)

cyclophosphamide (Cytoxan, **4**)

busulfan (Myleran, **5**)

The mechanism of action (MOA) of nitrogen mustards is via alkylation of the ring nitrogen (or exocyclic oxygen) atoms of DNA bases, leading to a nonreplicating form of DNA and death of the malignant cells. Mechlorethamine (**2**) forms an aziridinium cation intermediate **6** under physiological conditions. The basic amine atom on either adenine or guanine of DNA strain **7** would attack of aziridinium **6** to produce mono-alkylated DNA **8**, which could undergo an additional nucleophilic attack of aziridinium **6** to deliver interstranded DNA **9**. Both **8** and **9** cannot be replicated, causing death of the malignant cells.[9]

Regrettably, alkylating agents **1**–**5** are not selective against normal cells. Like most "carpet-bombing" chemotherapies, they cause severe toxicity by alkylating other fast-growing cells of hair follicles, bone marrow, lymph nodes, and epithelium. They inflict damage of inner lining of intestines, depress bone marrow, and cause hair loss. Therefore, we should become alert whenever there is presence of an alkylating group on a drug.[10]

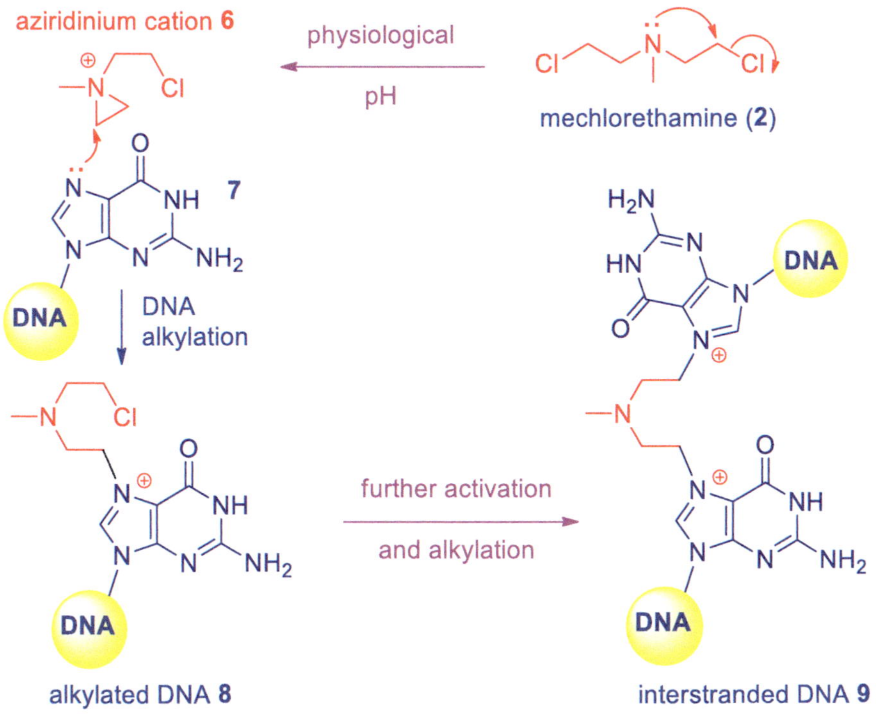

Similarly, a non-selective irreversible α receptor antagonist phenoxy-benzamine (Dibenzyline, **10**) also works as an alkylating agent. Under physiological conditions, phenoxybenzamine (**10**) exists in a more reactive form of aziridinium **11**, which is susceptible to attack by nucleophiles such as thiol, hydroxyl, and acetate groups, as well as off-target nucleophilic substitutions. The on-target covalent bond formation takes place from the nucleophilic attack by the cysteine at the 3.36 position in transmembrane helix 3 (TM3) of the α receptor to form a stable linkage on covalent adduct **12**.[11] Its

MOA as a nonselective irreversible alkylating agent also contributes to its toxicity such as reflex tachycardia.

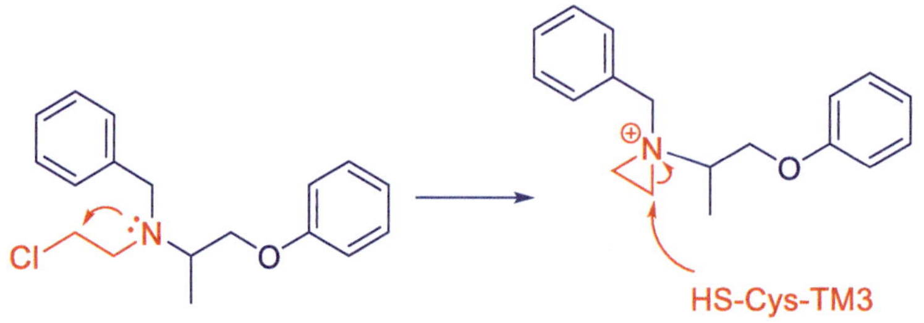

phenoxybenzamine (Dibenzyline, **10**)
non-selective irreversible
α receptor antagonist

aziridinium intermediate **11**

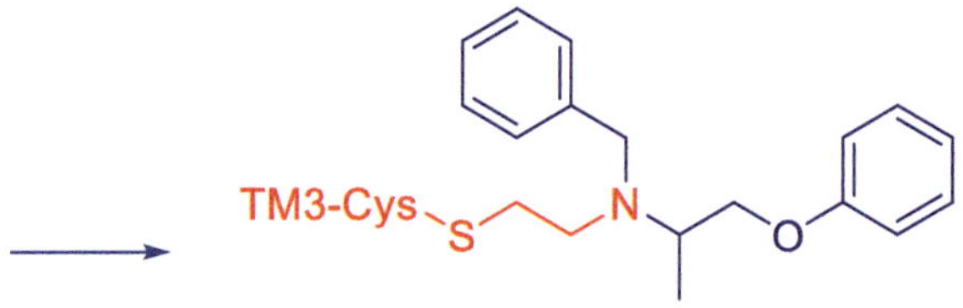

covalent adduct **12**

Parke–Davis′ chloramphenicol (Chloromycetin, **13**, 1949), a broad-spectrum antibiotic, possesses both nitrophenyl and dihaloalkane structural alerts. Its rare bone marrow toxicity may be a consequence of the nitrosophenyl metabolite derived from reduction of the nitrophenyl moiety; however, it could be the result of metabolism of the dichloromethyl substituent as well. Metabolically, chloramphenicol (**13**) is oxidized by CYP450 to give hydroxyl **14**, which loses a molecule of HCl to provide acid chloride **15**. Subsequently, the very reactive electrophile **15** may react with a number of nucleophiles. If the nucleophile happens to be water, the end-product would be oxamic acid **16**, which was detected in both rats and humans as one the major metabolites.[11] Incidentally, metabolism of chloroform follows a similar pathway to produce phosgene, a highly reactive chemical.

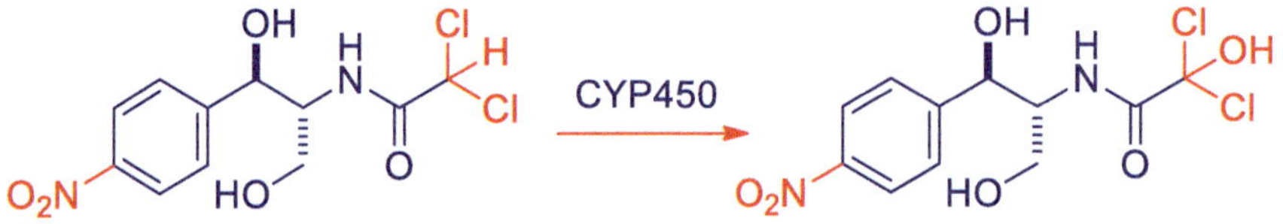

chloramphenicol (Chloromycetin, **13**)

hydroxyl metabolite **14**

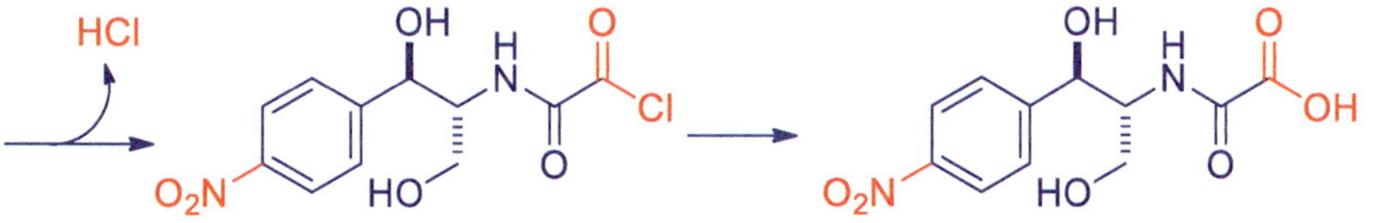

acyl chloride **15** oxamic acid **16**

Recently, 2-chloropropionamide was identified as a low-reactivity electrophile "warhead" for irreversible small-molecule probe identification. In particular, (*S*)-CW3554 (**17**) selectively labeled protein disulfide isomerase (PDI) thus inhibiting its enzymatic activity. Subsequent profiling against five diverse cancer cell lines revealed (*S*)-CW3554 (**17**)'s unique cytotoxicity in cells derived from multiple myeloma (MM), a cancer recently reported to be sensitive to PDI inhibition. This novel PDI inhibitor highlights the potential of 2-chloropropionamides as weak and stereochemically tunable electrophiles for covalent drug discovery. Interestingly, the analogous (*R*)-α-chloropropionamide specifically labeled a distinct protein, the aldehyde dehydrogenase (ALDH), underlining the role of the stereochemistry at the α-position of the warhead.[13] However, it was noted that PDI contains strongly nucleophilic active site cysteines and the potential of α-chloropropionamides in targeted covalent inhibitor (TCI) design remains to be demonstrated.[14]

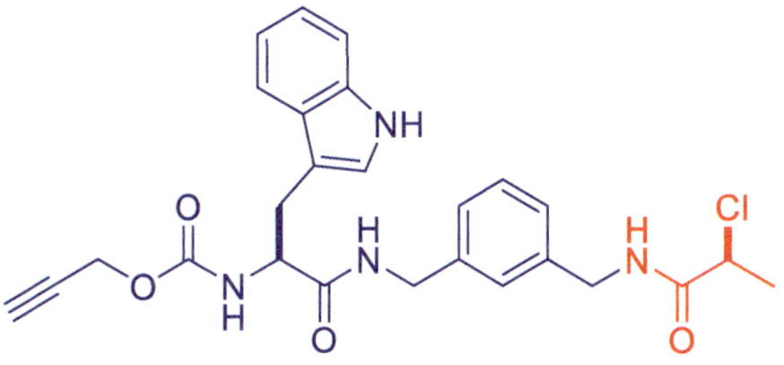

(*S*)-CW3554 (**17**)
selective PDI inhibitor
cytotoxic to multiple
myeloma cell lines

In addition to alkylating agents shown here, several other classes of chemotherapy function as alkylating agents as well.[13] They include monofunctional alkylating agents such as dacarbazine (**18**), procarbazine, streptozotocin, temozolomide, and triazene, as well as bifunctional alkylating agents such as aziridines, altretamine, mitomycin, and thiotepa.

The common toxicities associated with anticancer drug dacarbazine (**18**), which functions as a DNA methylating agent, include hepatic necrosis and hemopoietic depression. In terms of metabolism, demethylation of dacarbazine (**18**) provides demethylated **20** via the intermediacy of hydroxymethyl **19**. Triazene **21**, as the tautomer of **20**, loses a molecule of nitrogen to deliver aminoimidazole **22**, along with methyl carbocation. Although DNA methylation is the mechanism of action of dacarbazine (**18**), indiscriminant methylation of proteins by the methyl carbocation could trigger the hepatotoxicity associated with the drug.[15] In fact, dacarbazine (**18**)'s triazene motif is also a structural alert. See Section 5.3 for more information.

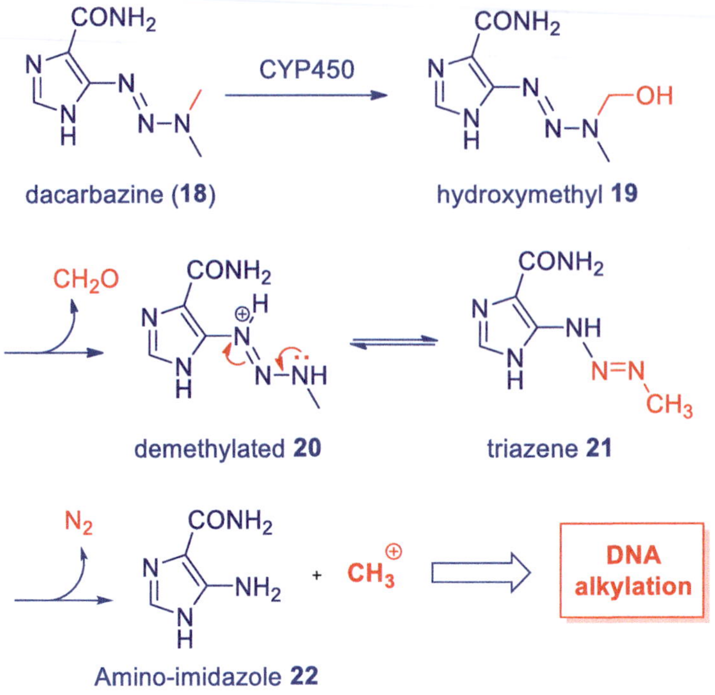

5.1.2. Michael Acceptors

What a difference a decade or two makes! In the 2000s, Michael acceptors were frowned upon by almost everyone. Today in 2019, six drugs with Michael acceptors are now on the market, mostly as anticancer drugs. Many more targeted covalent drugs are progressing at different stages in the pipeline.

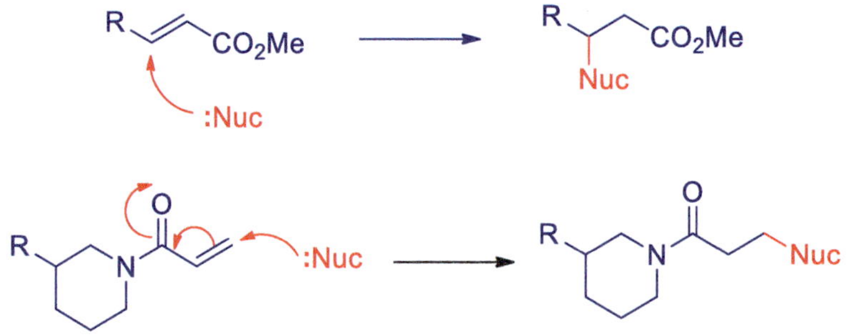

What changed? Mother Nature has not changed. It is our appreciation that has changed. We now understand that if a drug binds to the target tightly and selectively enough, a Michael acceptor as the "warhead" is not a liability, but an advantage. Section 1.2.3 showcases many drugs with Michael acceptors as warheads.

Michael acceptors as warheads in drugs existed long before they became fashionable. Aspirin, clopidogrel, and omeprazole are but a few older covalent drugs. Merck's finasteride (Proscar, Propecia, **23**) as a steroid 5α-reductase irreversible inhibitor possesses an α,β-unsaturated lactam as the Michael acceptor. As shown below, nicotinamide adenine dinucleotide phosphate (NADPH) delivers a hydride as the nucleophile, which adds to the α,β-unsaturated lactam on finasteride (**23**) to afford intermediate **24**. Enol **24** then reacts with the pyridinium intermediate from NADPH to form a covalent bond to deliver adduct **25**.[16] Therefore, finasteride (**23**) is a *bona fide* covalent inhibitor and its Michael acceptor as a structural alert does not damage its safety profile.

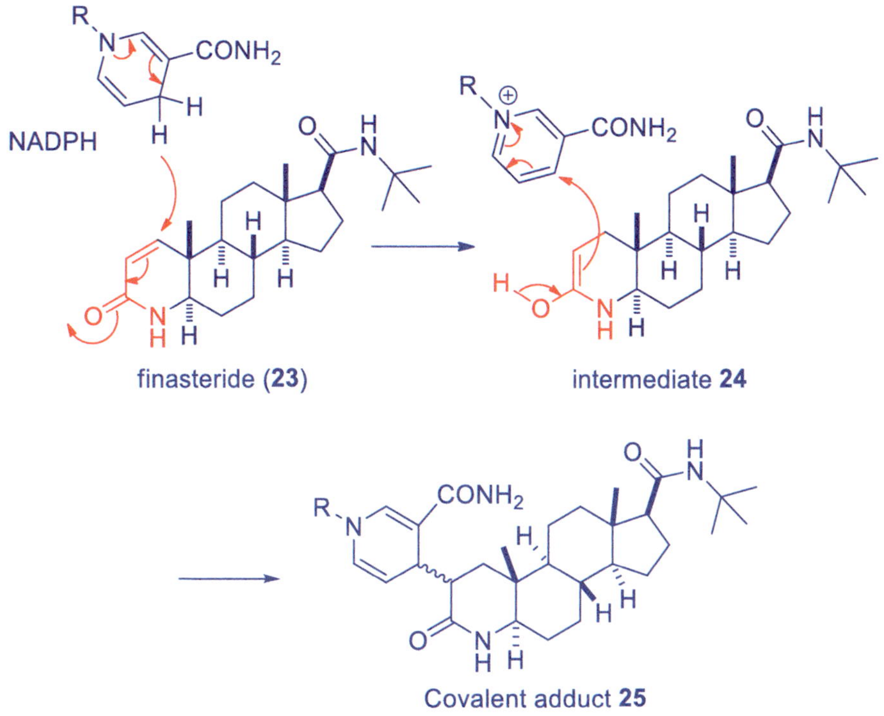

Finasteride (**23**) exhibits very slow offset kinetics from its target, the 5α-reductase enzyme. Although containing an α,β-unsaturated lactam as a Michael acceptor as a structural alert, it is selectively recognized by the two 5-reductase isoenzymes and off-target modification of other biomolecules is minimal.

It poses certain challenge when a Michael acceptor is "in disguise" as a metabolite. GSK's nucleoside reverse transcriptase inhibitor (NRTI) abacavir (Ziagen, **26**) is oxidized by aldehyde dehydrogenase (ALDH) to the corresponding β,γ-unsaturated aldehyde **27**, which converts to the thermodynamically more stable α,β-unsaturated aldehyde **28**. While the aldehyde functionality on both **27** and **28** may react with nucleophiles to form cyclic adducts, conjugated aldehyde **28**, as a Michael acceptor, is prone for conjugated addition to form covalent linkage. The Michael adducts are suspected to be partially responsible for abacavir (**26**)′s cases of hypersensitivity in approximately 4% of the patient population.[17]

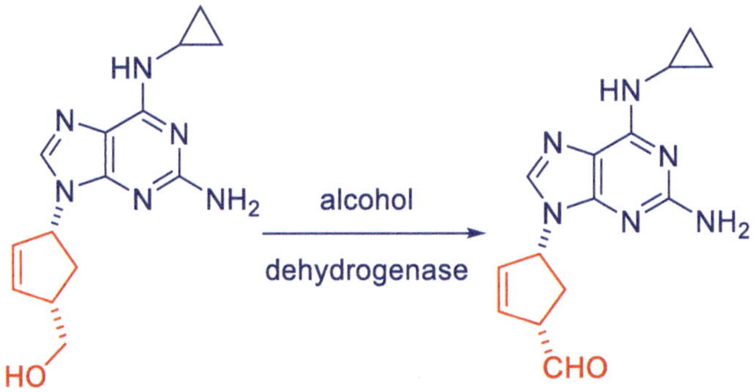

abacavir (Ziagen, **26**)
GSK, 1998, nucleoside
reverse transcriptase inhibitor

β,γ-unsaturated aldehyde **27**

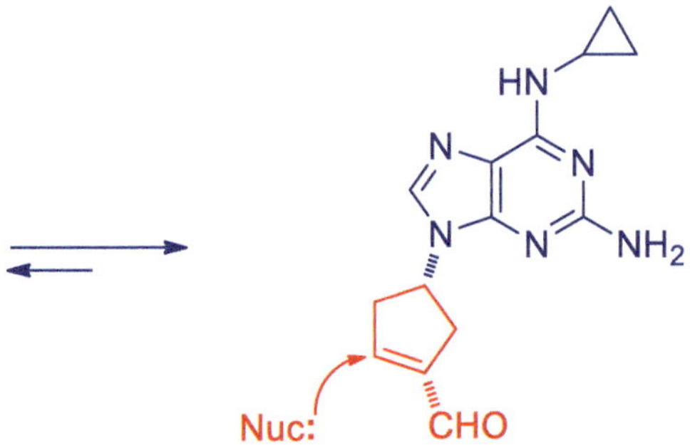

α,β-unsaturated aldehyde **28**

Below is a collection of some representative Michael acceptors as warheads for covalent inhibitors:

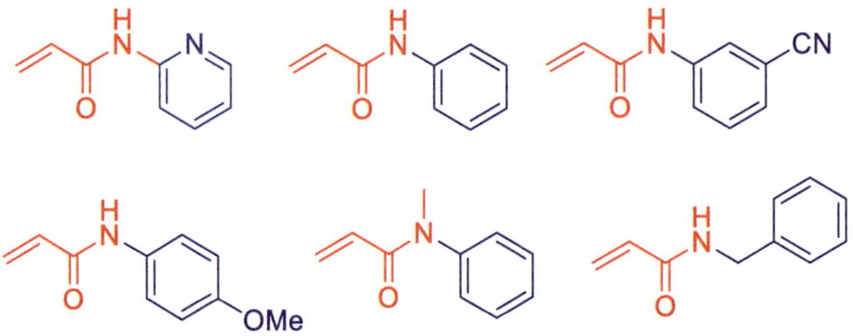

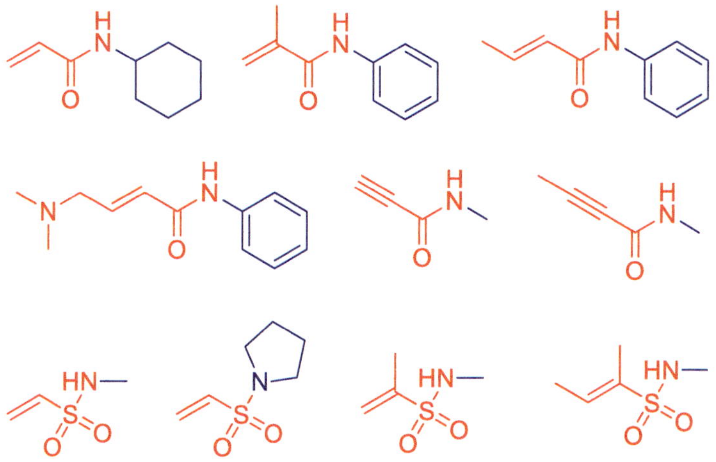

To conclude, a Michael acceptor is still a structural alert unless it serves as a warhead of a molecule that tightly and selectively binds to the target, accompanying minimal off-target binding.

5.1.3. Heteroaromatic Halides

The halides of either α- or γ-halopyridine readily undergo nucleophilic substitution on an aromatic ring (S_NAr), forming a covalent linkage with the nucleophile. If the reaction lacks selectivity for the intended target, the drug's off-target toxicity may manifest.

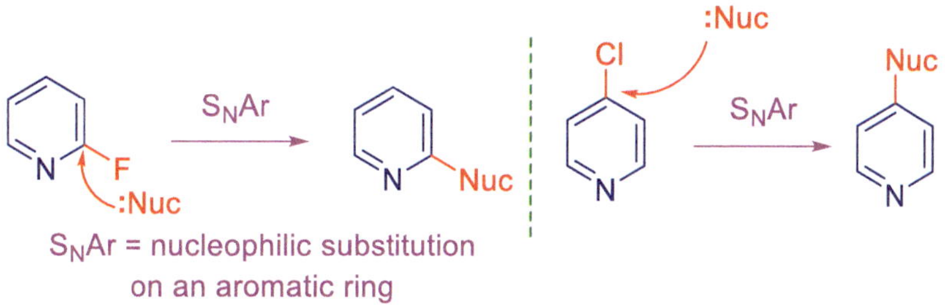

Not surprisingly, some α-haloheteroaromatic fragments such as α-chloropyridine and α-chloro-1,3,4-thiadiazole are now being investigated as warheads of covalent inhibitors:

But not all α-haloheteroaromatic fragments are bad. Leukemia drug cladribine (Leustatin, **30**), also known as 2-chloro-2′-deoxyadenosine (2CdA), functions as a purine analog antimetabolite. It possesses a 2-chlorine substituent on the pyrimidine ring. Unlike adenosine (**29**), cladribine (**30**)′s chlorine renders it partially resistant to breakdown by adenosine deaminase (ADA). When treated with glutathione and *N*-α-acetyl lysine, cladribine (**30**)′s daily covalent binding burden was found to be under 1 mg/d. Normally, if a compound has less than 10 mg/d estimated daily covalent binding burden, the compound is more likely to be safe. In fact, in literature, cladribine (**30**) is well tolerated with little drug-related hypersensitivity—a testimony to its selectivity for the biological target.[18]

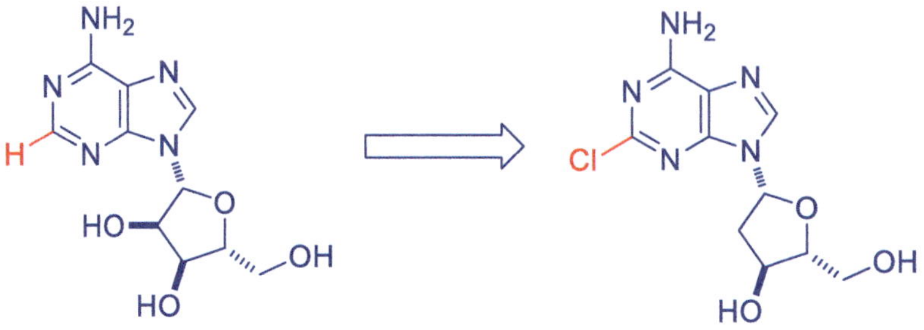

adenosine (**29**) cladribine (Leustatin, **30**)

In the same vein, angiotensin II (AT1) receptor antagonist losartan (Cozaar, **31**) is a quite safe medicine for treating hypertension. The α-chlorine atom on the imidazole ring poses little off-target toxicity.[19] Furthermore, S_NAr reaction of five-membered α-halo-heteroaromatics is not as facile as their six-membered counterparts. All other angiotensin II receptor antagonists (e.g., irbesartan, valsartan, candesartan, and telmisartan), for better or worse, steered away from the α-chloroimidazole structure.

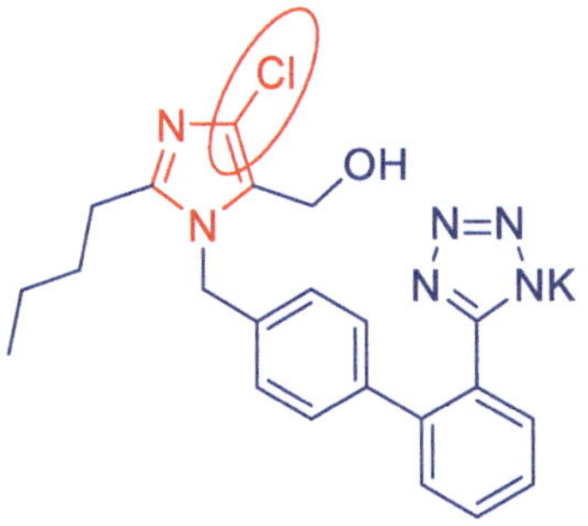

losartan (Cozaar, **31**)
BMS, 1995
angiotensin receptor blocker

Again, potency and selectivity matter. Even with the presence of an α-halohetero-aromatic structural alert, if there is little off-target binding, chances are the drug may not pose much toxicity issue.

5.1.4 Miscellaneous Reactive Electrophiles

Acid chlorides, acid anhydrides, isocyanates, and isothiocyanates are among the more reactive electrophiles. The challenge is to recognize them when they are generated as active metabolites via *in vivo* metabolism. For instance, one of the reactive metabolites of chloramphenicol (**10**) is acyl chloride **12**. Both hydroxamic acid and thiazolidinedione (TZD) may give rise to isocyanates as their reactive metabolites.

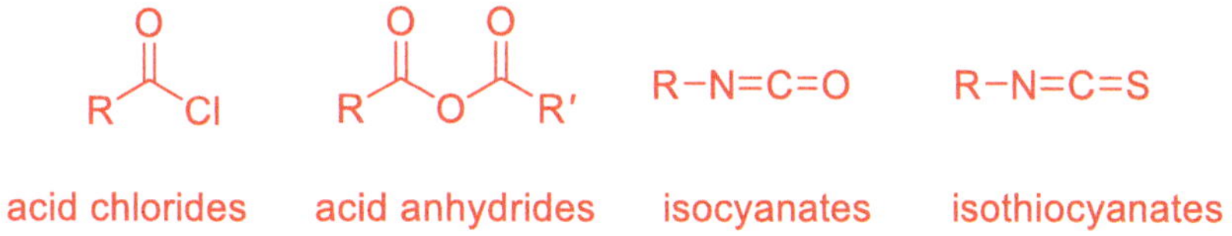

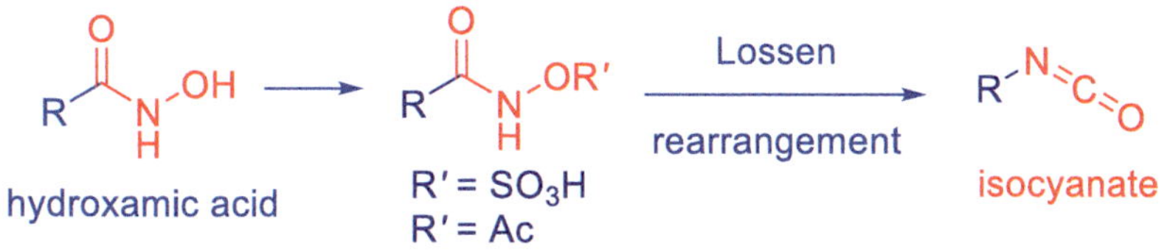

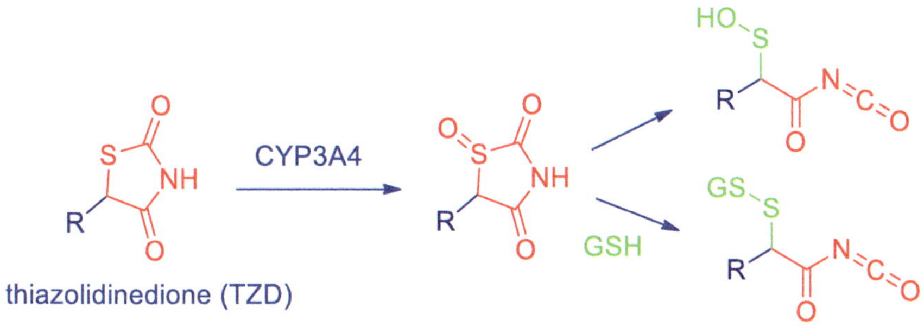

Sulfonyl fluorides[20] and fluorosulfates,[21] with potential of forming selective covalent inhibitors, have attracted much attention lately as chemical probes. Many additional reactive electrophiles such as carbodiimide, oxazolines, cyanoamines, and Woodward's reagent K, may also serve as warheads in designing covalent inhibitors.

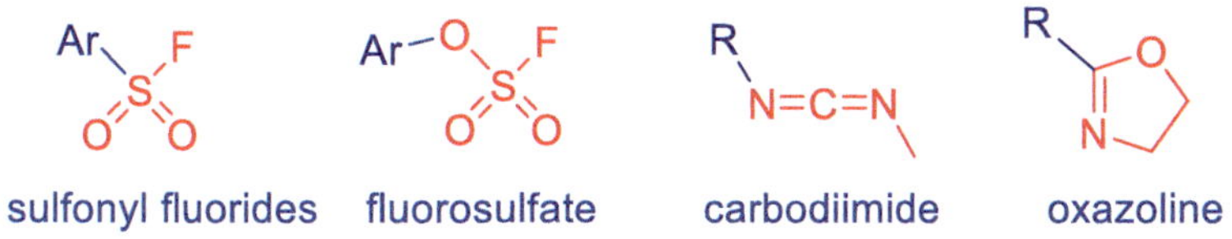

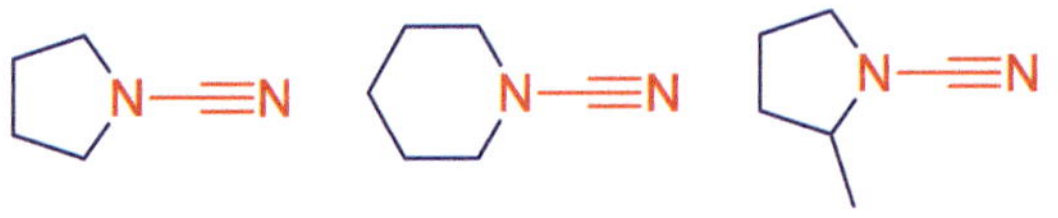

cyanoamines

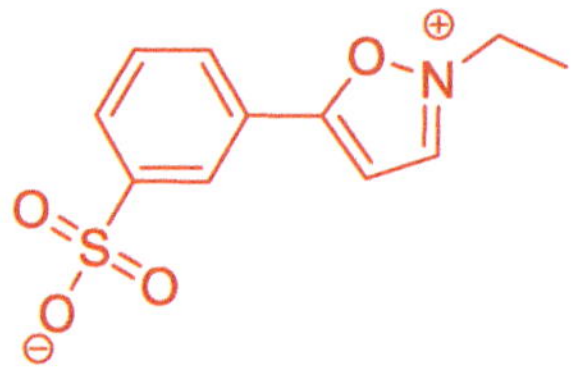

Woodward's reagent K

5.2 DNA Intercalators

Drugs targeting DNA include DNA alkylating agents and DNA intercalating agents. DNA intercalation is the MOA for several classes of drugs. There are six major modes for reversible binding of molecules with double-helical DNA: (i) electrostatic attractions with the anionic sugar-phosphate backbone of DNA; (ii) interactions with the DNA major groove; (iii) interactions with the DNA minor groove; (iv) intercalation between base pairs *via* the DNA major groove; (v) intercalation between base pairs *via* the DNA minor groove; and (vi) threading intercalation mode. The MOA of quinolone anti-malarial drugs such as quinine (**32**) and chloroquine (**33**) is intercalating malaria parasites′ DNA transcription. Platinum-containing cancer drugs cisplatin (Platinol, **34**), carboplatin (Paraplatin, **35**), and oxaliplatin (Eloxatin, **36**),[22] as well as additional cancer drugs *N*-(2-(dimethylamino)-ethyl)-acridine-4-carboxamide (DACA, **37**), camptothecin (**38**), and daunomycin (**39**) also function as DNA intercalators.[23] DACA (**37**) is actually a DNA intercalating dual topoisomerase I/II poison.

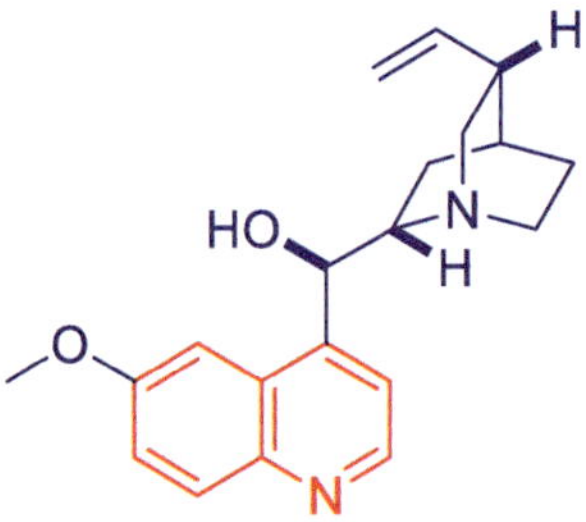

quinine (**32**)

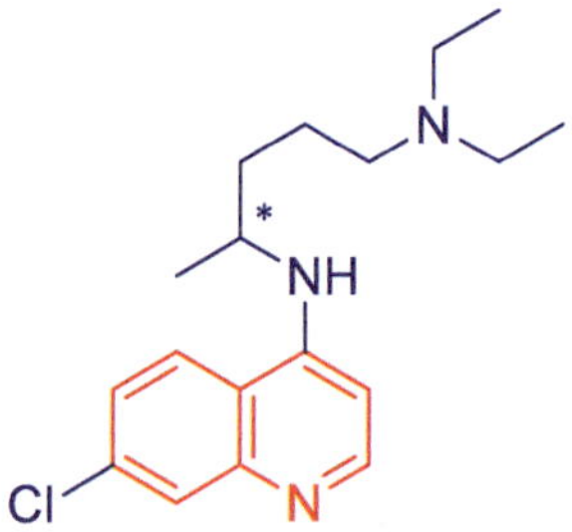

chloroquine (**33**)

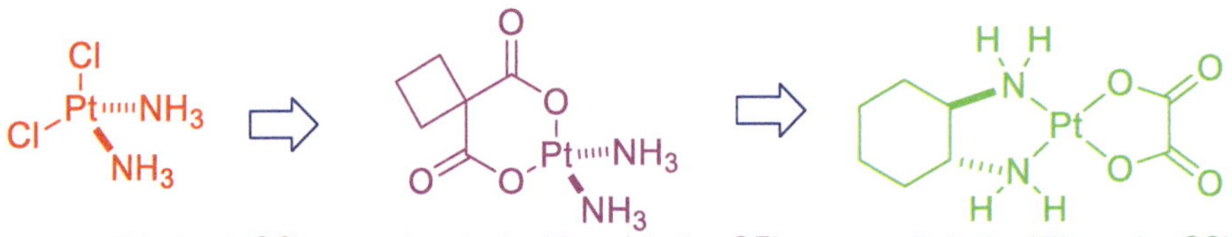

cisplatin (Platinol, **34**) renal damage

carboplatin (Paraplatin, **35**) much lower nephrotoxicity

oxaliplatin (Eloxatin, **36**) devoid of nephrotoxicity

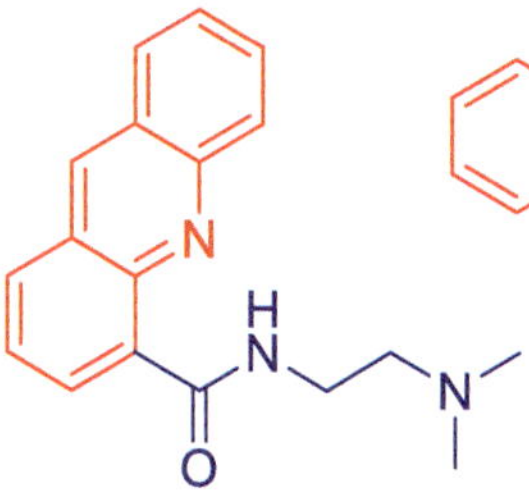
DACA (**37**)

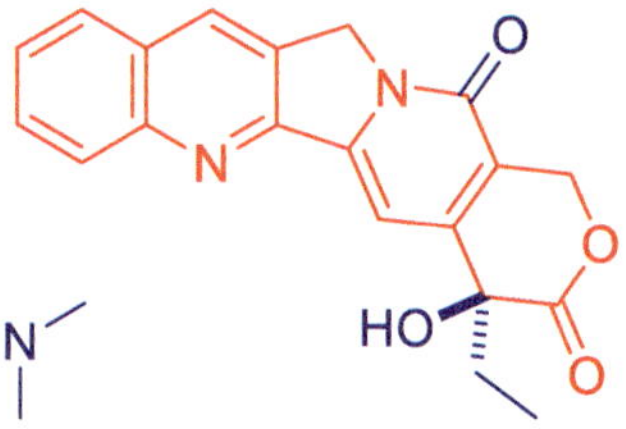
camptothecin (**38**)

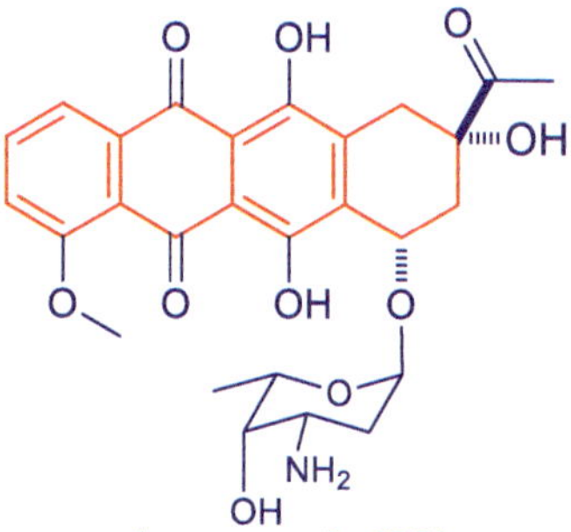
daunomycin (**39**)

DNA intercalators kill malignant cells while sometimes causing mutagenicity to normal cells at the same time. Their toxicities manifest as poisons of DNA topoisomerases, causing mutation and cancer.[24] Polycyclic aromatic hydrocarbons such as tricycles including psoralen (furocoumarin, **40**), fluorene (**41**), and carbazole (**42**) tend to be DNA intercalators that require special vigilance with regard to their toxicities.

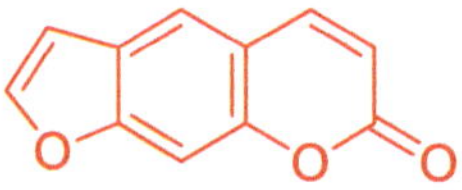
psoralen (furocoumarin, **40**)

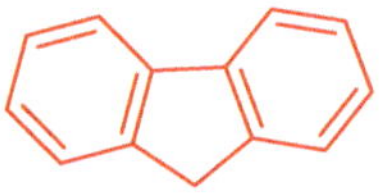
fluorene (**41**)

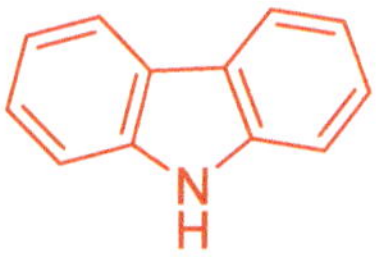
carbazole (**42**)

5.3 Carcinogens

Carcinogens are substances that cause cancer. *N*-Nitrosoamines (nitrosamines) and *N*-nitrosoamides are potent carcinogens. Meanwhile, *N*-nitrosoureas **43–45** are the early antineoplastic chemotherapies that are both carcinogenic and mutagenic. They are readily cleaved by nucleophiles to give *N*-nitrosoamines, which then lose water and nitrogen gas to yield strongly alkylating carbocations.

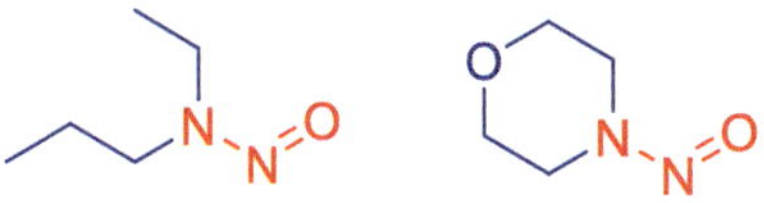
N-nitrosoamines

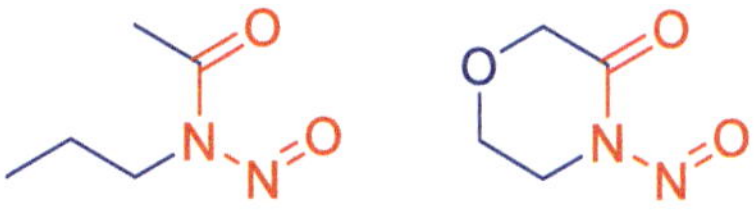
N-nitrosoamides

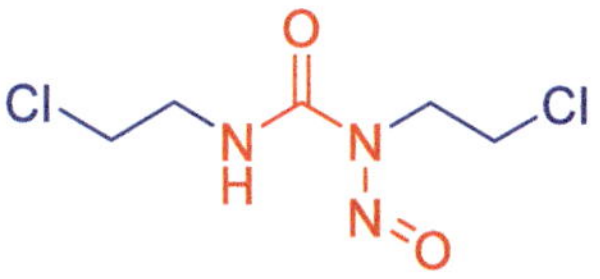
carmustine (BCNU, **43**)

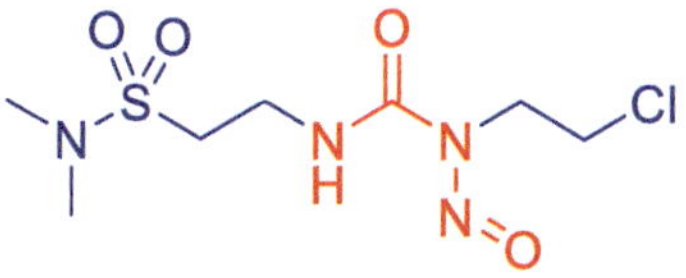
tauromustine (TCNU, **44**)

N-nitrosoureas

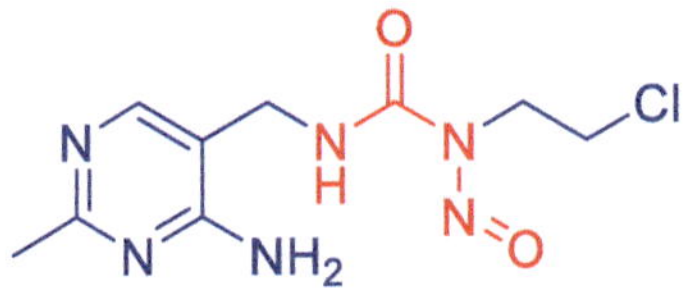

nimustine (ACNU, **45**)
N-nitrosourea

Like *N*-nitrosoamines and *N*-nitrosoamides, aflatoxins such as AFB1 and AFG2 are potent carcinogens to be avoided in designing drugs. Many polycyclic aromatic hydrocarbons are nongenotoxic carcinogens that do not cause direct DNA damage but induce cancer via other mechanisms.

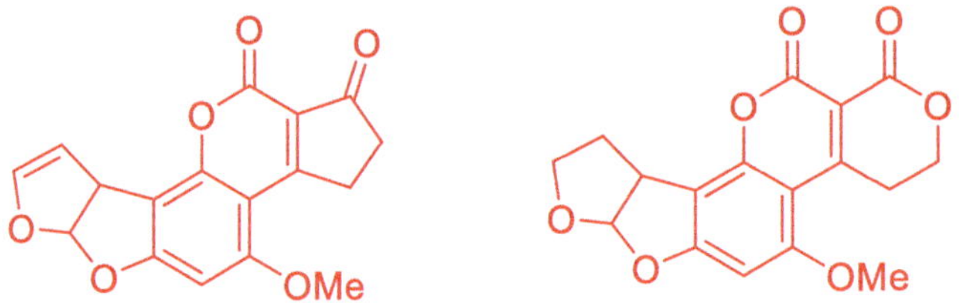

AFB1 AFG2

Carcinogenic structures *per se* are easy to recognize while carcinogenic metabolites are more challenging to discern. An experienced medicinal chemist becomes aware of drugs with potential carcinogenic metabolites. Diazo compounds, triazenes, and hydrazines may result in carcinogenic metabolites so they are considered as structural alerts as well.

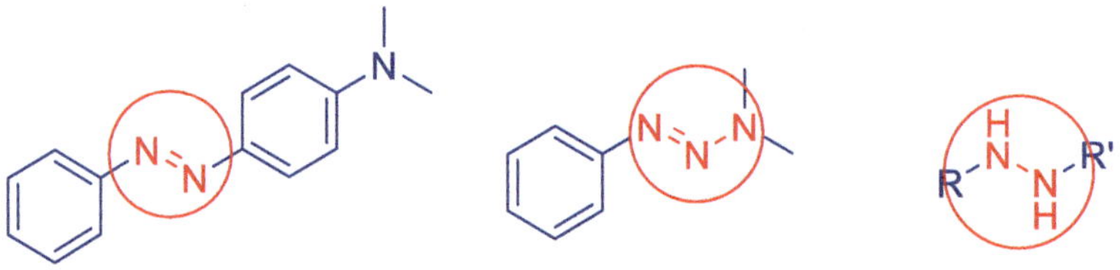

diazo dyes triazenes hydrazines

The best-known diazo drug is probably Prontosil (**46**) discovered by Domagk in 1932. It is a prodrug, and the diazo group is cleaved in the gut by intestinal bacteria. Another diazo drug is analgesic phenazopyridine (Pyridium, **47**). Triazene-containing drugs exemplified by antineoplastic dacarbazine (**18**)[15] and temozolomide (Temodar, **48**) are toxic as well. After possible liabilities of the diazo group became known, these older drugs are no longer widely used.

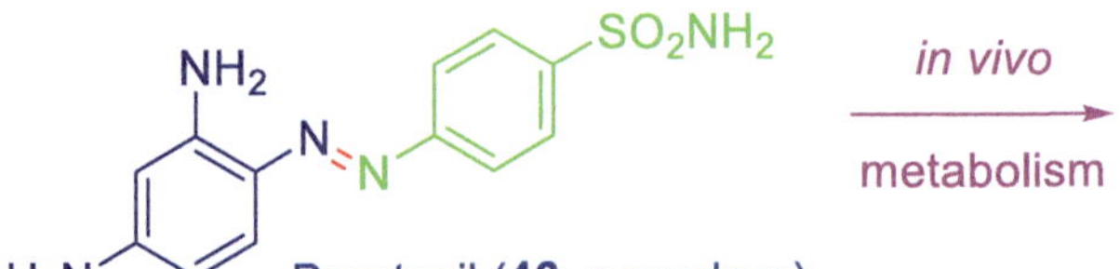

Prontosil (**46**, a prodrug)

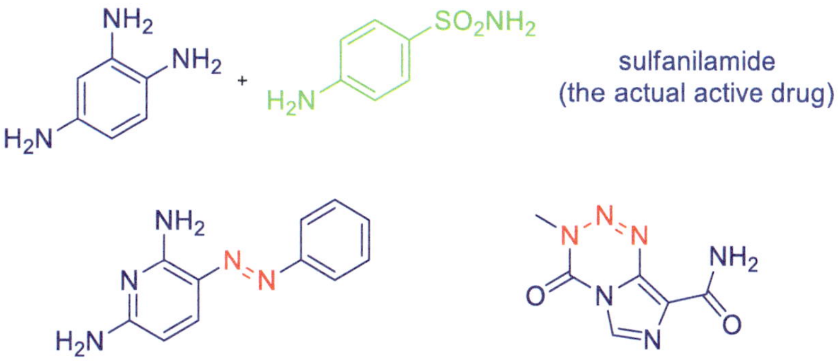

phenazopyridine (Pyridium, **47**) temozolomide (Temodar,**48**)

5.4 Metabolism Problematic Molecules

Cytochrome P450 enzymes are the major "engine" for drug metabolism. As our understanding deepens concerning drug metabolism, we now understand that some idiosyncratic toxicities are the consequence of reactive drug metabolites.[25]

5.4.1 Anilines and Anilides

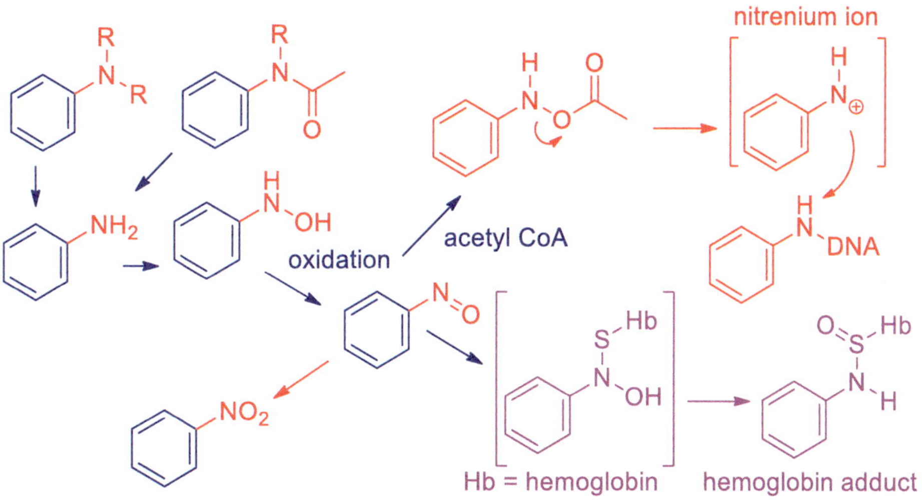

Drugs with anilines and anilides may be mutagenic, often causing methemoglobinemia (MetHb↑). Their reactive metabolites are largely activated by CYP450 oxidation. As shown above,[26] alkylaniline and anilides may be converted to primary aniline, which is oxidized to hydroxylamine by CYP 1A2 (and 1A1 to a lesser extent). The hydroxylamine intermediate itself is not chemically reactive, but it is readily further oxidized to nitrosobenzene, a known carcinogen and highly reactive. Once the carcinogenic nitrosobenzene metabolite is generated, three fates await: (i) it can be further oxidized to nitrobenzene; or (ii) reacts with acetyl CoA to form an *O*-acyl hydroxylamine, which is converted to a highly reactive nitrenium ion. The nitrenium ion is then trapped by endogenous biomolecules such as DNA to form adducts; or (iii) alternatively, the nitroso intermediate is trapped by a sulfur nucleophile such as the cysteine residue in hemoglobin to generate hemoglobin adducts.

Reactive metabolites from aniline metabolic activation render some anilines mutagens, although aniline itself, *per se*, is not mutagenic.[27] Due to the effects of reactive metabolites of aniline-containing drugs, some of them saw idiosyncratic toxicities including hypersensitivity, hepatotoxicity, and agranulocytosis. These aniline-containing drugs include leprosy treatment dapsone (with two anilines), sulfa drug sulfathiadiazine, and anti-arrhythmic agent procainamide. The fact that they are still useful as medicines may contribute to the fact that all three of them are attached to electron-withdrawing groups to their aniline moieties.

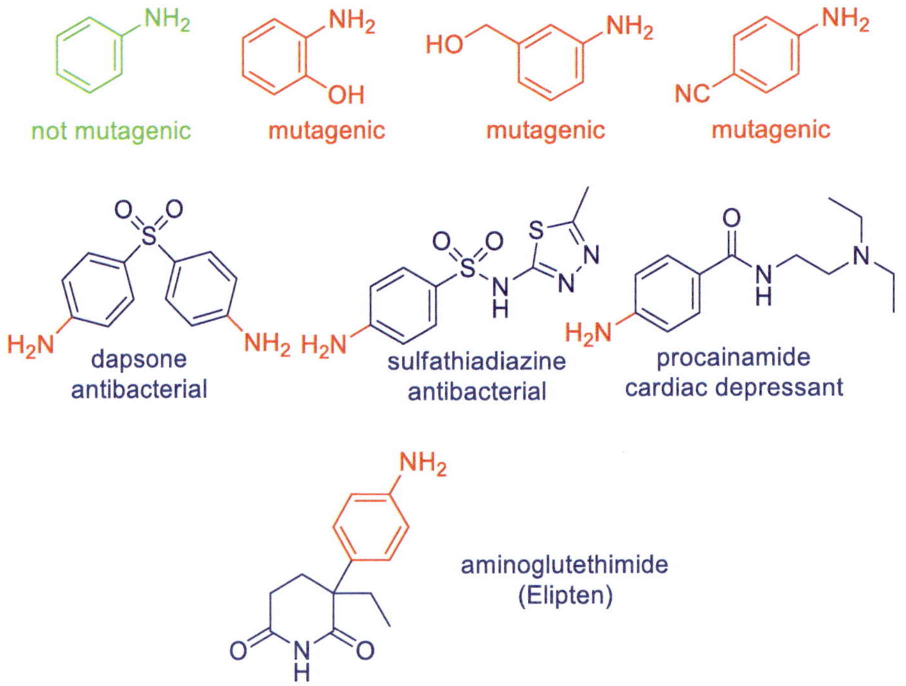

It is evident that diseases to treat and therapeutic indices also matter for drugs. Depending on the diseases a drug treats, some aniline-containing drugs are viable choices despite their potential liabilities. When aminoglutethimide (Elipten) was initially marketed as an anticonvulsant, it was withdrawn due to toxicities. However, as an aromatase inhibitor, its efficacy/safety profile is acceptable to treat advanced breast cancer. It is a primary aniline and has been used as a model to study aniline's hepatic effects. While agranulocytosis is a common idiosyncratic drug reaction (IDR) caused by aminoglutethimide, liver injury is not common. Therefore, it was speculated that the liver may be able to effectively deal with aminoglutethimide reactive metabolites and changes observed in the model study may be involved in adaptation.[28]

Recent examples of aniline-containing drugs to treat serious diseases such as AIDS and cancer include: HIV protease inhibitor amprenavir (Agenerase), non-nucleoside reverse transcriptase inhibitors (NNRTIs) etravirine (Intelence), Brc-Abl kinase inhibitor imatinib (Gleevec), hedgehog pathway inhibitor vismodegib (Erivedge), and mitogen-activated protein kinase (MEK)1/2 inhibitor cobimetinib (Cotellic). Last but not the least, histone deacetylase (HDAC) inhibitor chidamide (Epidaza), approved in China in 2015, has two aniline groups. Here the benzendiamine motif serves as a hydroxamate surrogate to chelate the catalytic zinc cation of the HDAC enzyme.

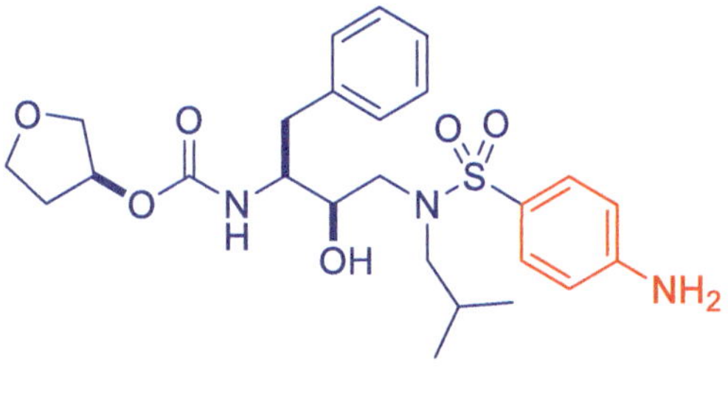

amprenavir (Agenerase)
GSK, 1999
HIV protease inhibitor

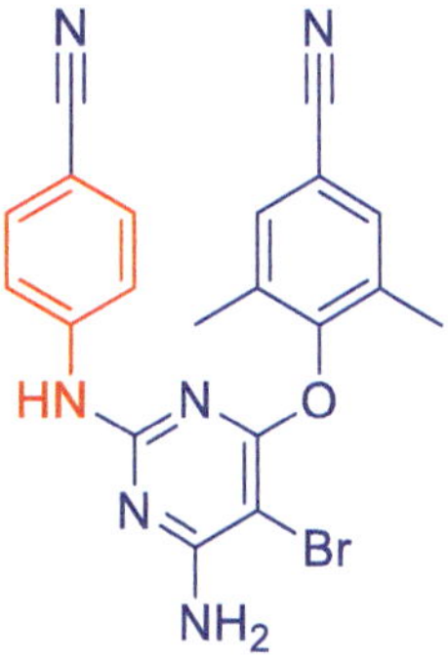

etravirine (Intelence)
Tibotec/J&J, 2008
NNRTI

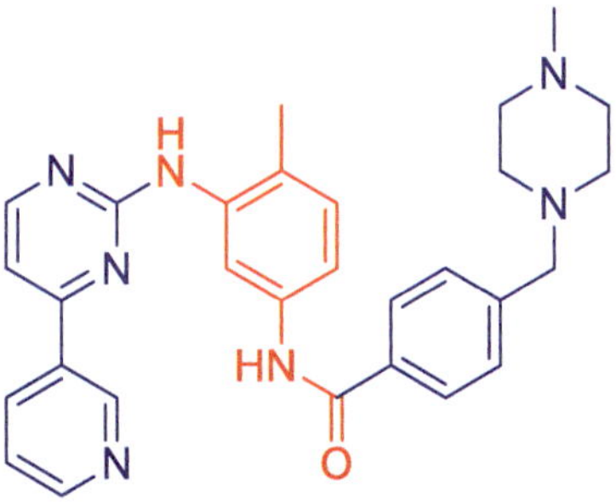

imatinib (Gleevec)
Novartis, 2011
Brc-Abl kinase inhibitor

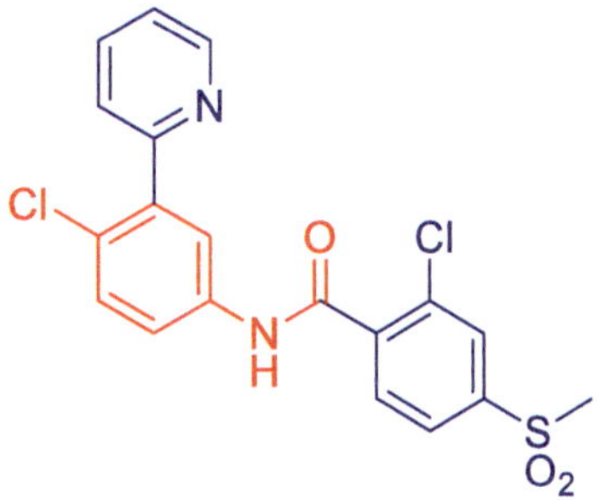

vismodegib (Erivedge)
Genentech, 2012
Hedgehog pathway inhibitor

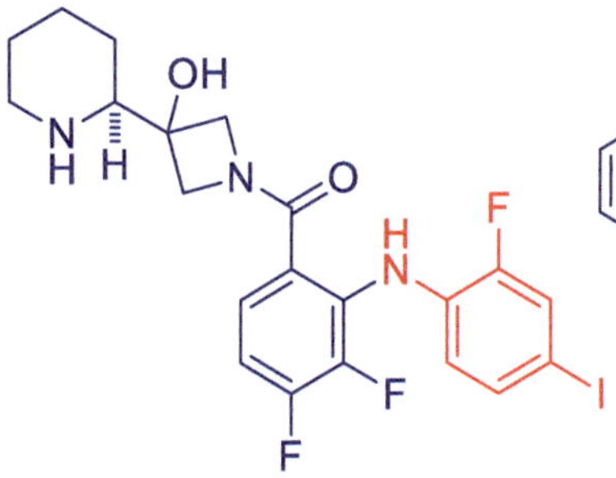

cobimetinib (Cotellic)
Exelixis/Genentech, 2015
MEK1/2 inhibitor

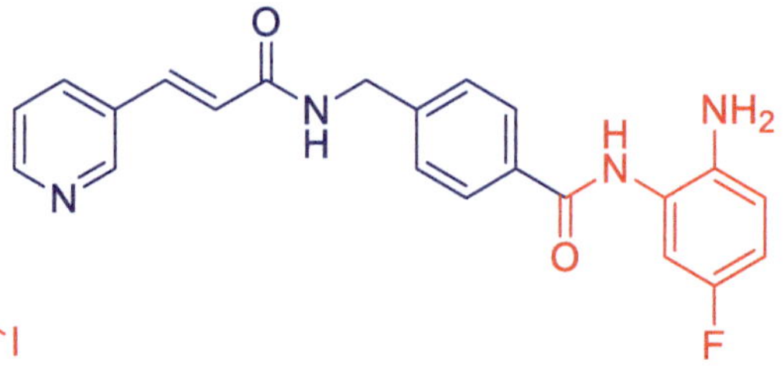

chidamide (Epidaza)
Huya (China), 2015
HDAC inhibitor

Over the years, many tactics have been developed to abrogate aniline′s liabilities. Attaching the amine functionality to an electron-deficient heterocycle makes the nitrogen atom less electron-rich, thus less prone to CYP oxidation. Pyridine, pyrazine, and indole (to a less extent since indole is electron-rich) rings all fit the bill. Phenol, benzylamine (a structural alert of its own right), and aliphatic amines may be explored as bioisosteres for anilines if they have with comparable activities.

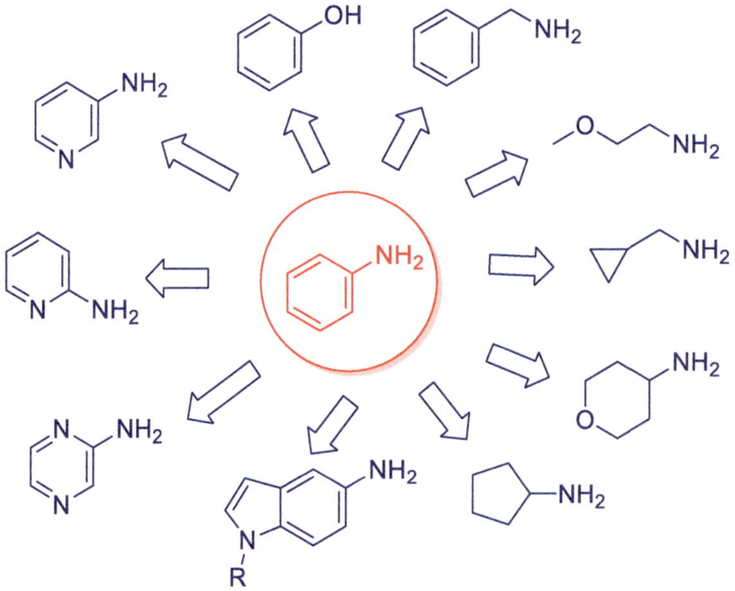

Same principles apply for anilides to further decrease the nitrogen atom′s electron density. Insertion of a methylene or replacement with a "retro-amide" has been successfully employed (see Chapter 4 for merits of retroamide atenolol vs. anilide practolol). Heterocyclic isosteres can replace anilides as well. Another well-trodden path to curb "naked" aniline′s metabolism is to design the amine in a ring as a cyclic amine.

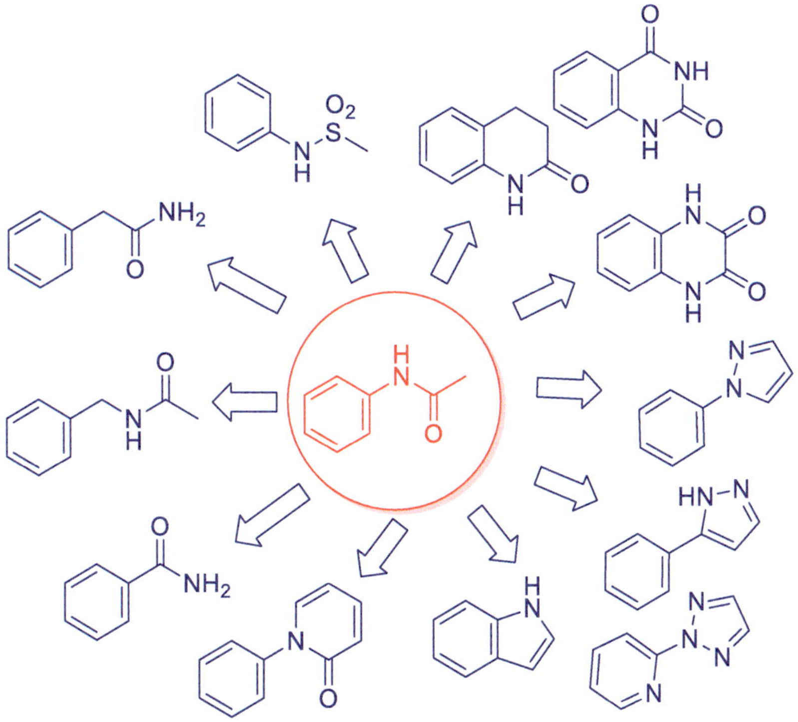

The amine group attached to an electron-deficient heterocycle reduces aniline's metabolism. For instance, Boehringer Ingelheim's BI 207524 (**49**) is a potent and selective thumb pocket HCV NS5B polymerase inhibitor, but its aniline metabolite 4-amino-2-ethoxycinnamic acid motif (in red) was found to have genotoxic liability. Replacing the aniline with nitrogen-containing isosteres led to identification of 2-aminopyridine analog **50**, which was tested negative in the Ames test and provided comparable genotype (GT)1a/1b potency to the prototype **49**.[29]

BI 207524 (**49**)
GT1a/1b replicon EC_{50} = 29/11 nM

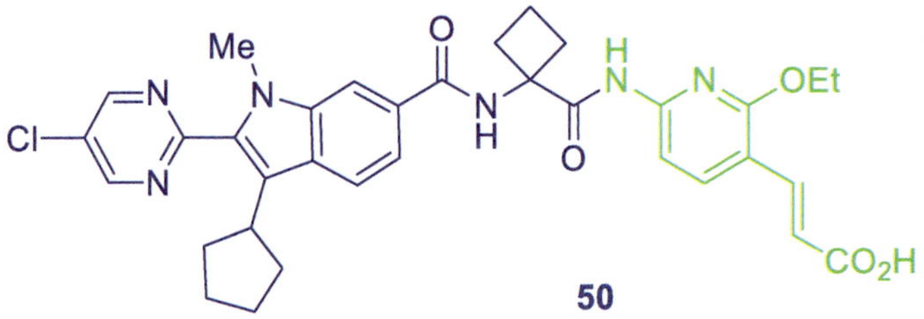

50

GT1a/1b replicon EC_{50} = 34/20 nM

The saga of aniline as a structural alert does not end here, sadly. Anilines and anilides can also be metabolically oxidized to the corresponding *p*- or *o*-iminoquinones.

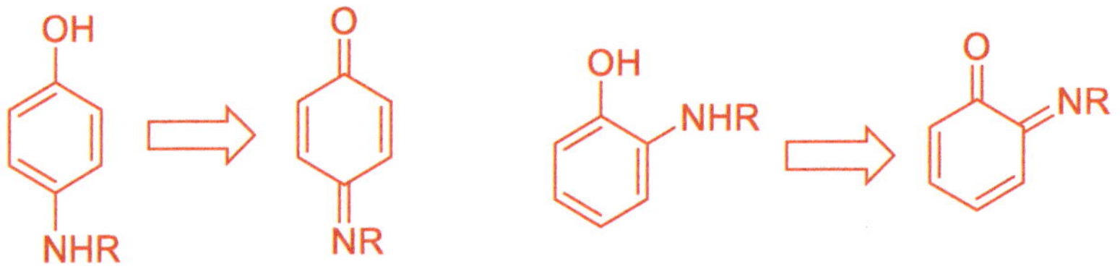

Nefazodone (**51**), a broad opioid receptor antagonist prescribed to treat alcohol dependence, has been associated with many cases of idiosyncratic hepatotoxicity at the therapeutic range of 200–400 mg qd, some so severe that liver transplantations or fatalities ensued. As shown below, the aniline moiety of nefazodone (**51**) is oxidized by CYP3A4 to the corresponding *para*-hydroxyl-nefazodone (**52**), which is further oxidized to the quinone–imine intermediate **53**. Needless to say, quinone–imine **53**, being such an excellent Michael acceptor, is vulnerable to nucleophilic attacks by either glucuronide (GSH) or water to form covalent bonds as the corresponding glucuronide or catechol, respectively.[30] If nucleophiles on DNA or proteins attack the quinone–imine intermediate **53**, it will cause toxicity.

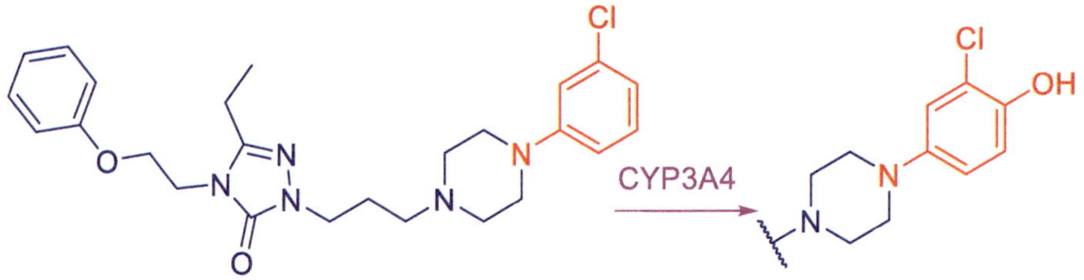

nefazodone (**51**) *para*-hydroxyl-nefazodone (**52**)

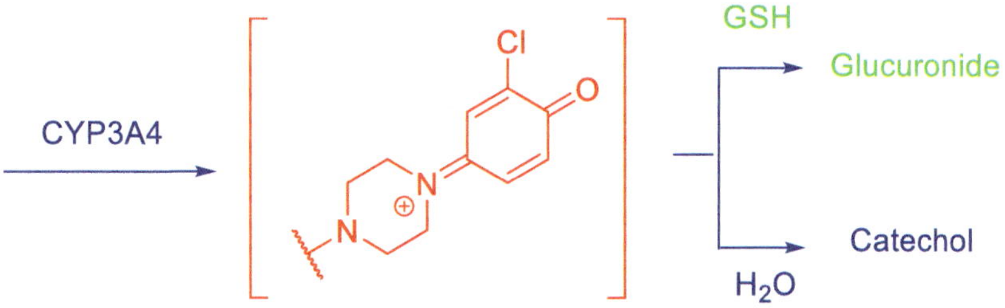

Quinone–imine (**53**)

In contrast, the metabolism of buspirone may serve as a teachable lesson on drug safety associated with reactive metabolites. One of buspirone's major circulating metabolites in humans is 5-hydroxyl-buspirone (**54**), which is impervious to further CYP3A4 oxidation to the ultra-reactive metabolite in the form of quinone–imine **55** because pyrimidine ring is electron-deficient. The absence of the quinone–imine reactive metabolite **55** may offer an explanation why buspirone (**20**) is not associated with idiosyncratic toxicity despite decades of clinical use.[31]

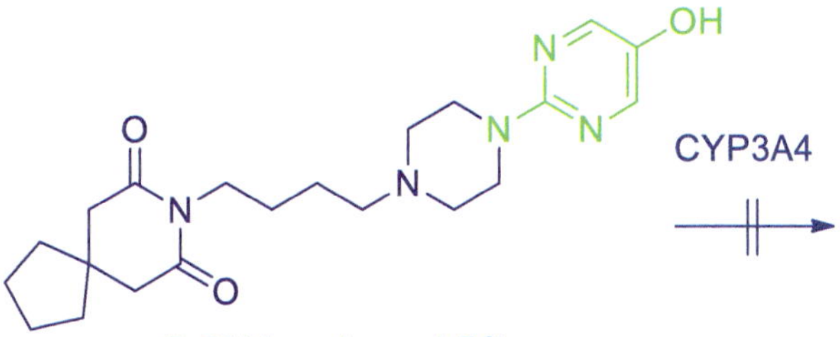

5-OH-buspirone (**54**)

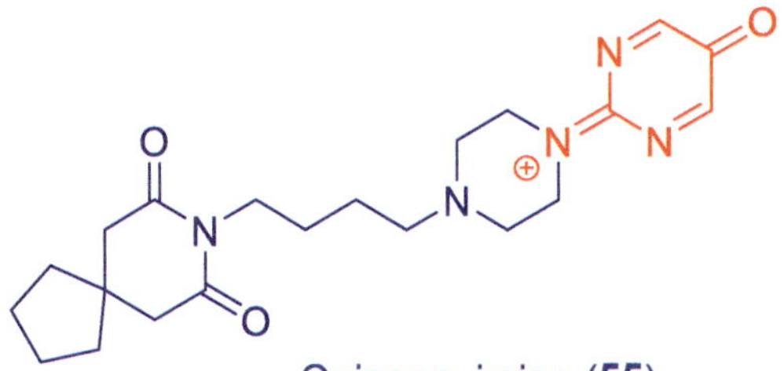

Quinone–imine (**55**)

The highly reactive quinone–imine metabolite **53** is generated from *para*-hydroxyl-nefazodone (**52**), which itself is an oxidative metabolite of nefazodone (**51**). For drugs already possessing an aniline/anilide group and a hydroxyl group at either para- or ortho-position, they are directly oxidized to the highly reactive quinone–imine metabolite. Acetaminophen is a classic example of this class of drugs.

Acetaminophen's liver toxicity may be explained by its reactive metabolites. With a phenol group, the drug itself may undergo Phase II metabolism by forming its corresponding sulfate and glucuronide conjugates. Aside from such a benign outcome, acetaminophen may be oxidized by CYP450 to the *N*-hydroxyl metabolite, which dehydrates to produce highly reactive metabolite *N*-acetyl-*p*-benzoquinone imine (NAPQI). Two fates await NAPQI: (i) it may react with the thiol of glutathione to form an innocuous adduct; or (ii) it may form conjugates with protein and nucleic acids thus leads to toxicity. This explains why severe hepatic toxicities ensue when overdosed with acetaminophen.[32] Meanwhile, Paracelsus′ sage words apply: "The dose makes poison." The larger the dose, the more toxicities one would observe for acetaminophen, or any other drug.

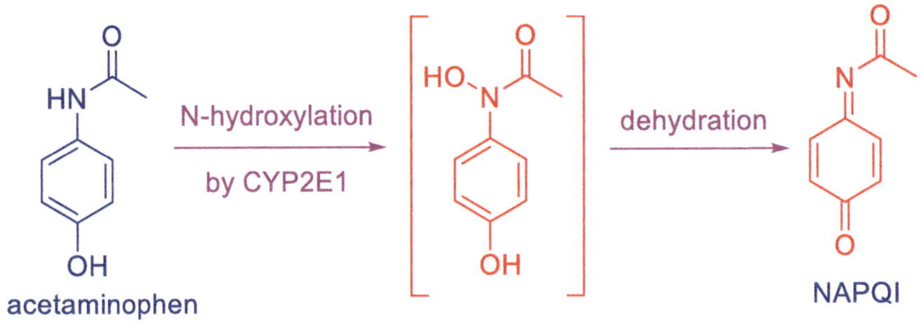

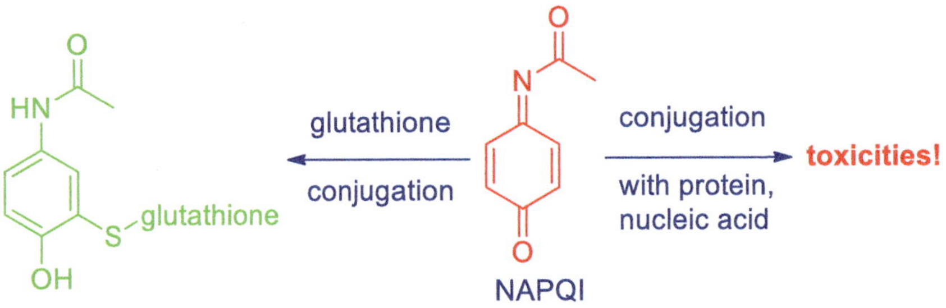

In the same vein, an old antimalarial amodiaquine has a similar *para*-hydroxylaniline motif. Formation of the iminoquinone reactive metabolite helps to explain its hepatotoxicity.

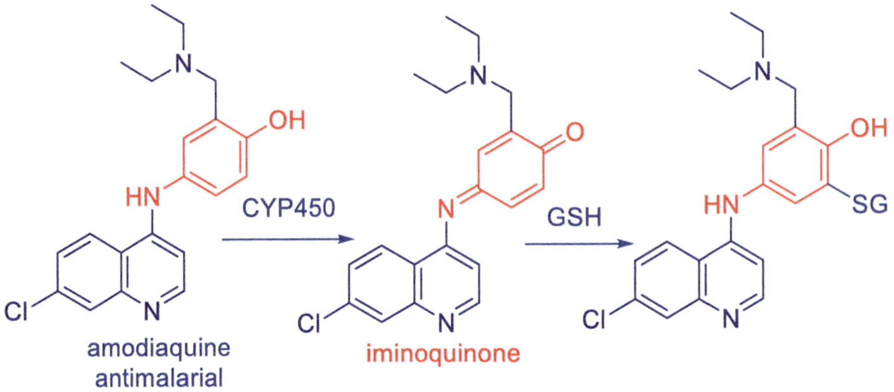

Aniline and anilide structural alerts, like all structural alerts, must be viewed in context of efficacy, safety, and therapeutic indices. While becoming "alert" when designing an aniline or an anilide fragment to the molecule is helpful, totally shying away from them would be a mistake and many life-saving medicines would have been missed. Case in point is that AstraZeneca's third-generation epidermal growth factor receptor (EGFR) inhibitor, osimertinib (Tagrisso, **56**) has two aniline and one anilide structural alerts on an electron-rich phenyl ring with a methoxyl substituent. Yet, it was safe enough to garner the FDA approval in 2015 for treating T790M positive patients with non-small cell lung cancer (NSCLC) with 80 mg qd dosing regimen. Scrutiny of osimertinib (**56**)'s *in vivo* metabolism revealed that demethylation metabolites **57** and **58** are the two major circulating metabolites. Oxidative metabolism of the three aniline and anilide structural alerts is not significant.[33]

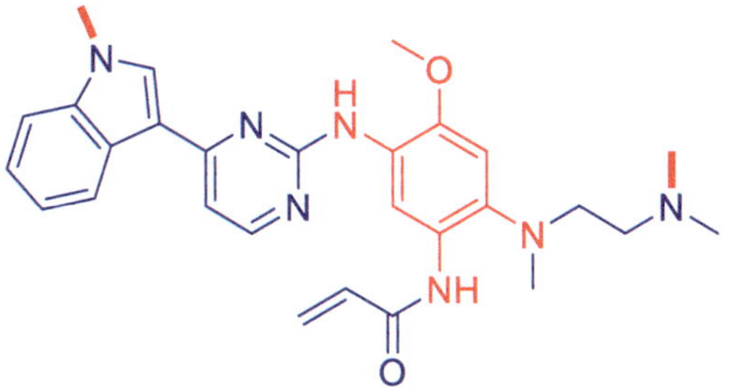

osimertinib (Tagrisso, **56**)
AZ, 2015
EGFR inhibitor (third generation)
overcome C797S mutation

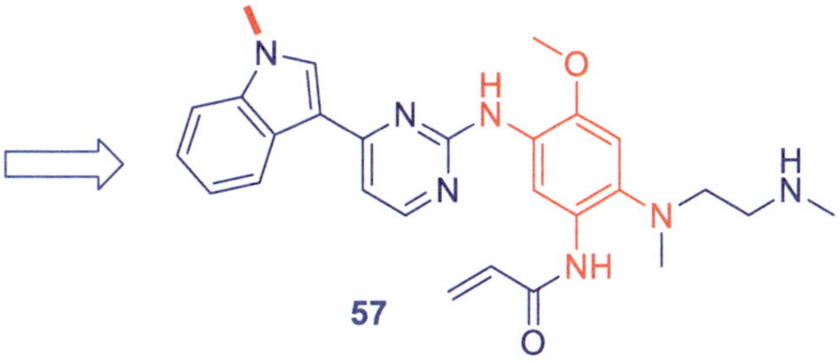

57

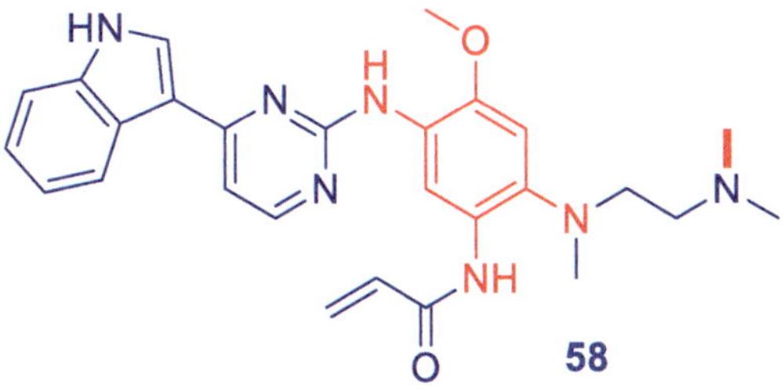

58

Would you have designed a drug with three nitrogen atoms and one oxygen atom on a phenyl ring?

5.4.2 Problematic Amines

Metabolism of osimertinib (**56**) producing demethylation metabolites **57** and **58** serves as a good segue to metabolism of aliphatic amines. An aliphatic amine is prone to α-hydroxylation, followed by demethylation or dealkylation as the "normal" metabolic pathways. But problems arise when potentially toxic reactive metabolites are generated, and those amines become problematic for drug hunters. Whereas many problematic amines exist, we focus on two frequently encountered classes: benzylamines and cyclopropylamines.

5.4.2.1 Benzylamines

Different from anilines, a benzylamine with an extra methylene moiety undergoes deamination to produce benzaldehyde as a major metabolite via the intermediacy of a hemiaminal. The oxidative deamination process is promoted by amine oxidases such as mitochondrial monoamine oxidase-B (MOA-B) and semicarbazide-sensitive amine oxidases (SSAO). The benzaldehyde metabolite itself, in turn, is rapidly oxidized to the corresponding carboxylic acid, which conjugates with glycine to form hippuric acid as another major metabolite.[34a]

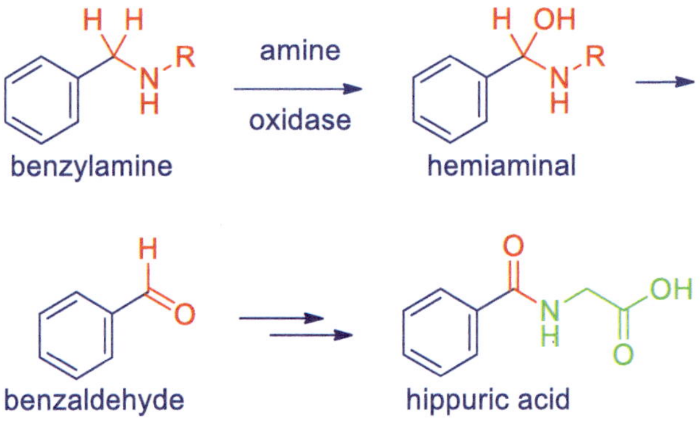

Besides monoamine oxidases (MAOs), CYP450 enzymes can wreak havoc on benzylamine-containing drugs as well. CYP450 oxidation of BMS's factor Xa inhibitor DPC-423 (**59**) leads to hydroxylamine **60**, which may be excreted after forming *O*-glucuronide **61**. Meanwhile, hydroxylamine **60** is readily further oxidized to the highly carcinogenic nitroso metabolite **62**. Tautomerization of nitroso **62** results in oxime **63**, which has been detected as a metabolite. Oxime **63** may be hydrolyzed to the corresponding carboxylic acid, or further oxidized to reactive nitrile *N*-oxide **64**. Both oxime **63** and nitrile *N*-oxide **64** are reactive enough to form GSH adduct **65**, which follows a metabolic cascade to generate more metabolites.[34b]

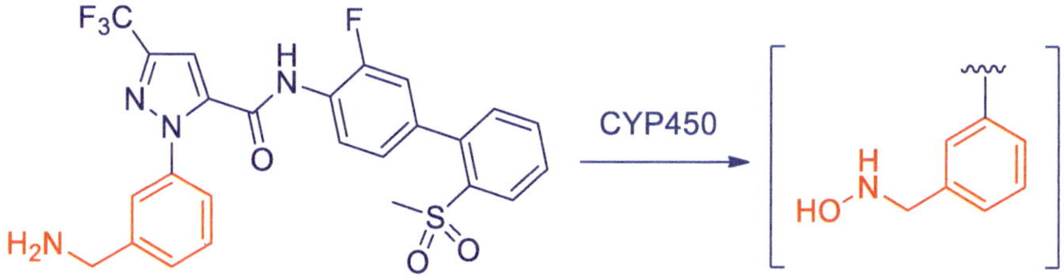

DPC 423 (**59**) hydroxylamine **60**

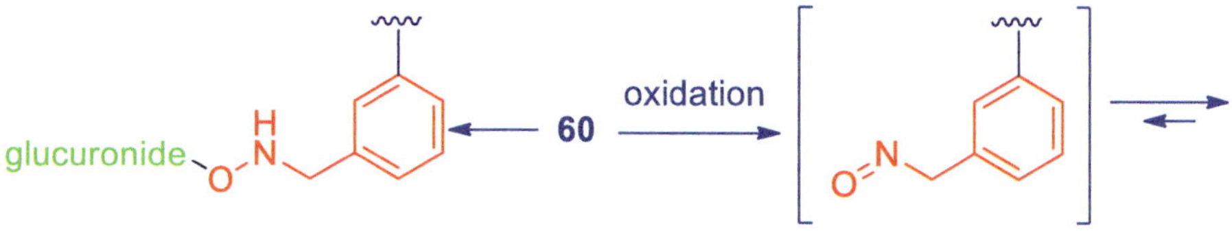

O-glucuronide **61** nitroso **62**

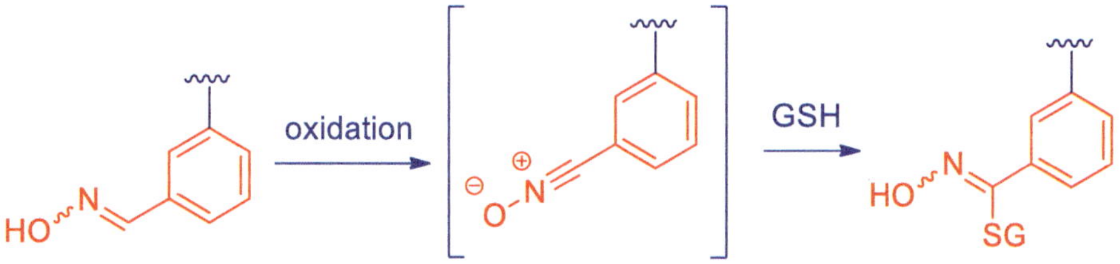

oxime **63** reactive nitrile oxide **64** GSH adduct **65**

A cyclic benzylamine is still a structural alert. Johnson and Johnson's pyrrolidine-substituted arylindenopyrimidine **66** is a potent dual adenosine A_{2A}/A_1 receptor antagonist with a potential as a treatment of Parkinson's disease. It contains a "disguised" benzylamine with the amine hidden in the pyrrolidine ring. Bioactivation of the benzylamine may explain why **66** is tested positive in the Ames test to show genotoxicity. After bio-activation, intermediate metabolites detected include endocyclic iminium ion **67**, amino aldehyde **68**, epoxide, and α,β-unsaturated ketone, all of them reactive intermediates to react with nucleophiles on DNA molecules. To minimize bio-activation leading to the major reactive intermediate iminium **67**, the pyrrolidine moiety of **66** was replaced with either 2,5-dimethylpyrrolidine to add steric hindrance, or pyridine-3-yl to eliminate the amine altogether, providing two analogs that were devoid of genotoxic liability.[35]

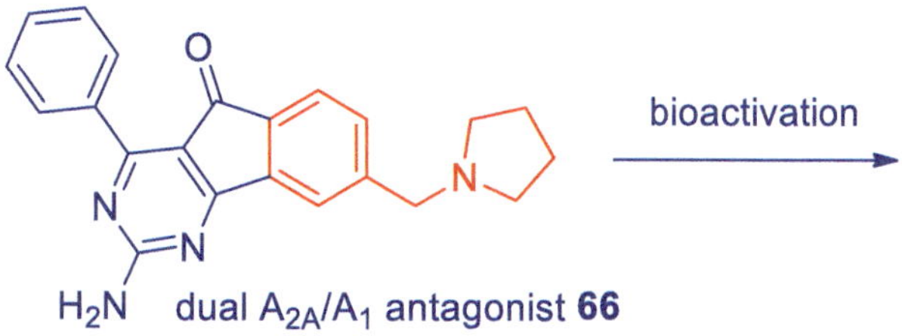

dual A_{2A}/A_1 antagonist **66**

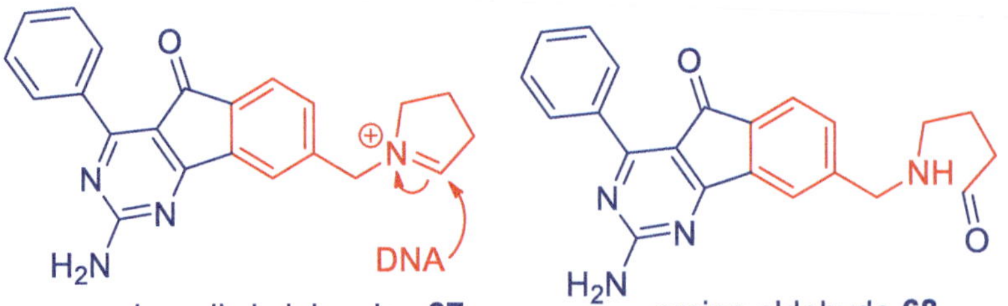

endocyclic iminium ion **67** amino aldehyde **68**

GSK's EGFR and human epidermal growth factor receptor 2 (HER2) dual kinase inhibitor lapatinib (Tykerb, **69**) contains a "benzylamine" (furan is an electron-rich isostere of benzene here) structural alert that is implicated for its black box warning for liver enzyme elevation and sporadic cases of hepatotoxicity. In addition to the "normal" deamination pathway, lapatinib (**69**) is also oxidized by CYP3A4 to secondary hydroxylamine **70**, which is further oxidized to imine *N*-oxide **71**. CYP3A4 cleaves the sulfone fragment on **71** and oxidizes it to the nitroso reactive metabolite **72**, which isomerizes to the corresponding oxime **73**.[36a–c]

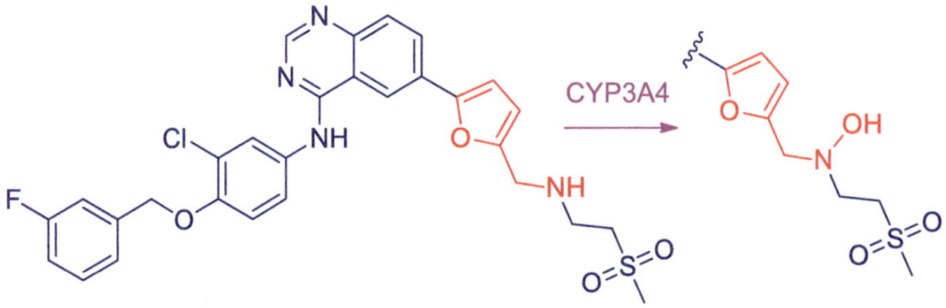

lapatinib (Tykerb, **69**)
GSK, 2007
EGFR and HER-2 inhibitor

secondary hydroxylamine **70**

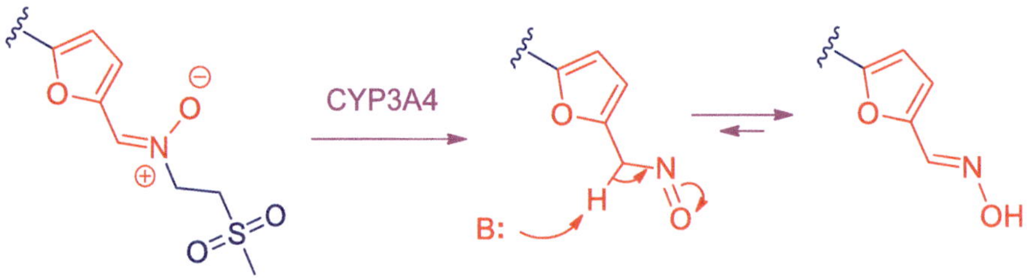

imine *N*-oxide **71** nitroso **72** oxime **73**

Nitroso intermediate **72** forms a complex with human CYP3A4 known as metabolite–intermediate (MI) complex. Although not a covalent bond, the MI complex between the nitroso reactive metabolite **72** and the CYP3A4 enzyme is so strong a coordination that it is virtually irreversible under physiological conditions. Another key metabolite, lapatinib quinone–imine from the chloroaniline ether metabolism, does form a covalent bond with CYP3A5 and is responsible for lapatinib (**69**)'s mechanism-based inactivation (MBI) of CYP3A5. Lapatinib (**69**)'s MBIs may be the culprit of the drug's hepatotoxicity. An MBI is also known as a suicide inhibitor of an enzyme, it binds to the target enzyme irreversibly, leading to permanent inhibition of its enzymatic function. Lapatinib (**69**)'s MBIs recapitulate the importance of our understanding of metabolism.[36d]

Even though benzylamine "hides" itself in a ring as a cyclic amine, it should be a structural alert as well since it metabolizes like normal benzylamines. Antidepressant nomifesine (**74**), a cyclic benzylamine, saw its phenyl ring oxidatively metabolized to phenol, catechol, and methoxyl catechol.[37a] The cyclic benzylamine portion is oxidatively metabolized to hydroxylamine **75**, which is readily further oxidized to dihydroisoquinolinium ion metabolite **76** by human myeloperoxidase, hemoglobin, monoamine oxidase A, and CYP450 enzymes.[37b]

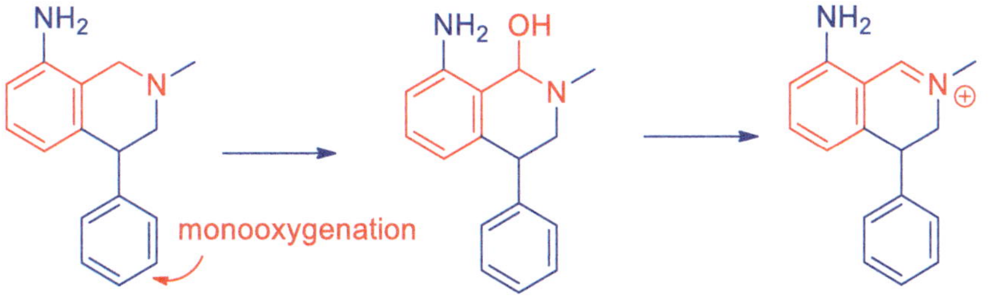

nomifesine (**74**) hydroxylamine **75** dihydroisoquinolinium **76**

Drugs containing the benzylamine structural alerts include donepezil (Aricept), sertraline (Zoloft), cetirizine (Zyrtec), and imatinib (Gleevec). Some of them are very safe. Therefore, we have to be philosophical about the benzylamine structural alert, just like all structural alerts.

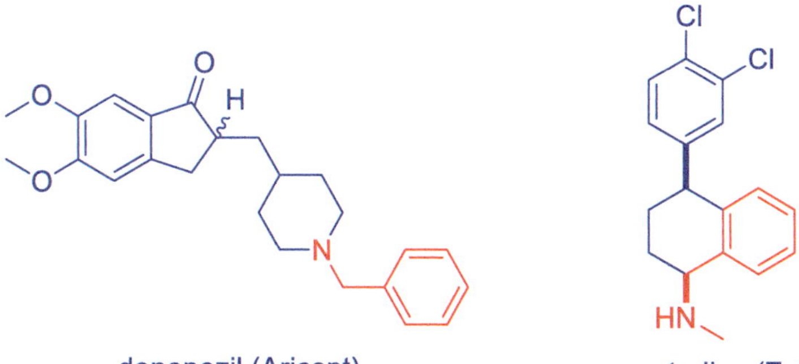

donepezil (Aricept)
Eisai/Pfizer, 1996
AChE Inhibitor for AD

sertraline (Zoloft)
Pfizer, 1997
SSRI

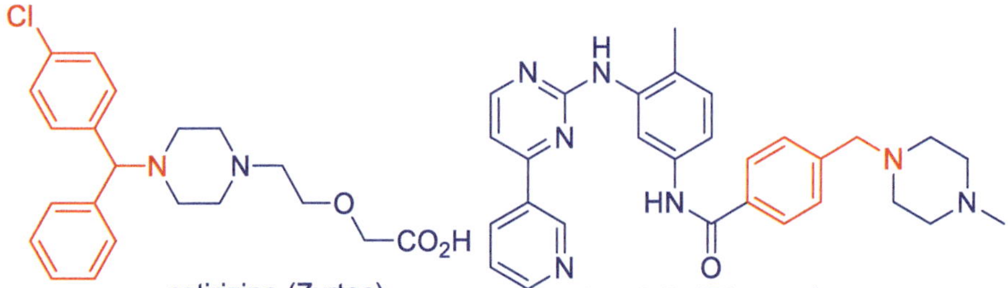

cetirizine (Zyrtec)
Pfizer, 1999
H_1 receptor antagonist

imatinib (Gleevec)
Novartis, 2001
bcr-abl kinase inhibitor

5.4.2.2 Cyclopropylamines

Drugs with the cyclopropylamine fragments have seen enough toxicities to warrant it a structural alert.

Tranylcypromine (Parnate, SK&F, 1961) is an old anti-depressant that works as an MAO inhibitor. The fact that its hippuric acid conjugate was isolated among its metabolites indicates that cinnamaldehyde is probably the intermediate and chances are that radical-mediated ring-opening is involved.[38] Subsequent studies showed that CYP450 enzymes, MAOs, and horseradish peroxide all can oxidize cyclopropylamine to a carbon-centered radical, which can be subsequently oxidized to a reactive α,β-unsaturated aldehyde.

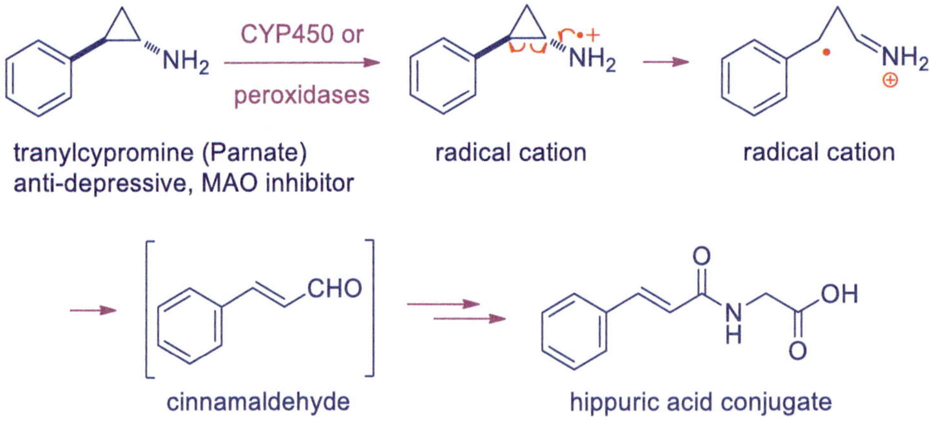

Whereas tranylcypromine's metabolism was not thoroughly investigated, that of Pfizer's fluoroquinolone antibacterial trovafloxacin (Trovan) was scrutinized employing a drug model (DM) compound to probe the culprit of its idiosyncratic hepatotoxicity. The purpose of using a DM is to exclude interferences from other possible oxidizable positions, especially the difluoroaniline moiety. Detection of the glutathione adduct suggests that α,β-unsaturated aldehyde is likely the metabolic intermediate. This is consistent with the proposed metabolic pathway involving the single-electron transfer (SET) mechanism.[39]

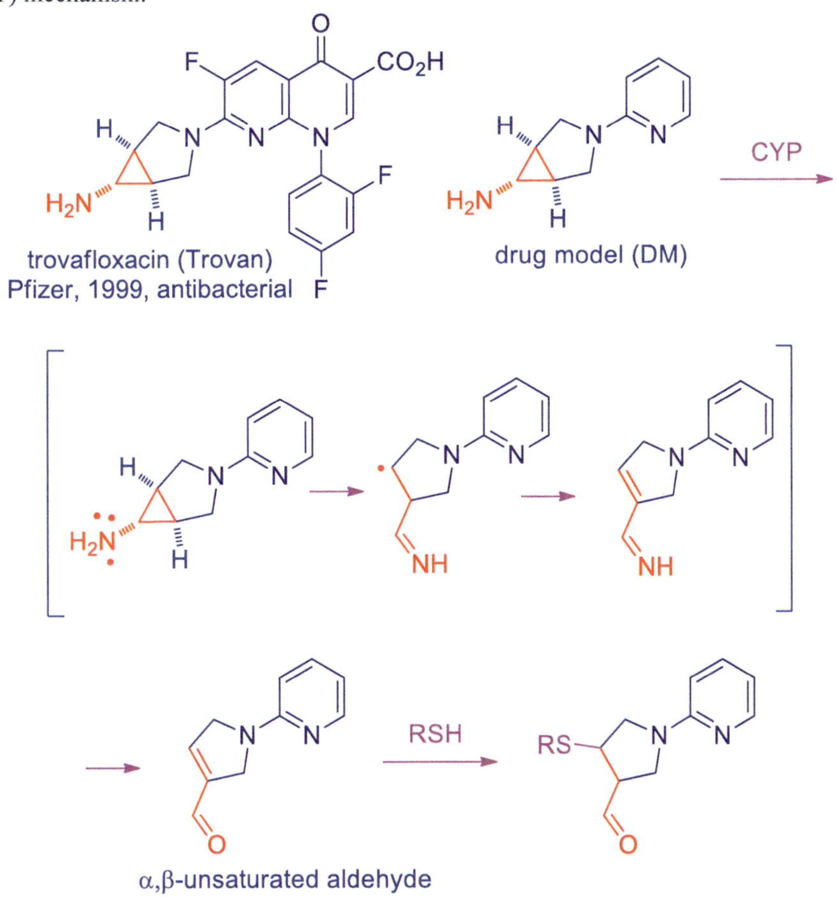

On the other hand, the cyclopropylamine fragment has done wonders for some drugs′ pharmacokinetic properties. The cyclopropyl moiety on cyclopropylamine fragment is more resistant to CYP450 metabolism thus less prone to dealkylation in comparison to simple alkylamines such as an ethylamine. Two good examples are GSK′s abacavir (Ziagen, **26**), a nucleoside reverse-transcriptase inhibitor (NRTI), and Boehringer Ingelheim′s nevirapine (Viramune), a non-nucleoside reverse-transcriptase inhibitor (NNRTI). Both of the drugs were selected over the corresponding ethylamine analogs because of cyclopropyl group′s better resistance to metabolism.[39] Ironically, nevirapine causes idiosyncratic episodes of liver toxicity. But the culprit is not the cyclopropyl group, rather it is the methyl group. Furthermore, the cyclopropyl fragments made appearance in successful drugs such as Bayer′s antibacterial ciprofloxacin (Cipro), GSK′s MEK1/2 inhibitor trametinib (Mekinist), and Eisai′s dual vascular endothelial growth factor receptor (VEGFR) and fibroblast growth factor receptor (FGFR) inhibitor lenvatinib (Lenvima).

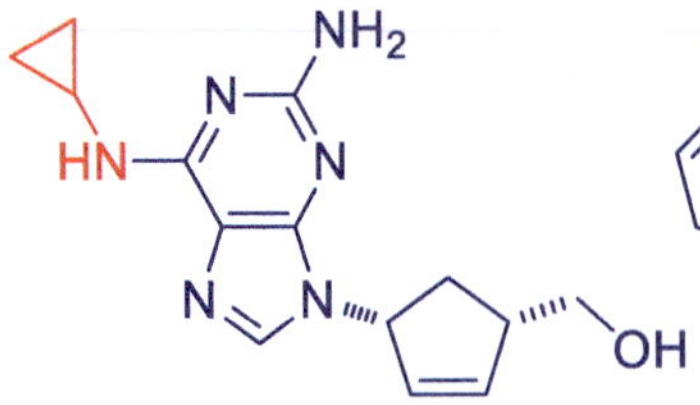

abacavir (Ziagen, **26**)
GSK, 1998
NRTI

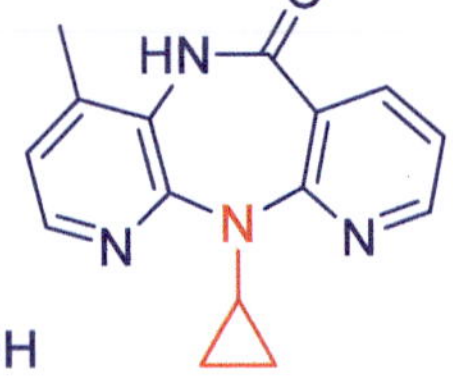

nevirapine (Viramune)
Boehringer Ingelheim, 1996
NNRTI

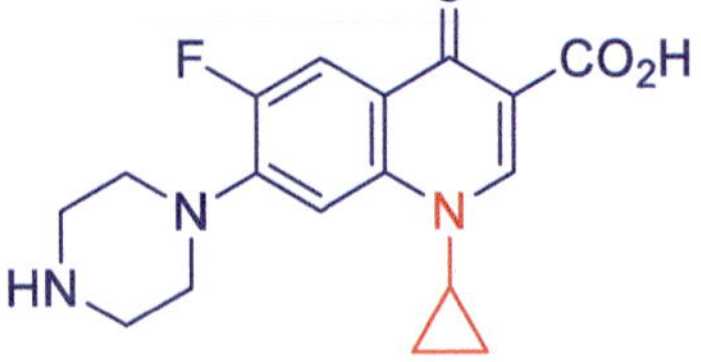

ciprofloxacin (Cipro)
Bayer, 1987
antibacterial

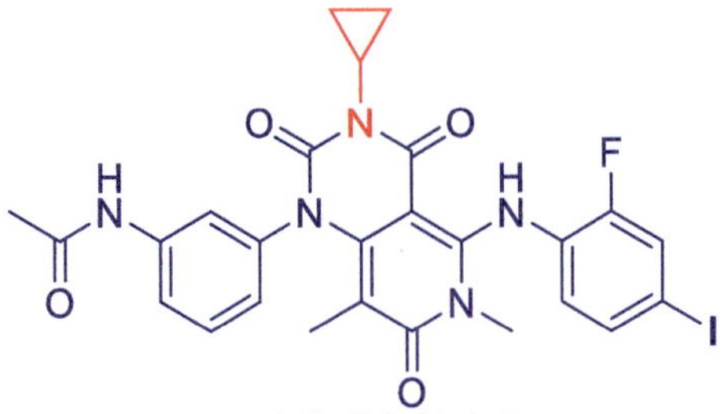

trametinib (Mekinist)
GSK, 2013
MEK1/2 inhibitor

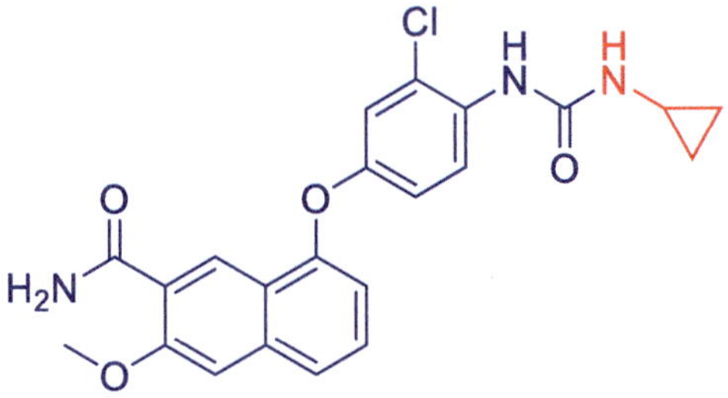

lenvatinib (Lenvima)
Eisai, 2015
VEGFR and FGFR inhibitor

5.4.3 *Nitroaromatics*

As discussed in Section 3.4.4, both nitrophenyl- and aniline-containing drugs may have nitrosophenyl as a fragment of their metabolites, which may be associated a broad spectrum of mutagenic, genotoxic, and carcinogenic properties:[40]

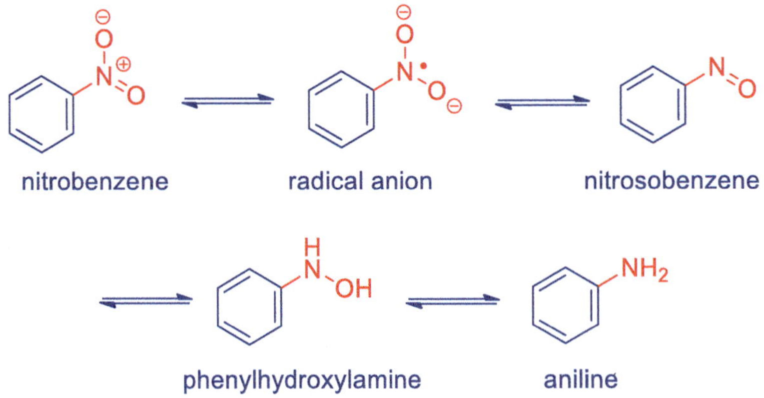

With the highly reactive nitrosophenyl metabolite in the mix, it is not all that surprising that a number of nitro-aryl drugs caused toxicities, especially liver toxicities. The label for an old skeletal muscle relaxant dantrolene has a liver toxicity warning. Tolcapone (Tasmar) and entacapone (Comtan) are selective and reversible nitrocatechol-type inhibitor of catechol-O-methyltransferase (COMT) for the treatment of Parkinson's disease. Both are known to have potential to cause hepatotoxicity.

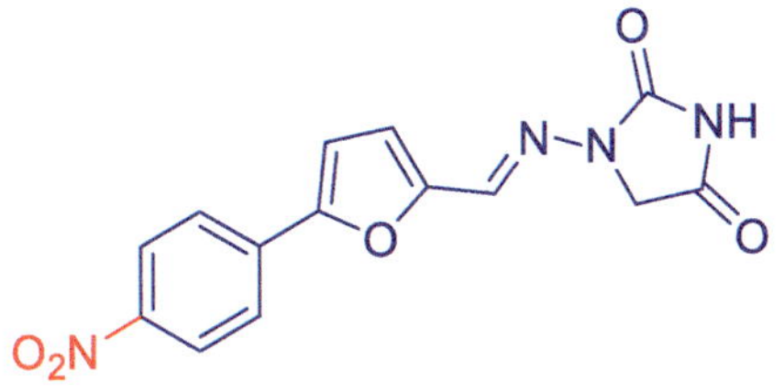

dantrolene
skeletal muscle relaxant
Liver warning

tolcapone
anti-parkinsonian
Liver warning

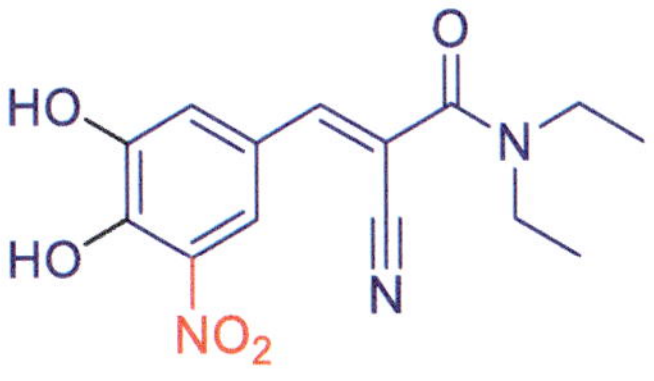

entacapone (Comtan)
anti-parkinsonian
Liver warning

Cyclooxygenase-2 (COX-2) inhibitor nimesulide, a nonsteroidal anti-inflammatory drug (NSAID), contains an aniline and a nitrophenyl structural alert, which may have contributed its rare, idiosyncratic but severe hepatotoxicity known as drug-induced liver failure (DILF). The nitro group may be converted to the amine functionality by nitro reductase. The resulting 1,4-diaminophenyl ether, as one of the major metabolites, is electron-rich and prone to CYP450 (2C19 and 1A2) or myeloperoxidase oxidation to the corresponding electrophilic metabolite diiminoquinone, similar to iminoquinones. GSH conjugate of the diiminoquinone has been isolated and identified.[41a] In 2015, an alternative metabolic bioactivation pathway for nimesulide was proposed.[41b]

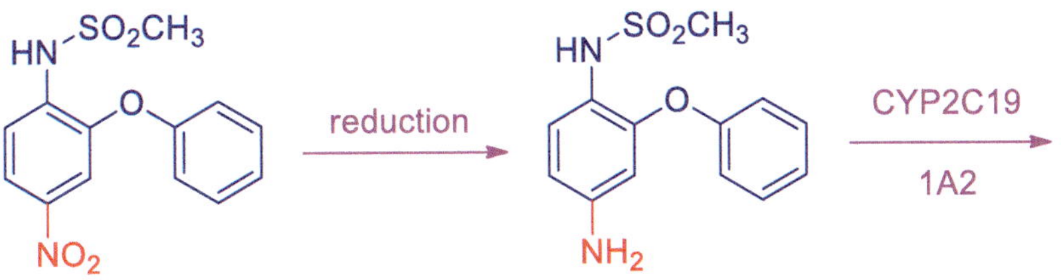

nimesulide, COX-2 inhibitor

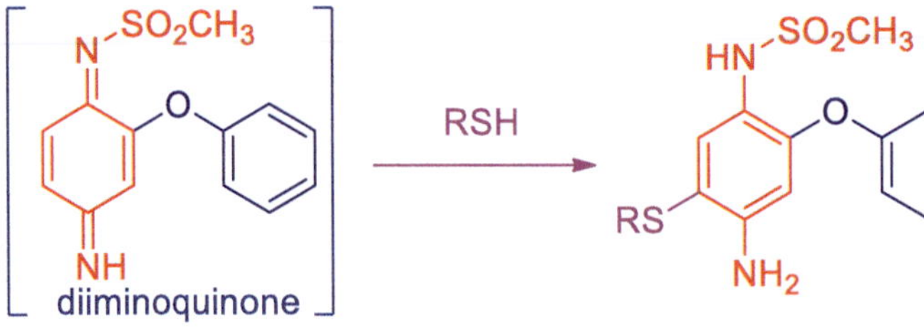

One of the more popular tactics in medicinal chemistry to abrogate nitrophenyl group's liability is replacing it with the corresponding pyridine. Pyridine isostere retains intrinsic potency and prevents metabolic activation by existing as a zwitterion. The pyridine analogs showed activity in an *in vivo* in a model of inflammation. It was further optimized by substitution NH for O and adding Br to eliminate the potential for di-iminoquinone formation.[42]

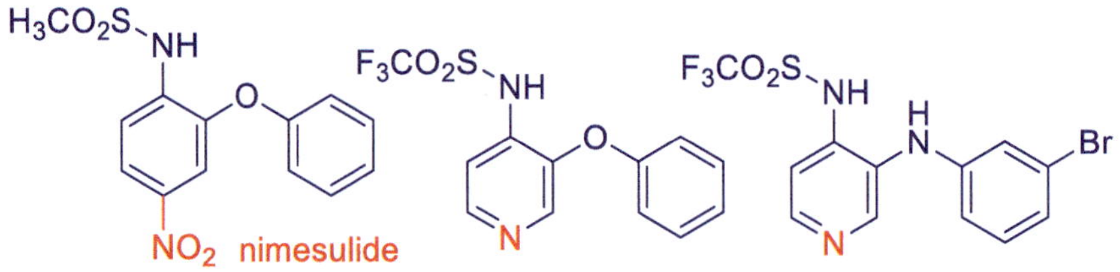

	nimesulide		
IC_{50} COX-1 (μM)	3.76	0.14	0.91
IC_{50} COX-2 (μM)	0.70	0.62	0.12
p*K*a	6.56	6.1	
Rat paw odema	58.0 @ 10 mpk	54.1 @ 30 mpk	

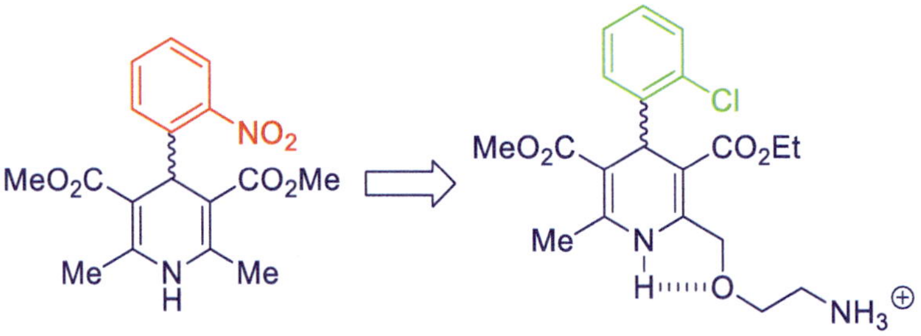

nifedipine (Adalat)
Bayer, 1981
calcium channel blocker

amlodipine (Norvasc)
Pfizer, 1990
calcium channel blocker

Another tactic to mitigate nitrophenyl group's potential toxicities is to replace the nitro group with halides such as fluorine atom(s) or chlorine atom(s). The most successful case is probably Pfizer's third-generation calcium channel blocker amlodipine (Norvasc). Bayer's nifedipine (Adalat), a first-generation calcium channel blocker, has several short-comings. The nitrophenyl fragment is likely the root cause of some its side effects. Pfizer third-generation calcium channel blocker amlodipine uses chlorine to replace nifedipine's nitro group, which was the impetus for initial drug design and may help minimizing its toxicity profile. Indeed, at 5 mg or 10 mg qd, amlodipine has proven to be both efficacious and safe.

Abbvie's wonder cancer drug Bcl-2 inhibitor venetoclax (Venclexta) serves as a good argument that completely shying away from the nitrophenyl structural alert is counter-productive. One of Abbvie's Bcl-2 inhibitors in clinical trials was navitoclax. Its trifluoromethylsulfonyl group was intentionally designed to replace the nitro group. Navitoclax was not selective, also binding to Bcl-xL and inducing a rapid, concentration-dependent decrease in the number of circulating platelets. This mechanism-based thrombocytopenia is the dose-limiting toxicity of single-agent navitoclax treatment in patients and limits the ability to drive drug concentrations into a highly efficacious range. In contrast, venetoclax is selective against Bcl-2 (TR FRET, K_i = 0.01 nM, Bcl-xL, K_i = 48 nM) thus platelet-sparing. Many factors contribute to venetoclax's success, but the nitrophenyl structural alert is not a deterrent to its superb efficacy and safety.[42]

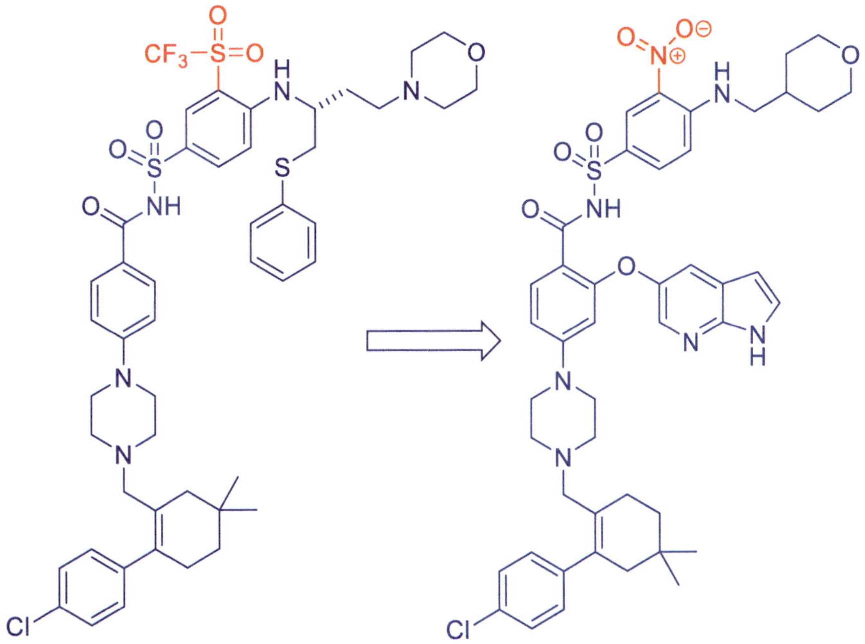

navitoclax
Bcl-2, TR FRET, K_i = 0.04 nM
Bcl-X_L, TR FRET, K_i = 0.05 nM

venetoclax (Venclexta)
Bcl-2, TR FRET, K_i = 0.01 nM
Bcl-xL, TR FRET, K_i = 48 nM

In 2017, the FDA approved two nitro-containing drugs: benznidazole (Rochagan), an anti-parasitic drug for treating Chagas disease; and secnidazole (Solosec) for the treatment of bacterial vaginosis.

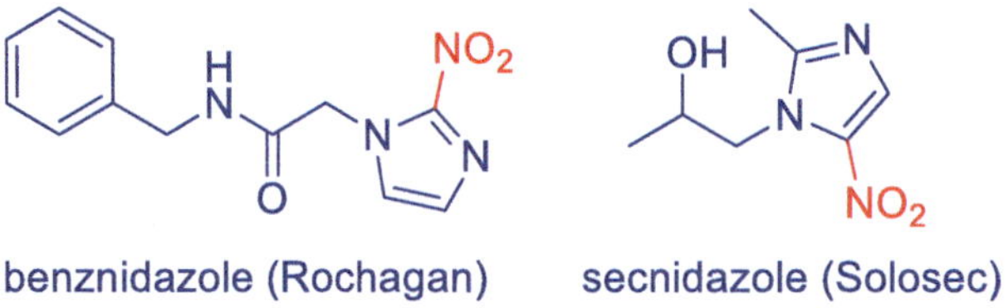

5.4.4 Quinones and Phenols

During bioactivation, hydroquinones and phenols are readily oxidized to the corresponding quinones and quinone methides, which are highly electrophilic and reactive as Michael acceptors.[43] Alkylating reactivity deceases as follows:

Quinone methides > iminoquinones > quinones

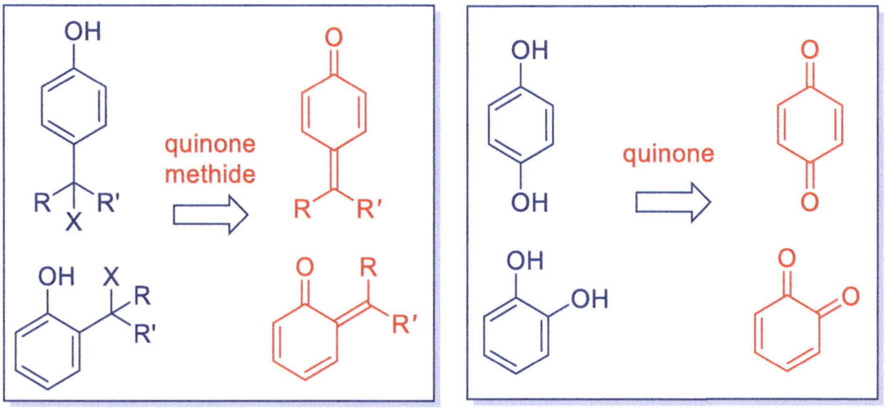

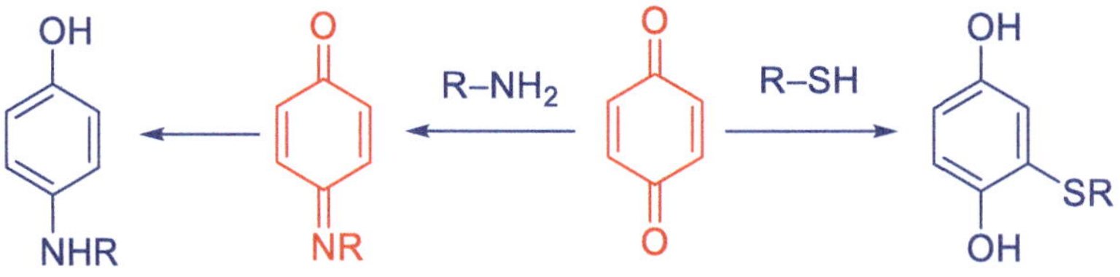

Antineoplastic agent doxorubicin (Adriamycin) works as a topoisomerase II poison. It has been plagued by several side effects including lipid peroxidation, cell damage, cardiac toxicity (in the form of a cumulative and irreversible cardiomyopathy), and drug-induced interference with cardiac mitochondrial calcium homeostasis.[44a] It is speculated that metabolic activation of the quinone–hydroquinone moiety could be important in the mechanism of cytotoxicity. Quinone–hydroquinone is readily reduced by a variety of enzyme systems, most notably, CYP450-reductase, via one-electron reduction, forming a semiquinone radical. As shown below, doxorubicin redox cycles liberate highly reactive free radical species of molecular oxygen, along with the hydroquinone radical from the anthraquinone chromophore. Free radicals liberated from doxorubicin redox cycling are thought to be responsible for many of the secondary effects of doxorubicin, including lipid peroxidation, the oxidation of both proteins and DNA, and the depletion of glutathione (GSH) and pyridine nucleotide reducing equivalents in the cell.[44b]

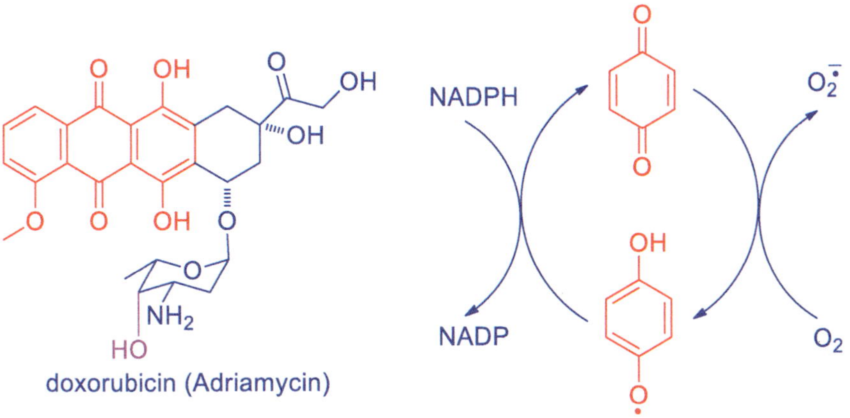

doxorubicin (Adriamycin)

Sankyo's troglitazone (Rezulin) is a peroxisome proliferator-activated receptor (PPAR)-γ agonist for the treatment of type 2 diabetes mellitus (T2DM). Regrettably, drug-induced idiosyncratic hepatotoxicity took place in 1.9% patients after its approval in 1997 and it was eventually pulled off the market 3 years later. One of the offending toxicophores is the thiazolidinedione (TZD, *vide infra*: Section 5.4.5. In addition, bioactivation of its phenol moiety on the chromane (benzodihydropyran) ring may contribute to its liver toxicity via apoptosis. One of the major metabolites, reactive quinone intermediate M3, is produced by CYP3A4 and CYP2C8 enzymes. Meanwhile, the corresponding quinone methide products were also observed *in vitro*.[45] The case of troglitazone's hepatotoxicity was a watershed for the pharmaceutical industry, which started screening reactive metabolites in the wake of its withdrawal in 2000.

troglitazone (Rezulin), antidiabetic, liver toxicities

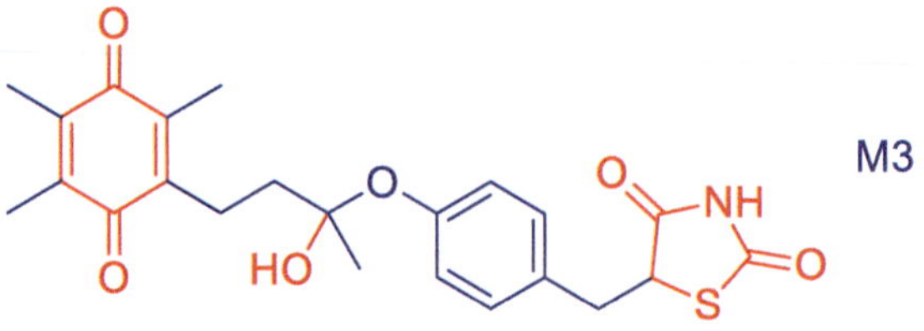

A detailed SET mechanism of troglitazone's chromane portion's metabolism is shown beneath:[46]

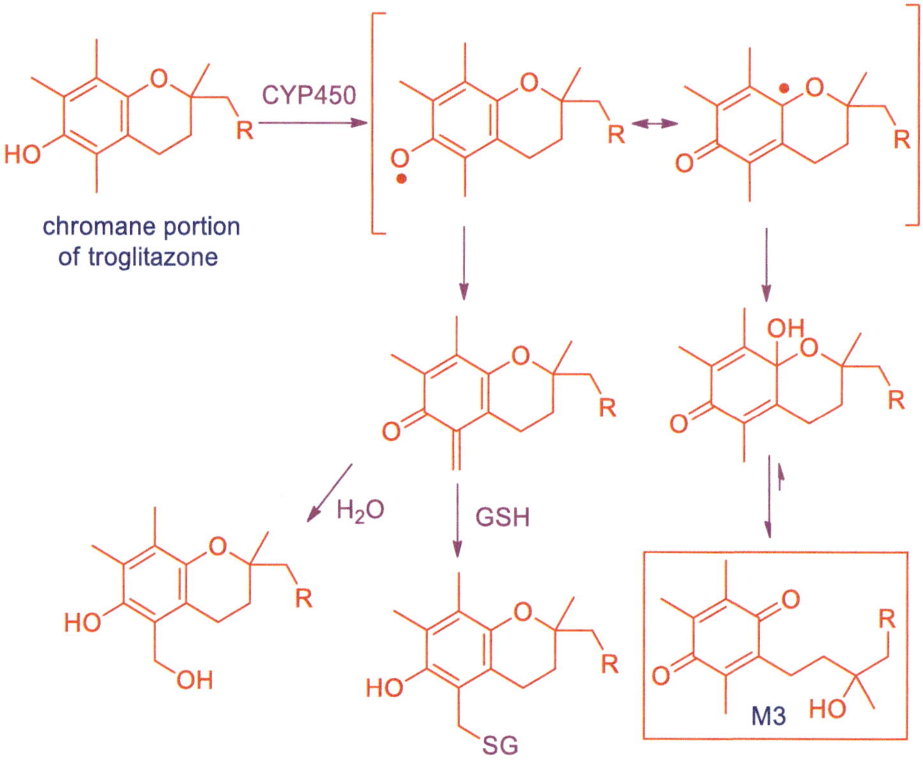

Even when the phenol functionality is masked with an alkyl ether group, oxidative metabolism to quinone or quinone methide may occur also because dealkylation takes place with ease.[47] Case in point is remoxipride (Roxiam), a selective D_2 receptor antagonist. Its major side effect, aplastic anemia, may be the consequence of its bio-activation. Under the influence of CYP450, α-oxidation (α-hydroxylation) and O-demethylation take place, giving rise to two metabolites: hydroquinone NCQ344 and an *o*-catechol (not shown) based on the regiochemistry of demethylation. Hydroquinone NCQ344 is further oxidized to *para*-quinone, a reactive electrophilic Michael acceptor, which is trapped by glutathione to provide the conjugate with concurrent release of HBr. In contrast, the *o*-catechol is reluctant to be oxidized either enzymatically or chemically to the corresponding *ortho*-quinone thus does not contribute to remoxipride's toxicities as much as NCQ344.[47]

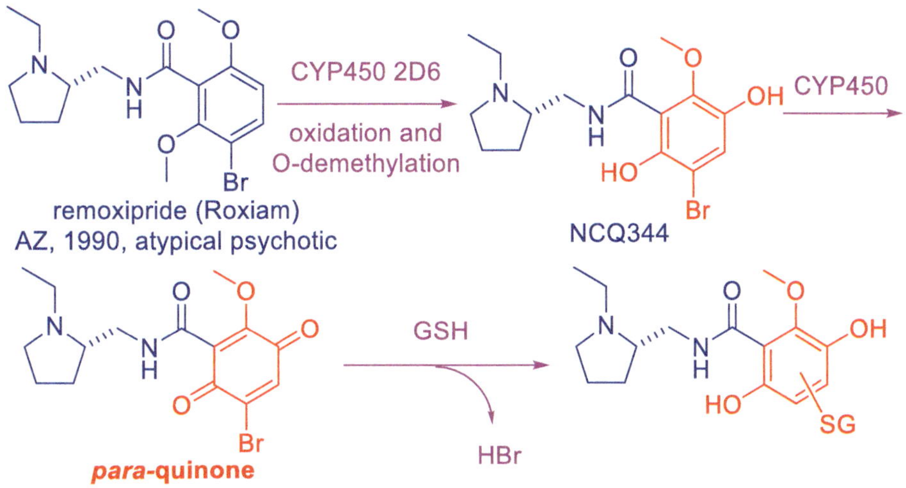

Selective estrogen receptor modulator (SERM) tamoxifen (Nolvadex) was the gold standard for the treatment of hormone-dependent breast cancer for many years after its appearance on the market. But its long-term usage has been linked to an increased risk of endometrial cancers in women and the three major electrophilic reactive metabolites, including tamoxifen cation (not shown here), two quinone methides and an *ortho*-quinone, may be the perpetrators.

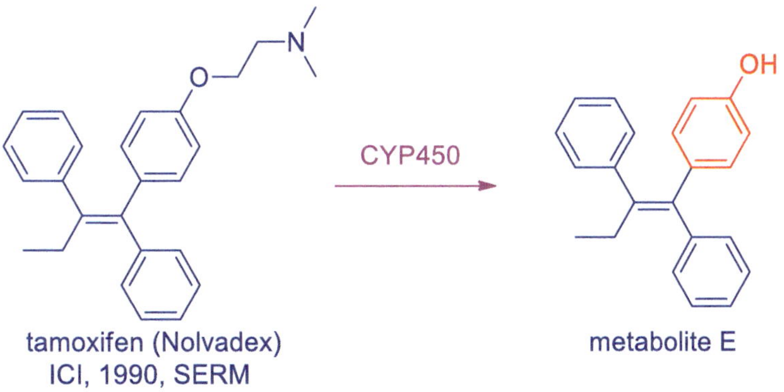

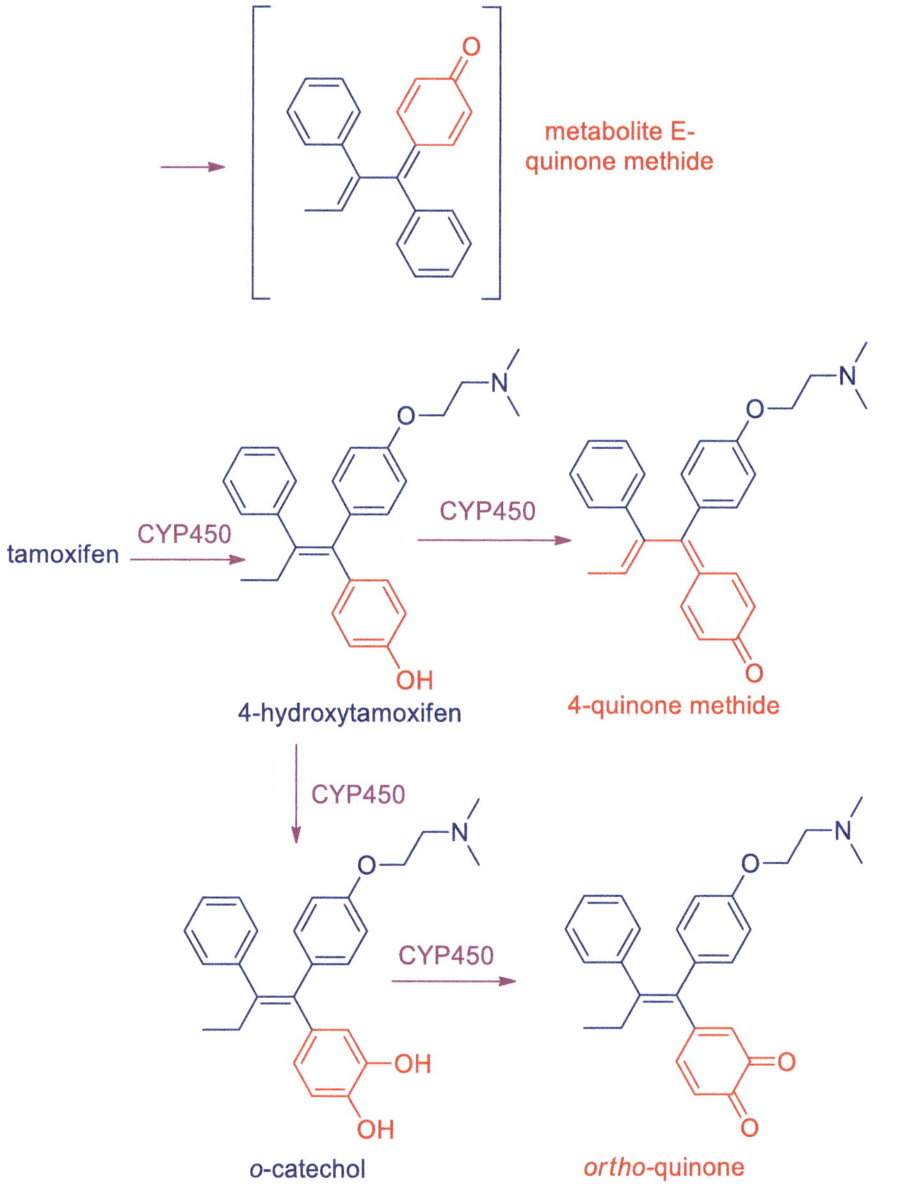

The tamoxifen sidechain, dimethylaminoethyl ether, may be dealkylated by CYP2B6 to generate metabolite E, which is even more active than the parent drug tamoxifen as an antiandrogen. Metabolite E then undergoes further oxidative metabolism to produce metabolite E-quinone methide.[48a] The mechanism for formation of the 4-quinone methide likely involves CYP2D6 (with contributions from CYP2C9 and 3A4)-catalyzed aromatic hydroxylation of tamoxifen, generating 4-hydroxytamoxifen, which undergoes a two-electron oxidation of the π-system, resulting in the 4-quinone methide.

The 4-quinone methide can react with DNA to form covalent adducts *in vitro*. Furthermore, 4-hydroxytamoxifen may be oxidized to the *ortho*-catechol, which is further oxidized to the corresponding *ortho*-quinone.[48b,c]

In addition to tamoxifen cation, all three aforementioned reactive metabolites: metabolite E-quinone methide, 4-quinone methide, and *ortho*-quinone, have potential to alkylate DNA and initiate the carcinogenic process.[48]

5.4.5 Sulfur-Containing Compounds

Most of the nearly 300 marketed drugs containing sulfur are reasonably safe. Among 285 sulfur-containing drugs, there are 72 sulfonamides, a dozen β-lactam antibiotics with the cephem core, 31 thioethers, 23 drugs have at least a thiazole core, as well as sulfonylureas, sulfonic acids, sulfamic acids, and others.[49]

Regrettably, several sulfur-containing functionalities have been implicated with idiosyncratic drug reactions (IDRs), some severe, making them structural alerts. They include: thiols, thiocarbonyls and thioureas, TZD, etc. They have been associated with hepatotoxicity, lung damage, bone-marrow depression, neoplasia, hormonal imbalance, and destroying CYP450 by suicide inactivation of hemoprotein.[50]

5.4.5.1 Thiols

Cysteine, glutathione, and coenzyme A all possess the thiol functionality, which is strongly nucleophilic and a strong reducing agent. *In vivo*, the thiol may be metabolized to sulfenic acid, sulfinic acid, sulfonic acid, and disulfide, as well as subsequent phase II metabolites.

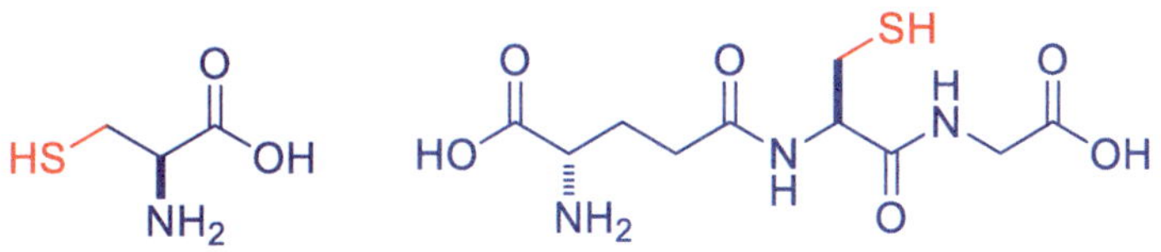

cysteine (Cys, C glutathione (GSH)

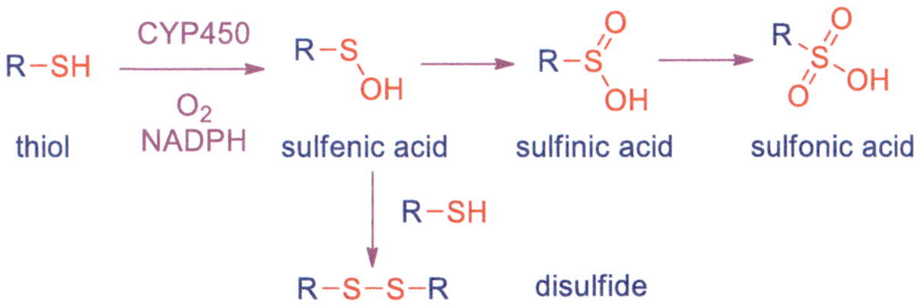

One of the early thiol-containing drugs was Wellcome's mercaptopurine (Purinethol) for treating cancer and autoimmune diseases. Its thiol group may be responsible for its adverse effects such as liver toxicity and bone marrow suppression.

The first orally active angiotensin-converting enzyme (ACE) inhibitor captopril (Capoten) as an antihypertensive has a thiol group as a chelator of the catalytic zinc cation of the enzyme. However, a trio of side effects have been associated with thiol: short-half life ($t_{1/2}$ = 2 h), rashes and loss of taste. Cysteamine (Cystaran) for treating cystinosis is one of the smallest drugs: 2-aminoethanethiol.

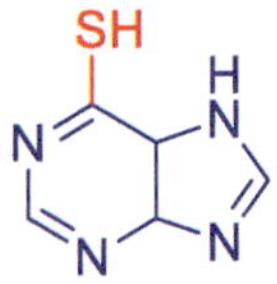

mercaptopurine (Purinethol)
Oncology
1953

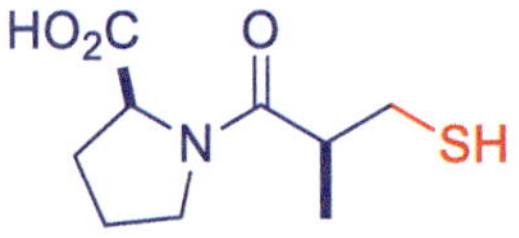

captopril (Capoten)
Squibb, 1977
ACE Inhibitor

cysteamine (Cystaran)
sensory organ/cystinosis
1994

Shown below are the four major isolated metabolites for captopril. The captopril–protein adduct may be responsible for its hypersensitivity. In addition, the disulfide linkage to macromolecules may trigger an immune response in addition to disruptive cellular functions. Finally, thiol′s reactive metabolites sulfenic acid, sulfinic acid, and sulfonic acid may challenge the body′s GSH defense system, which is the specified detoxification pathway.[50]

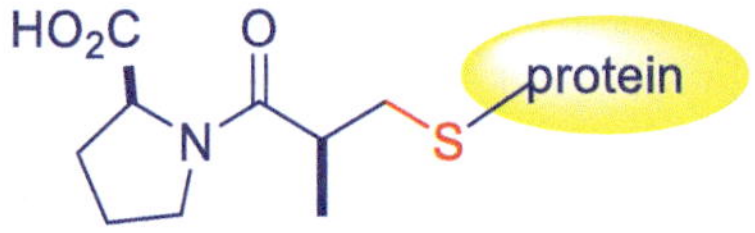

captopril–protein adduct

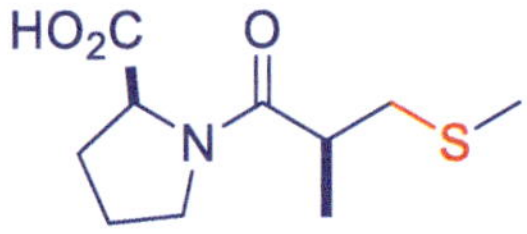

S-methyl-captopril

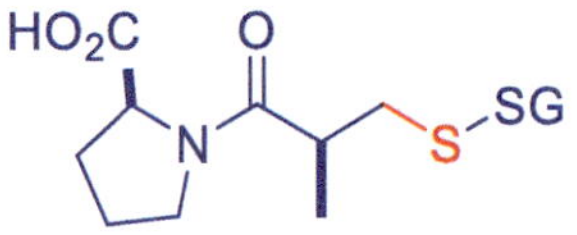

captopril–GSH adduct

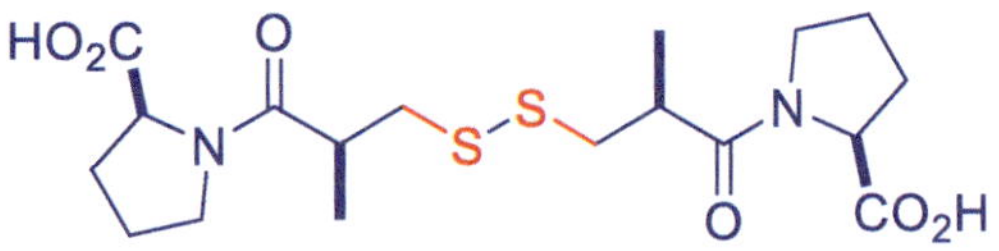

captopril disulfide dimer

5.4.5.2 Thioamides and Thioureas

The thiocarbonyl group is soft, highly polarizable, and easily oxidized. Thioamides and thioureas are readily oxidized by oxygenases such as flavin-containing monooxygenase (FMO) and CYP450 enzymes to their S-oxides, which are surprisingly stable both chemically and enzymatically thus can be and have been isolated. Further metabolic oxidation of the S-oxide intermediates leads to S,S-dioxides, which are extremely reactive thus not isolable as such. At least two major pathways await the reactive S,S-dioxide. It could isomerize to its tautomeric iminosulfinic acid form, which undergoes hemolytic scission to either acylium radical or iminium radical. Those radicals are then trapped by proteins. Another pathway for the S,S-dioxide involves nucleophilic attack by electrophiles such as an amine (e.g., lysine) to form an amidine or a guanidine.[51] We know by now that protein covalent binding of chemically reactive metabolites brings toxicities to these and many other small molecule drugs.

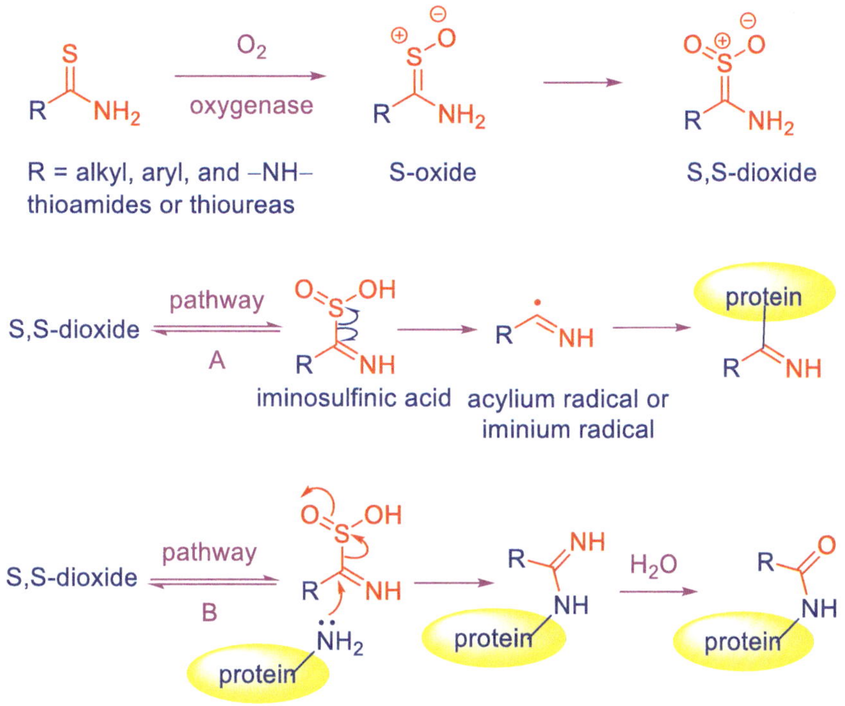

The aforementioned metabolic pathways of thioamides and thioureas explain the genesis of their potential toxicities. In fact, both thioureas propylthiouracil and methimazole for treating hyperthyroidism caused granulocytopenia, agranulocytosis, and a myriad of adverse effects.

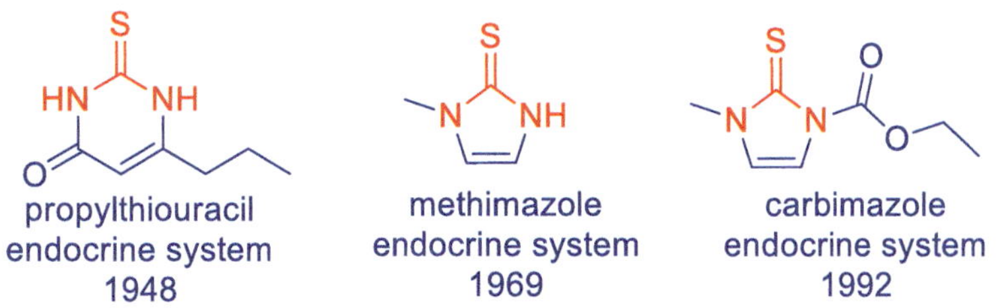

Carbimazole, as a prodrug of methimazole, has relatively few adverse effects. For methimazole, it is initially oxidized by CYP450 to give the 4,5-epoxide metabolite, which undergoes hydrolysis to afford the hemiaminal intermediate. Ring scission of the hemiaminal intermediate then gives rise to glyoxal and *N*-methylurea, which is oxidized to sulfenic acid and sulfinic acid, mainly by FMO. Those reactive metabolites can bind to proteins and cause toxicities. It has been shown that CYP450 and FMO work in tandem to cause methimazole's hepatotoxicties.[52]

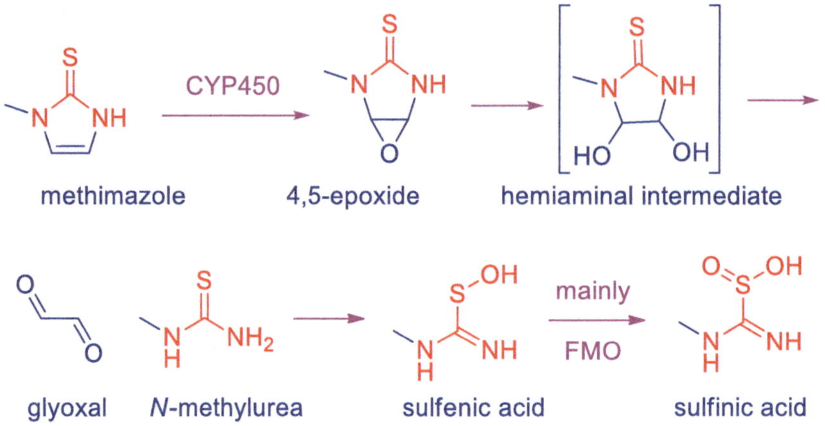

En route to the discovery of the first block-buster drug cimetidine (Tagamet), James Black's team at SmithKline & French prepared two thioureas. Whereas burimamide was not bioavailable, metiamide caused agranulocytosis. Even though metiamide also has a thioether functionality, thiourea was clearly the root cause of agranulocytosis by covalently binding to proteins with its reactive metabolites, presumably S-oxide and S,S-dioxide.

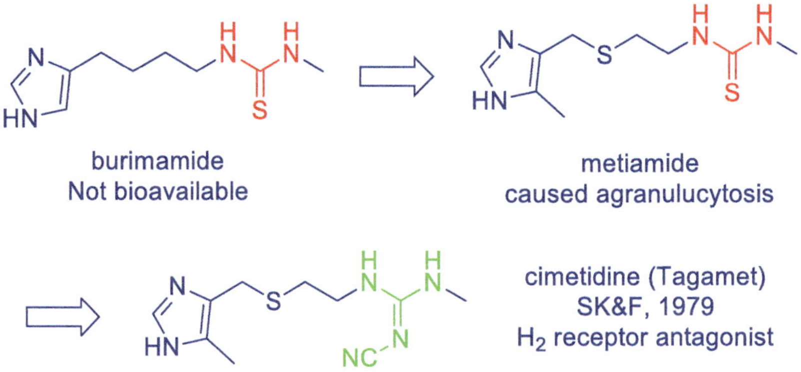

As always, a rigid view of structural alerts is dangerous. Jung discovered enzalutamide (Xtandi, 4 × 40 mg qd) and apalutamide (Erleada, 4 × 60 mg qd), two androgen receptor antagonists for the treatment of metastatic castration-resistant prostate cancer (CRPC). Each drug contains a thiourea moiety in a ring as a thiohydantoin, their risk/benefit profiles are robust enough to garner FDA's approval in 2012 and 2018, respectively.[53]

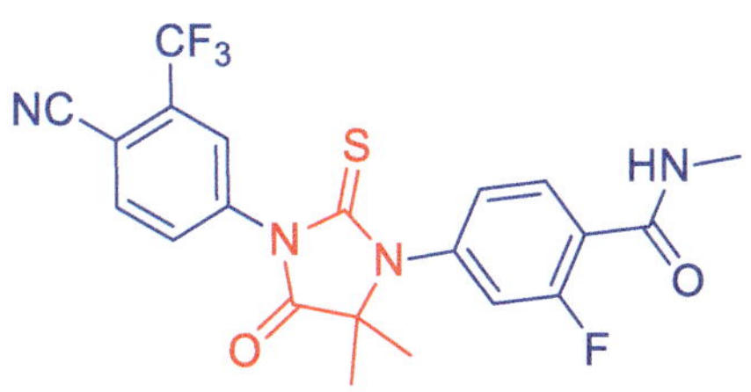

enzalutamide (Xtandi)
Medivation/Astellas, 2012
Anti-androgen

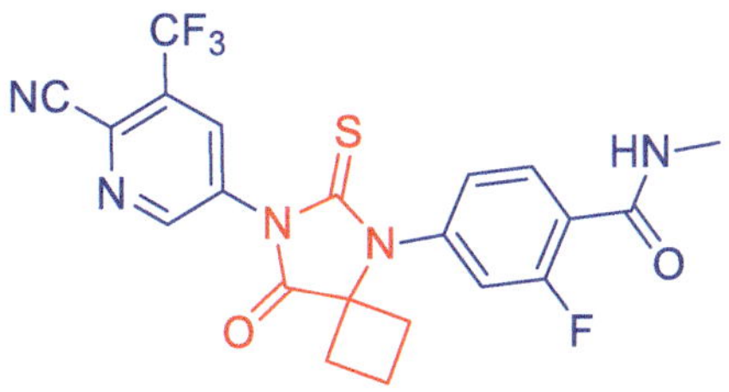

apalutamide (Erleada)
Janssen, 2018
Anti-androgen

5.4.5.3 Thiazolidinediones

In Section 5.4.4, we discussed reactive metabolites of the chromane portion of troglitazone (Rezulin, 200 mg qd). Its "me-too" drug rosiglitazone (Avandia, 2 mg qd) also contains the same pharmacophore TZD. Whereas rosiglitazone was withdrawn from the market in 2010 for myocardial infarction adverse effects, another "me-too" drug, pioglitazone (Actos, 4 mg qd) is still on the market despite the presence of the same TZD structural alert. Certainly, pioglitazone's 50-time lower dosage over that of troglitazone plays a key role in this discrepancy. Rosiglitazone has the lowest daily dosage among the three, yet it was plagued by cardiovascular adverse effects. Therefore, it is challenging to tease out how much TZD is responsible for the toxicities of both troglitazone and rosiglitazone, but pioglitazone's efficacy and safety justify its place in the market place.

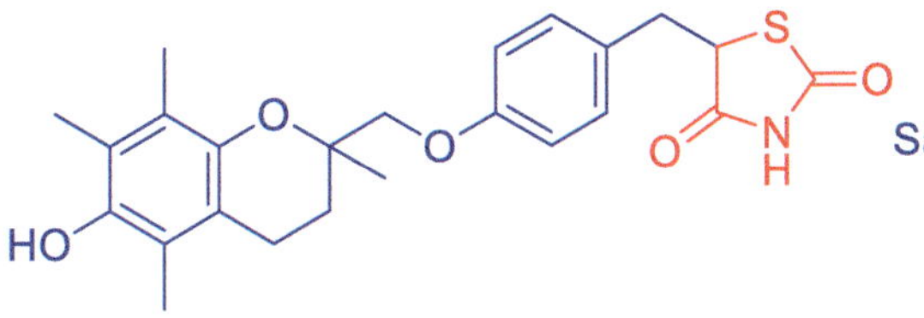

troglitazone (Rezulin)
Sankyo/Parke-Davis, 1997–2000
liver toxicities

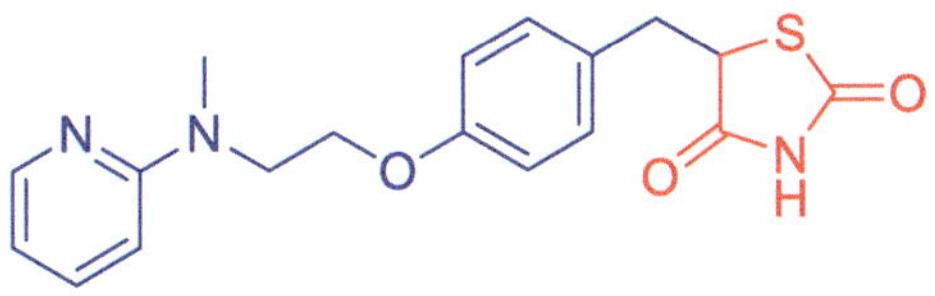

rosiglitazone (Avandia)
GSK, 1998–2010
cardiovascular toxicities

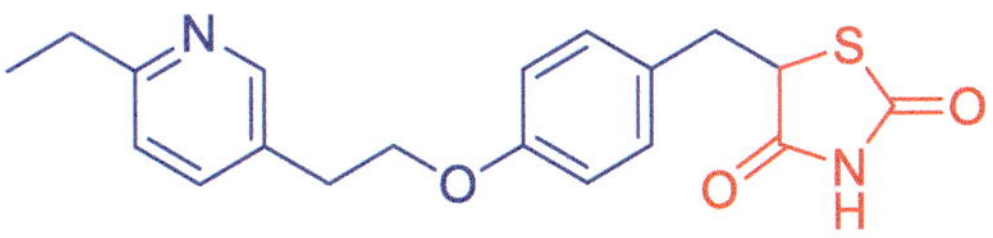

pioglitazone (Actos)
Takeda/Lilly, 1999
PPARγ agonist

The metabolism of the TZD functionality begins with CYP3A4 S-oxidation to yield the sulfoxide, which readily collapses to ring-opening products including reactive metabolites such as isocyanate and sulfine. GSH, DNA, and proteins can intercept and form covalent bonds with those very electrophilic reactive metabolites, which may be the culprit of liver and cardiovascular toxicities.[45]

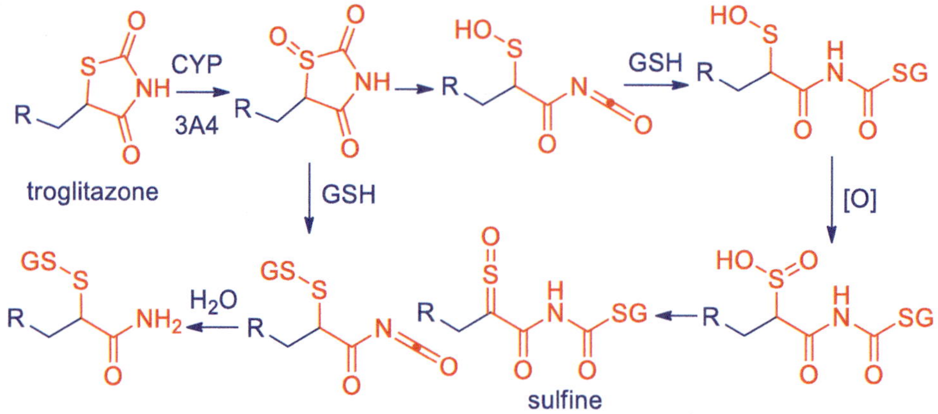

5.4.6 Hydrazines and Hydrazides

The demise of hydrazines and hydrazides was known as early as in the 1950s. Therefore, they became the early structural alerts and do not show up on drugs too often nowadays.

An old anti-hypertensive drug hydralazine contains a hydrazine structural alert that is responsible for its lupus-like erythematosus or rheumatoid arthritis adverse effects. It is metabolized by microsomal enzymes to metabolites capable of reacting covalently with macromolecules.[54a] An analogous drug dihydralazine is known to induce immunoallergic hepatitis. It was shown that its reactive metabolites from bioactivations covalently bind to CYP1A2 and CYP3A4 in human liver miscrosomes (HLMs) and trigger an immunological response as a neoantigen. The fact that chemically reactive metabolites bound to and inactivated the enzyme themselves suggest that dihydralazine is a mechanism-based inactivator (MBI) of CYP1A2 and CYP3A4.[54b] In addition, an old antidepressant phenelzine's hydrazine substituent may contribute to its adverse effects. The diazene phenylethylidenehydrazine is the putative metabolite intermediate that led to reactive metabolites to form covalent bonds with proteins.[54c]

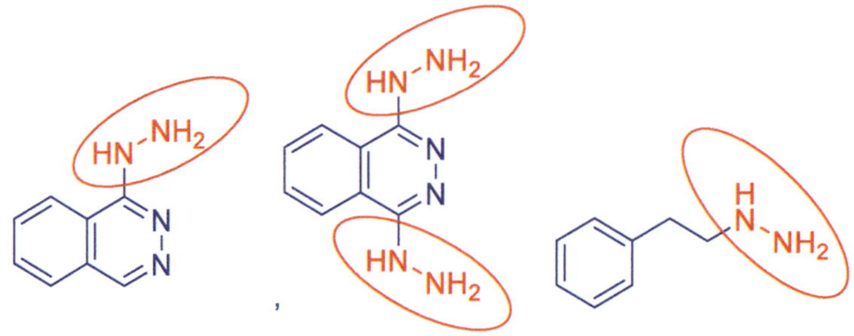

hydralazine anti-hypertensive | dihydralazine anti-hypertensive | phenelzine antidepressant

The hepatic and renal toxicities associated with hydrazine drugs including hydralazine and phenelzine treatment have been linked to free radical damage resulting from metabolism by CYP2E1.[54d] From experimental data, we could surmise a general metabolic pathway for hydrazine metabolism. At first, CYP450 dehydrogenation converts hydrazine to diazene, which is first oxidized to the corresponding diazo intermediate. Alternatively, the diazo intermediate may be arrived at from the diazene via the azoxy intermediate. Release of a molecule of nitrogen is accompanied by a radical that is responsible for forming covalent bonds with macromolecules such as proteins and DNAs and initiate toxic effects.

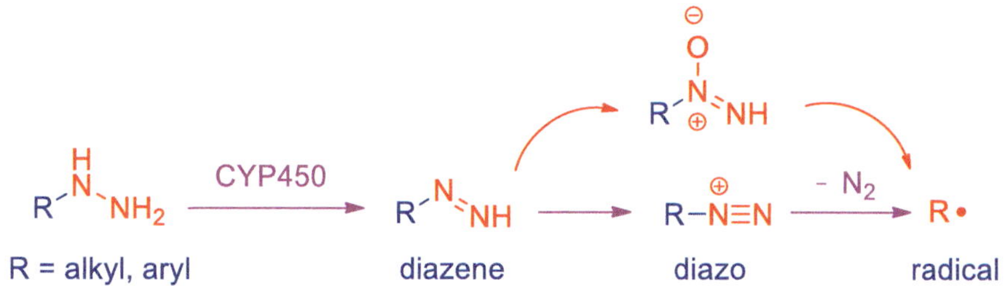

Hydrazide-containing drugs are as problematic as their hydrazine counterparts and they are considered as structural alerts as well. Isoniazid, iproniazid, and isocarboxazid are three representative hydrazides, and isoniazid is employed here to dissect their reactive metabolites.[55a]

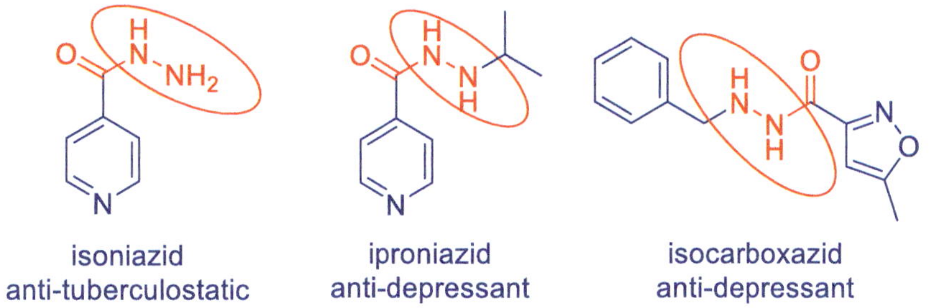

Shortly after its introduction to the market in 1952, tuberculosis (TB) drug isoniazid was recognized to cause rare cases of hepatitis (acute hepatocellular injury) known as idiosyncratic drug-induced liver injury (DILI). It received a black box warning in 1969. Surprisingly, isoniazid is still a widely used and effective first-line agent for treating TB even as of today. In contrast, its analog iproniazid, an MAO inhibitor anti-depressant, was withdrawn in 1956 due to severe hepatotoxicity. Another hydrazide-containing drug, isocarboxazid, also an MAO inhibitor anti-depressant, suffers many adverse effects as well. Since the emergence of selective serotonin reuptake inhibitors (SSRIs) in the 1990s, MAO inhibitor anti-depressants are no longer widely used due to their toxicity profiles.

Initially, it was believed that the polymorphisms of drug-metabolizing enzymes, *N*-acetyltransferase-2 (NAT2) and CYP2E1, played an important role. Thus, isoniazid is converted to *N*-acetylisoniazid, which is cleaved by amidase to produce isonicotinic acid and *N*-acetylhydrazine. Radiolabeled experiments showed that *N*-acetylhydrazine is the perpetrator of liver toxicity because its metabolites bind covalently to liver proteins. It is speculated that CYP450 enzymes, largely 2E1, oxidize *N*-acetylhydrazine to *N*-acetyldiazene, which collapses to afford, after losing a molecule of nitrogen, acetyl radical or acetylonium, both of which are highly reactive and bind covalently to proteins. But recently, the importance of human leukocyte antigen (HLA) has been increasingly recognized as possible culprit for causing isoniazid's DILI.[55b]

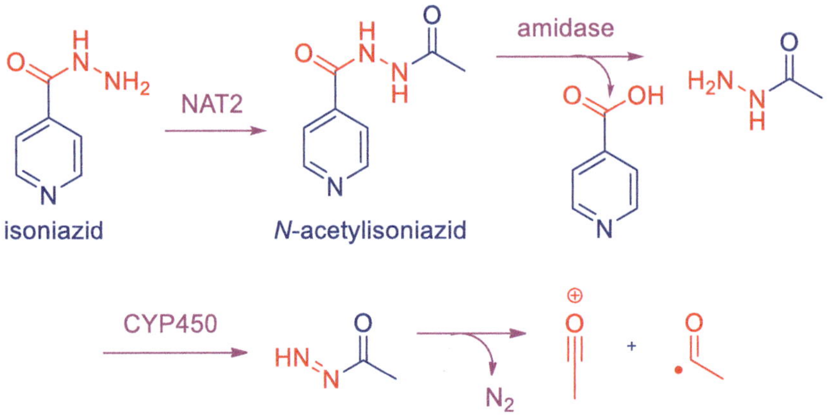

The demise of the hydrazine structural alert has not prevented its analog hydrazone from making appearances in marketed drugs although not too frequently, for a good reason.

Novartis' eltrombopag (Promacta) is an orally active thrombopoetin (Tpo) receptor agonist for the treatment of chronic idiopathic thrombocytopenia purpura. From high throughput screen (HTS), a deep purple diazonaphthalene dye SKF-56485 was identified as a hit with an EC_{50} of 200 nM for TpoR. But the diazo is a structural alert and is easily reduced *in vivo* by intestinal bacteria, resulting in destruction of the molecule. Inspired by orange food coloring tartrazine, which is an azopyrazole that is known to be resistant to azoreduction, the two molecules were fused to give SB-394725 (EC_{50} = 30 nM). In parallel, thiosemicarbazone (SB-450572, EC_{50} = 20 nM) was obtained via optimization of another HTS hit. Combination of SB-394725 and SB-450572 eventually led to eltrombopag with a hydrazone functionality after lead optimization.[56a]

The source of eltrombopag's hepatotoxicity and idiosyncratic reactions may not be the hydrazine structural alert. Radiolabeled absorption, distribution, metabolism, and excretion (ADME) investigations revealed that one of its major metabolites is the hydroxy-eltrombopag (M1), which is readily oxidized to reactive metabolite imine methide. Another reactive metabolite, the acylglucuronide (M2, not shown) attached to the carboxylic acid functionality is likely to contribute to its adverse effects as well.[56b]

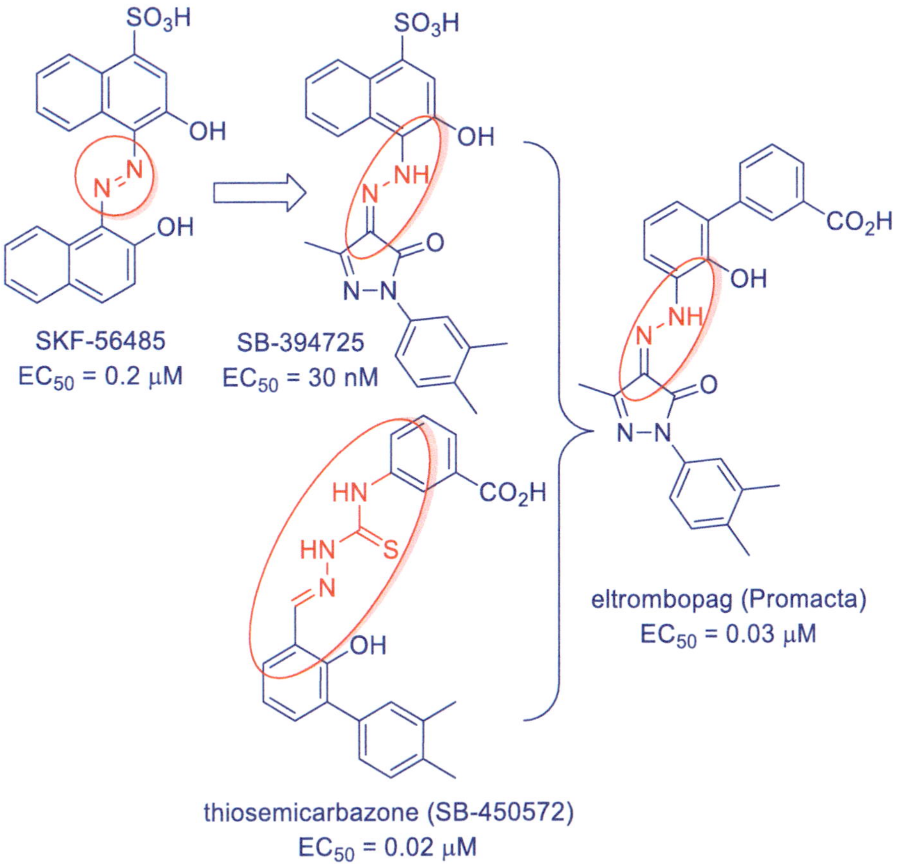

SO3H
OH
N=N
OH
SKF-56485
EC50 = 0.2 μM
SO3H
OH
N-NH
O
N-N
SB-394725
EC50 = 30 nM
HN
CO2H
HN
S
N
OH
thiosemicarbazone (SB-450572)
EC50 = 0.02 μM
CO2H
OH
N-NH
O
N-N
eltrombopag (Promacta)
EC50 = 0.03 μM

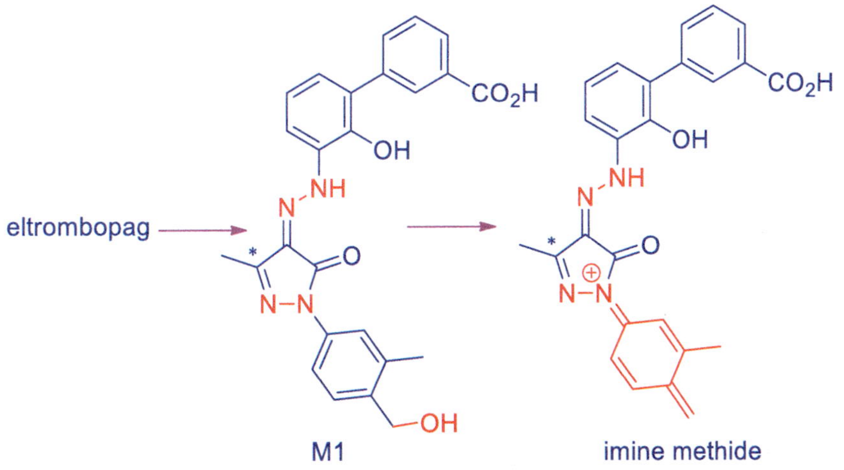

eltrombopag
CO2H
OH
N-NH
*
O
N-N
OH
M1
CO2H
OH
N-NH
*
O
N-N
imine methide

Metabolic cleavage of eltrombopag's hydrazine bond takes place to give two anilines, probably the products of gut microbial reductive biotransformations. The resulting aniline I and aniline II undergo Phase II metabolism, giving rise to their glucuronides and acetamides as M3, M4, and M8, respectively.[56c]

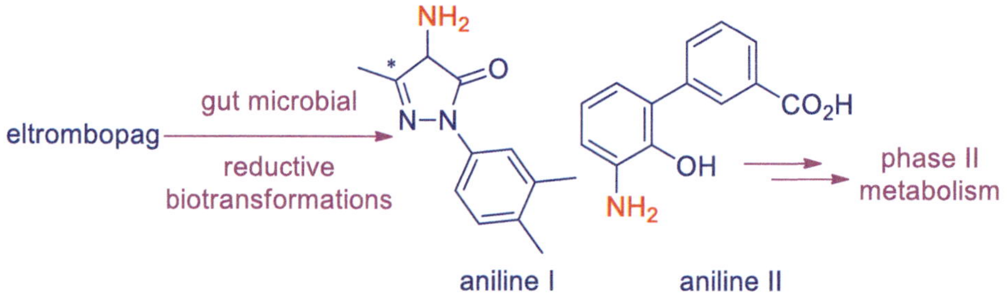

Another hydrazone-containing drug is dantrolene (Dantrium) for the treatment of malignant hyperthermia during anesthesia. It has a black box warning for potential to cause liver injury. The molecule is quite "ugly" and replete with a sizable collection of structural alerts: nitrophenyl, furan, hydrazone, and cyclic azaimide. Dantrolene's metabolism in human liver was investigated to narrow down the perpetrators of the hepatotoxicity.[57] One of dantrolene's major metabolites is 5-hydroxy-dantrolene catalyzed by CYP3A4. Since overexpression of CYP3A4 did not produce dantrolene cytotoxicity, 5-hydroxy-dantrolene can be ruled out.

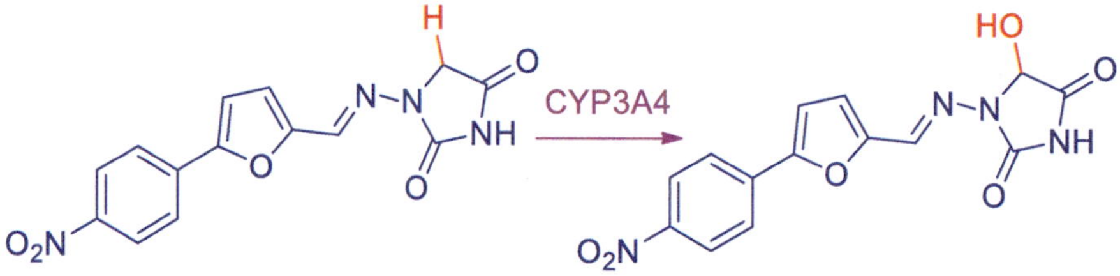

dantrolene (Dantrium)
Par Pharmaceuticals, 1974
muscle relaxant

5-hydroxydantrolene

It was shown that aldehyde oxidase-1 (AOX1) is responsible for reducing dantrolene to hydroxylamine-dantrolene, which is further reduced to amino-dantrolene, also by AOX1. Intermediate amino-dantrolene is then acetylated to acetylamino-dantrolene, which is one of the major metabolites. More relevantly, hydroxylamine-dantrolene is further oxidized to most likely nitroso-dantrolene, which can form covalent bonds with proteins and initiates toxicity events.

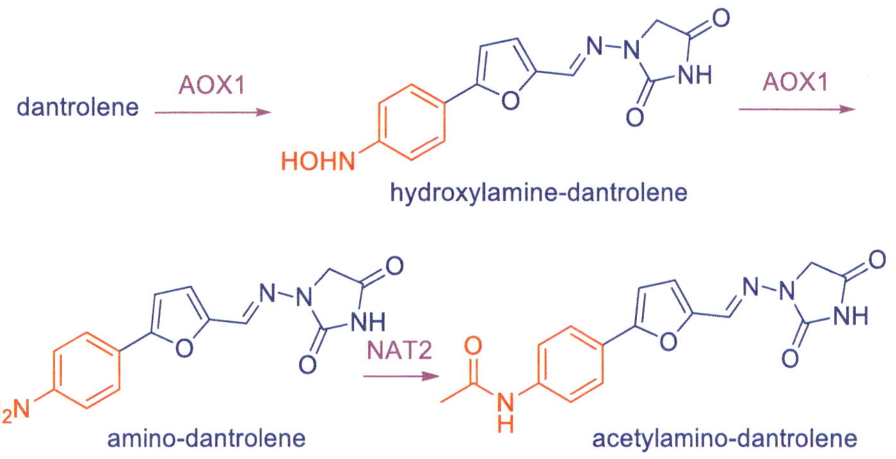

5.4.7 Methylenedioxyphenyl Moiety

Methylenedioxyphenyl fragment is a common functionality. Safrole with a methylenedioxyphenyl motif, is the principal ingredient of sassafras oil, isolated from nutmeg, cinnamon, black pepper, and root beer. The infamous recreational drug 3,4-methylenedioxymethamphetamine (MDMA, Ecstasy) also contains the methylenedioxyphenyl motif. In medicinal chemistry, the methylenedioxyphenyl substituent is frequently employed as a bioisostere of dimethyl *o*-catechol. Both GSK's paroxetine (Paxil, 20 mg qd) as an SSRI for treating depression and Icos/Lilly's phosphodiesterase-5 (PDE-5) inhibitor tadalafil (Cialis, 5–20 mg qd) for treating erectile dysfunction (ED) are remarkably safe despite the presence of the methylenedioxyphenyl structural alert.

safrole, sassafras oil (nutmeg, cinnamon, black pepper, root beer)

methylenedioxy-methamphetamine (MDMA, Ecstasy)

paroxetine (Paxil) GSK, 1992 SSRI

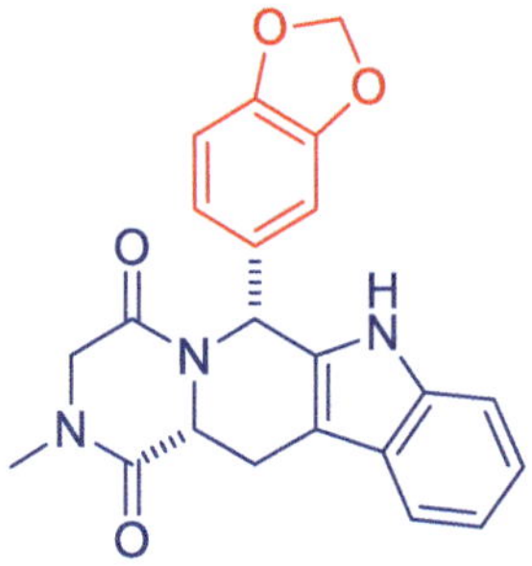

tadalafil (Cialis)
PDE 5 Inhibitor
Icos/Eli Lilly, 2003

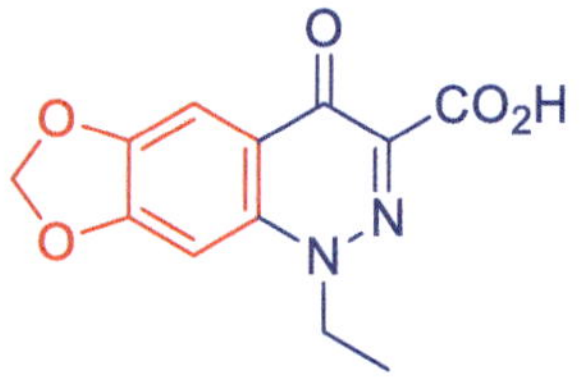

cinoxacin (Cinobac)
Eli Lilly, 1990
antibacterial

In contrast, Lilly's quinolone antibacterial cinoxacin (Cinobac) as a gyrase inhibitor suffered from gastrointestinal (GI) system and the central nervous system (CNS) adverse effects. Pfizer's endothelin antagonist sitaxsentan (Thelin) for treating pulmonary hypertension and congestive heart failure dosed at 100–500 mg qd was withdrawn in 2010 due to fatal liver toxicity. While appreciating that "the dose makes poison," let us take a look of methylenedioxyphenyl's metabolism and reactive metabolites to appreciate the molecular origin of potential toxicities.

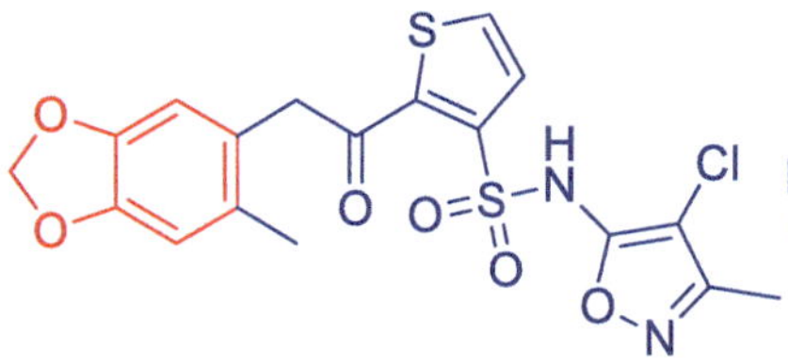

sitaxsentan
endothelin antagonist
pulmonary hypertension
congestive heart failure

MDMA's metabolism may be responsible for its neurotoxicity, possibly through formation of glutathione adducts. As shown beneath, its major metabolic pathways is O-demethylation to give 3,4-dihydromethamohetamine (*o*-catechol), mainly via CYP2D6. The *o*-catechol metabolite then undergoes Phase II metabolism including methylation, sulfation and glucuronization.[58] In addition, a carbene intermediate, generated via CYP2D6, may coordinate with the heme iron on the CYP enzyme to form a metabolic intermediate complex (MI complex or MIC). The MI complex results in an inactivation of CYP2D6.[59] Generally speaking, metabolic intermediate complex (MIC) can be monitored as a 455 nm absorbance.

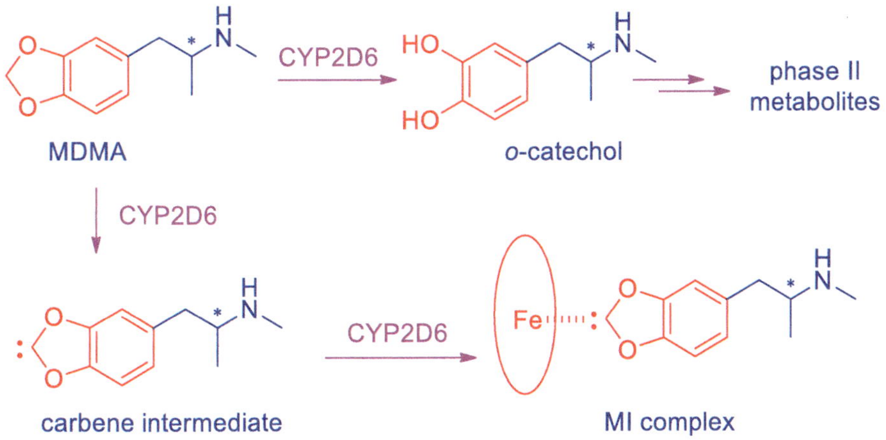

Like MDMA, paroxetine's methylenedioxyphenyl substituent is also metabolized by CYP2D6 to give O-demethylation metabolite, *o*-catechol-paroxetine, presumably produced via the intermediacy of hydroxyl-paroxetine. Two fates await *o*-catechol-paroxetine metabolite. (i) It may be methylated by COMT to deliver guaiacols as paroxetine's major metabolites. And (ii) it may be oxidized to the corresponding *o*-quinone intermediate, which is readily trapped by GSH. The GSH adducts help minimizing covalent binding to microsomes, thus promote detoxication. This process, in addition to small dosage (20 mg qd), is likely responsible for its good safety profile.[60]

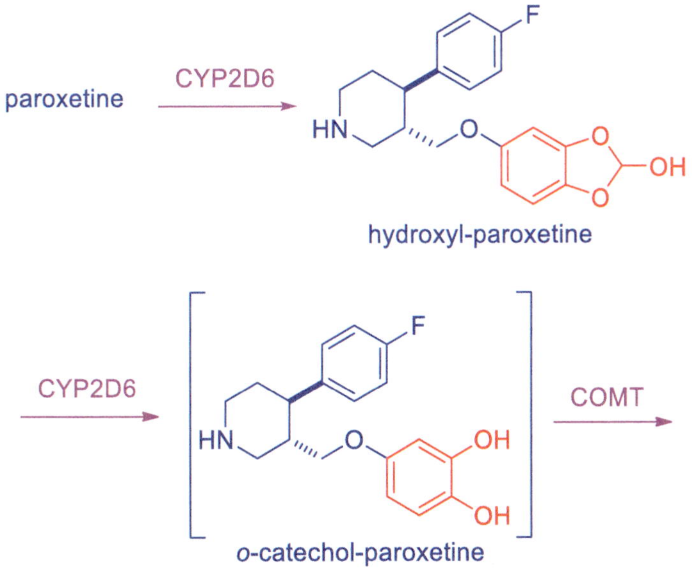

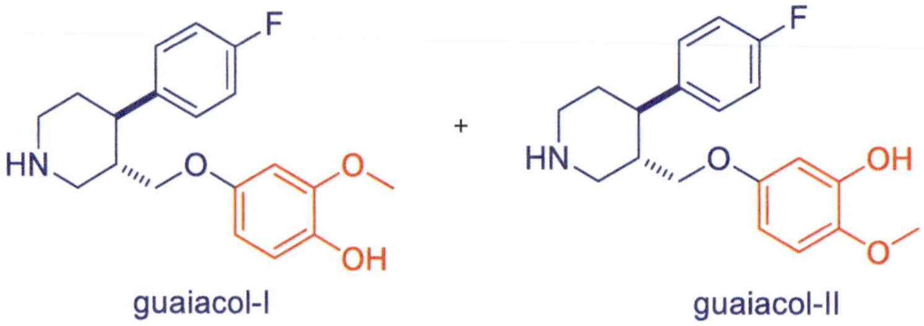

But paroxetine's methylenedioxyphenyl substituent is blamed to cause DDIs via inactivation of CYP2D6. Similar to MDMA, ring scission of the hydroxyl-paroxetine intermediate is aided by CYP2D6 to produce the highly active carbene intermediate, which readily coordinates with the heme iron on the CYP2D6 enzyme to form an MIC. This process is called mechanism-based inhibition (MBI).[3b] Formation of the MIC would result in DDIs with other drugs that are also CYP2D6 substrates such as desipramine, metoprolol, risperidone, and atomoxetine. MIC also explains paroxetine's nonstationary pharmacokinetics when co-administered with CYP2D6 extensive metabolizers.[61]

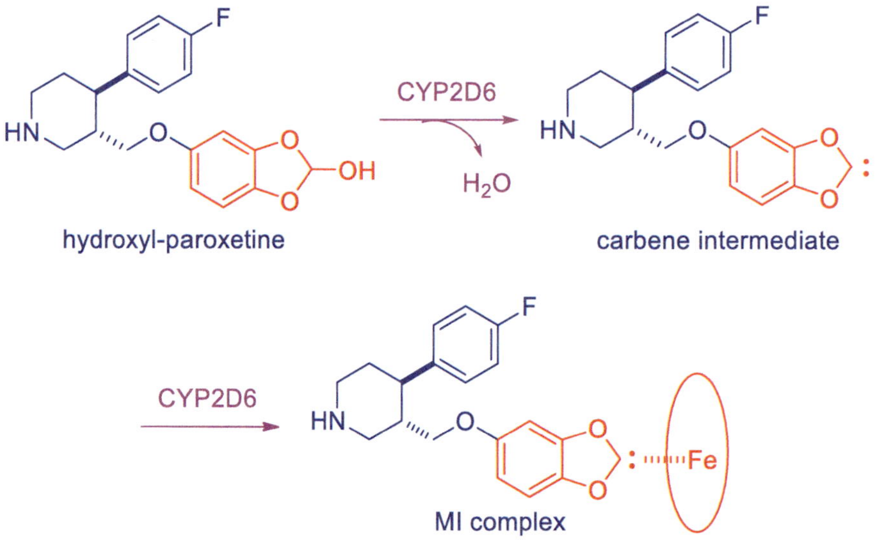

Whereas both MDMA and paroxetine are mainly metabolized by CYP2D6, the main driver of tadalafil's metabolism is CYP3A4, giving rise to the corresponding *o*-catechol-tadalafil. O-methylation of the *o*-catechol-tadalafil by COMT leads to dimethoxy-tadalafil as the major metabolite. The drug also undergoes mechanism-based inactivation (MBI) of CYP3A4.[62] There are no reported idiosyncratic toxicity and/or DDIs associated with tadalafil for the treatment of ED. The low daily dosage is most likely the reason behind its remarkable safety profile.

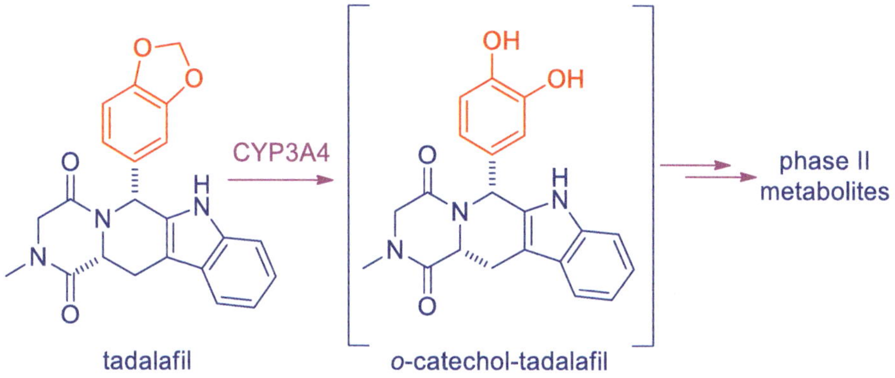

Since both cinoxacin and sitaxsentan were withdrawn from the market, not many investigations for their metabolism and their reactive metabolites have been published. But from the examples described before, it may be surmised that their MI complexes are mostly to blame for their respective adverse effects.

Metabolism of the methylenedioxyphenyl structural alert is summarized below.[63] The heme iron–carbene complex leads to P450 inhibition, which may be largely responsible for its toxicities. In addition, the OH radical, common by-product *in vivo*, can oxidize OCH_2O moiety to cause ring scission. Both pathways ultimately lead to the production of a catechol, which can undergo further oxidation to *ortho*-quinones, electrophilic reactive metabolites.

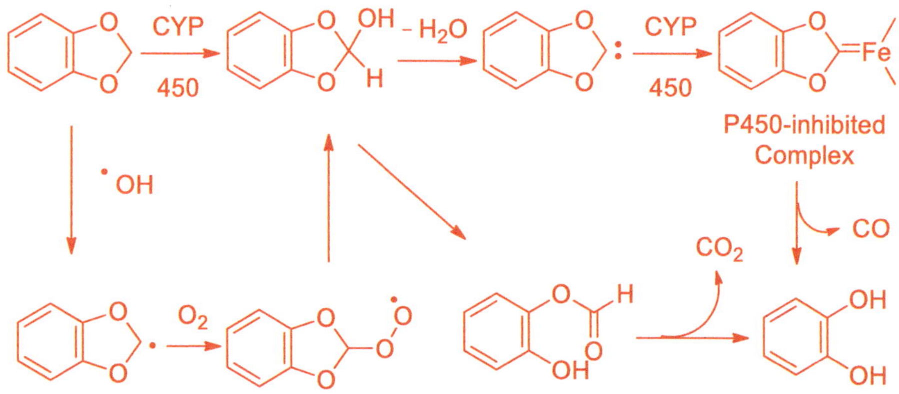

Several tactics are available to abrogate the methylenedioxyphenyl moiety's potential metabolic soft spot: the methylene sandwiched between the two oxygen atoms.[64] Replacing the vulnerable five-membered methylenedioxy with six-membered rings as shown beneath buttresses its resistance to CYP metabolism. Furthermore, blocking the two methylene protons of the methylenedioxy group with either two fluorine or two methyl groups also minimizes its metabolism.

Concert Pharmaceuticals prepared deuterated paroxetine (CTP-347) with its two methylene protons of the methylenedioxy group replaced with two deuterium atoms.

While CTP-347 is still as active as paroxetine biologically, *in vitro*, HLMs cleared CTP-347 faster than paroxetine as a result of decreased inactivation of CYP2D6. CTP-347 demonstrated little to no CYP2D6 MBI, apparently because of a dramatic reduction in the formation of the reactive carbene intermediate since C–D bonds are stronger than the C–H bonds. In Phase I clinical trials, CTP-347 was metabolized more rapidly in humans thus deuteration significantly reduced drug–drug interactions with tamoxifen and dextromethorphan. Concert's precision deuteration can improve the metabolism profiles of existing pharmacotherapies without affecting their intrinsic pharmacologies.[64]

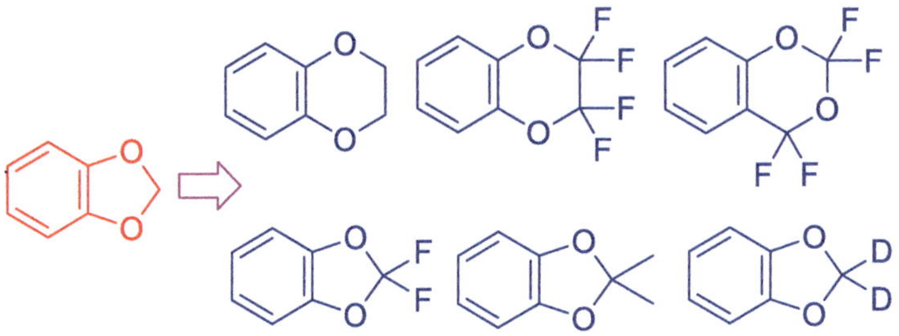

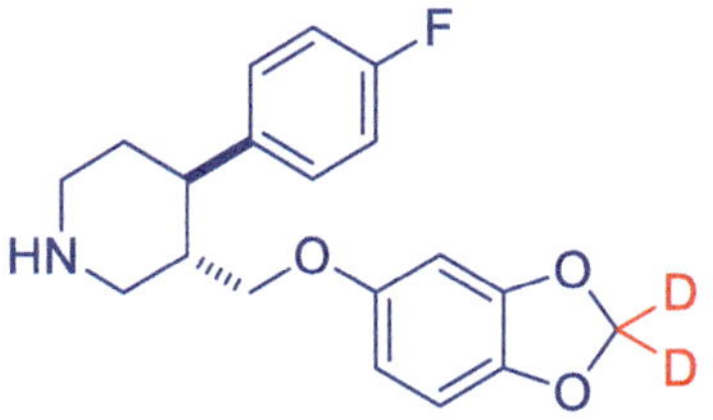

CTP-347

5.4.8 Electron-rich Heteroaromatics

Fundamental organic chemistry principles apply to drug metabolism in human body as well. As such, electron-rich heteroaromatic drugs are more prone to be oxidized by CYP enzymes. Some of them including pyrroles, indoles, furans, thiophenes, and thiazoles may result in reactive metabolites, making them structural alerts.

5.4.8.1 Pyrroles

Because the pyrrole ring is extremely electron-rich, pyrrole-containing drugs are easily oxidized by CYP-450 enzymes in the liver. The resulting metabolic oxidation products are prone to nucleophilic replacement by physiological nucleophiles such as the thiol group. The consequence is toxicities such as agranulocytosis, hepatotoxicity, and so on. An anti-hypertensive agent, mopidralazine (MDL-899), was extensively investigated with regard to the metabolic oxidation of its pyrrole ring in rats and dogs.[65–67] Isolation and characterization of mopidralazine's metabolites led to the hypothesis that the biotransformations of pyrrole may involve the introduction of molecular oxygen into the pyrrole ring. The intermediacy of 1,2-dioxetane explains that an oxidative cleavage of the pyrrole ring could provide all metabolites identified.

As one could imagine, the highly reactive intermediates from pyrrole could wreak havoc in the physiological system. Nucleophiles such as the thiol group could induce toxicities. As a result, the development of mopidralazine was subsequently discontinued.

Similar oxidative metabolism was observed for premazepam, an anti-anxiety drug, in the rat and the dog;[68] prinomide, an anti-inflammatory agent, in six species of laboratory animals;[69] and pyrrolnitrin, an anti-fungal agent, in rats.[70]

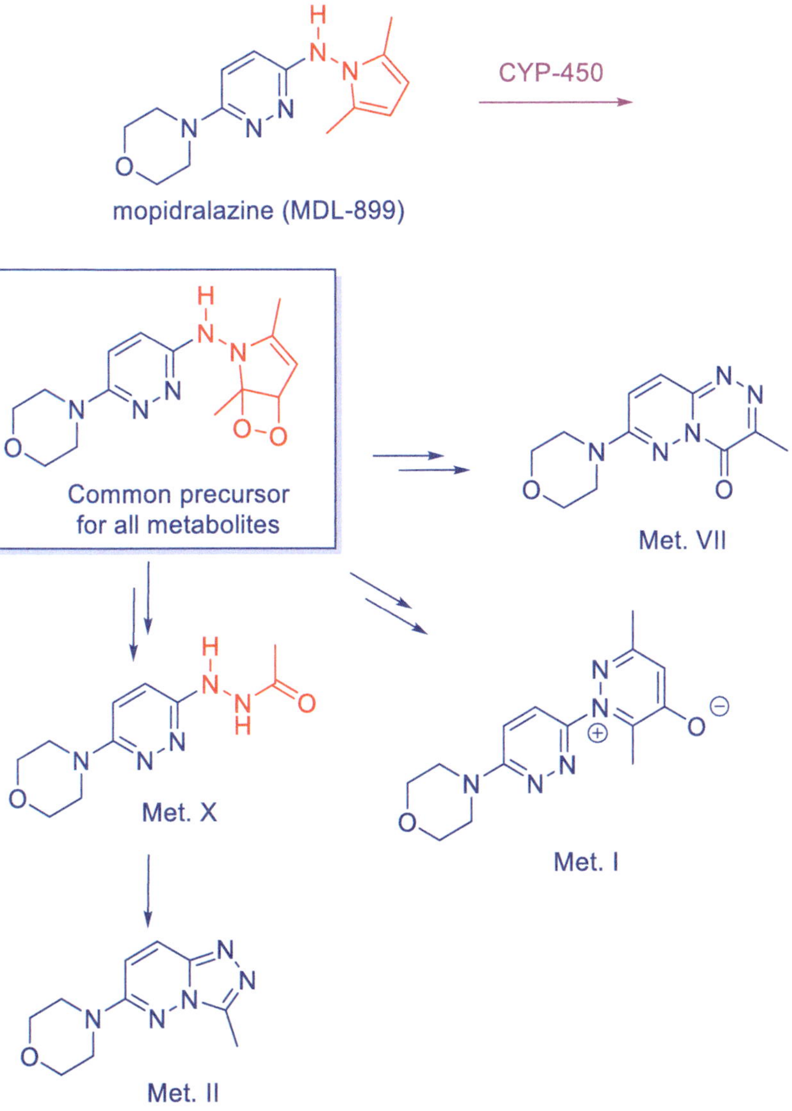

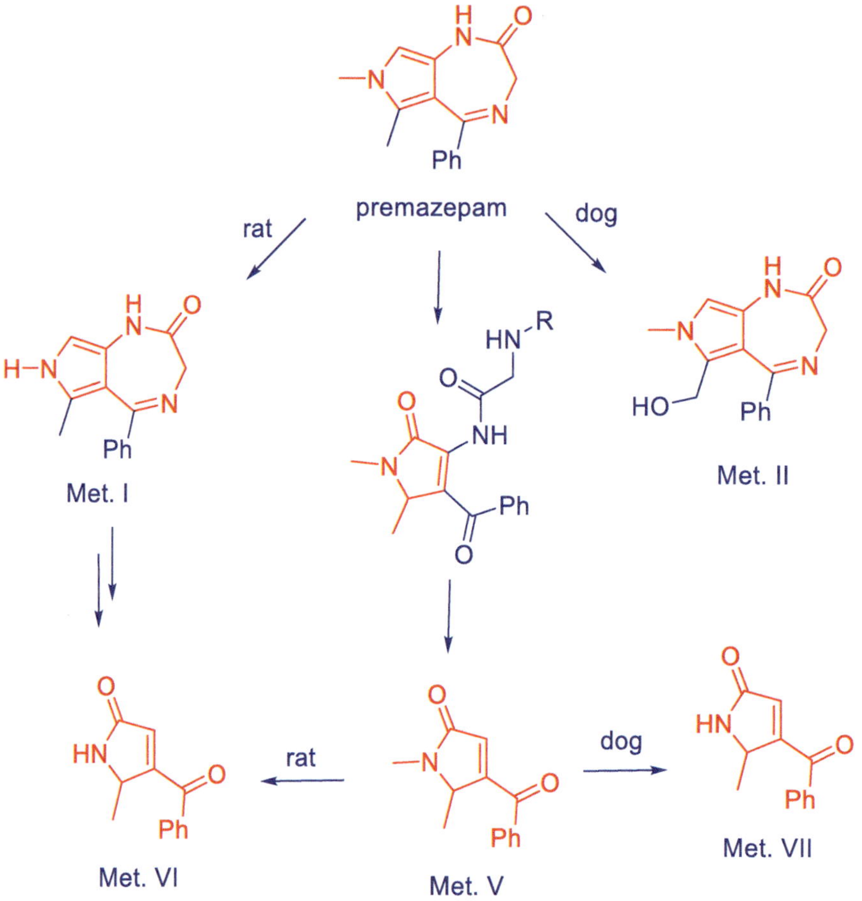

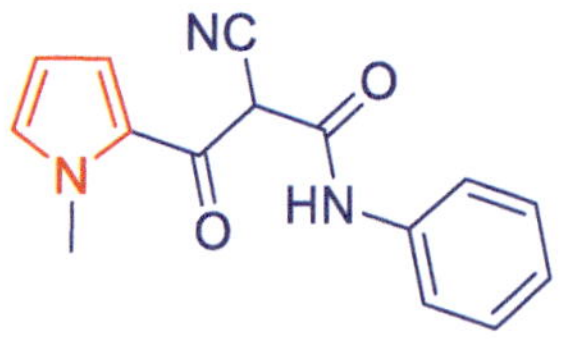

prinomide

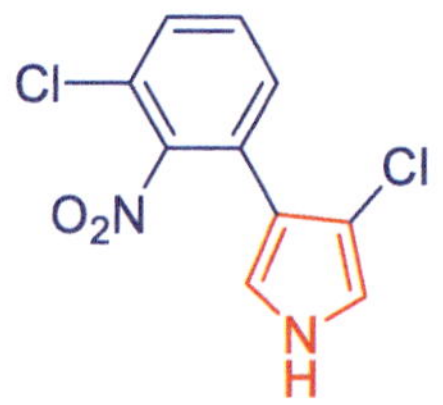

pyrrolnitrin

Nonetheless, if one assumes that all pyrrole-containing drugs are toxic, one would have missed atorvastatin (Lipitor). Atorvastatin is remarkably safe barring the mechanism-based safety concerns such as rhabdomyolysis that are associated with all HMG-CoA inhibitors.[71] Although atorvastatin contains the pyrrole ring, it has at least three factors that strongly attenuate its nucleophilicity. First, it is fully substituted at all possible positions—it is a penta-substituted pyrrole, thus, the steric hindrance would block CYP oxidation of the pyrrole ring. Second, two phenyl and one amide substitution form large delocalization to disperse the electronic density of the pyrrole ring. Third, *para*-fluorophenyl and amide are both electron-withdrawing, further diminishing the electronic density of the pyrrole ring.

In drug discovery, as in many things in life, there are always exceptions to the rules. Frequently, the safety and efficacy of a drug can only be determined by clinical trials as the gold standard.

5.4.8.2 Indoles

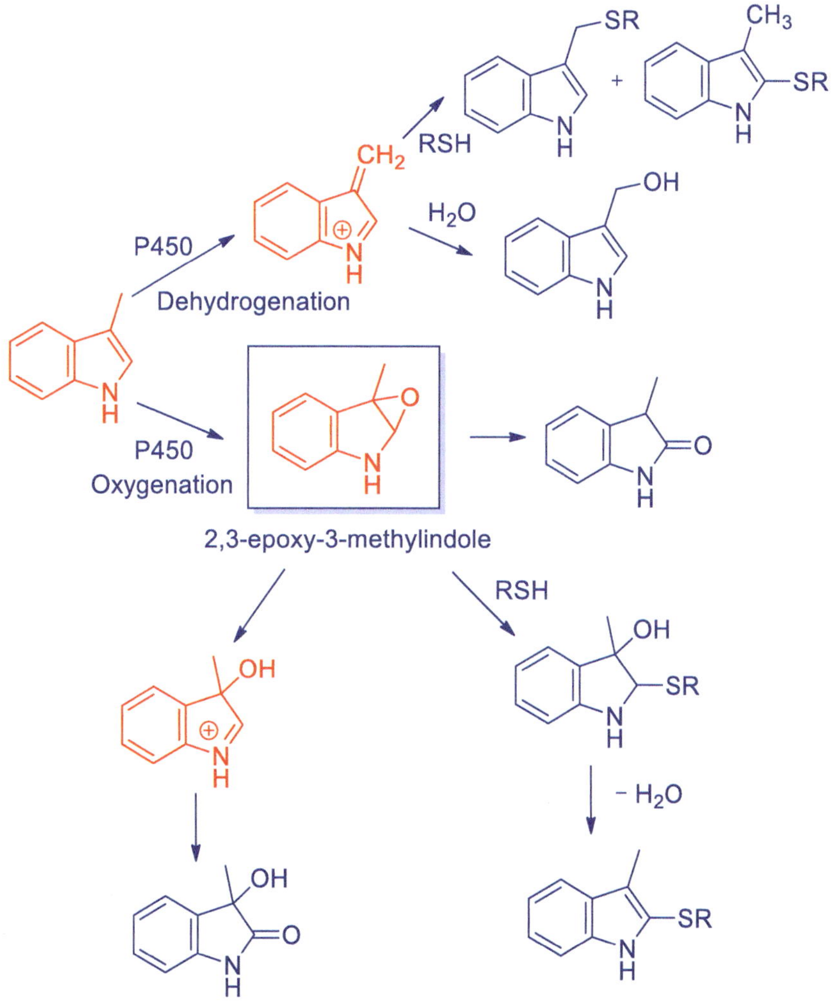

The indole-ring system exists in a plethora of endogenous amino acids, neurotransmitters, and drugs. The metabolic 2,3-oxidation of the indole ring by CYP450 takes place from time to time, but its correlation to *in vivo* toxicity is not often observed.

One particular indole, 3-methylindole, unfortunately, has been associated with higher risk of adverse outcomes, namely, pneumotoxin in animals. Evidence was found to support the formation of 2,3-epoxy-3-methylindoline as a reactive intermediate of the pneumotoxin 3-methylindole.[72] 3-Methylindole has been shown to form adducts with glutathione, proteins, and DNA using *in vitro* preparations.[73]

The CYP450-mediated bioactivation of 3-methylindole may be summarized below. Oxidation of the 3-methyl group occurs either directly via deoxygenation or via epoxidation of the 2,3-double bond leading to 2,3-epoxy-3-methylindole, the reactive intermediate that can be trapped by endogenous nucleophiles, such as glutathione.

The presence of a leaving group on the C3-methyl increases the likelihood of formation of electrophilic reactive intermediates.

Zafirlukast (Accolate) is a leukotriene antagonist indicated for the treatment of mild-to-moderate asthma, but the drug has been associated with occasional idiosyncratic hepatotoxicity. Structurally, zafirlukast is similar to 3-methylindole because it contains an *N*-methylindole moiety that has a 3-alkyl substituent on the indole ring. The results presented here describe the metabolic activation of zafirlukast via a similar mechanism to that described for 3-methylindole. NADP(H)-dependent biotransformation of zafirlukast by hepatic microsomes from rats and humans afforded a reactive metabolite, which was detected as its GSH adduct.[74] The formation of this reactive metabolite in HLMs was shown to be exclusively catalyzed by CYP3A enzymes. Evidence for *in vivo* metabolic activation of zafirlukast was obtained when the same GSH adduct was detected in bile of rats given an intravenous (IV) or oral dose of the drug.

The observation of *in vitro* metabolic activation of the 3-benzylindole moiety in zafirlukast to give the glutathione adduct is an indication that the 3-methyl-indole activation pathway applies to other activated 3-alkyl indoles as well.[7]

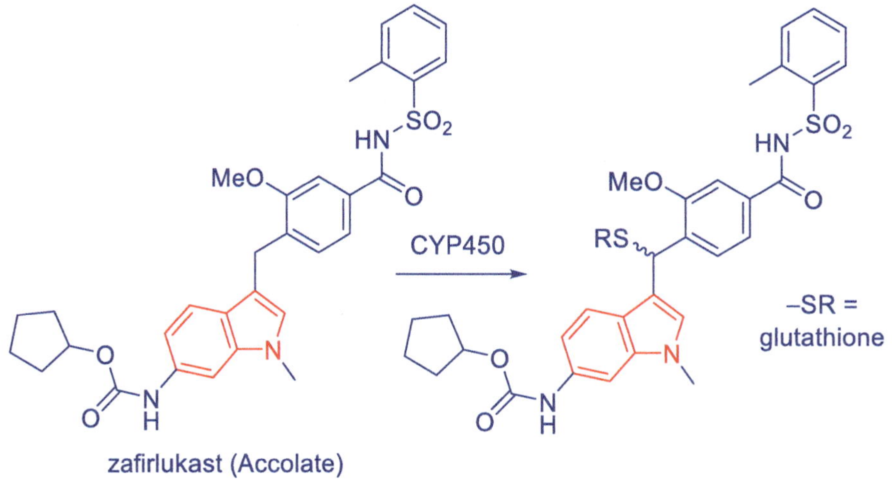

5.4.8.3 Furans and Thiophenes

Electron-rich furan and thiophene ring systems are susceptible to oxidation by cytochrome P-450 (CYP450) enzymes. The oxidized products are then capable of reacting with various biological nucleophiles; the resulting metabolites can lead to toxicity, typically hepatotoxicity.[75–78]

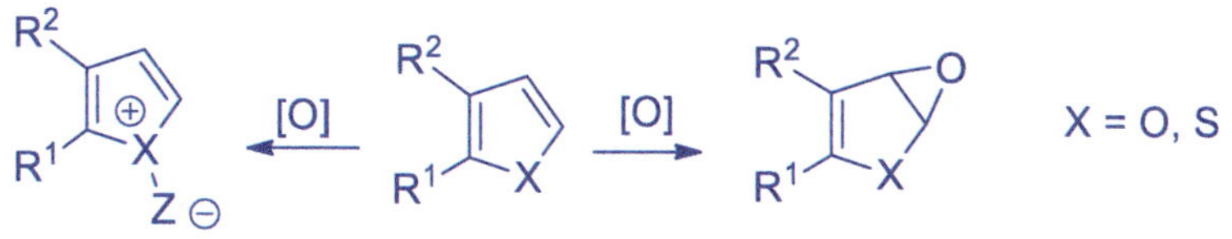

In general, the furan ring system appears to be far less reactive than the thiophene system, and it is therefore much less toxic. This is most likely due to the higher electronegativity of the oxygen atom, which reduces the reactivity of the ring toward oxidation.

Of furan-containing drugs on the market, furosemide, a diuretic, has been shown to cause hepatic necrosis in mice.[79] The mechanism involves metabolic activation of furosemide by oxidation of the furan ring by CYP450 followed by conjugation to glutathione to produce a furosemide–glutathione conjugate. Despite these results, furosemide has not been shown to present significant toxicity to humans.

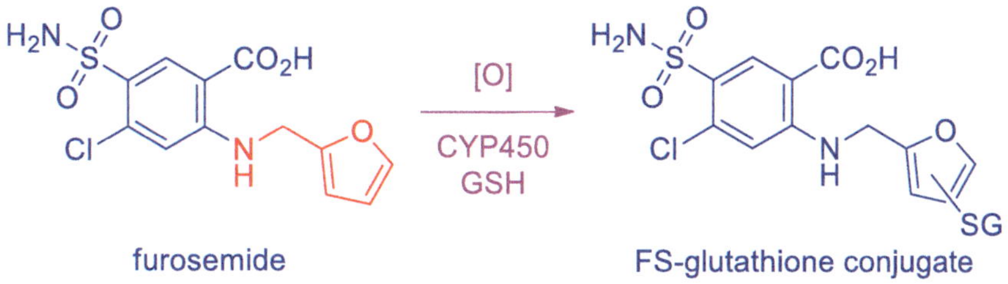

Tienilic acid, a thiophene-based diuretic used to treat hypertension, has been shown to cause hepatotoxicity. Oxidation of the thiophene and subsequent reactions of the activated product with nucleophilic proteins is responsible for the observed pathology.[80] Tienilic acid was withdrawn from the market shortly after there was evidence of drug induced hepatitis. Additionally, tienilic acid was found to be a "suicide" inhibitor for the cytochrome P-450 enzyme (CPY2C9) to which it became covalently linked.[81]

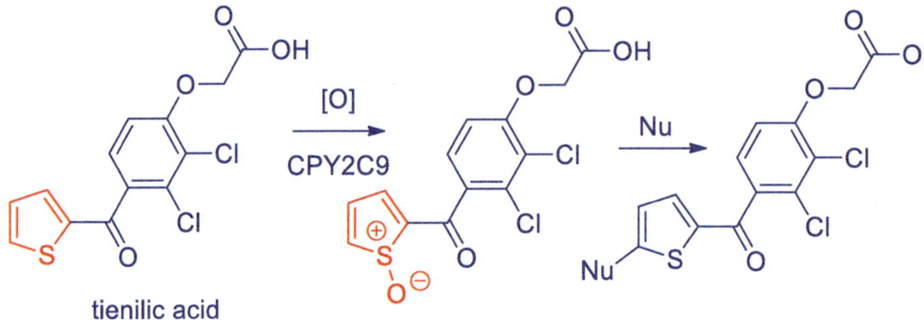

Although the electron-rich thiophene may lead to toxicity, the metabolic chemistry of thiophene can also lead to desirable therapeutic effects as in the case of clopidogrel (Plavix). The parent compound is oxidized by cytochrome P-450, and further oxidation in the presence of water opens the thiophene ring to produce an electrophilic sulfenic acid.[80] This electrophilic intermediate is susceptible to a nucleophilic thiol found on the $P2Y_{12}$ receptor.[82] Creation of this disulfide bond modifies the receptor and inhibits platelet aggregation, leading to the desired therapeutic effect.

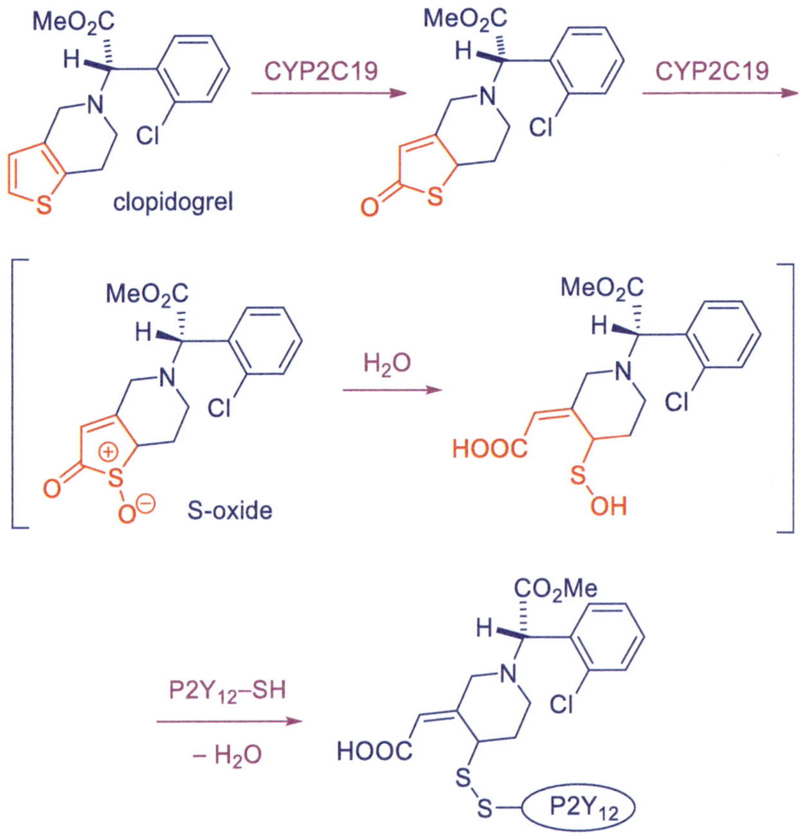

Potential toxicity arising from the furan and thiophene ring structures is a concern in designing potential drug leads, but potential leads should not be eliminated on those grounds alone. The success of current pharmaceuticals that contain furan and thiophene moieties is a clear indication that furan- and thiophene-containing therapies can be made safe, and even exploited as in the case of clopidogrel.

5.4.8.4 Thiazoles

Commercially outsourced libraries as well as pharmaceuticals often have 1,3-thiazoles or benzothiazoles.[83] Yet thiazoles especially 2-aminothiazoles are considered as structural alerts and often excluded when considering the design of new drug candidates. The risk associated with structural alerts like these can be reduced by inducing an alternative metabolic pathway or simply low clinical exposure, but such an approach cannot accurately predict the human clinical response. An understanding of the "metabolic purpose" and the tendency to from reactive metabolites is therefore necessary. This is especially relevant from a position concerning hit triage and follow-up strategies. The function of most cytochrome P450 (or CYP) enzymes, the major enzymes involved in drug metabolism, is to catalyze the oxidation of organic substances. 1,3-Thiazoles are prone to oxidative metabolism and typically undergo epoxidation. This happens at the 4,5-double bond and causes the formation of α-dicarbonyl metabolites and thioamide derivatives, such as thioamides, thioureas, or acylated thioureas. Both types of metabolites are capable of undergoing further metabolism to form reactive intermediates. For example, compounds with documented adverse reactions due to associated downstream reactive intermediates have been observed.[84]

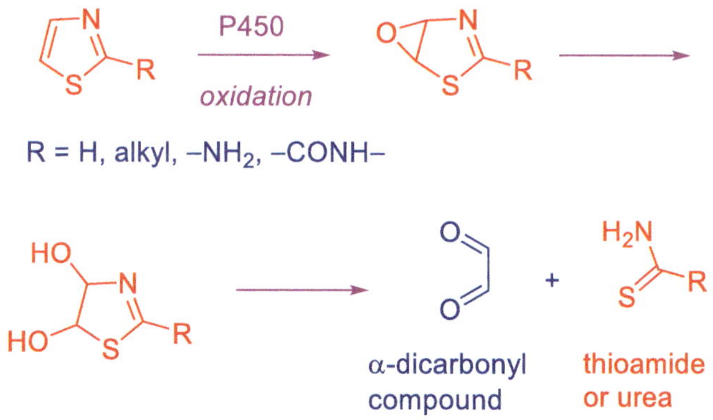

The existence of substituents at the 4- or 5-carbon can, however, delay this oxidative pathway as seen in some examples like meloxicam shown below. Meloxicam and sudoxicam are NSAIDs and belong to the enol–carboxamide category. They are structurally very similar; the only difference is the presence of an additional methyl group on the five-carbon in the thiazole ring. Here, the 5-methyl group in meloxicam undergoes oxidative metabolism, which prevents the oxidative ring opening of the 2-amidothiazole. Sudoxicam, which has an unsubstituted 2-amidothiazole, has been observed to form oxidative ring opened products *in vivo*.[85] Thus, it has been associated with severe hepatotoxicity precluding its further use, while meloxicam has been on the market for more than a decade showing much less hepatotoxicity.[86]

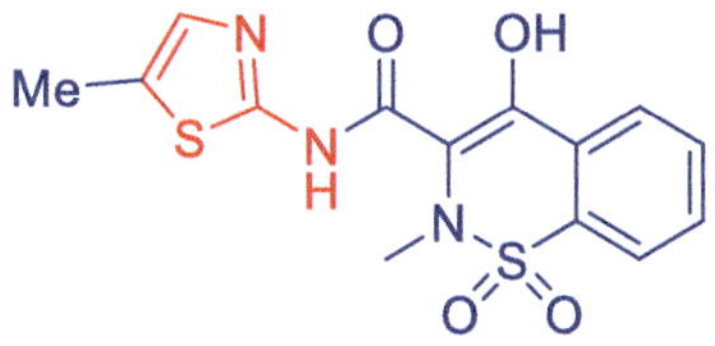

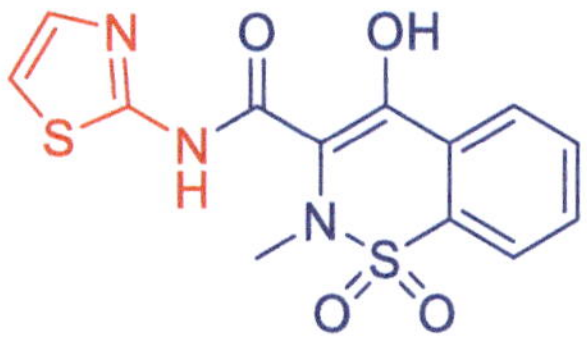

Another example is that of the HIV protease inhibitor ritonavir that contains two thiazole-rings and involves the oxidative ring-opening of thiazole to reactive intermediates. The oxidation appears to be a rate-limiting step in the mechanism based inactivation (MBI) of the cytochrome P450 enzyme CYP3A4. Non-thiazole-containing HIV protease inhibitors like indinavir, nelfinavir, or saquinavir are metabolized by CYP3A4. The inhibition of CYP3A4 by ritonavir has been found to result in a decreased metabolism of simultaneously administered protease inhibitors such as saquinavir or indinavir, thereby causing them to be cleared from the body more slowly.[87] Thus, ritonavir can boost the efficacy of other protease inhibitors, enabling the clinician to lower their dosing frequency.[88]

5.5 PAINS

PAINS stands for Pan Assay INterference compounds. Frequent, false, or promiscuous hits or colloidal aggregators in HTS had been known but not taken too seriously. Similarly, invalid metabolic panaceas (IMPs) may be false positive hits, too. Since Baell and Holloway first proposed new substructure filters for removal of PAINS from screening libraries and for their exclusion in bioassays in 2010,[89] the demise of PAINS has become well-known enough to warrant an editorial entitled "*The Ecstasy and Agony of Assay Interference Compounds*" in 2017 by editors-in-chief of eight ACS journals.[90] Therefore, PAINS are now official structural alerts.

The mechanisms of action (MOA) for PAINS are many. They may be reactive, chelators, redox active, or colored so as to interfere with biological assays. Here we focus on discussing several well-known classes of PAINS. We should pay attention to aggregators with colloidal behaviors as well when evaluating assay results. As Baell and Nissink summarized,[91] major MOAs of PAINS are various and include:

1. Aggregators, physicochemical interference such as micelle formation;
2. Chemical reactivity with biological and bioassay nucleophiles such as thiols and amines that can bind to protein covalently;
3. Redox cycling and redox activity;
4. Impurity interference. For instance, metal chelation that can interfere with proteins, assay reagents, or through bringing in heavy metal contaminants;
5. Fluorescence interference, photoreactivity with any protein functionality; having photochromic properties that might interfere with typically used

assay signaling such as absorption and fluorescence (λ_{ex} = 680 nM or λ_{em} = 520–620 nM; and

6. Assay-specific interference.

5.5.1 Unsaturated Rhodanines

Alkylidene rhodanines are the most widely reported PAINS. Rhodanine itself is colorful and interfere with assays, but alkylidene-rhodanine derivatives and arylidene-rhodanines are even worse as frequent hitters. This implied that their colors may play a role in their promiscuity in addition to their reactivity (with thiol for example) and chelating ability. Like azo compounds and quinones, arylidene-rhodanines could interfere in some assay technologies where absorption of light in 570–620 nM and could interfere in signaling.[89] Alkylidene rhodanines are designated as "ene_rhod_A" (unsaturated rhodanines) in several computational programs available to screen PAINS.

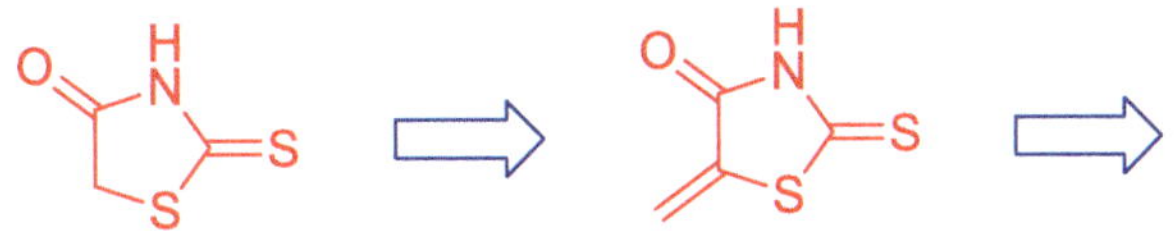

rhodanine ene rhodanine

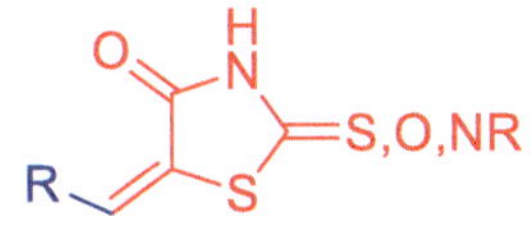

alkylidene-rhodanine derivatives

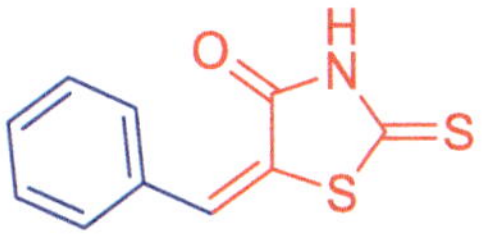

arylidene-rhodanine

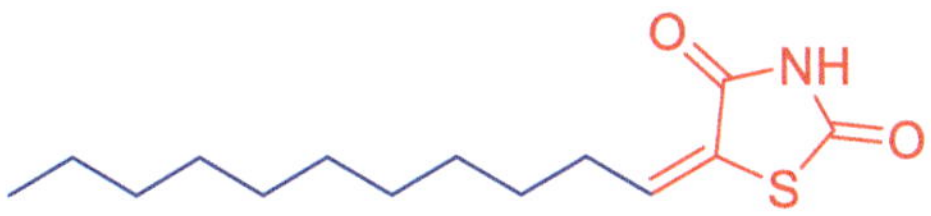

(*E*)-5-undecylidenethiazolidine-2,4-dione

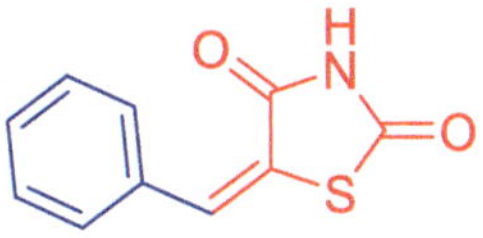

(*E*)-5-benzylidene-thiazolidine-2,4-dione

While investigating thiazolidinones as tumor necrosis factor receptor-1 (TNFRc1) inhibitors, BMS scientists noticed that several of them displayed "photochemically enhanced" binding to their targets. When light is present, these compounds form covalent bonds to their protein targets. All of the compounds as exemplified by IV560 and IW927 exhibiting this light-sensitive reactivity possess extended π systems.[91]

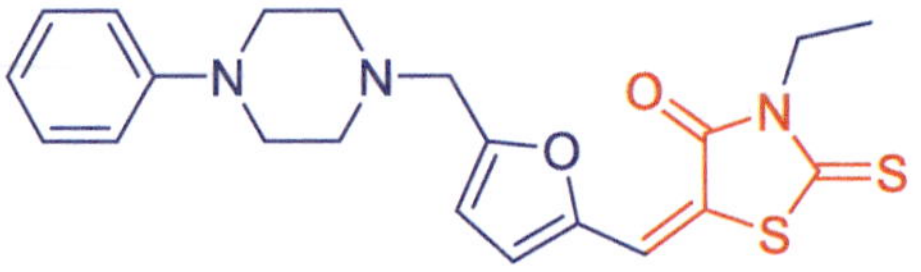

IV560

IW927

Later on, they concluded that 5-arylidene-2-thioxo-dihydro-pyrimidine-4,6(1*H*,5*H*)-diones and 3-thioxo-2,3-dihydro-1*H*-imidazo-[1,5-*a*]indol-1-ones are light-dependent TNFRc1antagonists.[92]

Kiessling′s group ran into similar 5-arylidene-4-thiazolidinone structures in their quest of chemical probes of UDP-galactopyranose mutase (UGM). It was found that 5-arylidene-4-thiazolidinones can serve as electrophiles and undergo conjugate addition reactions with nucleophiles to afford adducts as shown below. Upon addition of dithiothreitol (DTT) as the nucleophile, a significant decrease of absorbance of the maximum peak (380 nm in this particular case) was observed in less than a minute. All similar compounds underwent rapid reaction, indicating that conjugate addition readily took place and loss of extended chromophore immediately ensued.[93] This experience serves as a cautionary tale that 5-arylidene-4-thiazolidinones are PAINS that could be from their colors or as Michael acceptors, and more likely, both.

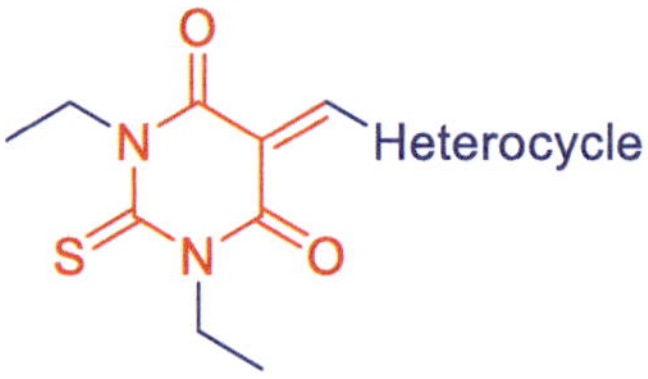

5-arylidene-2-thioxodihydro-pyrimidine-4,6(1*H*,5*H*)-diones

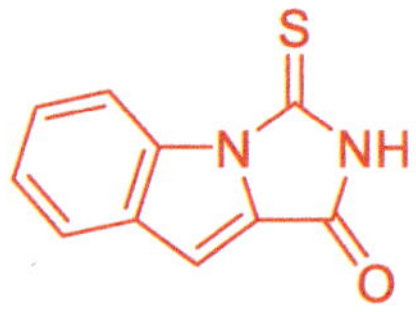

3-thioxo-2,3-dihydro-1*H*-imidazo[1,5-*a*]indol-1-one

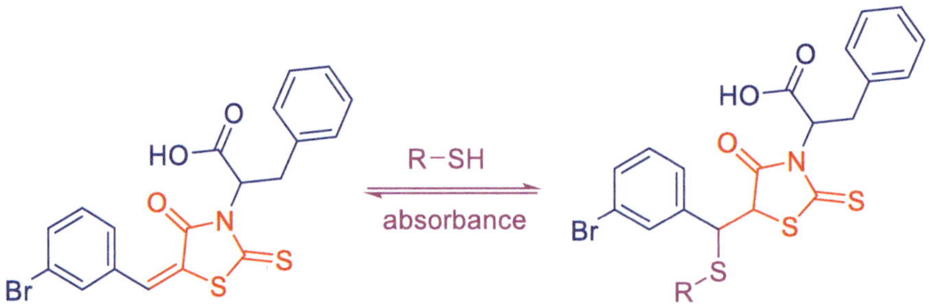

PAINS found their ways to the field of protein–protein interactions. In 2001, a group at Harvard Medical School reported identification of BH3I-1 as a small-molecule inhibitor of interaction between the BH3 domain and Bcl-x$_L$.[94] In 2006, BH3I-1 was tested inactive as both Bcl-2 and Bcl-w inhibitors with IC_{50} of >50 and 100 μM, respectively, and its weak affinity was determined by solution competition assays with an optical biosensor.[95] Regardless, BH3I-1 has continued attracting attentions as Bcl-x$_L$ inhibitor, although it is selective against Mcl-1.[96]

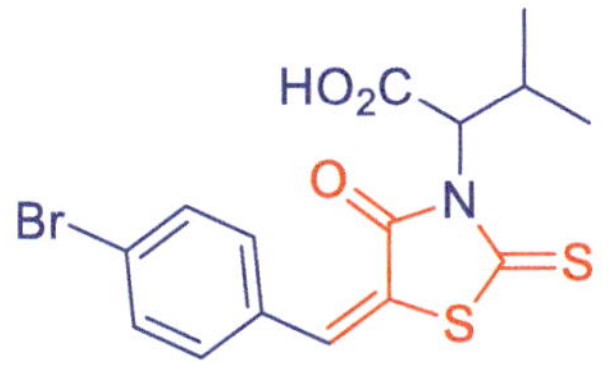

BH3I-1
Bcl-X$_L$, IC_{50} = 91 μM[96]
Mcl-1, IC_{50} = 9,110 μM[96]

Having learned alkylidene rhodanines as the most prominent PAINS, it is stunning to peruse literature and realize that so many research papers exist in literature reporting discoveries of their pharmacological activities.[97] They are equivalent of "fool's gold" in drug discovery.

5.5.2 Phenolic Mannich Bases

Phenolic Mannich bases interfere with biological assays because of their reactivity,[98] ability to chelate, and cytotoxicity.[99] As shown below, 1-hydroxybenzylamines as phenolic Mannich bases are locked by intramolecular hydrogen bonds. Sometimes, even under physiological conditions, they could undergo elimination reaction (where the amine group serves as a leaving group) to produce *ortho*-quinone methides (*o*-QMs), which are excellent electrophilic Michael acceptors. *o*-QMs may be trapped with many nucleophiles such as thiol, hydroxyl, and amino groups on protein to form covalent bonds although the reaction is reversible. Even an "inverse electron-demand Diels–Alder" reaction between *o*-QMs and ethyl vinyl ether (EVE) took place readily to produce the adduct.[98]

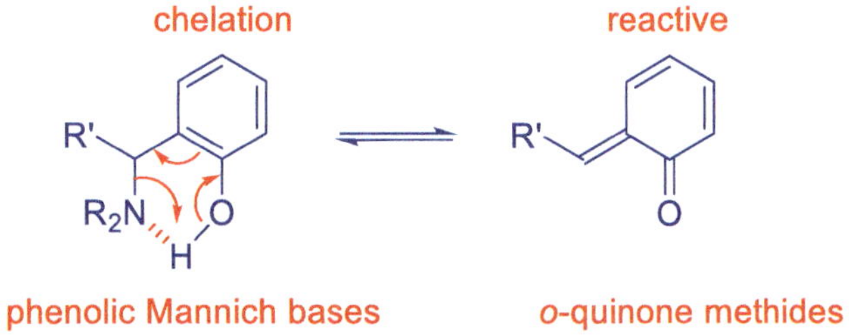

The phenol and the Mannich base do not have to be *ortho*- to each other to elicit such reactivity. *para*-Phenolic Mannich bases are prone to decomposition to afford *para*-quinone methides (*p*-QMs). Take 5-hydroxy-1-aminoindan as an example, even in the absence of a base, their intrinsic instability renders its facile decomposition to the corresponding *p*-quinone methide.[98a] In addition, when a hydroxyl- or an amino group situates at the para-position of the benzyl alcohol-carbamate of doxorubicin prodrugs, a self-immolative mode of action of decomposition takes place via a similar mode of action to produce *para*-quinone methide or *para*-iminoquinone methide.[98a]

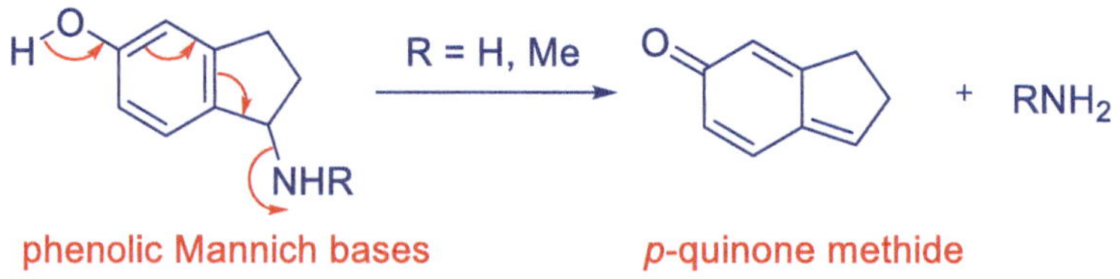

Based on X-ray crystal structures of macrophage migration inhibitory factor (MIF) tautomerase, Sanofi–Aventis carried out virtual screening and identified several phenolic Mannich bases as inhibitors. However, biochemical and X-ray crystallographic studies revealed that the hydroxyquinone derivatives were actually covalent inhibitors of the MIF tautomerase. Adducts were formed by N-alkylation of the Pro-1 at the catalytic domain with a loss of an amino group of the inhibitor:[99]

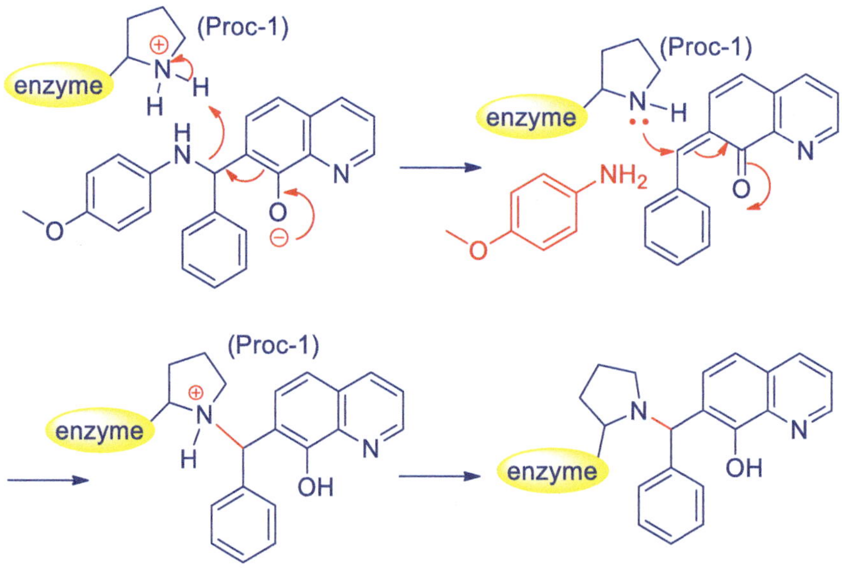

A closely related class of PAINS are hydroxyphenylhydrazones. On the one hand, both *o*- and *p*-hydroxyphenylhydrazones may interfere with biological assay because they are both colorful and reactive. On the other hand, *o*-hydroxyphenylhydrazones, in particular, also interfere with biological assay because they are chelators, in a manner similar to *ortho*-phenolic Mannich bases.

color reactive

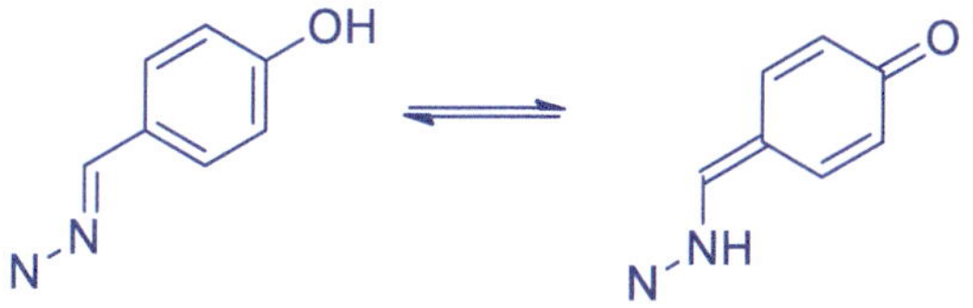

p-hydroxyphenylhydrazones *p*-hydrazone-quinones

chelation, color reactive

o-hydroxyphenylhydrazones *o*-hydrazone-quinones

5.5.3 Invalid Metabolic Panaceas

Invalid metabolic panaceas, or IMPs, have given some "hits" and natural products bad names, for the rightful reasons because they are misleading and could have wasted invaluable resources.

Curcumin is a constituent (~5%) of the traditional Chinese medicine (TCM) turmeric. The amount of research, publications, and clinical trials on curcumin is astonishing, greater than 15,000 manuscripts averaging >50 published per week! It is active in many biological assays including histone acetyltransferase (HAT) p300, HDAC8, tau and amyloid fibril formation, cystic fibrosis transmembrane conductance regulator (CFTR), and CB1. It has been investigated in 120 clinical trials for colon and pancreatic cancer, Alzheimer's disease, erectile dysfunction, and any other diseases under the Sun. No wonder many dubbed it a panacea. But as a 2017 review by Walters and coworkers revealed, curcumin should be classified under PAINS and IMPs.[100] First of all, curcumin is unstable, degrades to several fragments even in physiological conditions. Its physiochemical properties are poor, forming chemical aggregates (colloids) under common biochemical assay conditions. And finally, its ADMET properties are far from ideal despite tremendous amount of efforts devoted to improve them. The authors threw cold water to research on curcuminoids.[100]

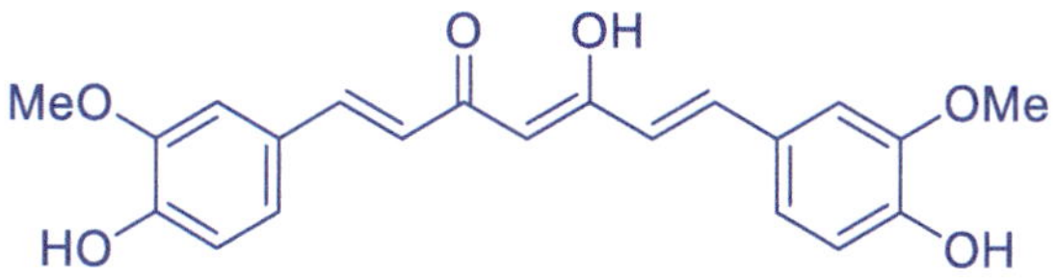

curcumin

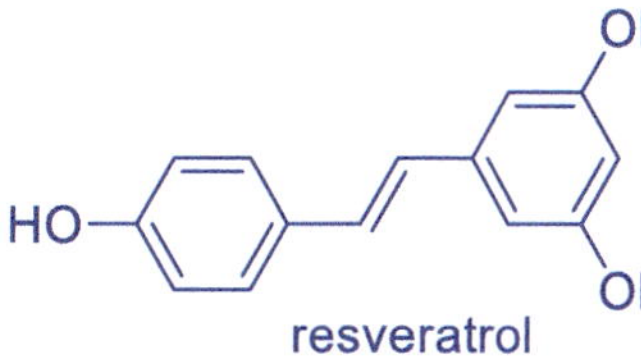

resveratrol

A few other natural products listed under this category of IMPs are ginsenosides, genistein, quercetin, apigenin, nordihydroguaiaretic acid, resveratrol, kaempferol, and fistein.[101,102]

A letter to the editor in 2017 offered a counter argument. The authors speculated that curcumin works its "magic" via mechanisms not appreciated by current well-accepted tenets of medicinal chemistry. They cautioned that "summary dismissal of an entire area of research is like throwing the baby out with bath water."[103]

5.5.4 Alkylidene Barbiturates etc.

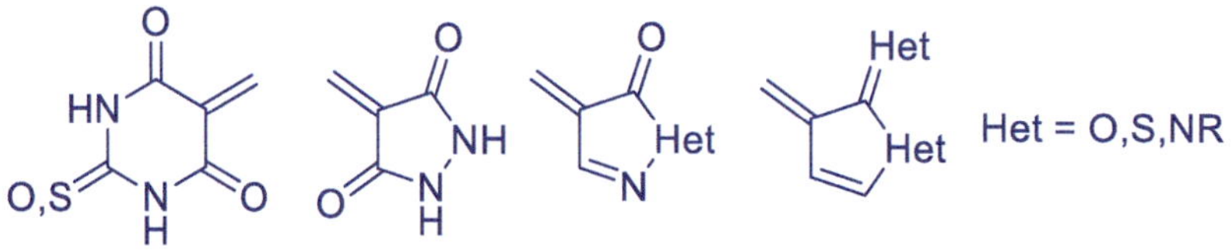

alkylidene barbiturates

A few excellent reviews have been published. Several years passed by after the 2010 JMC paper by Baell and Holloway. Some initial PAINS are no longer PAIN-full, having moved out of the PAINS space. Some new PAINS have been added to the list.[104]

One class of PAINS are alkylidene barbiturates. So are the three five-membered heterocycles shown here. Another frequently encountered class of PAINS are dialkylamines.[105]

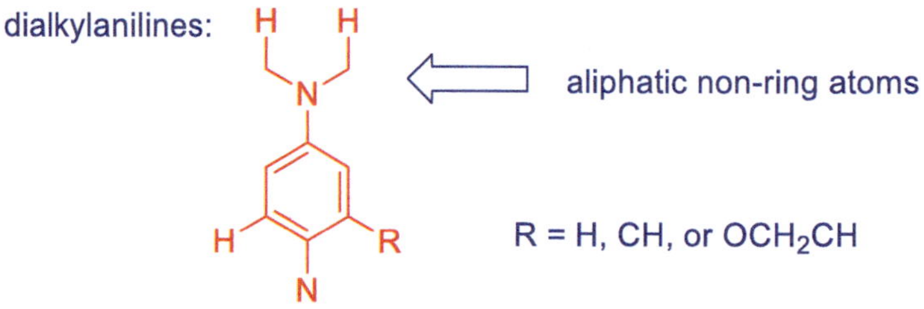

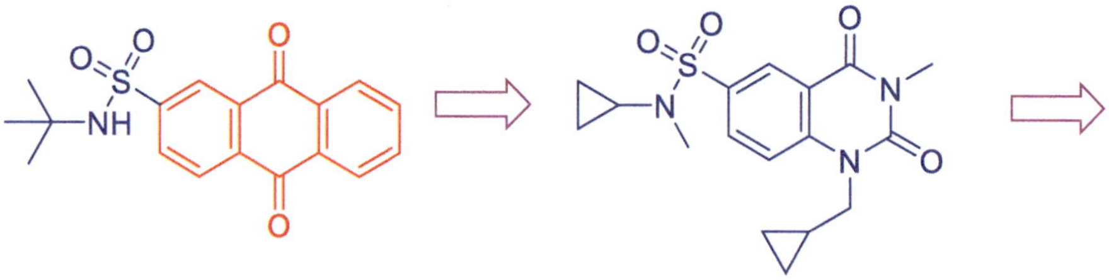

PAINS chemotype

biochemoical EC_{50} = 3.9 μM
cell POM EC_{50} = 1.8 μM

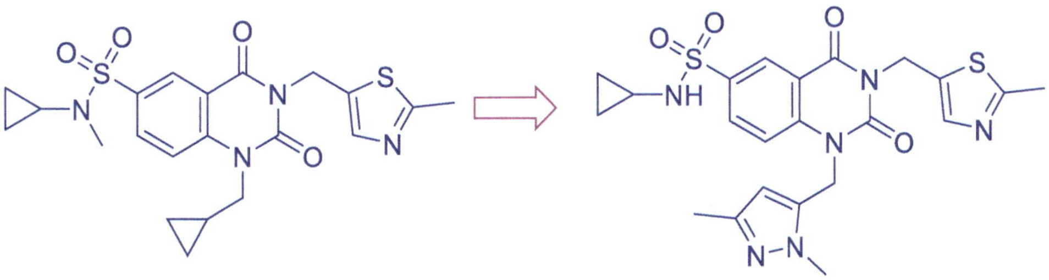

chemical EC_{50} = 0.019 μM
cell POM EC_{50} = 0.028 μM
limited solubility

non-PAIN chemotype
chemical EC_{50} = 0.026 μM
cell POM EC_{50} = 0.037 μM
in vitro tool compound

To gain a better perspective of PAINS, it helps to recognize that over 60 FDA-approved and worldwide drugs (~5%) contain PAINS chemotypes, and about the same number have been shown to aggregate. In addition, even though the initial hit is one of the PAINS, it is possible to design the drug out of the PAINS space. From screening of a library of 1,400,000 compounds for chemical probes against poly(ADP-ribose) glycohydrolase (PARG) inhibit DNA repair with differential pharmacology to olaparib, a group at Manchester only found one hit, which is a anthraquinone sulfonamide. While some quinones are classified under PAINS, not all of them are. This particular hit, however, showed cytotoxicity (after 72 h) at doses comparable to those required for PARG inhibition. This finding was not unexpected, given the flat aromatic structure, which is typical of a DNA-intercalating chemotype. Surface plasmon resonance (SPR) and structural biology demonstrated credible and stoichiometric binding, leading to *in silico* scaffold hopping and clear structure-activity relationship (SAR). They successfully identified a chemotype that is not PAINS.[106]

5.6 Conclusions

1. *The dose makes poison. For the same structural alert, efficacious drugs with lower dosages have less chances of causing toxicities;*
2. *Clear-cut is a rarity in determining "druggability";*
3. *Experience and understanding of the issues are valuable for the "gray" areas;*
4. *Avoiding these "ugly" functional groups saves time, energy, and resources; and*
5. *When in doubt, let data speak.*

5.7 References

1. Limban, C.; Nuta, D. C.; Chirita, C.; Negres, S.; Arsene, A. L.; Goumenou, M.; Karakitsios, S. P.; Tsatsakis, A. M.; Sarigiannis, D. A. *Toxicol. Rep.* **2018**, *5,* 943–953.
2. Claesson, A.; Minidis, A. *Chem. Res. Toxicol.* **2018**, *31,* 389–411.
3. (a) Kalgutkar, A. S. *Chem. Res. Toxicol.* **2017**, *30*, 220–238. (b) Orr, S. T. M.; Ripp, S. L.; Ballard, T. E.; Henderson, J. L.; Scott, D. O.; Obach, R. S.; Sun, H.;

Kalgutkar, A. S. *J. Med. Chem.* **2012**, *55,* 4896–4933. (c) Leung, L.; Kalgutkar, A. S.; Obach, R. S. *Drug Metab. Rev.* **2012**, *44,* 18–33. (d) Kalgutkar, A. S.; Didiuk, M. T. *Chem. Biodivers.* **2009**, *6*, 2115–2137. (e) Kalgutkar, A. S.; Gardner, I.; Obach, R. S.; Shaffer, C. L.; Callegari, E.; Henne, K. R.; Mutlib, A. E.; Dalvie, D. K.; Lee, J. S.; Nakai, Y.; et al. *Curr. Drug Metab.* **2005**, *6,* 161–225.

4. Garcia-Serna, R.; Vidal, D.; Remez, N.; Mestres, J. *Chem. Res. Toxicol.* **2015**, *28*, 1875–1887.
5. Kalgutkar, A. S.; Dalvie, D. *Annu. Rev. Pharmacol. Toxicol.* **2015**, *55*, 35–54.
6. Stepan, A. F.; Walker, D. P.; Bauman, J.; Price, D. A.; Baillie, T. A.; Kalgutkar, A. S.; Aleo, M. D. *Chem. Res. Toxicol.* **2011**, *24*, 1345–1410.
7. Blagg, J. Structural Alerts for Toxicity. In Abraham, D. J.; Rotella, D. P., Eds *Burger's Medicinal Chemistry, Drug Discovery, and Development, 7th Edition.* Wiley: New York, NY, **2010**, pp. 301–334.
8. Alves, V. M.; Muratov, E. N.; Capuzzi, S. J.; Politi, R.; Low, Y.; Braga, R. C.; Zakharov, A. V.; Sedykh, A.; Mokshyna, E.; Farag, S.; et al. *Green Chem.* **2016**, *18,* 4348–4360.
9. Polavarapu, A.; Stillabower, J. A.; Stubblefield, S. G. W.; Taylor, W. M.; Baik, M.-H. *J. Org. Chem.* **2012**, *77,* 5914–5921.
10. (a) Fu, D.; Calvo, J. A.; Samson, L. D. *Nat. Rev. Cancer* **2012**, *12*, 104–120. (b) Cravedi, J. P.; Perdu-Durand, E.; Baradat, M.; Alary, J.; Debrauwer, L.; Bories, G. *Chem. Res. Toxicol.* **1995**, *8*, 642–648.
11. Frang, H.; Cockcroft, V.; Karskela, T.; Scheinin, M.; Marjamäki, A. *J. Biol. Chem.* **2001**, *276*, 31279–31284.
12. Corpet, D. E.; Bories, G. F. *Drug Metab. Dispos.* **1987**, *15*, 925–927.
13. Allimuthu, D.; Adams, D. J. *ACS Chem. Biol.* **2017**, *12*, 2124–2131.
14. Gehringer, M.; Laufer, S. A. *J. Med. Chem.* **2019**, *62*, 5673–5724.
15. Reid, J. M.; Kuffel, M. J.; Miller, J. K.; Rios, R.; Ames, M. W. *Clin. Cancer Res.* **1995**, *5*, 2192–2197.
16. Aggarwal, S.; Thareja, S.; Verma, A.; Bhardwaj, T. R.; Kumar, M. *Steroids* **2010**, *75*, 109–153.
17. Charneira, C.; Godinho, A. L. A.; Oliveira, M. C.; Pereira, S. A.; Monteiro, E. C.; Marques, M. M.; Antunes, A. M. M. *Chem. Res. Toxicol.* **2011**, *24*, 2129–2141.
18. Dahal, U. P.; Obach, R. S.; Gilbert, A. M. *Chem. Res. Toxicol.* **2013**, *26*, 1739–1745.
19. (a) Naik, P.; Murumkar, P.; Giridhar, R.; Yadav, M. R. *Bioorg. Med. Chem.* **2010**, *18*, 8418–8456. (b) Schmidt, B.; Schieffer, B. *J. Med. Chem.* **2010**, *46*, 2261–2270. (c) Timmermans, P. B.; Duncia, J. V.; Carini, D. J.; Chiu, A. T.; Wong, P. C.; Wexler, R. R.; Smith, R. D. *J. Hum. Hypertens.* **1995**, *9(Suppl 5)*, S3–S18.
20. Chinthakindi, P. K.; Arvidsson, P. I. *Eur. J. Org. Chem.* **2018**, 3648–3666.
21. Jones, L. H. *ACS Med. Chem. Lett.* **2018**, *9,* 584–586.
22. Cheff, D. M.; Hall, M. D. *J. Med. Chem.* **2017**, *60,* 4517–4532.
23. Rescifina, A.; Zagni, C.; Varrica, M. G.; Pistara, V.; Corsaro, A. *Eur. J. Med. Chem.* **2014**, *74,* 95–115.
24. Snyder, R. D.; Ewing, D.; Hendry, L. B. *Mutat. Res.* **2006**, *609*, 47–59.
25. (a) Kalgutkar, A. S.; Dalvie, D.; Obach, R. S.; Smith, D. A. *Reactive Drug Metabolites*. Wiley-VCH: Weinheim. **2012**. (b) Kalgutkar, A. S.; Soglia, J. R. *Exp. Opin. Drug Metab. Toxicol.* **2005**, *1,* 91–142.

26. Famulok, M.; Boche, G. *Angew. Chem. Int. Ed. Engl.* **1989**, *28*, 468–469.
27. Shamovsky, I.; Börjesson, L.; Mee, C.; Nordén, B.; Hasselgren, C.; O'Donovan, M.; Sjö, P. *J. Am. Chem. Soc.* **2011**, *133*, 16168–16185.
28. Ng, W.; Metushi, I. G.; Uetrecht, J. *J. Immunotoxicol.* **2015**, *12*, 24–32.
29. Beaulieu, P. L.; Bolger, G.; Duplessis, M.; Gagnon, A.; Garneau, M.; Stammers, T.; Kukolj, G.; Duan, J. *Bioorg. Med. Chem. Lett.* **2015**, *25*, 1140–1145.
30. Mahmood, I.; Sahajwalla, C. *Clin. Pharmacokinet.* **1999**, *36*, 277–287.
31. Bauman, J. S.; Frederick, K. S.; Sawant, A.; Walsky, R. L.; Cox, L. M.; Obach, R. S.; Kalgutkar, A. S. *Drug Metab. Dispos.* **2008**, *36*, 1016–1029.
32. Chen, W.; Koenigs, L. L.; Thompson, S. J.; Peter, R. M.; Rettie, A. E.; Trager, W. F.; Nelson, S. D. *Chem. Res. Toxicol.* **1998**, *11*, 295–301.
33. (a) Cheng, H.; Planken, S. *ACS Med. Chem. Lett.* **2018**, *9*, 861–863. (b) Cheng, H.; Nair, S. K.; Murray, B. W. *Bioorg. Med. Chem. Lett.* **2016**, *26*, 1861–1868.
34. (a) Mutlib, A. E.; Dickenson, P.; Chen, S.-Y.; Espina, R. J.; Daniels, J. S.; Gan, L.-S. *Chem. Res. Toxicol.* **2002**, *15*, 1190–1207. (b) Mutlib, A. E.; Chen, S.-Y.; Espina, R.J; Shockcor, J.; Prakash, S. R.; Gan, L.-S. *Chem. Res. Toxicol.* **2002**, *15*, 63–75.
35. Lim, H.-K.; Chen, J.; Sensenhauser, C.; Cook, K.; Preston, R.; Thomas, T.; Shook, B.; Jackson, P. F.; Rassnick, S.; Rhodes, K.; et al. *Chem. Res. Toxicol.* **2011**, *124*, 1012–1030.
36. (a) Castellino, S.; O'Mara, M.; Koch, K.; Borts, D. J.; Bowers, G. D.; MacLauchlin, C. *Drug Metab. Dispos.* **2012**, *40*, 139–150. (b) Takakusa, H.; Wahlin, M. D.; Zhao, C.; Hanson, K. L.; New, L. S.; Chan, E. C. Y.; Nelson, S. D. *Drug Metab. Dispos.* **2011**, *39*, 1022–1030. (c) Teng, W. C.; Oh, J. W.; New, L. S.; Wahlin, M. D.; Nelson, S. D.; Ho, H. K.; Chan, E. C. Y. *Mol. Pharmacol.* **2010**, *78*, 693–703. (d) Ho, H. K.; Chan, J. C. Y.; Hardy, K. D.; Chan, E. C. Y. *Drug Metab. Rev.* **2015**, *347*, 21–28.
37. (a) Yu, J.; Brown, D. G.; Burdette, D. *Drug Metab. Dispos.* **2010**, *38*, 1767–1778. (b) Obach, R. S.; Dalvie, D. K. *Drug Metab. Dispos.* **2006**, *34*, 1310–1316.
38. Alleva, J. J. *J. Med. Chem.* **1963**, *6*, 621–624.
39. (a) Sun, Q.; Zhu, R.; Foss, F. W.; Macdonald, T. L. *Chem. Res. Toxicol.* **2008**, *21*, 711–719. (b) Sun, Q.; Zhu, R.; Foss, F. W.; Macdonald, T. L. *Bioorg. Med. Chem. Lett.* **2007**, *17*, 6682–6686.
40. Nepali, K.; Lee, H.-Y.; Liou, J.-P. *J. Med. Chem.* **2019**, *62*, 2851–2893.
41. (a) Li, F.; Chordia, M. D.; Huang, T.; Macdonald, T. L. *Chem. Res. Toxicol.* **2009**, *22*, 72–80. (b) Zhou, L.; Pang, X.; Xie, C.; Zhong, D.; Chen, X. *Chem. Res. Toxicol.* **2015**, *28*, 2267–2277.
42. Souers, A. J.; Leverson, J. D.; Boghaert, E. R.; Ackler, S. L.; Catron, N. D.; Chen, J.; Dayton, B. D.; Ding, H.; Enschede, S. H.; Fairbrother, W. J.; et al. *Nat. Med.* **2013**, *19*, 202–208.
43. Klopčič, I.; Dolenc, M. S. *Chem. Res. Toxicol.* **2019**, *32*, 1–34.
44. (a) Wallace, K. *Cardiovasc. Toxicol.* **2007**, *7*, 101–107. (b) Sinha, B. K.; Mason, R. P. *J. Drug Metab. Toxicol.* **2015**, *6*, 186(1–8).
45. (a) Patel, H; Sonawane, Y.; Jagtap, R.; Dhangar, K.; Thapliyal, N.; Surana, S.; Noolvi, M.; Shaikh, M. S.; Rane, R. A.; Karpoormath, R. *Bioorg. Med. Chem.*

Lett. **2015**, *25,* 1938–1946. (b) Ikeda, T. *Drug Metab. Pharmacokinet.* **2011**, *26,* 60–70. (c) Masubuchi, Y. *Drug Metab. Pharmacokinet.* **2006**, *21,* 347–356.
46. Chadha, N.; Bahia, M. S.; Kaur, M.; Silakari, O. *Bioorg. Med. Chem.* **2015**, *23,* 2953–2974.
47. Erve, J. C. L.; Svensson, M. A.; von Euler-Chelpin, H.; Klasson-Wehler, E. *Chem. Res. Toxicol.* **2004**, *17,* 564–571.
48. (a) Fan, P. W., Bolton, J. L. *Drug Metab. Dispos.* **2001**, *29,* 891–896. (b) Crewe, H. K., Notley, L. M., Wunsch, R. M., Lennard, M. S., Gillam, E. M. *Drug Metab. Dispos.* **2002**, *30,* 869–874. (c) Fan, P. W., Zhang, F., Bolton, J. L. *Chem. Res. Toxicol.* **2000**, *13,* 45–52.
49. Scott, K. A.; Njardarson, J. T. *Top. Curr. Chem.* **2018**, *376*, 1–34.
50. Zuniga, F. I.; Loi, D.; Ling, K. H. J.; Tang-Liu, D. D.-S. *Exp. Opin. Drug Metab. Toxicol.* **2012**, *8,* 467–485.
51. (a) Nishida, C. R.; Ortiz de Montellano, P. R. *Chem. Biol. Interact.* **2011**, *192,* 21–25. (b) Ji, T.; Ikehata, K.; Koen, Y. M.; Esch, S. W.; Williams, T. D.; Hanzlik, R. P. *Chem. Res. Toxicol.* **2007**, *20,* 701–708. (c) Chilakapati, J.; Shankar, K.; Korrapati, M. C.; Hill, R. A.; Mehendale, H. M. *Drug Metab. Dipos.* **2005**, *33,* 1877–1885.
52. Mizutani, T.; Yoshida, K.; Murakami, M.; Shirai, M.; Kawazoe, S. *Chem. Res. Toxicol.* **2000**, *13,* 170–176.
53. Jung, M. E.; Ouk, S.; Yoo, D.; Sawyers, C. L.; Chen, C.; Tran, C.; Wongvipat, J. *J. Med. Chem.* **2010**, *53*, 2779–2796.
54. (a) Streeter, A. J.; Timbrell, J. A. *Drug Metab. Dispos.* **1983**, *11,* 179–183. (b) Masubuchi, Y.; Horie, T. *Chem. Res. Toxicol.* **1999**, *12*, 1028–1032. (c) Parent, M. B.; Master, S.; Kashlub, S.; Baker, G. B. *Biochem. Pharmacol.* **2002**, *63*, 57–64. (d) Runge-Morris, M.; Feng, Y.; Zangar, R. C.; Novak, R. F. *Drug Metab. Dispos.* **1996**, *24,* 734–737.
55. (a) Boelsterli, U. A.; Lee, K. K. *J. Gastroenterol. Hepatol.* **2014**, *29*, 678–687. (b) Polasek, T. M.; Elliot, D. J.; Somogyi, A. A.; Gillam, E. M. J.; Lewis, B. C.; Miners, J. O. *Br. J. Clin. Pharmacol.* **2006**, *61*, 570–584.
56. (a) Duffy, K. J.; Erickson-Miller, C. L. The Discovery of Eltrombopg, An Orally Bioavailable TpoR Agonist. In Metcalf, B. W.; Dillon, S., eds. *Target Validation in Drug Discovery.* Academic Press: Cambridge, MA, **2011**, pp. 241–254. (b) Deng, Y.; Rogers, M.; Sychterz, C.; Talley, K.; Qian, Ya.; Bershas, D.; Ho, M.; Shi, W.; Chen, E. P.; Serabjit-Singh, C.; et al. *Drug Metab. Dispos.* **2011**, *39,* 1747–1754. (c) Deng, Y.; Madatian, A.; Wire, M. B.; Bowen, C.; Park, J. W.; Williams, D.; Peng, B.; Schubert, E.; Gorycki, F.; Levy, M.; et al. *Drug Metab. Dispos.* **2011**, *39,* 1734–1746.
57. Amano, T.; Fukami, T.; Ogiso, T.; Hirose, D.; Jones, J. P.; Taniguchi, T.; Nakajima, M. *Biochem. Pharmacol.* **2018**, *152*, 69–78.
58. Steuer, A. E.; Schmidhauser, C.; Schmid, Y.; Rickli, A.; Liechti, M. E.; Kraemer, T. *Drug Metab. Dispos.* **2015**, *43,* 1864–1871.
59. O'Mathuna, B.; Farre, M.; Rostami-Hodjegan, A.; Yang, J.; Cuyas, E.; Torrens, M.; Pardo, R.; Abanades, S.; Maluf, S.; Tucker, G. T.; et al. *J. Clin. Psychopharmacol.* **2008**, *28,* 525–531.

60. Zhao, S. X.; Dalvie, D. K.; Kelly, J. M.; Soglia, J. R.; Frederick, K. S.; Smith, E. B.; Obach, R. S.; Kalgutkar, A. S. *Chem. Res. Toxicol.* **2007**, *20,* 1649–1657.
61. Bertelsen, K. M.; Venkatakrishnan, K.; Von Moltke L. L.; Obach, R. S.; Greenblatt, D. J. *Drug Metab. Dispos.* **2003**, *31,* 289–293.
62. Ring, B. J.; Patterson, B. E.; Mitchell, M. I; Vandenbranden, M.; Gillespie, J.; Bedding, A. W; Jewell, H.; Payne, C. D.; Forgue, S. T.; Eckstein J.; et al. *Clin. Pharmacol. Ther.* **2005**, *77,* 63–75.
63. Yu, H.; Balani, S. K.; Chen, W.; Cui, D.; He, L.; Griffith, H. W.; Mao, J.; George, L. W.; Lee, A. J.; Lim, H.-K.; et al. *Drug Metab. Dispos.* **2015**, *43,* 620–630.
64. Uttamsingh, V.; Gallegos, R.; Liu, J. F.; Harbeson, S. L.; Bridson, G. W.; Cheng, C.; Wells, D. S.; Graham, P. B.; Zelle, R.; Tung, R. *J. Pharmacol. Exp. Ther.* **2015**, *354,* 43–54.
65. Assandri, A.; Perazzi, A.; Baldoli, E.; Ferrari, P.; Ripamonti, A.; Bellasio, E.; Tuan, G.; Zerilli, L. F.; Tarzia, G. *Xenobiotica* **1985**, *15*, 1069–1087.
66. Assandri, A.; Perazzi, A.; Bellasio, E.; Ciabatti, R.; Tarzia, G.; Ferrari, P.; Ripamonti, A.; Tuan, G.; Zerilli, L. F. *Xenobiotica* **1985**, *15*, 1089–1102.
67. Assandri, A.; Tarzia, G.; Bellasio, E.; Ciabatti, R.; Tuan, G.; Ferrari, P.; Zerilli, L.; Lanfranchi, M.; Pelizzi, G. *Xenobiotica* **1987**, *17*, 559–573.
68. Assandri, A.; Barone, D.; Ferrari, P.; Perazzi, A.; Ripamonti, A.; Tuan, G.; Zerilli, L. *Drug Metab. Dispos*. **1984**, *12*, 257–263.
69. Egger, H.; Itterly, W.; John, V.; Shimanskas, C.; Stancato, F.; Kapoor, A. *Drug Metab. Dispos*. **1988**, *16*, 568–575.
70. Murphy, P. J.; Williams, T. L. *J. Med. Chem.* **1972**, *15,* 137–139.
71. Walsh, K. M.; Albassam, M. A.; Clarke, D. E. *Toxicol. Pathol.* **1996**, *24,* 468–476.
72. Skordos, K.; Skiles, G. L.; Laycock, J. D.; Lanza, D. L.; Yost, G. S. *Chem. Res. Toxicol*. **1998**, *11*, 741–749.
73. Regal, K. A.; Laws, G. M.; Yuan C, Yost, G. S.; Skiles, G. L. *Chem. Res. Toxicol.* **2001**, *14*, 1014–1024.
74. Kassahun, K.; Skordos, K.; McIntosh, I.; Slaughter, D.; Doss, G. A.; Baillie, T. A.; Yost, G. S. *Chem. Res. Toxicol*. **2005**, *18*, 1427.
75. Mansuy, D.; Valadom, P.; Erdelmeier, I.; Lopez-Garcia, P.; Amar, C.; Girault, J.-P.; Dansette, P. M. *J. Am. Chem. Soc.* **1991**, *113*, 7825–7826.
76. Valadon, P.; Dansette, P. M.; Girault, J.-P.; Amar, C.; Mansuy, D. *Chem. Res. Toxicol*. **1996**, *9*, 1403–1413.
77. Treiber, A.; Dansette, P. M.; Amri, H. E.; Girault, J.-P.; Ginderow, D.; Mornon, J.-P.; Mansuy, D. *J. Am. Chem. Soc*. **1997**, *119*, 1565–1571.
78. Blagg, J. Structural Alerts for Toxicity. In Abraham, D. J.; Rotella, D. P., Eds *Burger's Medicinal Chemistry, Drug Discovery, and Development, 7th Edition*. Wiley: New York, NY, **2010**, pp. 301–334.
79. Williams, D. P.; Antoine, D. J.; Butler, P. J.; Jones, R.; Randle, L.; Payne, A.; Hoard, M.; Gardner, I.; Blagg, J.; Park, B. K. *J. Pharmacol. Exp. Ther.* **2007**, *322*, 1208–1220.
80. Stepan, A. F.; Walker, D. P.; Bauman, J.; Proce, D. A.; Baillie, T. A.; Kalgutkar, A. S.; Aleo, M. D. *Chem. Res. Toxicol*. **2001**, *24*, 1345–1410.
81. Bonierbale, E.; Valadon, P.; Pons, C.; Desfosses, B.; Dansette, P. M.; Mansuy, D. *Chem. Res. Toxicol.* **1999**, *12*, 286–296.

82. Dansette, P. M.; Libraire, J.; Bertho, G.; Mansuy, D. *Chem. Res. Toxicol.* **2009**, *22*, 369–373.
83. Kalgutkar, A. S. Metabolic Activation of Organic Functional Groups Utilized in Medicinal Chemistry. In Lee, M. S.; Zhu, M., eds. *Mass Spectroscopy in Drug Metabolism and Disposition: Basic Principles and Applications*. Wiley: Hoboken, NJ. **2010**, pp. 43–82.
84. Abraham, D. J.; Rotella, D. P., eds. *Burger's Medicinal Chemistry, Drug Discovery and Development, Seventh Edition*. Wiley: Hoboken, NJ, **2010**, pp. 301–334.
85. Hobbs, D. C.; Twomey, T. M. *Drug Metab. Dispos.* **1977**, *5*, 75–81.
86. Obach, R. S.; Kalgutkar, A. S.; Ryder, T. F.; Walker, G. S. *Chem. Res. Toxicol.* **2008**, *21*, 1890–1899.
87. Zeldin, R. K.; Petruschke, R. A. *J. Antimicrob. Chemother.* **2004**, *53*, 4–9.
88. Merry, C.; Barry, M. G.; Mulcahy, F.; Ryan, M.; Heavey, J.; Tjia, J. F.; Gibbons, S. E.; Breckenridge, A. M.; Back, D. J. *AIDS* **1997**, *11,* F29.
89. Baell, J. B.; Holloway, G. A. *J. Med. Chem.* **2010**, *53*, 2719–2740.
90. Aldrich, C.; Bertozzi, C.; Georg, G. I.; Kiessling, L.; Lindsley, C.; Liotta, D.; Merz, K. M., Jr.; Schepartz, A.; Wang, S. *J. Med. Chem.* **2017**, *60*, 2165–2168.
91. Voss, M. E.; Carter, P. H.; Tebben, A. J.; Scherle, P. A.; Brown, G. D.; Thompson, L. A.; Xu, M.; Lo, Y. C.; Yang, G.; Liu, R.-Q.; Strzemienski, P.; Everlof, J. G.; Trzaskos, J. M.; Decicco, C. P. *Bioorg. Med. Chem. Lett.* **2003**, *13*, 533–538.
92. Carter, P. H.; Scherle, P. A.; Muckelbauer, J. A.; Voss, M. E.; Liu, R.-Q.; Thompson, L. A.; Tebben, A. J.; Solomon, K. A.; Lo, Y. C.; Li, Z.; et al. *Proc. Natl. Acad. Sci. USA* **2001**, *98*, 11879–11884.
93. Carlson, E. E.; May, J. F.; Kiessling, L. L. *Chem. Biol.* 2006, *13*, 825–837.
94. Degterev, A.; Lugovsky, A.; Cardone, M.; Mulley, B.; Wagner, G.; Mitchison, T.; Yuan, J. *Nat. Cell Biol.* **2001**, *3*, 173–182.
95. van Delft, M. F.; Wei, A. H.; Mason, K. D.; Vandenberg, C. J.; Chen, L.; Czabotar, P. E.; Willis, S. N.; Scott, C. L.; Day, C. L.; Cory, S.; et al. *Cancer Cell* **2006**, *10*, 389–399.
96. (a) Bernardo, P. H.; Sivaraman, T.; Wan, K.-F.; Xu, J.; Krishnamoorthy, J.; Song, C. M.; Tian, L.; Chin, J. S. F.; Lim, D. S. W.; Mok, H. Y. K.; et al. *Pure Appl. Chem.* **2011**, *83*, 723–731. (b) Stucki, D.; Brenneisen, P.; Reichert, A. S.; Stahl, W. *Toxicol. Lett.* **2018**, *295*, 369–378.
97. (a) Kaminskyy, D.; Kryshchyshyn, A.; Lesyk, R. *Eur. J. Med. Chem.* **2017**, *140*, 542–594. (b) Holota, S.; Kryshchyshyn, A.; Trufin, Y.; Demchuk, I.; Derkach, H.; Gzella, A.; Grellier, P.; Lesyk, R. *Bioorg. Chem.* **2019**, *86*, 126–136.
98. (a) Herzig, Y.; Lerman, L.; Goldenberg, W.; Lerner, D.; Gottlieb, H. E.; Nudelman, A. *J. Org. Chem.* **2006**, *71*, 4130–4140. (b) Weinert, E. E.; Dondi, R.; Colloredo-Melz, S.; Frankenfield, K. N.; Mitchell, C. H.; Freccero, M.; Rokita, S. E. *J. Am. Chem. Soc.* **2006**, *128*, 11940–11947.
99. (a) McLean, L. R.; Zhang, Y.; Li, H.; Li, Z.; Lukasczyk, U.; Choi, Y.-M.; Han, Z.; Prisco, J.; Fordham, J.; Tsay, J. T.; Reiling, S.; Vaz, R. J.; Li, Y. *Bioorg. Med. Chem. Lett.* **2009**, *23*, 6717–6720. (b) Cisneros, J. A.; Robertson,

M. J.; Valhondo, M.; Jorgensen, W. L. *Bioorg. Med. Chem. Lett.* **2016**, *26*, 2764–2767.
100. Nelson, K. M.; Dahlin, J. L.; Bisson, J.; Graham, J.; Pauli, G. F.; Walters, M. A. *J. Med. Chem.* **2017**, *60*, 1620–1637.
101. Bisson, J.; McAlpine, J. B; Chen, S.-N.; Graham, J.; Pauli, G. F; Friesen, J. B. *J. Med. Chem.* **2016**, *59*, 1671–1690.
102. Baell, J. B. *J. Nat. Prod.* **2016**, *79*, 616–628.
103. Padmanaban, G.; Nagaraj, V. A. *ACS Med. Chem. Lett.* **2017**, *8*, 274.
104. (a) Gilberg, E.; Guetschow, M.; Bajorath, J. *J. Med. Chem.* **2018**, *61*, 1276–1284. (b) Gilberg, E.; Stumpfe, D.; Bajorath, J. *RSC Adv.* **2017**, *7*, 35638–35647. (c) Baell, J. B.; Nissink, J. W. M. *ACS Chem. Biol.* **2018**, *13*, 36–44.
105. Vidler, L. R.; Watson, I. A.; Margolis, B. J.; Cummins, D. J.; Brunavs, M. *ACS Med. Chem. Lett.* **2018**, *9*, 792–796.
106. James, D. I.; Smith, K. M.; Jordan, A. M.; Fairweather, E. E.; Griffiths, L. A.; Hamilton, N. S.; Hitchin, J. R.; Hutton, C. P.; Jones, S.; Kelly, P. *ACS Chem. Biol.* **2016**, *11*, 3179–3190.

Index

A

D

E

F

G

H

I

N

O

P

Q

R

S

V

W

X

Z

Printed and bound by CPI Group (UK) Ltd, Croydon, CR0 4YY